Lecture Notes in Computer Science 10307

Commenced Publication in 1973
Founding and Former Series Editors:
Gerhard Goos, Juris Hartmanis, and Jan van Leeuwen

Editorial Board

More information about this series at http://www.springer.com/series/7407

Jarkko Kari · Florin Manea
Ion Petre (Eds.)

Unveiling Dynamics and Complexity

13th Conference on Computability in Europe, CiE 2017
Turku, Finland, June 12–16, 2017
Proceedings

 Springer

Editors
Jarkko Kari
University of Turku
Turku
Finland

Ion Petre (ID)
Åbo Akademi University
Turku
Finland

Florin Manea
Christian-Albrechts-University of Kiel
Kiel
Germany

ISSN 0302-9743   ISSN 1611-3349 (electronic)
Lecture Notes in Computer Science
ISBN 978-3-319-58740-0   ISBN 978-3-319-58741-7 (eBook)
DOI 10.1007/978-3-319-58741-7

Library of Congress Control Number: 2017940625

LNCS Sublibrary: SL1 – Theoretical Computer Science and General Issues

Printed on acid-free paper

This Springer imprint is published by Springer Nature
The registered company is Springer International Publishing AG
The registered company address is: Gewerbestrasse 11, 6330 Cham, Switzerland

Preface

CiE 2017: Unveiling Dynamics and Complexity
Turku, Finland, June 12–16, 2017

The conference Computability in Europe (CiE) is organized yearly under the auspices of the Association CiE, a European association of mathematicians, logicians, computer scientists, philosophers, physicists, and others interested in new developments in computability and their underlying significance for the real world. CiE promotes the development of computability-related science, ranging from mathematics, computer science, and applications in various natural and engineering sciences, such as physics and biology, as well as related fields, such as philosophy and history of computing. CiE 2017 had as motto "Unveiling Dynamics and Complexity," emphasizing in this way two important broad research directions within the CiE community.

CiE 2017 was the 13th conference in the series, and the first one to take place in Finland, in the city of Turku. The conference was jointly organized by the Department of Mathematics and Statistics, University of Turku, and the Department of Computer Science, Åbo Akademi University. The 12 previous CiE conferences were held in Amsterdam (The Netherlands) in 2005, Swansea (Wales) in 2006, Siena (Italy) in 2007, Athens (Greece) in 2008, Heidelberg (Germany) in 2009, Ponta Delgada (Portugal) in 2010, Sofia (Bulgaria) in 2011, Cambridge (UK) in 2012, Milan (Italy) in 2013, Budapest (Hungary) in 2014, Bucharest (Romania) in 2015, and Paris (France) in 2016. CiE 2018 will be held in Kiel, Germany. Currently, the annual CiE conference is the largest international meeting focused on computability theoretic issues. The proceedings containing the best submitted papers as well as extended abstracts of invited, tutorial, and special session speakers, for the CiE conferences are published in the Springer series *Lecture Notes in Computer Science*.

The CiE conference series is coordinated by the CiE Conference Series Steering Committee consisting of: Arnold Beckmann (Swansea), Alessandra Carbone (Paris), Natasha Jonoska (Tampa FL), Benedikt Löwe (Amsterdam), Florin Manea (Kiel, chair), Klaus Meer (Cottbus), Mariya Soskova (Sofia), Susan Stepney (York), and ex-officio members Paola Bonizzoni (Milan, president of the Association CiE) and Dag Normann (Oslo).

The Program Committee of CiE 2017 was chaired by Jarkko Kari (University of Turku) and Ion Petre (Åbo Akademi University, Turku). The committee selected the invited and tutorial speakers and the special session organizers, and coordinated the reviewing process of all submitted contributions.

The Program Committee invited six speakers to give plenary lectures:

- Scott Aaronson (University of Texas at Austin)
- Karen Lange (Wellesley College)
- Ludovic Patey (Université Paris Diderot)
- Nicole Schweikardt (Humboldt-Universität zu Berlin)
- Alexander Shen (Université de Montpellier)
- Moshe Vardi (Rice University)

The conference also had two plenary tutorials by

- Denis R. Hirschfeldt (University of Chicago)
- Daniel M. Gusfield (University of California, Davis)

CiE 2017 had six special sessions, listed here. Speakers in these special sessions were selected by the respective special session organizers and were invited to contribute a paper to this volume. The History and Philosophy of Computing session was focused on a special topic this year: history and foundations of recursion, in memory of Rósza Péter (1905–1977).

Algorithmics for Biology

> *Organizers:* Paola Bonizzoni and Veli Mäkinen
> *Speakers:* Tobias Marschall (Max-Planck-Institut für Informatik), Fabio Vandin (University of Padova), Gregory Kucherov (University Paris-Est Marne-la-Vallée), Gianluca Della Vedova (University of Milano-Bicocca)

Combinatorics and Algorithmics on Words

> *Organizers:* Tero Harju and Dirk Nowotka
> *Speakers:* Stepan Holub (Charles University in Prague), Pascal Ochem (Université de Montpellier), Svetlana Puzynina (Sobolev Institute of Mathematics and École Normale Supérieure de Lyon), Narad Rampersad (University of Winnipeg)

Computability in Analysis, Algebra, and Geometry

> *Organizers:* Julia Knight and Andrey Morozov
> *Speakers:* Saugata Basu (Purdue University), Margarita Korovina (University of Aarhus), Alexander Melnikov (University of California, Berkeley), Russell Miller (Queens College, City University of New York)

Cryptography and Information Theory

> *Organizers:* Delaram Kahrobaei and Helger Lipmaa
> *Speakers:* Jean-Charles Faugère (Université Pierre et Marie Curie), Elham Kashefi (University of Edinburgh, Université Pierre et Marie Curie), Aggelos Kiayias (University of Edinburgh), Ivan Visconti (Università degli Studi di Salerno)

Formal Languages and Automata Theory

Organizers: Juhani Karhumäki and Alexander Okhotin
Speakers: Kai Salomaa (Queen's University at Kingston), Matrin Kutrib (Justus-Liebig-Universität Gießen), Thomas Colcombet (Université Paris Diderot), Artur Jez (University of Wrocław)

History and Philosophy of Computing

Organizers: Liesbeth De Mol and Giuseppe Primiero
Speakers: Juliette Kennedy (University of Helsinki), Jan von Plato (University of Helsinki), Giovanni Sommaruga (Université de Fribourg), Hector Zenil (University of Oxford and Karolinska Institute)

The members of the Program Committee of CiE 2017 selected for publication in this volume and for presentation at the conference 25 of the 52 non-invited submitted papers. Each paper received at least three reviews by the Program Committee and their subreviewers. In addition to the accepted contributed papers, this volume contains 12 invited full papers. The production of the volume would have been impossible without the diligent work of our expert referees, both Program Committee members or subreviewers. We would like to thank all of them for their excellent work.

Springer generously funded this year a Best Student Paper Award, awarded to a paper authored solely by students. The winner of this award was the paper *"Towards Computable Analysis on the Generalized Real Line,"* by Lorenzo Galeotti and Hugo Nobrega. All authors who contributed to this conference were encouraged to submit significantly extended versions of their papers, with additional unpublished research content, to *Computability. The Journal of the Association CiE.*

The Steering Committee of the conference series CiE is concerned about the representation of female researchers in the field of computability. In order to increase female participation, the series started the Women in Computability (WiC) program in 2007. In 2016, after the new constitution of the Association CiE allowed for the possibility of creating special interest groups, the Special Interest Group Women in Computability was established. Also since 2016, the WiC program has been sponsored by ACM's Women in Computing. This program includes a workshop, the annual WiC diner, the mentorship program, and a grant program for young female researchers. In 2017, the Women in Computability Workshop, coordinated by Liesbeth De Mol, invited the following speakers: Juliette Kennedy (University of Helsinki, Finland), Karen Lange (Wellesley College, USA), Ursula Martin (University of Oxford, UK)

The symposium Magic in Science was co-located with CiE 2017. It took place immediately after the conference and celebrated the 75th birthday of Grzegorz Rozenberg (University of Leiden, The Netherlands and University of Colorado at Boulder, USA), one of the world leaders in research on theoretical computer science and natural computing.

The organizers of CiE 2017 would like to acknowledge and thank the following entities for their financial support (in alphabetical order): the Academy of Finland, the Association for Symbolic Logic (ASL), the City of Turku, the European Association for Theoretical Computer Science (EATCS), Federation of Finnish Learned Societies, Finnish Academy of Science and Letters, Springer, Turku Centre for Computer

Science, Turku Complex Systems Institute, Turku University Foundation, University of Turku, Åbo Akademi University, Åbo Akademi University Foundation. We would also like to acknowledge the support of our non-financial sponsor, the Association Computability in Europe (CiE).

March 2017   Jarkko Kari
                                                              Florin Manea
                                                                  Ion Petre

Organization

Program Committee

Andrew Arana	University of Illinois at Urbana-Champaign, USA
Arnold Beckmann	Swansea University, UK
Paola Bonizzoni	University of Milano-Bicocca, Italy
Olivier Bournez	LIX and Ecole Polytechnique, France
Vasco Brattka	Universität der Bundeswehr München, Germany
Cristian S. Calude	University of Auckland, New Zealand
Ann Copestake	University of Cambridge, UK
Liesbeth De Mol	CNRS UMR8163 Savoirs, Textes, Language Université de Lille 3, France
Ekaterina Fokina	Vienna University of Technology, Austria
Tero Harju	University of Turku, Finland
Emmanuel Jeandel	Laboratoire Lorrain de Recherche en Informatique et ses Applications (LORIA), France
Emil Jebek	Institute of Mathematics of the Czech Academy of Sciences, Czech Republic
Natasha Jonoska	University of South Florida, USA
Jarkko Kari	University of Turku (Co-chair), Finland
Viv Kendon	Durham University, UK
Takayuki Kihara	University of California, Berkeley
Florin Manea	Institut für Informatik, Christian-Albrechts-Universität
Klaus Meer	Brandenburgische Technische Universität Cottbus Senftenberg, Germany
Russell Miller	Queens College and The Graduate Center, CUNY, USA
Bernard Moret	EPFL, Switzerland
Rolf Niedermeier	TU Berlin, Germany
Dag Normann	University of Oslo, Norway
Dirk Nowotka	Christian-Albrechts-Universität zu Kiel, Germany
Isabel Oitavem	CMAF
Ion Petre	Abo Akademi (Co-chair), Finland
Kai Salomaa	Queens University
Reed Solomon	University of Connecticut, USA
Mariya Soskova	Sofia University, Bulgaria
Susan Stepney	University of York, UK
Peter Van Emde Boas	ILLC-FNWI-Universiteit van Amsterdam (Emeritus), The Netherlands
Philip Welch	University of Bristol, UK
Damien Woods	California Institute of Technology, USA

Additional Reviewers

Alten, Clint Van
Atanasiu, Adrian
Azimi, Sepinoud
Berger, Ulrich
Beyersdorff, Olaf
Bollig, Beate
Bonizzoni, Paola
Calvert, Wesley
Carl, Merlin
Carroy, Raphael
Cho, Da-Jung
Csima, Barbara
Darwiche, Adnan
Day, Adam
Diener, Hannes
Durand-Lose, Jérôme
Dzhafarov, Damir
Filmus, Yuval
Franklin, Johanna
Georgiev, Ivan
Gibbons, Jeremy
Golovnev, Alexander
Gregoriades, Vassilios
Hansen, Kristoffer
    Arnsfelt
Hamkins, Joel David
Hirschfeldt, Denis

Hoyrup, Mathieu
Hlzl, Rupert
Ikegami, Daisuke
Ishmukhametov, Shamil
Jalonen, Joonatan
Kahle, Reinhard
Kilinç, Görkem
Knight, Julia
Ko, Sang-Ki
Komusiewicz, Christian
Kopra, Johan
Lecomte, Dominique
Lempp, Steffen
Lenzi, Giacomo
Löwe, Benedikt
Marcone, Alberto
Martin, Barnaby
Mauro, Luca San
Mercas, Robert
Metcalfe, George
Michel, Pascal
Mol, Liesbeth de
Morozov, Andrey
Mundhenk, Martin
Mäkinen, Veli
Ng, Timothy
Niwinski, Damian

Patey, Ludovic
Patterson, Murray
Pich, Ján
Porreca, Antonio E.
Porter, Chris
Poulsen, Bøgsted Danny
Previtali, Marco
Primiero, Giuseppe
Quinn, Sara
Rojas, Cristobal
Rossegger, Dino
Salo, Ville
Schmid, Markus L.
Schweber, Noah
Stephenson, Jonny
Thapen, Neil
Toth, David
Tzameret, Iddo
Törmä, Ilkka
Verlan, Sergey
Vikas, Narayan
Vink, Erik De
Zenil, Hector
Zetzsche, Georg
Zizza, Rosalba

Local Organization

Mikhail Barash
Minna Carla
Christel Engblom

Joonatan Jalonen
Jarkko Kari
Ion Petre

Susanne Ramstedt
Ville Salo

Contents

Contributed Papers

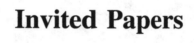

Invited Papers

Character-Based Phylogeny Construction and Its Application to Tumor Evolution

Gianluca Della Vedova[✉], Murray Patterson, Raffaella Rizzi, and Mauricio Soto

DISCo, University of Milano–Bicocca, Milan, Italy
gianluca.dellavedova@unimib.it

Abstract. Character-based Phylogeny Construction is a well-known combinatorial problem whose input is a matrix M and we want to compute a phylogeny that is compatible with the actual species encoded by M.

In this paper we survey some of the known formulations and algorithms for some variants of this problem. Finally, we present the connections between these problems and tumor evolution, and we discuss some of the most important open problems.

1 Introduction

A *phylogeny* is a common representation of any evolutionary history: a labeled tree whose leaves are the extant species or taxa, or individuals, or simply biological data that we are currently able to analyze [16]. The construction of a phylogeny by character-based methods have always found a wide interest among the most theoretically inclined researchers, thanks to its connections with combinatorics (and graph theory in particular). In this paper we will survey some of the main known results on character-based phylogenies, as well as its recent application in cancer genomics.

In this context the main units of study are the *species* and the *characters*: each species is described by a set of attributes, called characters, where each character is inherited independently and can assume one of a finite set of values, called *states*. Notice that species must not be interpreted literally, as they can also be single individuals or entire populations: the only relevant fact is that they can be thought of as a set of characters. Recent applications, indeed, show that such models can be applied to analyze the evolution of data related to various genomic information, such as haplotyping [2,5,13] protein domains [33], markers in tumors [24] or single-nucleotide variants (SNV) [34].

In the following we will denote by C, S, and Q respectively the characters, the species and the possible states that we are considering. Moreover, we will denote by n, m, and k respectively the number of such characters, species and possible states. Without loss of generality, we can assume that there is a special state, represented by 0, the *initial* state of each character. In this paper we will consider only rooted trees, where the root has this initial state for all characters.

© Springer International Publishing AG 2017
J. Kari et al. (Eds.): CiE 2017, LNCS 10307, pp. 3–13, 2017.
DOI: 10.1007/978-3-319-58741-7_1

In a character-based (rooted) tree T, each edge $e = (x, y)$ of T represents an evolutionary event (or *mutation*) and is labeled by some character-state pairs (c, q), where c is a character of M and q is a possible state for c. More precisely, if the edge e is labeled by (c, q), then the mutation occurring in e is the change of state of character c, which assumes state q. This notion allows to extend the idea of state from characters to nodes of the tree T. Let (x, y) be an edge of T. Then the state of y for the character c, denoted by $l_y(c)$ is equal to q if the pair (c, q) labels the edge (x, y), while $l_y(c) = l_x(c)$ otherwise.

Given an $n \times m$ matrix M and a tree T, we will say that T explains M (or that M is consistent with T) if, for each row s of M, there exists a node x of T whose state $l_x(\cdot)$ is equal to the row s. These notions lead to a natural computational problem where the input is a matrix M, while the output is a tree T whose root r has state 0 for each character, and T explains the matrix M — this problem corresponds to computing a putative evolutionary history consistent with the available data (the matrix M). We will call such problem the CHARACTER COMPATIBILITY (CC) problem.

Restricting the set of possible mutations that appear in a tree T results in different variants of CC, each corresponding to different models of evolution, and hence with different possible biological applications.

2   Perfect Phylogeny and Variants

Classically, the most widely studied CC version is PERFECT PHYLOGENY, where no character-state pair (c, q) can label two edges of T. Since the Perfect Phylogeny problem is NP-hard [4], we need to focus on restrictions: for example on the number of possible states, that is, bounding the value of k.

BINARY PERFECT PHYLOGENY considers the case when Q contains only two states, 0 and 1. In this case, state 1 represents the fact that a species possesses a character (for example that a species has wings, or an individual has blood type A). In this problem, not only are there only two states, but each character can label only one edge of the tree T. More formally, given the matrix M, we want to compute a rooted tree T with vertices V_T such that:

(1) For each species $s \in S$ there is a vertex v_s of T with $l_{v_s}(\cdot) = s$.
(2) For every $c \in C$ and $q \in q$, the set $\{v \in V_T \mid l_v(c) = q\}$ induces a connected component of T.

This formulation implies that the infinite sites assumption holds: in fact no mutation can involve twice the same character. While this assumption is very restrictive (in fact it is too restrictive to be applied in several contexts), there are some applications where it is a useful model, such as haplotyping [2] or tumoral phylogeny [14]. From a combinatorial point of view, there is a rich literature, starting from [11], with a simple characterization of the matrices that can be explained by a tree T. Let M be a binary matrix. Then there exists a tree T explaining M if and only if there does not exist two characters c_1 and c_2 and three species s_{01}, s_{10}, s_{11} such that $M[s_{01}, c_1] = M[s_{10}, c_2] = 0$

and $M[s_{01}, c_2] = M[s_{10}, c_1] = M[s_{11}, c_1] = M[s_{11}, c_2] = 1$. Such condition is also called the 4-gamete test, since it consists of finding two characters c_1 and c_2 inducing all four possible pairs $(0,0)$, $(0,1)$, $(1,0)$, and $(1,1)$ in any tree explaining M — notice that the previous definition does not check for the pair $(0,0)$ since the root of T must have state 0 for all characters.

The Binary Perfect Phylogeny problem has a well-known linear-time algorithm [21,22], as well as an $O(nm \log^2(n+m))$-time algorithm when the matrix M has missing entries [32]. While the first algorithm frames the problem as sorting the rows of M, the latter paper describes the problem in a graph-theoretic framework, by introducing a bipartite graph G associated to M. More precisely the vertex set of G is $C \cup S$ and an edge of G connects the species s and the character c iff $M[s, c] = 1$ — with a slight abuse of notation we consider the graph G and the matrix M equivalent. In this context, the 4-gamete test is a forbidden subgraph test (i.e., G can be explained by a tree T iff G does not contain a certain induced subgraph).

Whenever the number k of states is larger than 2, the problem becomes more complex. There have been several papers presenting polynomial-time algorithms for $k \in \{3, 4\}$, culminating with a polynomial-time algorithm for any constant k [1], later improved in [25].

Generalizing the model. A missing entry $M[s, c]$ in a binary matrix M means that we do not know if the species s possesses the character c. As soon as we move away from binary matrices, we can have partial uncertainty on the states that can be assumed by species s and character c, that is, some states are possible but others are not. To model this situation, we can introduce the notion of *generalized character* [3], where each entry $M[s, c]$ is a subset of the set Q of all possible states.

Another form of partial information regards the possible transitions between the states of characters, that is, for each edge (x, y) labeled by (c, q), the relation between the states of c in x and in y. For example, in the case of Binary Perfect Phylogeny, the only possible transition is $0 \rightarrow 1$, but for Perfect Phylogeny in general, only the transitions ending in 0 are forbidden — all other transitions are allowed. Therefore we can model the possible transitions as a directed graph G_Q with vertices Q (Fig. 1).

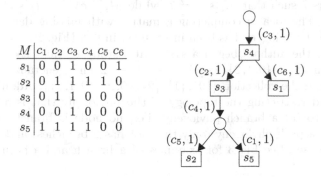

M	c_1	c_2	c_3	c_4	c_5	c_6
s_1	0	0	1	0	0	1
s_2	0	1	1	1	1	0
s_3	0	1	1	0	0	0
s_4	0	0	1	0	0	0
s_5	1	1	1	1	0	0

Fig. 1. A 5×6 binary matrix M that is explained by a perfect phylogeny.

These observations lead to the GENERALIZED CHARACTER COMPATIBILITY (GCC) problem [3] which is to find a perfect phylogeny for a set S of species on a set C of generalised characters, that is, to find a rooted tree $T = (V_T, E)$ that is a perfect phylogeny and, additionally, for every $c \in C$, that the graph $G(c)$ is an induced subgraph of G_Q, where $G(c)$ is the graph obtained from T by contracting all edges that are not labeled by c. Essentially, this new condition means that the state transitions for each character c must respect G_Q. Note that in [3], where the idea of restricting the transitions has originated, they focused on the case when G_Q is a tree. In Binary Perfect Phylogeny with missing entries [32], for example, $Q = \{0, 1\}$ and $G_Q = 0 \to 1$, where missing entries correspond to the subset $\{0, 1\}$ of Q.

From a biological point of view, the most interesting case is when $Q = \{0, 1, 2\}$ and G_Q consists of the two arcs $0 \to 1$, $1 \to 2$ — representing the progression from absent to present to dormant [3] — the case that spurred this study [3] to begin with. Unfortunately, this case is NP-complete [3], motivating the search for different variants of the problem. The first algorithmic results in this context is a polynomial-time algorithm for the case when the set of states in each entry of M induces a directed path on G_Q [3]. Note that Binary Perfect Phylogeny with missing entries is such a case, and so the above result (also) implies that it is polynomial-time solvable.

A subcase of this case of the GCC has also been studied under the name of the PERSISTENT PHYLOGENY problem [6,7,23]. This problem allows exactly one edge to be labeled by $(c, 1)$ or $(c, 0)$, for each character c. The edge labeled by $(c, 1)$ represents the loss (or the transition to the dormant state) of the character c, which is deemed to be possible exactly once in the entire tree T.

Recently there have been some important advances on the Persistent Phylogeny problem, most notably an exhaustive combinatorial algorithm [6], an ILP algorithm [23], and a polynomial-time algorithm [9]. All these algorithms are based on the notions of extended matrix.

The *extended matrix* M_e associated to the input matrix M, is obtained by replacing each column c of M with two columns (c^+, c^-). Moreover for each row s of M, $M_e[s, c^+] = 1$ and $M_e[s, c^-] = 0$ whenever $M[s, c] = 1$, while $M_e[s, c^+] = M_e[s, c^-] = ?$ otherwise. Solving the instance M corresponds to completing the extended matrix M_e [6], that is, replacing the question marks with 0 or 1, obtaining a new matrix M_f which is equal to M_e for all species s and characters c such that $M_e[s, c] \neq ?$, while $M_f[s, c^+] = M_f[s, c^-]$ whenever $M_e[s, c] = ?$. The idea of completing a matrix with missing data in order to obtain a perfect phylogeny has been introduced in [32] (Fig. 2).

In [28,29], the authors began a systematic study of this case of the GCC problem when $Q = \{0, 1, 2\}$, G_Q is $0 \to 1 \to 2$ and generalized characters are chosen from the collection $\{\{0\}, \{1\}, \{2\}, \{0, 2\}, \{0, 1, 2\}\}$. Additionally, the study involved restricting the topology of the phylogeny that is a solution to the problem to (a) a branch phylogeny (i.e., no node is branching, even the root) and (b) a path phylogeny (only the root has 2 branches) and (c) the tree (the general case). See Fig. 3 for examples of a branch and a path phylogeny.

Table 1. Complexity of all cases of the GCC Problem when $Q = \{0, 1, 2\}$, G_Q is $0 \rightarrow 1 \rightarrow 2$ and set of states chosen from the collection $\mathcal{Q} \subseteq \{\{0\}, \{1\}, \{2\}, \{0, 2\}, \{0, 1, 2\}\}$. A ? means that the case remains open.

	$\mathcal{Q}$\soln	Branch	Path	Tree
1	$\mathcal{Q} \subseteq \{\{0\}, \{1\}, \{2\}\}$	P [1]	P [1]	P [1,3]
2	$\{\{0, 1, 2\}\} \subseteq \mathcal{Q} \subseteq \{\{0\}, \{1\}, \{2\}, \{0, 1, 2\}\}$; $|\mathcal{Q}| \leq 2$	trivial	trivial	trivial
3	$\{\{0, 1, 2\}\} \subseteq \mathcal{Q} \subseteq \{\{0\}, \{1\}, \{2\}, \{0, 1, 2\}\}$; $|\mathcal{Q}| \geq 3$	P [29]	NP-c [29]	P [3]
4	$\mathcal{Q} \subseteq \{\{0\}, \{0, 2\}, \{0, 1, 2\}\} \vee \mathcal{Q} \subseteq \{\{2\}, \{0, 2\}, \{0, 1, 2\}\}$	trivial	trivial	trivial
5	$\{\{1\}, \{0, 2\}\}$	P [28]	P [28]	P [9]
6	$\{\{0\}, \{1\}, \{0, 2\}\}$	P [28]	NP-c [29]	?
7	$\{\{0\}, \{2\}, \{0, 2\}\} (\cup \{\{0, 1, 2\}\})$	P [29]	NP-c [29]	P [3]
8	$\{\{1\}, \{2\}, \{0, 2\}\}$	P [28]	P [29]	?
9	$\{\{0\}, \{1\}, \{2\}, \{0, 2\}\}$	P [28]	NP-c [29]	?
10	$\{\{1\}, \{0, 2\}, \{0, 1, 2\}\} \subseteq \mathcal{Q}$	NP-c [28]	NP-c [29]	NP-c [3]

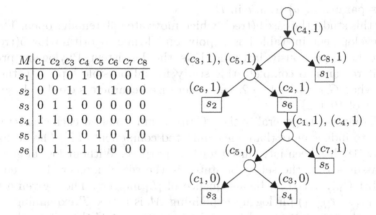

Fig. 2. A binary matrix M of size 6×8 that is explained by the persistent phylogeny on the right. The boldfaced entries of M correspond to two characters (c_2 and c_4) inducing the states $(0, 1)$ $(1, 0)$ and $(1, 1)$, hence M cannot be explained by a perfect phylogeny.

The reason for studying the restrictions to branch and path phylogies is that 70% of real instances of human genotype data that admit a tree phylogeny also admit a path phylogeny [19]. The results of this study are summarized in Table 1.

The tractable cases of Table 1 can be summarized as follows. First of all, 1(branch–tree) are cases of [1] — 1(tree) also implied by the algorithm in [3]. In [28], the authors show that 5(branch–path) are equivalent to the linear-time solvable *Consecutive-Ones Property* (C1P) problem [10]. The C1P problem is, given an $n \times m$ binary matrix M, to decide if there is an ordering of the columns of M in such a way that for each row, the set of columns that are 1 in the row appear consecutively — such an ordering is called a *consecutive-ones ordering*.

The C1P can be solved in time $O(n+m+f)$ [10], where f is the number of non-zero entries in M, by giving an algorithm that builds a so-called PQ-tree [10], a structure that encodes all consecutive-ones orderings of M. In fact, 6(branch), 8(branch) and 9(branch) are based on algorithm that uses the PQ-tree that runs in time $O(nm^4 + f)$. Then, in [29], the follow-up to [28], the authors show that 8(path) can be reduced to solving case 8(branch). While 3(tree) and 7(tree) are tractable given the algorithm of [3], in [29], the authors give an algorithm based on this one for the case where the solution must be a branch and the set of states in each entry of M induces a directed path on G_Q. Problem 6(tree) has also been called Constrained Persistent Phylogeny [8].

For completeness, 10(tree) was shown to be NP-complete in [3]. The remaining cases in Table 1 have been shown to be NP-complete [28,29] by reduction from different versions of the NP-complete *Path Triple Consistency* (PTC) problem of [28]. The PTC problem is, given a set $V = \{1, \ldots, n\}$ and a collection of triples $\{a, b, c\}$ from V, does there exist a path $P = (V, E)$ (an ordering) on the elements of V such that for each triple, there exists an edge $e \in E$ of P whose removal separates $\{a, b\}$ from c in P.

After this study, the case 9(tree), which motivated [3] remains open. The most recent development in Table 1 is a polynomial-time algorithm for 5(tree) [9], which we have previously introduced as the Persistent Phylogeny problem. Finally, it remains to complete this study for all possible inputs to the GCC problem when $G_Q = 0 \to 1 \to 2$, i.e., generalized characters can be chosen from any subset of $\{0, 1, 2\}$.

A different way to generalize the CC problem is to allow each character-state pair (c, q) to induce more than one connected component of T. The notion of ℓ-phylogeny [18] has been introduced for this purpose. Without loss of generality, we can assume that the set Q of states is the set of integers between 0 and $|L| - 1$. Let $(\ell_0, \ldots, \ell_{|L|-1})$ be a sequence of $|L|$ integers. Then, given a matrix M, an $(\ell_0, \ldots, \ell_{|L|-1})$-phylogeny explaining M is a tree T explaining M such that for each state $q \in Q$ and for each character $c \in C$ the subgraph induced

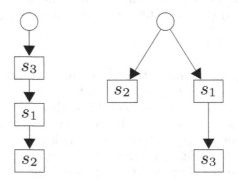

Fig. 3. Examples of a branch phylogeny (left), and a path phylogeny (right) on some set $S = \{s_1, s_2, s_3\}$ of species

by the set $l_c^{-1}(q)$ has at most at most ℓ_q connected components, where $l_c^{-1}(q)$ is the set of nodes of T where the character c has state q. In other words, for each character each state q can be acquired at most ℓ_q times in the tree T. Unfortunately, this direction has led mostly to NP-complete problems, such as finding an $(\ell, \ldots, \ell)$-phylogeny when $\ell > 1$ [18].

Two cases that have been left open in [18] are finding a $(2,1)$-phylogeny and a $(2,2)$-phylogeny. Notice that the Persistent Phylogeny problem is equivalent to finding a $(2,1)$-phylogeny [36], hence that problem has been settled [9].

3 Tumor Evolution

An important application of phylogenetic trees is the evolutionary reconstruction of the history of tumor cells. Cancer can be seen as an uncontrolled evolutionary process of somatic mutations of tumor cells from a single founder cell [20] creating a diverse set of subpopulations [12,27,37]. From this point of view, representing tumor progression could be stated as a phylogeny reconstruction problem by considering a particular subpopulation of mutated cells as the species and mutations as characters. Nowadays, single cell sequencing is still a daunting process [31]. Instead, the most common procedure in cancer patients is to extract and sequence multiples samples of the tumor tissue. Each of these samples contains millions of cells which come from multiple tumor subpopulations called clones. From sequencing data we can obtain the variant allele frequency (VAF) which corresponds to the proportion of cells in the sample containing a somatic mutation.

The natural problem in this framework is to infer, only from VAF data, the mutations present in the clone subpopulations and their history, that is, a phylogeny that can explain the input VAF data. Formally speaking, VAF information for p samples and m somatic mutations can be stored in a $p \times m$ (frequency) matrix F where $F[t,j]$ represents the proportion of cells in sample t that have the mutation j. If n clones are supposed to be present in the samples, they can be described by a binary $n \times m$ matrix B where $M[i,j]$ represents the presence or absence of mutation j in the i-th clone: hence B is a matrix that can be explained by a perfect phylogeny. Finally, the $p \times n$ (usage) matrix U describes the relative proportion of clones in each sample, that is, $U[t,i]$ corresponds to the proportion of the cells in the sample belonging to the subpopulation i. From these definitions, given the frequency matrix F, our problem can be stated as the following matrix factorization problem $F = \frac{1}{2}UB$ [14], where clonal matrix B respects a phylogenetic evolutionary model (usually a perfect phylogeny) — the $\frac{1}{2}$ is a technical consequence of the fact that each human being has two copies of each chromosome and a mutation affects only one of these two copies. Figure 4 presents an instance of the problem and a solution.

Many works have been proposed under different assumptions like a single sample [30,35] or considering different underlying evolutionary processes. From a theoretical perspective, the problem of the reconstruction of a perfect phylogenetic history for multiple samples is NP-complete [14] even for binary trees [24].

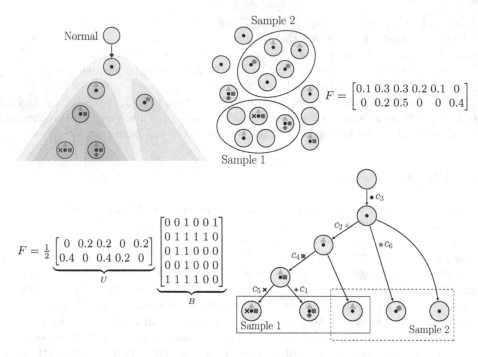

Fig. 4. Example of the phylogenetic clonal reconstruction for the tumor composition problem. On the top, we have (on the left) the actual unknown evolutionary history and the actual unknown clonal subpopulations, where each colored dot is a mutation; (center) the cells in the tumor tissue and in the samples; (on the right) the resulting VAF matrix that is the instance of our problem. On the bottom, we have (on the left) the solution of the instance expressed as the matrices U and B; (on the right) the solution of the instance expressed as the perfect phylogeny.

Nevertheless, the authors of [14] propose a mixed tools strategy of combinatorial and integer linear programming approaches to find a feasible solution for the reconstruction problem. The multi-state generalization [17] of the binary model is studied in [15] where the authors provided an algorithm to enumerate all possible evolutionary trees explaining the observed proportion under the multi-state perfect phylogenetic model.

4   Conclusions

In this survey we have briefly introduced the current status of character-based phylogeny. We conclude the paper with a short discussion on the open problems. In Table 1 we have classified several problems, describing only if they are NP-complete or if they have a polynomial-time algorithm. Some of the known algorithms are not optimal, hence designing faster algorithms is a possible direction.

The study of approximation or fixed-parameter algorithms in this field is only in its infancy. Since all problems presented here are decision problems, the best choice of an objective function is not clear.

Finally, we leave the reader two problems that have been left open from [18] — more than two decades ago — on a matrix M: (1) finding an $(\ell, \ell, \ldots, \ell)$-phylogeny explaining M when $k > 2$ is a constant and $\ell > 1$, and (2) finding a $(2, 2)$-phylogeny explaining M.

Acknowledgments. We acknowledge the support of the MIUR PRIN 2010–2011 grant "Automi e Linguaggi Formali: Aspetti Matematici e Applicativi" code 2010LYA9RH, of the Cariplo Foundation grant 2013–0955 (Modulation of anti cancer immune response by regulatory non-coding RNAs), of the FA grants 2013-ATE-0281, 2014-ATE-0382, and 2015-ATE-0113.

References

1. Agarwala, R., Fernández-Baca, D.: A polynomial-time algorithm for the perfect phylogeny problem when the number of character states is fixed. SIAM J. Comput. **23**(6), 1216–1224 (1994)
2. Bafna, V., Gusfield, D., Lancia, G., Yooseph, S.: Haplotyping as perfect phylogeny: a direct approach. J. Comput. Biol. **10**(3–4), 323–340 (2003)
3. Benham, C., Kannan, S., Paterson, M., Warnow, T.: Hen's teeth and whale's feet: generalized characters and their compatibility. J. Comp. Biol. **2**(4), 515–525 (1995)
4. Bodlaender, H.L., Fellows, M.R., Warnow, T.J.: Two strikes against perfect phylogeny. In: Kuich, W. (ed.) ICALP 1992. LNCS, vol. 623, pp. 273–283. Springer, Heidelberg (1992). doi:10.1007/3-540-55719-9_80
5. Bonizzoni, P.: A linear-time algorithm for the perfect phylogeny haplotype problem. Algorithmica **48**(3), 267–285 (2007)
6. Bonizzoni, P., Braghin, C., Dondi, R., Trucco, G.: The binary perfect phylogeny with persistent characters. Theor. Comput. Sci. **454**, 51–63 (2012)
7. Bonizzoni, P., Carrieri, A.P., Della Vedova, G., Rizzi, R., Trucco, G.: A colored graph approach to perfect phylogeny with persistent characters. Theor. Comput. Sci. **658**, 60–73 (2016)
8. Bonizzoni, P., Carrieri, A.P., Della Vedova, G., Trucco, G.: Explaining evolution via constrained persistent perfect phylogeny. BMC Genomics **15**(6), S10 (2014)
9. Bonizzoni, P., Della Vedova, G., Trucco, G.: Solving the persistent phylogeny problem in polynomial time. CoRR, abs/1611.01017 (2016)
10. Booth, K.S., Lueker, G.S.: Testing for the consecutive ones property, interval graphs, and graph planarity using PQ-tree algorithms. J. Comput. Syst. Sci. **13**(3), 335–379 (1976)
11. Buneman, P.: The recovery of trees from measures of dissimilarity. In: Hodson, F.R., Kendall, D.G., Tautu, P. (eds.) Mathematics in the Archaelogical and Historical Sciences. Edinburgh University Press, Edinburgh (1971)
12. Ding, L., Raphael, B.J., Chen, F., Wendl, M.C.: Advances for studying clonal evolution in cancer. Cancer Lett. **340**(2), 212–219 (2013)
13. Ding, Z., Filkov, V., Gusfield, D.: A linear-time algorithm for the perfect phylogeny haplotyping (PPH) problem. J. Comput. Biol. **13**(2), 522–553 (2006)

14. El-Kebir, M., Oesper, L., Acheson-Field, H., Raphael, B.J.: Reconstruction of clonal trees and tumor composition from multi-sample sequencing data. Bioinformatics **31**(12), i62–i70 (2015)
15. El-Kebir, M., Satas, G., Oesper, L., Raphael, B.: Inferring the mutational history of a tumor using multi-state perfect phylogeny mixtures. Cell Syst. **3**(1), 43–53 (2016)
16. Felsenstein, J.: Inferring Phylogenies. Sinauer Associates, Sunderland (2004)
17. Fernandez-Baca, D.: The perfect phylogeny problem. In: Du, D.Z., Cheng, X. (eds.) Steiner Trees in Industries. Kluwer Academic Publishers, Dordrecht (2000)
18. Goldberg, L.A., Goldberg, P.W., Phillips, C.A., Sweedyk, E., Warnow, T.: Minimizing phylogenetic number to find good evolutionary trees. Discrete Appl. Math. **71**(1–3), 111–136 (1996)
19. Gramm, J., Nierhoff, T., Sharan, R., Tantau, T.: Haplotyping with missing data via perfect path phylogenies. Discrete Appl. Math. **155**, 788–805 (2007)
20. Greaves, M., Maley, C.C.: Clonal evolution in cancer. Nature **481**(7381), 306–313 (2012)
21. Gusfield, D.: Efficient algorithms for inferring evolutionary trees. Networks **21**, 19–28 (1991)
22. Gusfield, D.: Algorithms on Strings, Trees and Sequences: Computer Science and Computational Biology. Cambridge University Press, Cambridge (1997)
23. Gusfield, D.: Persistent phylogeny: a galled-tree and integer linear programming approach. In: Proceedings of the 6th ACM BCB Conference, pp. 443–451 (2015)
24. Hajirasouliha, I., Mahmoody, A., Raphael, B.J.: A combinatorial approach for analyzing intra-tumor heterogeneity from high-throughput sequencing data. Bioinformatics **30**(12), i78–i86 (2014)
25. Kannan, S., Warnow, T.: A fast algorithm for the computation and enumeration of perfect phylogenies. SIAM J. Comput. **26**(6), 1749–1763 (1997)
26. Kollar, E., Fisher, C.: Tooth induction in chick epithelium: expression of quiescent genes for enamel synthesis. Science **207**, 993–995 (1980)
27. Lawrence, M.S., Stojanov, P., et al.: Mutational heterogeneity in cancer and the search for new cancer-associated genes. Nature **499**(7457), 214–218 (2013)
28. Maňuch, J., Patterson, M., Gupta, A.: On the generalised character compatibility problem for non-branching character trees. In: Ngo, H.Q. (ed.) COCOON 2009. LNCS, vol. 5609, pp. 268–276. Springer, Heidelberg (2009). doi:10.1007/978-3-642-02882-3_27
29. Maňuch, J., Patterson, M., Gupta, A.: Towards a characterisation of the generalised cladistic character compatibility problem for non-branching character trees. In: Chen, J., Wang, J., Zelikovsky, A. (eds.) ISBRA 2011. LNCS, vol. 6674, pp. 440–451. Springer, Heidelberg (2011). doi:10.1007/978-3-642-21260-4_41
30. Miller, C.A., et al.: Sciclone: inferring clonal architecture and tracking the spatial and temporal patterns of tumor evolution. PLoS Comput. Biol. **10**(8), e1003665 (2014)
31. Navin, N.E.: Cancer genomics: one cell at a time. Genome Biol. **15**(8), 452 (2014)
32. Pe'er, I., Pupko, T., Shamir, R., Sharan, R.: Incomplete directed perfect phylogeny. Siam J. Comput. **33**(3), 590–607 (2004)
33. Przytycka, T., Davis, G., Song, N., Durand, D.: Graph theoretical insights into evolution of multidomain proteins. J. Comput. Biol. **13**(2), 351–363 (2006)
34. van Rens, K.E., Mäkinen, V., Tomescu, A.I.: SNV-PPILP: refined SNV calling for tumor data using perfect phylogenies and ILP. Bioinf. **31**(7), 1133–1135 (2015)
35. Roth, A., Khattra, J., et al.: Pyclone: statistical inference of clonal population structure in cancer. Nat. Methods **11**(4), 396–398 (2014)

36. Steel, M.A.: Phylogeny: Discrete and Random Processes in Evolution. CBMS-NSF Regional Conference Series in Applied Mathematics. SIAM, Philadelphia (2016)
37. Vogelstein, B., Papadopoulos, N., Velculescu, V.E., Zhou, S., Diaz, L.A., Kinzler, K.W.: Cancer genome landscapes. Science **339**(6127), 1546–1558 (2013)

Is there any Real Substance to the Claims for a 'New Computationalism'?

Alberto Hernández-Espinosa[1], Francisco Hernández-Quiroz[1],
and Héctor Zenil[2,3,4(✉)]

[1] Departamento de Matemáticas, Facultad de Ciencias, UNAM, Mexico City, Mexico
albertohernandezespinosa@gmail.com, fhq@ciencias.unam.mx
[2] Department of Computer Science, University of Oxford, Oxford, UK
[3] Information Dynamics Lab, Karolinska Institutet, Stockholm, Sweden
[4] Algorithmic Nature Group, LABORES, Paris, France
hector.zenil@algorithmicnaturelab.org

Abstract. *Computationalism* is a relatively vague term used to describe attempts to apply Turing's model of computation to phenomena outside its original purview: in modelling the human mind, in physics, mathematics, etc. Early versions of computationalism faced strong objections from many (and varied) quarters, from philosophers to practitioners of the aforementioned disciplines. Here we will not address the fundamental question of whether computational models are appropriate for describing some or all of the wide range of processes that they have been applied to, but will focus instead on whether 'renovated' versions of the *new computationalism* shed any new light on or resolve previous tensions between proponents and skeptics. We find this, however, not to be the case, because the *new computationalism* falls short by using limited versions of "traditional computation", or proposing computational models that easily fall within the scope of Turing's original model, or else proffering versions of hypercomputation with its many pitfalls.

Keywords: Computationalism · Classical computation · Natural computation · Computability · Turing machine model

1 Classical vs. Non-classical Computation

The simplest view of the Turing machine model (TM) construes it as a decision problem solver, tackling such questions as whether a certain string represents a prime number or whether a certain other string belongs to a context-free language. Of course, this view is rather restrictive, as there are many interesting questions that cannot be answered with a simple "yes" or "no". But TMs can be viewed as mechanisms for calculating functions, with the input string representing the argument(s) of the function and the string left on the tape at halting time

---

Invited contribution to Computability in Europe (CiE) 2017 – 'Unveiling Dynamics and Complexity'.

© Springer International Publishing AG 2017
J. Kari et al. (Eds.): CiE 2017, LNCS 10307, pp. 14–23, 2017.
DOI: 10.1007/978-3-319-58741-7_2

representing the result. Given the easy correspondence between natural numbers and finite strings in an alphabet, a TM can be said to calculate a function from natural numbers to natural numbers. Decision problems can be viewed as special cases of functions from natural numbers to natural numbers.

A basic set-theoretical argument tells us that there are many more functions from natural numbers to natural numbers than there are possible TMs and, ergo, that most functions cannot be computed by TMs. The halting problem is one such function.

Within the field of classical computation, and indeed coeval with the introduction of classical computation, certain forms of non-classical computation were devised, such as the oracle machine, which was introduced by Turing himself [31].

Here we do not aim to add to the already lengthy list of possible objections to hypercomputation, which claims the feasibility of computational models that may go beyond the Turing limit in theory but not in practice. Instead we offer an analysis and criticism of supposedly new models of computation that claim to be different from and even to exceed (regardless of whether or not they can be classified as hypercomputation) the classical Turing model in their ability to describe how nature works and–so it is claimed–compute in radical or innovative ways.

2 A Brief Roadmap to Computationalism

While there is no current consensus as to the validity of attacks on *classical computationalism*, nowadays many researchers in different fields seem to agree that new models of computation are needed in order to overcome such objections (for a summary of which see [12]).

In this paper we will refer to the former type of computationalism as *classical computationalism* and to the latter type as *new computationalism*. The new computationalist wave is a highly varied mix which encompasses both rejections of Turing's model and appeals to "natural" computation.

In the decades following Turing's introduction of his formalization of effective procedure (as defined in [14]) in his seminal paper [30, 31], and especially after the widespread and profound success of electronic computers in science and engineering (now universally regarded as incarnations of Turing's mathematical model), there was a strong impulse to not only use computers in every field within sight but also to view them as models of how things really are, the *computational model of the human mind* being quite probably one of the most, if not the most, iconic instance of this tendency [24]. The process of encoding (or rewriting/or reinterpreting) a problem as a finite sequence of symbols which could be manipulated mechanically by Turing machines in order to solve it is what came to be referred to as *classical computationalism*.

Very soon dissenting voices raised objections based on (controversial) interpretations of Gödel's theorems [17, 23], failures to close the gap between mechanical processing of information and real understanding of it [25, 26], and the obvious differences between the way brains process information and the particular operation of a Turing machine. The crisis in Artificial Intelligence in the 1980s [4] did not help to advance the cause of computationalism, as some early efforts to apply computers to (seemingly) not very complex human abilities like language translation or vision failed miserably. That the objectors to computationalism were not able to present better models of the human mind did not lead them to demur.

The time was ripe for bold proposals to overcome the impasse. Among the most popular were *hypercomputation* and some forms of *natural computing*, together with what we will classify as 'other models of computation' based upon variations of the operation of classical models.

3   The Uninstantiation of Hypercomputation

A mechanism more powerful than any TM must be able to compute more functions than a TM can. If it merely calculates what a TM does, only (finitely) faster, or more intuitively or with less hassle for its creator, then we cannot say it is computationally more powerful, as it can be simulated by a TM. The Church-Turing thesis states that any formalism capturing what an effective procedure is will be equivalent to a TM [15]. The Church-Turing thesis is much maligned among neo- and hyper-computationalists, but as Sieg [27] has shown (following ideas first advanced by Gandy [9]), it can be reduced to two very basic principles: boundedness and locality conditions. The former implies that a computing device can immediately recognize only a bounded number of configurations, the latter that a computing device can change only immediately recognizable configurations. In (perhaps) oversimplified terms, in order to overcome the TM's limits, the device must be able to either access an infinite amount of information or must act upon places that are not immediately accessible in a finite number of operations. Sieg's formulation does not imply a proof of Church's thesis, but instead establishes a mathematical baseline for the kind of device needed to violate the so-called *Turing limit*.

Such devices may exist or may eventually be created, but they must act very differently from our current computers and cannot be based on trivial variations of classical models.

Some models for hypercomputation include the Oracle Turing Machine model [32], analog recurrent neural networks [29], and analog computation [22].

However, these models are just theoretical constructs, and not only are there no actual devices based on them or physical processes which correspond to them (as far as we know), but there is no prospect of turning them into concrete, viable tools for research in the foreseeable future (even Turing did not have such an eventuality in mind). See Davis' paper on hypercomputation [5] for a critique of those who think otherwise. As we consider Davis' analysis quite complete and well founded, we shall dwell no further on this issue and we will conclude this section by saying that non-existence and breach of physical laws (mainly the 2nd. law of thermodynamics) are good reasons to overlook hypercomputation as a meaningful alternative to computation.

More recently, Maldonado [18, 19] has offered a defense of a form of *biological hypercomputation*, claiming that: "[...] life is not a standard Turing Machine", but rather that living systems hypercompute, and that an understanding of life is reached not by grasping what life is but what it does.

It has even been suggested that phenomena such as death can be sources of a sort of uncomputability, due to the alleged incapability of information theory to describe death or to have it programmed into a system as a desirable property so as to provide meaning to artificial life–just as it does in the case of natural life [8]. Molecular biology, however, can explain death, using straightforward analogies to computation and reprogramming contextualized within information theory [37].

However, that science has not yet fully explained life and death, among other things, does not mean that it will not do so in the future. Thus we consider the claim that neither computationalism nor information theory can explain death (and hence life, according to [8]) shortsighted, if not simply incorrect. Since the discovery of DNA we have known that developmental and molecular biology (and thus biology) are mostly information theoretic, and the more we explore these fields the more we find to confirm our sense that this is indeed the case.

4   Natural Computation

Nature is a rich source of ideas and there has lately been a turn toward *natural computation* in the literal sense. Of course, there is nothing wrong with looking to natural phenomena for inspiration. Wolfram takes a very pragmatic approach in his epistemological treatise, a non-classical exploration of the classical computational universe, finding qualitative parallels between nature and computation, with nature harnessing the power of classical computation as a natural source of algorithmic creativity [35].

Others, however, have gone further, offering a divergent notion of computation by attacking classical computation, alleging that a set of constraints that have been in place from the inception of the classical model of computation have handicapped not only the model but the scientific and technological progress of computation as such. The common idea behind most, if not all of these objections to the classical model is that nature does not operate like a Turing machine–because, e.g., nature works in parallel over analog information [7], because nature does more than solve problems [18], and because, it is claimed, there is no way to construct a machine with an infinite tape [7].

Perhaps the most puzzling aspect of the arguments of a group of researchers looking for new notions or types of computation is that while openly accepting that (natural) computing is about information processing (as is classical computation) [6], and also that nature certainly computes (because computers exist in and within nature), they posit a different kind of computation than the classical one [19] while using nature as evidence for non-Turing computation. At the same time none of them specifies exactly what makes this kind of computation distinctive, beyond stressing its difference from a Turing machine (or a trivial modification of a Turing machine, e.g., a non-terminating one, such as a cellular automaton, which can hardly be classified as non-classical). Such a line of reasoning eventuates in a trivialized natural computation thesis.

Another objection to this view is that by generalizing the notion of computation to any process that transforms its environment, it renders the concept vacuous. The specificity of symbolically encoding problems and solving them by a set of finite, formal rules is then lost. Mechanics, process, transformation *and* computation are all synonyms. Again there is nothing wrong with this per se, but in practical terms this trend does not point out how to attack problems with our current computers, which happen to be (less than ideal) Turing machines. In other words, an extension of the concept of computation should require an enhancement of computers. At present it is highly debatable where this enhancement will come from, but this line of thought definitely takes us back to hypercomputational ground.

5   Not More but Equally Powerful

Finally there is an abundance of computational models that have sometimes been touted as more powerful alternatives to Turing machines, but on inspection turn out to be mathematically equivalent to Turing's model. Of course, a different model may give us a new, powerful insight into an aspect of computation obscured by the rigidity of Turing machines (mind you, they were supposed to be rigid in the extreme). This is the case with numerous models of concurrent and distributed computation [21]. But this does not mean that these models solve problems a Turing machine cannot. Being the good theoreticians they are, the people behind concurrent or distributed models have not made any such claim.

Of course, the mathematical equivalence between the TM model and many others (programming languages, concurrent computation, etc.) does not mean that the latter are superfluous. They were introduced to solve real, important problems for which TMs did not provide a clear or manageable way of expressing the actual questions. If, for instance, you are trying to capture the fundamental issues of communication and synchronization dealt with by the π-calculus [20] you will not get very far by encoding them as a string to be manipulated by a sequential TM with its mind-boggling (and mostly irrelevant) details, even if this were theoretically possible. In other words, we are not challenging the utility of alternatives to the TM, only the claim that some models can do what no TM can *in principle*.

Among other models is the *Interaction machine* model that "extend[s] the Turing Machine model by allowing *interaction*, i.e. input and output actions (read and write statements) determined by the environment at each step of the computation" [33,34], π-calculus, "a mathematical model of processes whose interconnections change as they interact" [20,21]. Scott Aaronson offers a whimsical characterization of the Interactive model [1]. It is puzzling that interactive computations cannot simply be viewed as independent classical computations.

Moreover, software such as operating systems are implementations of highly interactive programs. They were introduced early in the development of the first computers, and concurrent computation is an active area of research where these kinds of questions are addressed within a very classical–so to speak–framework. Nothing in Turing's model prevents an external observer or machine from interacting with the working tape of the original machine, thus effectively interacting with the machine itself.

Another model is the *Inductive computing* model in the context of what the author has called 'super-recursive algorithms [2]. In many respects, the inductive machine model is not comparable to classical computation, but there is one respect in which it is not that far removed from a certain form of such computation. Inductive computation does not produce a definitive output and is thus similar to transducers, and like cellular automata they do not terminate, suggesting a transducer or cellular automaton-type of computation that supposedly generalizes the classical Turing machine model.

Many of the objections based upon models such as that of Gurevich [10], even when deployed with intent to disprove the Church-Turing Thesis [11]–in what constitutes a clear misunderstanding of the philosophical basis and content of said thesis [28]–are based upon, for instance, the argument that Turing machines cannot deal with structures other than strings on tapes, even when trivial modifications that preserve all of their classical properties have been made, modifications that in no way imply the invalidity of the original TM model [16] (for example, models of Turing machines operating on grids preserving, say, algorithmic information properties [36]. This does not mean, however, that such models cannot be useful, a case in point being high-level descriptions of classical computation, with Gurevich's [10] model being put to very practical use nowadays in software engineering, as a tool for software modelling.

6   Old Dogs, New Tricks

A standard for surpassing the Turing model and disproving the Church-Turing thesis must entail something far more stringent than trivial modifications of classical computation. Of course, the main question is what constitutes a nontrivial modification of the classical model, a modification that does more than simply introduce an infinite element which merely takes the purportedly feasible new model into the *hypercomputation* category.

We consider the use of the expression "more powerful" as merely metaphorical unless specifics are provided as regards what makes a given model more powerful. Likewise, if as soon as such a model is instantiated it merely becomes as powerful as or else not comparable to classical computation.

An example of a trivial modification (to modern eyes) that has been accepted as not leading to more computational power other than speed-up is the use of additional machine tapes.

In the same fashion then, when it is claimed that concurrent computation is more powerful than TMs as TMs are crippled by their 'sequentiality' (bearing in mind that though concurrency can be properly simulated in sequential

TMs, doing so is very cumbersome), we do not consider such an objection to be an objection in principle, as it is not related to the inability of the model to undergo minor changes without changing anything more than the details of its operation. Or that object oriented programming is more powerful because it is heuristically superior to clumsy TMs (although again TMs can simulate object oriented programming—with overhead). Similarly, objections concerning speed, illustrated by, e.g., quantum computing, do not fall into the category of fundamental challenges to the model, having to do only with its operation.

A model claimed to be more powerful than classical models is the so-called 'actor' model that, according to its originator, was inspired by physics, including general relativity and quantum mechanics. It was also influenced by the programming languages Lisp, Simula and early versions of Smalltalk, as well as capability-based systems and packet switching. Its development was "motivated by the prospect of highly parallel computing machines consisting of dozens, hundreds, or even thousands of independent microprocessors, each with its own local memory and communications processor, communicating via a high-performance communications network." [3]

According to Hewitt [13], "concurrency extends computation beyond the conceptual framework of Church, Gandy, Gödel, Herbrand, Kleene, Post, Rosser, Sieg, Turing, etc. because there are effective computations that cannot be performed by Turing Machines. . . . [and where] computation is conceived as distributed in space where computational devices communicate asynchronously and the entire computation is not in any well-defined state. (An *actor* can have stable information about what it was like when it received a message.) Turing's Model is a special case of the Actor Model." [13]

It appears trivial to most computer scientists that these models can be simulated by classical models (e.g. by dovetailing on parallel computations on different inputs stored in different tapes) as long as there are not an infinite number of interactions or an infinite number of actors acting at the same time that would violate the boundedness and locality principles of feasible models [27, 28].

7   Conclusion

This paper does not attempt to disprove the existence of ways of overcoming the limitations of the traditional Turing machine model or to provide a survey of models of computation purported to go beyond the Turing model (whether claiming the status of hypercomputation or not). Instead, it attempts to be a reminder of what those limitations are and how far some claims have gone in trying to establish a new type of computationalism, claims that are often, if not always, (mistakenly) predicated on the apparent weakness of classical models, weaknesses that are in fact only weaknesses of orthodox interpretations of their operating details.

There are very good theoretical models of what life looks like that purport to surpass the Turing machine model, but we are still far from being able to put any of these models into practice, assuming it will ever be possible to do so.

Every few years we see a claim of this sort, and its technical merits should be assessed in order to (most improbably) accept it or (as has been usual hitherto) debunk it.

Clearly we have not gone in for clever new theorems or innovations attempting to analyze specific proposals that have been floated, which you may find disappointing (just good old theory). For our part we are even more disappointed at not being able to acknowledge the appearance of novel and more solid ideas and have not felt compelled to spend time producing a theorem to show that a classical Turing machine can, for example, simply be extended to operate on grids and other structures and still preserve its classical nature by virtue of preserving all the theory of computation derived for it, respecting hierarchies and at most achieving speed-up gains.

The common denominator of all these attacks on classical computation, including Church's thesis, is the impression they create of refuting an opponent's argument though the arguments refuted are not ones that have actually been advanced by anyone– what is called a *straw man fallacy*. In effect disputes are generated where there are none. For example, no serious researcher has ever suggested that the mind, nature or the universe operates or is a mechanical incarnation of a (universal) Turing machine.

In order to build sound objections against classical computation and computationalism, we conclude that it is thus necessary to represent it in its full spectrum, and not to adopt an old, abstract, symbol- manipulation view of computation that is out of date or else has been oversimplified for other purposes.

References

1. Aaronson, S.: The Toaster-Enhanced Turing Machine, Blog entry. http://www. scottaaronson.com/blog/?p=1121. Accessed 11 Feb 2017
2. Burgin, M.: Super-Recursive Algorithms. Monographs in Computer Science. Springer, New York (2005)
3. Clinger, W.D.: Foundations of actor semantics. AITR-633 (1981)
4. Crevier, D.: AI: The Tumultuous History of the Search for Artificial Intelligence. Basic Books, New York (1993)
5. Davis, M.: The myth of hypercomputation. In: Teuscher, C. (ed.) Alan Turing: Life and Legacy of a Great Thinker, pp. 195–211. Springer, Heidelberg (2004)
6. Dodig-Crnkovic, G.: Significance of models of computation, from turing model to natural computation. Minds Mach. **21**(2), 301–322 (2011)
7. Fresco, N.: Physical Computation and Cognitive Science. Studies in Applied Philosophy, Epistemology and Rational Ethics, vol. 12. Springer, Heildeberg (2014)
8. Froese, T.: Life is precious because it is precarious: Individuality, mortality, and the problem of meaning. In: Dodig-Crnkovic, G., Giovagnoli, R. (eds.) Representation and Reality: Humans, Animals and Machines, Springer (in press)
9. Gandy, R.: Church's thesis and the principles for mechanisms. In: Barwise, H.J., Keisler, H.J., Kunen, K. (eds.) The Kleene Symposium, pp. 123–148. North-Holland Publishing Company (1980)
10. Gurevich, Y.: Sequential abstract state machines capture sequential algorithms. ACM Trans. Comput. Log. **1**(1), 77–111 (2000)

11. Gurevich, Y., Dershowitz, N.: A natural axiomatization of computatbility and proof of the church's thesis. Bull. Symbolic Logic. **14**(3), 299–350 (2008)
12. Hernández-Espinosa, A., Hernández-Quiroz, F.: Does the principle of computational equivalence overcome the objections against computationalism? In: Dodig-Crnkovic, G., Giovagnoli, R. (eds.) Computing Nature Turing Centenary Perspective. Studies in Applied Philosophy, Epistemology and Rational Ethics, pp. 225–233. Springer, Heidelberg (2013)
13. Hewitt, C.: What is computation? Actor model versus turing's model. In: A Computable Universe, Understanding Computation and Exploring Nature as Computation. World Scientific Publishing Company/Imperial College Press, Singapore (2013)
14. Hilbert, D.: Neubegrndung der Mathematik: Erste Mitteilung. Abhandlungen ausdem Seminar der Hamburgischen Universität **1**, 157–177 (1922)
15. Kleene, S.C.: Introduction to Metamathematics. North-Holland, Amsterdam (1952)
16. Langton, C.G.: Studying artificial life with cellular automata. Physica D: Nonlinear Phenom. **22**(1–3), 120–149 (1986)
17. Lucas, J.R.: Minds, machines and Gödel. Philosophy **36**(137), 112–127 (1961)
18. Maldonado, C.E., Gómez-Cruz, N.A.: Biological hypercomputation: A concept is introduced (2012). arXiv preprint arXiv:1210.4819
19. Maldonado, C.E., Gómez-Cruz, N.A.: Biological hypercomputation: A new research problem in complexity theory. Complexity **20**(4), 8–18 (2015)
20. Milner, R.: Communicating, Mobile Systems: The Pi Calculus. Cambridge University Press, Cambridge (1999)
21. Milner, R., Parrow, J., Walker, D.: A calculus of mobile processes. Inf. Comput. **100**(1), 1–40 (1992)
22. Mycka, J., Costa, J.F.: A new conceptual framework for analog computation. Theoret. Comput. Sci. **374**(1–3), 277–290 (2007)
23. Penrose, R., Mermin, D.: The emperor's new mind: Concerning computers, minds, and the laws of physics. Am. J. Phys. **58**(12), 1214–1216 (1990)
24. Putnam, H.: Representation and Reality. A Bradford Book, Cambridge (1988)
25. Searle, J.R.: Minds, brains, and programs. Behav. Brain Sci. **3**(3), 417–424 (1980)
26. Searle, J.R.: Is the brain a digital computer? Proc. Addresses Am. Philos. Assoc. **64**(3), 21–37 (1990). American Philosophical Association
27. Sieg, W.: Church without dogma: Axioms for computability. In: Cooper, S.B., Löwe, B., Sorbi, A. (eds.) New Computational Paradigms, pp. 139–152. Springer, New York (2008)
28. Sieg, W.: Axioms for computability: Do they allow a proof of church's thesis? In: Zenil, H. (ed.) A Computable Universe: Understanding and Exploring Nature as Computation. World Scientific Publishing Press, Singapore (2013)
29. Siegelmann, H.T.: Recurrent neural networks and finite automata. Comput. Intell. **12**(4), 567–574 (1996)
30. Turing, A.M.: On Computable Numbers, with an Application to the Entscheidungsproblem. Proc. London Math. Soc. **2**(42), 230–265 (1936)
31. Turing, A.M.: On Computable Numbers, with an application to the Entscheidungsproblem: A correction. Proc. London Math. Soc. vol. 2 (published 1937). **43**(6), 544–546 (1938)
32. Turing, A.M.: Systems of logic based on ordinals. Proc. London Math. Soc. **2**(1), 161–228 (1939)
33. Wegner, P.: Why interaction is more powerful than algorithms. Commun. ACM **40**(5), 80–91 (1997)

34. Wegner, P.: Interactive foundations of computing. Theoret. Comput. Sci. **192**(2), 315–351 (1998)

35. Wolfram, S.: A New Kind of Science. Wolfram Media, Champaign (2002)

36. Zenil, H., Soler-Toscano, F., Delahaye, J.P., Gauvrit, N.: Two-dimensional kolmogorov complexity and validation of the coding theorem method by compressibility. Peer J. Comput. Sci. **1**, 23 (2015)

37. Zenil, H., Schmidt, A., Tegnér, J.: Causality, information and biological computation. In: Walker, S.I., Davies, P.C.W., Ellis, G. (eds.) Information and Causality: From Matter to Life. Cambridge University Press (in press)

Formalizing a Fragment
of Combinatorics on Words

Štěpán Holub[1]([⊠]) and Robert Veroff[2]

[1] Department of Algebra, Charles University, Prague, Czech Republic
holub@karlin.mff.cuni.cz
[2] Computer Science Department, University of New Mexico, Albuquerque, USA
veroff@cs.unm.edu

Abstract. We describe an attempt to formalize some tasks in combinatorics on words using the assistance of Prover9, an automated theorem prover for first-order and equational logic.

Keywords: Formalization · Periodicity · Combinatorics on words · Automated theorem proving

1 Motivation

In this paper we discuss a formalized approach to some tasks in combinatorics on finite words. Formalization of mathematical knowledge classically has two rather different motivations. One is Automated Theorem Proving, where one hopes to develop methods to find (possibly difficult, or just tedious) proofs automatically. The second motivation is Formalization of Mathematics, that aims at human-assisted computer verification of (human originated) parts of mathematics.

A prominent example of the formalization approach has evolved around the proof of Kepler's conjecture announced by Thomas Hales in 1998 and subsequently reviewed by 13 reviewers of *Annals of Mathematics* for three years without a conclusive verdict [10]. The situation in combinatorics on words is certainly less dramatic, but there are some similar features. As an illustrative example, we are looking at the classification of binary equality words [2,3,5,7,8]. In addition to important concepts, this project requires a lot of detailed case analysis which is arduous to make, tedious and unrewarding to read and check, and therefore possibly unreliable as to its correctness and/or completeness. Many of the arguments are repetitive. All of this leads to a conclusion that formalization might be a good idea. Moreover, the project is not yet completed, including more than two hundred undecided cases that may have to be dealt with separately. In view of this example, we want to keep in mind both possible goals of the formalization. One also can point out that artificial intelligence may blur the sharp distinction between them [14].

A third, tangential interest in this paper is to see what assumptions are needed in order to prove certain results. Looking at words through the lens of a limited set of tools yields interesting insights.

Š. Holub—Supported by the Czech Science Foundation grant number 13-01832S.

J. Kari et al. (Eds.): CiE 2017, LNCS 10307, pp. 24–31, 2017.
DOI: 10.1007/978-3-319-58741-7_3

2 Formalization

The proofs presented in this paper were found with the assistance of Prover9 [12], an automated theorem prover for first-order and equational logic. Problems are represented with a set of first-order formulas in clause form [1] that includes a set of axioms and a problem statement posed for proof by contradiction. Prover9 searches for a proof by applying inference rules to clauses until either a contradiction is found or some processing limit is reached. The search is guided by heuristics for selecting clauses, applying inference rules, and managing the growing set of derived clauses. Full details for the computation are accessible on the support web page [9].

Our formalization is based on several decisions that are motivated partly by theoretical considerations and partly by features of Prover9.

2.1 Semigroup

We consider words as *semigroup elements*. There is no explicit use of the fact that the semigroup is free; in particular, words are not seen as sequences. This is partly motivated by the fact that formal representations of lists typically are not handled well by theorem provers such as Prover9. Moreover, we are mainly interested in proofs showing that a certain relation on words forces periodicity, which is a property rather algebraic in nature. Nevertheless, the combinatorial complexity reappears in a nontrivial use of associativity; there are an exponential number of ways to associate any given expression.

We do not allow the empty word. Existence of the empty word has both advantages and disadvantages, and we decided to avoid it to simplify the language and theory.

2.2 No Arithmetic

We do not use natural numbers; in particular, we have no strong concept of length. The main reason for this is that we want to avoid computation and reasoning about inductively defined objects. Here too, the motivation is to avoid weaknesses of the theorem prover. An important consequence is that there is no uniform way to deal with arguments that typically would be inductive in nature. Length is partly substituted with a weaker *length comparison* compatible with the semigroup operation.

2.3 Equidivisibility

We assume the *equidivisibility* property of the semigroup: if $xy = uv$, then there is an element w such that either $x = uw$ and $v = wy$, or $u = xw$ and $y = wv$. This is a property of a free semigroup, also called Levi's lemma. Levi [11] proved that an equidivisible semigroup S is free if and only if it is graded, that is, if it is endowed with a semigroup homomorphism $\varphi : S \to (\mathbb{N}_+, +)$ (see also [13, p. 26]). Levi's lemma is thus a kind of measure of the distance of our axioms from the free semigroup.

2.4 Power

In addition to semigroup multiplication and length comparison, we use *power* as a primitive concept. The choice stresses that the main feature of words we are interested in is periodicity. Note that the concept of power becomes a nontrivial extension of multiplication precisely in the absence of natural numbers, since the expression w^n, understood as $w \cdot w \cdots\cdots w$, is an expression in a meta-language.

The properties of power that we formalized can be seen in the list of axioms below. We want to stress the axiom claiming that if both y and uyv are powers of x, then all the words u, y and v are powers of some common z. This is in a sense the only nontrivial fact about powers that we are using for now. (cf. Sect. 3).

Axioms

The above decisions lead to the following formal theory. The logical symbols in Prover9 notation are & for the logical *and*, | for the logical *or*, the minus sign for *negation*, and != for *non-equal*. The existential quantifier is explicitly stated as ∃ (it has a verbal form exists in the computer code); non-quantified variables are implicitly universally quantified.

The non-logical symbols are

– binary operation *
– binary relations Power and Shorter

The standard interpretation of Power(y,x) is that y is a power of x; that is, $y \in x^+$. The standard interpretation of Shorter(x,y) is that x is shorter than y; that is, $|x| < |y|$.

The axioms are as follows:

`(x * y) * z = x * (y * z).`	(associativity)
`(x * y = x * z) -> y = z.`	(left cancellation)
`(x * y = z * y) -> x = z.`	(right cancellation)
`x * y != x.`	(no right unit)
`x * y != y.`	(no left unit)
`x * y = u * v -> (x = u	∃w (x * w = u | w * y = v)).`	
	(equidivisibility)

`-Shorter(x,x).`	(non-reflexive)
`Shorter(x,y) -> -Shorter(y,x).`	(anti-symmetric)
`Shorter(x,y) & Shorter(y,z) -> Shorter(x,z).`	(transitive)
`Shorter(x,x * y).`	(compatible with *)
`Shorter(x,y * x).`	
`Shorter(x,y) <-> Shorter(x * z,y * z).`	(cancelation of length)
`Shorter(x,y) <-> Shorter(z * x,z * y).`	
`Shorter(x,y) <-> Shorter(x * z,z * y).`	
`Shorter(x,y) <-> Shorter(z * x,y * z).`	

```
Power(x,x).                                     (reflexivity of power)
Power(x,y) & Power(y,z) -> Power(x,z).          (transitivity of power)
Power(y,x) & Power(z,x) -> Power(y * z,x).      (compatibility with *)
Power(y,x) & Power(z,x) -> ( y = z |  ∃u ( Power(u,x) &
  ((y = z * u & y = u * z)|(z = y * u & z = u * y)) ) ) )
                                                (cancellation of powers)
Power(y * z,x) -> (Power(y,x) & Power(z,x)) |
  (∃u ∃v (u * v = x &
         (y = u | ∃y1 (y = y1 * u & Power(y1,x)) &
         (z = v | ∃z1 (z = v * z1 & Power(z1,x)) ) ) (breaking a power)
(Power(y,x) & Power(u * (y * v), x)) ->
    ∃z (Power(y,z) & Power(u,z) & Power(v,z)). (no nontrivial shift)
```

3   Commutation

In combinatorics on words, the fact that two words commute if and only if they are powers of the same root is probably the most elementary fact (up to considering the existence of the common root to be the very definition of commutation). In our formalization, however, the formula

$$x * y = y * x \rightarrow \exists z \, (\text{Power}(x,z) \, \& \, \text{Power}(y,z)).$$

is not an axiom but the first fact we would like to prove (let us call it the "Commutation lemma"). It turns out that this is an easy task for Prover9, hence we witness maybe the very first theorem in this field ever obtained by Automated Theorem Proving. Its "difficulty" at best corresponds to the characterization offered by Thomas Hales in 2008 [6, p. 1377]:

> Overall, the level today of fully automated computer proof [...] remains that of undergraduate homework exercises...

The proof output from Prover9 is given in Fig. 1. The constants c_1 and c_2 appearing in the proof are Skolem constants [1] coming from the existentially quantified variables in the negation of the theorem (for proof by contradiction).

The proof consists of a unique (humanly) nontrivial observation implied by the commutativity of x and y: $(xy)(xy) = x(yx)y = x(xy)y$. Now the consequent follows from the (no nontrivial shift) axiom.

4   Conjugation and Missing Induction

An obvious candidate claim to be proven next is the following theorem, actually just a part of a well known characterization of conjugate words.

Theorem 1. *If $xz = zy$, then there are words u and v such that $x = uv$ and $y = vu$.*

```
=============================== PROOF ===============================

% Proof 1 at 0.03 (+ 0.00) seconds.
% Length of proof is 21.
% Level of proof is 8.
% Maximum clause weight is 23.000.
% Given clauses 69 (8.625 givens/level).

5 Power(y,x) & Power(z,x) -> Power(y * z,x) # label(non_clause).  [assumption].
8 Power(y,x) & Power(u * (y * v),x) -> (exists z (Power(y,z) & Power(u,z) & Power(v,
z))) # label(non_clause).  [assumption].
29 x * y = y * x -> (exists z (Power(x,z) & Power(y,z))) # label(non_clause) #
label(goal).  [goal].
39 (x * y) * z = x * (y * z).  [assumption].
45 Power(x,x).  [assumption].
47 -Power(x,y) | -Power(z,y) | Power(x * z,y) # label(non_clause).  [clausify(5)].
64 -Power(x,y) | -Power(z * (x * u),y) | Power(z,f7(y,x,z,u)).  [clausify(8)].
65 -Power(x,y) | -Power(z * (x * u),y) | Power(u,f7(y,x,z,u)).  [clausify(8)].
89 c2 * c1 = c1 * c2.  [deny(29)].
90 c1 * c2 = c2 * c1 # label("Goal 1").  [copy(89),flip(a)].
91 -Power(c1,x) | -Power(c2,x).  [deny(29)].
97 -Power(x,y) | Power(x * x,y).  [factor(47,a,b)].
143 -Power(x * y,z) | -Power(u * (x * (y * w)),z) | Power(u,f7(z,x * y,u,w)).
[para(39(a,1),64(b,1,2))].
146 -Power(x * y,z) | -Power(u * (x * (y * w)),z) | Power(w,f7(z,x * y,u,w)).
[para(39(a,1),65(b,1,2))].
264 c1 * (c2 * x) = c2 * (c1 * x).  [para(90(a,1),39(a,1,1)),rewrite([39(4)]),flip(a)].
302 Power(x * x,x).  [resolve(97,a,45,a)].
325 Power(x * (y * (x * y)),x * y).  [para(39(a,1),302(a,1))].
398 Power(c2 * (c2 * (c1 * c1)),c2 * c1).  [para(90(a,1),325(a,1,2,2)),rewrite([264(7),
264(6),90(10)])].
431 Power(c2,f7(c2 * c1,c2 * c1,c2,c1)).  [resolve(143,b,398,a),unit_del(a,45)].
461 -Power(c1,f7(c2 * c1,c2 * c1,c2,c1)).  [ur(91,b,431,a)].
474 $F.  [resolve(146,b,398,a),unit_del(a,45),unit_del(b,461)].

=============================== end of proof ===============================
```

Fig. 1. Proof of the Commutation lemma

Consider the following simple classical proof.

Proof. Proceed by induction on $|z|$. If $|z| \leq |x|$, then there is a word v such that $x = zv$ and $y = vz$. Therefore, we are done with $u = z$.

Assume that $|z| > |x|$. Then $z = xz' = z'y$ with $|z'| < |z|$, and the proof is completed by induction.

This proof cannot be formalized with our current axioms, since we do not have induction. Specifically, **Shorter** is not a well-founded relation.

In fact, the problem is not just with *this* proof, since the formula

$$x * z = z * y \rightarrow (x = y) \mid \exists u \exists v \ (x = u * v \quad \& \quad y = v * u) \quad \text{(conj.)}$$

cannot be proven with our axioms. This follows from the following semigroup in which all axioms hold, but Theorem 1 does not.

Let $\langle A \rangle$ be a semigroup generated by $A = \{a, b\} \cup \{c_i \mid i \in \mathbb{N}\}$ and defined by the set of relations $c_i = ac_{i+1} = c_{i+1}b$, $i \in \mathbb{N}$. The **Power** relation is interpreted in the natural way as in A^*. We define the "length" semigroup homomorphism $\ell : \langle A \rangle \rightarrow (\mathbb{Q}, +)$ by $\ell(a) = \ell(b) = 1$ and $\ell(c_i) = 2^{-i}$. The relation **Shorter**(x, y) is interpreted as $\ell(x) < \ell(y)$. Note that by Levi's lemma (see above), there cannot exist any semigroup homomorphism $\langle A \rangle \rightarrow (\mathbb{N}, +)$.

This example shows that our rudimentary axiomatic system must be extended by the (conj.) formula if we wish to prove anything about conjugate words.

5   Periodicity Lemma

The Periodicity lemma, often called the Fine and Wilf theorem [4], is a fundamental tool when dealing with periodicity. It states when a word can have two different periods p and q in a nontrivial way, where nontrivial means not having a period dividing both p and q. This formulation of the claim apparently depends strongly on arithmetical properties of periods, namely divisibility. Nevertheless, in this case we are able to prove the following version of the Periodicity lemma that is only slightly weaker than the full version:

Theorem 2. *Let u with $|u| \geq |xy|$ be a prefix of both x^ω and y^ω. Then x and y commute.*

For the sake of completeness, we recall that the full version of the Periodicity lemma has a weaker assumption $|u| \geq |xy| - \gcd(|x|, |y|)$, and moreover, it claims that the bound is optimal.

We have formulated Theorem 2 in a way that fits our formalization. Namely, periods of u are defined by its *periodic roots* x and y. Moreover, the infinite power is used, reminding us that we do not care about the exponent (since we haven't got the means needed). In order to make our formulas more intuitive, it may be convenient to enrich our language with binary relations `Prefix` and `Period`, defined by axioms

$$\texttt{Prefix(x,y) <->} \exists \texttt{z} \quad \texttt{(x * z = y).}$$
$$\texttt{Period(x,y) <->} \exists \texttt{z} \quad \texttt{(Power(z,x)} \quad \texttt{\&} \quad \texttt{Prefix(y,z)).}$$

Theorem 2 now has the following simple form:

$$\texttt{(Period (x,u) \& Period(y,u) \& -Shorter(u,x * y)) -> x * y = y * x.}$$

An "informal" proof of Theorem 2 using only accepted axioms is the following.

Proof. Let uu_1 be a power of x and let uu_2 be a power of y. Then $uu_1 = xx_1$ and $uu_2 = yy_1$ with $x_1 \in x^+$ and $y_1 \in y^+$. Then $u = xu_3 = yu_4$, where $u_3u_1 \in x^+$ and $u_4u_2 \in y^+$. Since $uu_1, u_3u_1x \in x^+$ and $|uu_1| = |u_3u_1x|$, we deduce $uu_1 = u_3u_1x$. Now,

$$uu_1x = xu_3u_1x = xuu_1 = xyu_4u_1,$$

and xy is a prefix of u. Similarly, we obtain that yx is a prefix of u, which concludes the proof.

This is a typical example of a very simple proof which is at the same time quite unpleasant to read and verify. Of course, the same argument can be made with an appeal to the intuition of the character of the periodicity. However, such an intuition is hardly preserved throughout more complex proofs.

To date, Prover9 has not found a proof entirely on its own. To get a fully formalized proof, we split the argument into several steps. Specifically, we first proved four auxiliary lemmas:

$$(x * y = u * v) \to (\text{Prefix}(x,u) \mid \text{Prefix}(u,x) \mid x = u). \tag{L1}$$

$$\text{Prefix}(x,y) \to \text{Shorter}(x,y) \ . \tag{L2}$$

$$-\text{Shorter}(u, x * y) \to (u * z \mathbin{!=} x * y) \ . \tag{L3}$$

$$((u * u1 = x * x1) \ \& \ -\text{Shorter}(u,x * y)) \to (\exists u3 \quad (u = x * u3)) \ . \tag{L4}$$

They are just tiny, humanly natural reformulations of existing axioms and definitions which nevertheless help to point the Prover9 search in the right direction. The use of (L4) is clear from the reformulation of Theorem 2 below. Note that the lemma says: if u and x are prefixes of the same word and $|u| \geq |xy|$ (for an arbitrary y), then x is a prefix of u. The lemma would be more natural if $|x| < |u|$, that is $\text{Shorter}(x,u)$, were used instead of $|u| \geq |xy|$ (that is $-\text{Shorter}(u,x * y)$). However, the latter being an explicit assumption of Theorem 2, the present form is one more little hint for the automated proof.

We then reformulated the task as

$$\begin{aligned} (\text{Power}(u * u1, \ x) \ \& \ u = x * u3 \ \& \\ \text{Power}(u * u2, \ y) \ \& \ u = y * u4 \ \& \\ -\text{Shorter}(u,x * y)) \\ \to x * y = y * x. \end{aligned}$$

Here $\exists u1 \ \text{Power}(u * u1, \ x)$ can be proved, or it can be considered as a different definition of $\text{Period}(x,u)$ (similarly for $\text{Period}(y,u)$). The claims $\exists u3$ $u = x * u3$ and $\exists u4$ $u = y * u4$ were proved separately.

The proof of Theorem 2 now splits into two cases. (1) $u = xy$ or $u = yx$; (2) $u \neq xy$ and $u \neq yx$. By symmetry of x and y, the first case can be reduced to $u = xy$. Note that this is a meta-argument sparing us one of two formal proofs identical up to exchange of x and y.

The case $u = xy$ was proved automatically when we suggested (L1) and (L2) to Prover9. For the case $u \neq xy$ and $u \neq yx$, we let Prover9 first prove intermediate conclusions (identical up to symmetry of x and y):

$$\exists u5 \ (u = (x * y) * u5) .$$
$$\exists u5 \ (u = (y * x) * u6) .$$

6   Conclusion

A text in combinatorics on words usually contains three dots (like $a_1 \cdots a_n$) somewhere on the first few lines. Experts on automated theorem proving quickly become skeptical when seeing those dots, since computers refuse to understand what they mean. The original intention of our research was therefore to break this skepticism and to show the very possibility of a formal approach to words. As in practically all other areas of mathematics, there is little hope (or fear) that computers will replace mathematicians in the near future. From our recent experience reported in this text, the realm of fully automated proving ends somewhere between the Commutation lemma and the Periodicity lemma. On the other hand, a vision of a computer assisted proof verification or search for individual steps in proofs seems more realistic. We wish to leave open for a further enquiry whether this or some modified formalization attempt can bring about something substantial.

References

1. Chang, C.-L., Lee, R.C.-T.: Symbolic Logic and Mechanical Theorem Proving. Academic Press, New York (1973)
2. Culik II, K., Karhumäki, J.: On the equality sets for homomorphisms on free monoids with two generators. RAIRO ITA 14(4), 349–369 (1980)
3. Czeizler, E., Holub, Š., Karhumäki, J., Laine, M.: Intricacies of simple word equations: An example. Int. J. Found. Comput. Sci. **18**(6), 1167–1175 (2007)
4. Fine, N.J., Wilf, H.S.: Uniqueness theorems for periodic functions. Proc. Am. Math. Soc. **16**(1), 109–109 (1965)
5. Hadravová, J.: Structure of equality sets. Ph.D. thesis, Charles University (2007)
6. Hales, T.C.: Formal proof. Not. AMS **55**(11), 1370–1380 (2008)
7. Holub, Š.: Binary equality sets are generated by two words. J. Algebra **259**(1), 1–42 (2003)
8. Holub, Š.: A unique structure of two-generated binary equality sets. In: Ito, M., Toyama, M. (eds.) DLT 2002. LNCS, vol. 2450, pp. 245–257. Springer, Heidelberg (2003). doi:10.1007/3-540-45005-X_21
9. Holub, Š., Veroff, R.: Formalizing a fragment of combinatorics on words (web support) (2017). http://www.cs.unm.edu/veroff/CiE2017/
10. Lagarias, J.C.: The Kepler Conjecture: The Hales-Ferguson Proof. Springer, New York (2011)
11. Levi, F.W.: On semigroups. Bull. Calcutta Math. Soc. **36**, 141–146 (1944)
12. McCune, W.: Prover9, version 02a (2009). http://www.cs.unm.edu/mccune/prover9/
13. Sakarovitch, J.: Elements of Automata Theory. Cambridge University Press, New York (2009)
14. Urban, J., Vyskočil, J.: Theorem proving in large formal mathematics as an emerging AI field. In: Bonacina, M.P., Stickel, M.E. (eds.) Automated Reasoning and Mathematics. LNCS, vol. 7788, pp. 240–257. Springer, Heidelberg (2013). doi:10.1007/978-3-642-36675-8_13

Turing's 1949 Paper in Context

Cliff B. Jones[✉]

School of Computing Science, Newcastle University, Newcastle upon Tyne, UK
cliff.jones@ncl.ac.uk

Abstract. Anyone who has written one knows how frustratingly difficult it can be to perfect a computer program. Some of the founding fathers of computing set out ideas for reasoning about software — one would say today 'techniques for proving that a program satisfies its specification'. Alan Turing presented a paper entitled *Checking a Large Routine* that laid out a workable method for reasoning about programs. Sadly his paper had little impact. Understanding the problem faced, Turing's proposal and what followed provides insight into how ideas evolve. Comparing three contributions from the 1940s with the current state of the art clarifies a problem that still costs society a fortune each year.

1 Introduction

In the 1940s, there were at least three published outlines of how it might be possible to reason about computer programs; Herman Goldstine and John von Neumann [GvN47], Alan Turing [Tur49] and Haskell Curry [Cur49]. Frustratingly these early insights appear to have then lain dormant only to be reinvented (and developed enormously) starting two decades later.[1] It is interesting to speculate why the early papers were only studied after the work of Bob Floyd, Tony Hoare and others had been published.

It is Alan Turing's 1949 paper that has the clearest similarity with what followed, for example in Floyd's [Flo67], and Sect. 3 gives enough detail of Turing's 1949 paper to support comparisons. Here, briefer comments are made on the other two contributions.

The report [GvN47] contains a long account of the process of designing programs that achieve a mathematical task. Looked at today, it is even tempting to say that the account is somewhat rambling but it must be remembered that in 1947 no one had any experience of writing computer programs. A distinction is made between parameters to a program which are likened to *free variables* and those locations whose values change during a computation that are compared to

---

[1] Connections other than the citations between these three pioneering pieces of work are difficult to trace. The current author has tried to determine whether there is a direct link between von Neumann's assertion boxes and Turing's annotations. It is known that Turing visited the US during the war but there is no evidence that the urgent business of cryptography left any time to discuss program development ideas.

© Springer International Publishing AG 2017
J. Kari et al. (Eds.): CiE 2017, LNCS 10307, pp. 32–41, 2017.
DOI: 10.1007/978-3-319-58741-7_4

bound variables. The notion of constructing a flowchart is, from today's stand-point, handled in a rather pedestrian description. However, crucially, a distinction is made between *operation boxes* (that change the values of bound variables) and *assertion boxes* that can be used to describe logical relations between values.

Haskell Curry's [Cur49] cites, and is clearly influenced by, [GvN47]. Curry however puts more emphasis on constructing programs from components or *transformations.* In a subsequent paper from 1950, Curry applies his proposals to constructing a program for *inverse interpolation.* Curry's emphasis on topics such as (homomorphic) transformations is not surprising given his background in combinatory logic.[2]

2   The Problem

Users of computers have to endure the situation that most software has bugs; those who purchase software are frustrated by the fact that it comes, not only without guarantees, but with explicit exclusions of any liabilities on the provider for losses incurred by the purchaser. It was claimed in a 2002 NIST report that software 'maintenance' cost US industry over $50Bn per year.

The problem for the programmer is the literal nature of the indefatigable servant called hardware. If one told a human servant to do anything nonsensical, there is at least a chance that the instruction would be queried. If, however, a computer program is written that continues to subtract one from a variable until it reaches zero, it will do just that — and starting the variable with a negative value is unlikely to yield a useful result.

A simplified expression of the question that has occupied many years of research – and which is of vital importance – is 'how can one be sure that a program is correct?'. In fact, this form of the question is imprecise. A better formulation is 'how can we be sure that a program satisfies an agreed specification?'.

Once one is clear that this property must apply to 'all possible inputs', a natural idea is to look at mathematical proof. The position argued in [Jon03] is that the desirability of reasoning about correctness was clear to the early pioneers and a key impediment to adoption has been a search for tractable methods of decomposing the reasoning so as to deal with large programs.

3   The Paper *Checking a Large Routine*

Alan Turing made seminal contributions related to software. In fact, it can be argued that 'Turing Machines' provide the first thought-through idea of what constitutes 'software'. Turing's way of showing that the *Entscheidungsproblem* was unsolvable was to propose an imaginary universal computer and then to

---

[2] Acknowledgements to Curry's influence are present in both the choice of name for the Haskell (functional) programming language and the term *Curry-Howard correspondence* (for proofs as programs).

prove that there were results which no program could compute. The Turing machine language was minimal but just rich enough to describe any step-by-step process that could be envisaged.

In 1949 the EDSAC computer in Cambridge became the world's second 'electronic stored program computer' to execute a program.[3] A conference to mark this event was held in Cambridge from 22–25 June 1949. Many people who became famous in the early history of European computing attended the event; among those who gave papers was Alan Turing. His paper [Tur49] is remarkable in several ways. Firstly, it is only three (foolscap) pages long. Secondly, much of the first page is taken up with one of the best motivations ever given for program verification ideas. Most interestingly, Turing presents the germ of an idea that was to lay dormant for almost 20 years; a comparison with Bob Floyd's 1967 paper – which became the seed of an important area of modern computer science research – is made below.

It is worth beginning, as does the 1949 paper, with motivation. The overall task of proving that a program satisfies its specification is, like most mathematical theorems, in need of decomposition. Turing made the point that checking the addition of a long series of numbers is a monolithic task that can be split into separate sub-tasks by recording the carry digits. His paper displays five (four-digit) numbers whose sum is to be computed and he makes the point that checking the four columns can be conducted separately if the carry digits are recorded (and a final addition including carries is another separable task). Thus the overall check can be decomposed into five independent tasks that could even be conducted in parallel.

Turing's insight was that annotating the flow chart of a program with claims that should be true at each point in the execution can also break up the task of recording an argument that the complete program satisfies its specification (a claim written at the exit point of the flowchart).

A program correctness argument is in several respects more difficult than the arithmetic addition: these, and how the 1949 paper tackled them, are considered one by one. Turing's example program computes factorial ($n!$); it was presented as a flowchart in which elementary assignments to variables were written in boxes; the sequential execution of two statements was indicated by linking the boxes with a (directed) line. In the original, test instructions were also written in rectangular boxes (they are enclosed diamond shaped boxes in Fig. 1) with the outgoing lines indicating the results of the tests and thus the dynamic choice of the next instruction.

Suppose a 'decorating claim' is that the values of the variables are such that $r < n$ and $u = (r + 1) * r!$. Consider the effect if the next assignment to be executed changes r to have a value one greater than the previous value of that same variable (in some programming languages this would be written $r := r + 1$). Then it is easy to check that after this assignment a valid decoration is $r \leq n$ and $u = r!$. Reasoning about tests is similar. Suppose that the decorating assertion

---

[3] The world's first embodiment of an 'electronic stored-program computer' to run was the Manchester 'Baby' that executed its first program on midsummer's day 1948.

before a test is $s - 1 \leq r < n$ and $u = s * r!$; if the execution of the test indicates that the current values of the variables are such that $(s - 1) \geq r$ then, on the positive path out of the test, a valid decoration is $r < n$ and $u = (r + 1) * r!$ (which expression can be recognised from above).

The flowchart in the 1949 paper represented what today would be written as assignments (e.g. $r := r + 1$ from above) by $r' = r + 1$. Furthermore, Turing chose to mark where decorating claims applied by a letter in a circle and to record the decorations in a separate table. There is the additional historical complication that his decorations were associated with numerical machine addresses. For these reasons – and those of space – the actual 1949 figures are not given here but exactly the same annotated program is presented in Fig. 1 in a modern form.

In today's terms, it could be said that Turing's programming language was so restricted that the meaning (semantics) of its constructs was obvious. This point is also addressed when subsequent developments are described below.

By decorating flowcharts, Turing – and some subsequent researchers – finessed some delicate issues about loops which are just represented by the layout of the flowchart.[4] In the case of Turing's example, factorial is computed by successive multiplication; and, in fact, the envisaged machine was so limited that even multiplication was not in the instruction set and the actual program has a nested inner loop that computes multiplication by successive addition.

There is a delicate and important issue about loops that Turing addressed formally: the need to argue about their termination. When a computer locks up (and probably has to be restarted) one possible cause is that a program is in a loop that never terminates. The 1949 paper contains a suggestion that loop termination can be justified by showing the reduction of an ordinal number which Turing commented would be a natural argument for a mathematician; he added however that a 'less highbrow' argument could use the fact that the largest number representable on the machine he was considering was 2^{40}.

The fascinating thing about the 1949 paper was the early recognition of the need for something more mathematical than the execution of test cases to support the claim that a program satisfies its specification. Of course, the example given in a (very short) paper is small but it is clear from the title of the paper that the intention was to apply the proposal to 'large routines'.

Turing gave a workable method for recording such reasoning. Lockwood Morris and the current author had a debate with Maurice Wilkes about the comparison with Floyd's approach but it should be clear that a major intellectual step had been made beyond the acceptance that 'write then debug' was the only way to achieve correct software.

---

[4] In fact, some form of looping concept is central to the power of general purpose programs. One could say that they make it possible to compute properties that are not in the basic instructions of the programming language. This point is illustrated with recursive functions in [Pét66, Chap. 1] and applied to interesting data structures in the appendix of Rózsa Péter's book. It is interesting to recall Tony Hoare's comment that he couldn't write Quicksort as a program until he learned a language that provided recursive procedures.

4   The Floyd/Hoare Approach

Turing's wonderfully brief and clear 1949 paper both identifies the issue of reasoning about programs and contains a clear proposal as to how it might be undertaken for numerical examples. Considering this early recognition of the issue, it is remarkable that the key paper on which so much subsequent research on program reasoning is based did not appear until the late 1960s. The landmark talk by Bob Floyd in 1967 (published as [Flo67]) appears to have been written in complete ignorance of Turing's 1949 paper. Some possible reasons for this are put forward in Sect. 6. The similarities might indicate the degree to which the idea of separating complex arguments into smaller steps is inevitable; the differences between Turing's and Floyd's approaches are more interesting.

Floyd also annotated flowcharts and Turing's factorial example can be presented in Floyd's style as in Fig. 1 (the two deduction steps traced in the previous section appear in the lower part of this flowchart). For Turing's example, arithmetic relations suffice for the decorating assertions; Floyd explicitly moved to the formal language of first-order predicate calculus for his assertion language. This decision made it possible for Floyd to be precise about what constitutes a valid argument. In fact, Floyd explored what were later called 'healthiness conditions' by Edsger Dijkstra.

Like Turing, Floyd offered formal ways of reasoning about termination. Not all later authors achieved this. The idea of showing that a program computes some desirable result *if it terminates* is sometimes misnamed 'partial

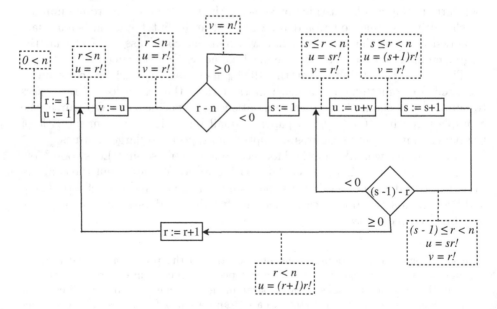

Fig. 1. A presentation of Turing's example in the Floyd style

correctness'; one can however take the view that, if a program is expected to terminate, this requirement is a part of its specification and needs to be justified.[5]

A crucial further step – still in ignorance of Turing's 1949 paper – was made by Tony Hoare who proposed an 'axiomatic' view of program semantics. In [Hoa69], he separated program reasoning from the flowchart view by proposing – what have come to be called – 'Hoare triples' as terms in an extended logic containing pre conditions, program constructs and post conditions. Both pre and post conditions are logical expressions that characterise sets of states. Valid triples record that, for all states satisfying the pre condition, the execution of the program text will result in states satisfying the post condition. What is important about this change of viewpoint is that it moves the programmer away from thinking in terms of program tracing and prompts viewing programs and assertions as combined terms in an extended logical system. A further advantage became clear in a second paper [Hoa71] by Hoare. The title of *Proof of a Program: FIND* betrays the fact that Hoare initially wrote a paper to provide a *post facto* proof of the correctness of a non-trivial sorting program (of his own invention). A first version of this proof was sent to referees one of whom was the current author. In the absence of the sort of mechanical theorem provers that are in use today, it was difficult to build confidence in the long and detailed proof. Hoare realised this and rewrote the paper (but did not change the title) to embody a stepwise development of the program with layers of abstraction that were much easier to grasp and reason about. Hoare's axiomatic approach became the foundation of a huge field of research on the formal design of software.

Hoare's original paper did not formalise termination arguments; of course, he was aware of the importance of showing that programs always terminated in any state satisfying the pre condition but the topic was not included in the initial set of rules. Hoare reasoned about termination (and in fact the relationship to the initial states) separately and informally.

Because of its importance below, it is worth recording that Hoare was generous in his credits to earlier researchers mentioning Floyd, Naur[6] and van Wijngaarden (whose role is explored in Sect. 6).

---

[5] John McCarthy – who did much to promote formal methods – used the following imaginary scenario to highlight the need for termination arguments 'an algorithm that might appear to ensure that someone becomes a millionaire: the person should walk along the street picking up any piece of paper — if it is a cheque made out to that person for one million dollars, take it to the bank; if not, discard the piece of paper and resume the process'.

[6] Peter Naur proposed in a paper in the journal *BIT* 'general snapshots' as a way of annotating a program text to reason about its correctness. Naur's system was based on comments in a program and was less formal than Floyd's but it is another indication that the idea of decomposing an argument about correctness had reached its moment in time.

5   Current Situation

Research on reasoning about programs (and designs of complex hardware) has not only been a major topic for academic research, it is now an essential approach used by industry in cases where life or business is at risk.

Jim King was one of Floyd's students and built Effigy [Kin71] which was an early system in which programs could be annotated with assertions (using first-order predicate calculus as in Floyd's paper). Today, powerful theorem proving assistant software such as Isabelle, PVS and Coq use heuristics from research on Artificial Intelligence and greatly reduce the human effort in creating completely formal proofs. Several notations have been developed for specifying both over-all systems and components that arise during design. Integrated systems that generate 'proof obligations' from design steps link to theorem proving assistants.

Perhaps the development that would cause the founding figures most surprise is today's emphasis on using abstract forms of data in specification and design. With the limited store sizes of the early machines, vectors – or perhaps multi-dimensional arrays – constituted the extent of data abstraction. Now computer software manipulates huge and interlinked data structures whose representation is itself a major design challenge. Fortunately, the ideas of data abstraction and reification are also handled in many modern design support systems.

The use of 'formal methods' for hardware design accelerated after Intel reportedly took a \$475M loss because of an undetected design flaw in the Pentium chip. Many researchers emphasise a 'stack' of verified components from programs, through compilers to the hardware designs themselves. As mentioned, the earliest uses of formal methods were for 'safety critical' software in which errors could put lives at risk. In some ways, the more interesting development is that some organisations use formal methods even where there is no requirement to do so, the argument being that they provide a cost-effective way of creating predictable and maintainable software. A useful review of the current state of deployment of formal methods is [WLBF09] (which is currently being updated).

Even where completely formal (machine checked) proofs are not considered necessary, an approach sometimes called 'formal methods light' offers an engineering approach founded on mathematics.

One area of research that is of strong interest today concerns 'concurrency'. This is important because hardware designers are putting ever more processors on a single chip in order to continue to provide the exponential speed increases that have revolutionised computing. Of particular interest here is the development of special logics such as Temporal Logic or Separation Logic to reason about concurrency.

6   Why Did Turing's Paper Not Have Immediate Impact?

Reasoning about programs is today seen as the only way to ensure the correctness of so-called 'safety critical' software and as a cost-effective way of achieving high quality software in a predictable process. It is, therefore, interesting to try to understand the lack of impact of Turing's 1949 gem on what has become such an important area of computing science.

First, it is worth noting that Turing was not careful in the production of printed materials especially for what he might have seen as a rather minor contribution. The reproduced version of the 1949 paper is a nightmare to interpret.[7] The identifiers used in the program could hardly have been more badly chosen for someone with unclear handwriting and little time for proofreading: u, v, n and r are mistyped 10 times. This is compounded by the fact that Turing chose – rather than $n!$ – to write factorial as a box around n; a perfectly acceptable notation but in the printed paper this symbol is missing in several places presumably because it should have been added by hand but was not.[8]

The 'typos' were, however, certainly not the main reason for the lack of impact. Neither Floyd nor Hoare were old enough to have attended the 1949 conference. It would also appear that few people who were at that meeting were inclined to follow Turing's rigorous approach to reasoning about software. There was one tantalising attendee: Aad van Wijngaarden is listed as a participant in the Cambridge meeting. He went on to publish an important paper [vW66] entitled *Numerical Analysis as an Independent Science*. The contribution of this paper was to provide axioms of finite computer arithmetic in which, for example, adding one to the largest number representable in a single word might yield a negative result. As mentioned above, Hoare acknowledges this contribution because it is one aspect of his 1969 'axiomatic' approach. In some sense then, van Wijngaarden had in his hands all of the pieces of Hoare's approach but failed to put them together.[9] The current author had the good fortune to know many of the main players (but not Turing). In particular, Aad van Wijngaarden was always a charming and interesting dinner companion; it is most unfortunate that Turing's 1949 paper only came to hand after van Wijngaarden's death. Neither first hand questions, nor recollections from closer colleagues have exposed any record of what van Wijngaarden thought of Turing's paper.[10]

Another consideration when looking for paths to impact is the language in which Turing's programs were written. Programs are inevitably written in a language. The intended meaning of that formal language (its 'semantics') is an essential assumption for any reasoning about programs. Typically, a programming language is translated into the machine code of a specific type of hardware. Were the translation not to reflect the intended semantics, any effort to reason about programs would be undermined. Fortunately this can be seen as decomposing the overall challenge of 'program correctness': ways are needed, on the

---

[7] A full review of Turing's 1949 paper with indications of the necessary typographical corrections is contained in [MJ84].

[8] For example: 'with n in line 29 we shall find a quantity in line 31 when the machine stops which is n'.

[9] More detail on the history of reasoning about programs can be found in [Jon03].

[10] Gerard Alberts has however made two important points (private communication March 2017): it is important to remember that van Wijngaarden's background was numerical analysis rather than logic; furthermore, Alberts has evidence that van Wijngaarden 'had little affinity' with Turing and his work.

one hand, for reasoning about programs written in a specific 'high level' language and, on the other hand, for showing that the translation of that language into machine code is correct. One suggestion for the lack of progress over almost two decades (1949–1967) is that programmers were so preoccupied with the many other developments their attention was diverted from the crucial issue of whether programs satisfied their specifications. Those 18 years also saw the development from machine code to ('high level') programming languages in which large programs can be written.

Another possible explanation is one that still plagues the software industry: to mathematicians like Turing and von Neumann the notion of proof was bread and butter, but many who took up the role of programmer were not conscious of the certainty to be gained from presenting careful logical arguments.

It is tempting to speculate what might have happened had Turing's paper been more widely read and understood at an earlier point in time. The story about van Wijngaarden ought give pause to anyone who suggests that, had Turing's paper been more widely known, the subject of program reasoning would have been automatically advanced by two decades. Who knows when someone of Tony Hoare's disposition and ability would have come along to make the crucial step to an axiomatic presentation? It is interesting to note that Hoare was actually looking for ways to record the meaning of programming languages when he proposed his axiomatic approach; so there is the initial requirement that the sought after person might have needed to have struggled with the specific thorny question of how to describe the semantics of a language precisely but to leave some aspects 'under determined'.

A more compelling avenue for speculation is to wonder what influence awareness of Turing's paper might have had on Bob Floyd. Floyd's paper, like Hoare's, is generous in its acknowledgement of previous work. In fact, Don Knuth's obituary note suggests excessive modesty in Floyd's statement that 'These modes of proof of correctness and termination are not original; they are based on ideas of Perlis and Gorn, and may have had their earliest appearance in an unpublished paper by Gorn.'[11]

The facts are simple: Turing's [Tur49] is another example of a paper from a brilliant mind; the paper was ahead of its time; sadly, this contribution appears to have gone unrecognised until after Floyd's independent invention of a scheme that went beyond Turing's had been published and taken up by other researchers.

Ideas have their time and independent inventions are not unusual; one part of the question of timing is that receptiveness can depend on people having struggled with preparatory problems;[12] for many years Turing's contributions were somewhat undervalued (a state of affairs that has recently been handsomely redressed); however his *Checking a large routine* had regrettably little impact. It is also important not to go too far in explaining scientific progress as a sequence of landmark papers.

---

[11] Repeated attempts to track down this elusive paper have so far yielded nothing: [Gor61] does not appear to be the missing document and [Gor68] is too late.

[12] A similar conclusion is drawn in Priestly's interesting book [Pri11, p. 302].

Acknowledgements. I should like to dedicate this paper to Lockwood Morris (1943–2014) who was a great but under-appreciated scientist. We worked together on [MJ84] when we were both in Oxford and he subsequently spent his sabbatical with me in Manchester. As well as many happy memories, Lockwood did me a great personal favour in December 2013 which was sadly the last time we met.

I am extremely grateful to Liesbeth de Mol for bringing Curry's work to my attention and to Gerard Alberts for interesting input on van Wijngaarden. Comments from anonymous referees have also helped improve the paper although some of their suggestions will not fit in the page ration. My funding for this research comes from the EPSRC *Strata* Platform Grant.

References

[Cur49] Curry, H.B.: On the composition of programs for automatic computing. Naval Ordnance Laboratory Memorandum 9806, 19-8, 52 pp. (1949)

[Flo67] Floyd, R.W.: Assigning meanings to programs. In: Proceedings of Symposium in Applied Mathematics. Mathematical Aspects of Computer Science, vol. 19, pp. 19–32. American Mathematical Society (1967)

[Gor61] Gorn, S.: Specification languages for mechanical languages and their processors a baker's dozen: a set of examples presented to ASA x3.4 subcommittee. Commun. ACM **4**(12), 532–542 (1961)

[Gor68] Gorn, S.: The identification of the computer, information sciences: their fundamental semiotic concepts and relationships. Found. Lang. **4**, 339–372 (1968)

[GvN47] Goldstine, H.H., von Neuman, J.: Planning and coding of problems for an electronic computing instrument. Technical report, Institute of Advanced Studies, Princeton (1947)

[Hoa69] Hoare, C.A.R.: An axiomatic basis for computer programming. Commun. ACM **12**(10), 576–580, 583 (1969)

[Hoa71] Hoare, C.A.R.: Proof of a program: FIND. Commun. ACM **14**, 39–45 (1971)

[Jon03] Jones, C.B.: The early search for tractable ways of reasoning about programs. IEEE Ann. Hist. Comput. **25**(2), 26–49 (2003)

[Kin71] King, J.C.: A program verifier. In: Freiman, C.V. (ed.) Proceedings of the Information Processing, IFIP 1971, pp. 234–249. North-Holland (1971)

[MJ84] Morris, F.L., Jones, C.B.: An early program proof by Alan Turing. Ann. Hist. Comput. **6**(2), 139–143 (1984)

[Pét66] Péter, R.: Recursive Functions. Academic Press, New York (1966)

[Pri11] Priestley, M.: A Science of Operations: Machines, Logic and the Invention of Programming. Springer Science & Business Media, London (2011)

[Tur49] Turing, A.M.: Checking a large routine. In: Report of a Conference on High Speed Automatic Calculating Machines, pp. 67–69. University Mathematical Laboratory, Cambridge, June 1949

[vW66] Wijngaarden, A.: Numerical analysis as an independent science. BIT **6**, 66–81 (1966)

[WLBF09] Woodcock, J., Larsen, P.G., Bicarregui, J., Fitzgerald, J.: Formal methods: practice and experience. ACM Comput. Surv. **41**(4), 1–36 (2009)

Gödel's Reception of Turing's Model of Computability: The "Shift of Perception" in 1934

Juliette Kennedy[✉]

University of Helsinki, Helsinki, Finland
juliette.kennedy@helsinki.fi

1 Introduction

The emergence of the mathematical concept of computability in the 1930 s was marked by an interesting shift of perspective, from viewing the intuitive concept, "human calculability following a fixed routine" in terms of calculability in a logic, to viewing the concept as more adequately expressed by Turing's model.[1] This happened in spite of, or in parallel with, *confluence*, as Gandy called it in his [1], namely the proven extensional equivalence of the models of computability which had been given prior to Turing's model.

In this talk we consider this shift—one in which Gödel was a key figure—in relation to Gödel's philosophical outlook subsequently. On the way we consider a question that Kripke has asked recently [2]: why did Gödel not see that the *Entscheidungsproblem* is an *immediate* consequence of the Completeness and Incompleteness Theorems? Kripke's analysis depends upon viewing computability in terms of calculability in a logic. We thus suggest that Kripke's own explanation for Gödel's purported blindness to the fact of having solved what was arguably viewed as the single most important problem in logic remaining open at the time,[2] be complemented by the story of the difficulties logicians had with viewing computability in this sense.

In particular we will suggest that the inherent circularity which characterises such a conception, may have contributed to a sense of reluctance on the part of the Princeton group of logicians at the time to embrace it: for if computability is understood in terms of calculability in a logic, then it must be the case that not any logic will serve, rather the logic in question must be given effectively. But then, how to understand the concept "given effectively", as applied to the logic in question?

What follows is an abbreviated history of the development of computability in the 1930 s, from the point of view of this shift in perspective.[3]

Juliette Kennedy: I thank Liesbeth de Mol for the invitation to address a special session of CiE2017, and to contribute this extended abstract to its proceedings.

[1] Gandy quote. See below.

[2] See, e.g. Sieg [3], p. 387.

[3] Some of the text below is adapted from the author's [4].

© Springer International Publishing AG 2017
J. Kari et al. (Eds.): CiE 2017, LNCS 10307, pp. 42–49, 2017.
DOI: 10.1007/978-3-319-58741-7_5

2   Isolating the Concept of "Computable Function"

Gödel defined the class of primitive recursive functions in his 1931 [5], while Church at the same time had developed the λ-calculus together with Kleene.

Church's first presentation of the λ-calculus in 1932 [6], which embeds that calculus in a *deductive formalism*, in the Hilbert/Bernays terminology, was found to be inconsistent in 1934. Two years later Church published a, in Gandy's words, logic-free definition of the λ-calculus, based on the primitives "function" and "iteration":[4]

> When it began to appear that the full system is inconsistent, Church spoke out on the significance of λ-definability, *abstracted from any formal system of logic*, as a notion of number theory.[5]

In the period between these two versions of the λ-calculus Church suggested his *thesis*, namely the suggestion to identify the λ-definable functions with those which are effectively computable.[6] The plausibility of the thesis became especially clear when Church, Kleene and Rosser established *confluence*, proving the equivalence of λ-definability with computability in the sense of the Herbrand-Gödel equational calculus, in 1935.[7]

Interestingly, Gödel was not persuaded of its adequacy, when he was told of Church's suggestion to view intuitive or effective computability in terms of λ-definability in early 1934, finding the proposal "thoroughly unsatisfactory."[8] Church described Gödel's then suggestion to instead take an axiomatic approach to the problem, in a letter to Kleene:

> His [Gödel's] only idea at the time was that it might be possible, in terms of effective calculability as an undefined notion, to state a set of axioms which would embody the generally accepted properties of this notion, and to do something on that basis. ... At that time he did specifically raise the question of the connection between recursiveness in this new sense and effective calculability, but said he did not think that the two ideas could be satisfactorily identified "except heuristically."[9]

Church may have been influenced by Gödel's negative view of the adequacy of λ-calculus, as in his lecture on his Thesis to the American Mathematical Society in 1935, Church used the Herbrand-Gödel equational calculus as a model of

---

[4] [1], Sect. 14.8. Following Gandy we use the term "effectively computable," or just "effective,"to mean "intuitively computable.".

[5] See [7], which relies on Church's [8]. See also [9]. Emphasis the author's.

[6] The phrase "Church's Thesis" was coined by Kleene in 1943. See his [10].

[7] The proof of the equivalence developed in stages. See Davis, [11] and Sieg, [12].

[8] Church, letter to Kleene November 29, 1935. Quoted in Sieg, op. cit., and in Davis [11].

[9] Church, op.cit. In this talk and elsewhere we distinguish the axiomatic from the logical method, viewing the former as an informal notion, and the latter as involving a formalism.

effective computation, i.e. recursiveness in the "new sense," rather than the λ-calculus.

In fact Church presented two approaches to computability in the AMS lectures and in his subsequent 1936 [13], based on the lectures: Firstly *algorithmic*, based still on what is now known as the untyped λ-calculus, i.e. the evaluation of the value fm of a function by the step-by-step application of an algorithm—and secondly *logical*, based on the idea of calculability in a logic:

> And let us call a function F of one positive integer *calculable within* the logic if there exists an expression f in the logic such that $f(\mu) = \nu$ is a theorem when and only when $F(m) = n$ is true, μ and ν being the expressions which stand for the positive integers m and n.[10]

In order to view computability in terms of calculability in a logic, one must first restrict the class of formal systems representing those computable functions. As we noted, not just any formalism will serve! For Church this meant that the theorems of the formal system should be recursively enumerable.[11] A number-theoretic function is taken to be effective, then, if its values can be computed in such an effectively given formalism.

The argument appears to be circular.[12] In Hilbert and Bernays' 1939 *Grundlagen der Mathematik II*, the computable functions are also presented in terms of a logical calculus, but here effectivity is now reduced to primitive recursion. In his [14], Sieg has this to say about Hilbert and Bernays' improvement of Church's system:

> In this way they provided the mathematical underpinnings for ... Church's argument, but only *relative* to the recursiveness conditions: the crucial one requires the proof predicate of deductive formalisms, and thus the steps informal calculations, to be primitive recursive!

We now come to one of the topics of this note, the shift in perspective in 1934 noted by Gandy, writing: "... in 1934 the interest of the group shifted from systems of logic to the λ-calculus and certain mild extensions of it: the $\lambda - \kappa$ and the $\lambda - \delta$ calculi."[13] Gandy does not speculate on the deeper reasons for this shift; and indeed it is interesting that there are so few direct references to

---

[10] [13], p. 357.

[11] In detail, recursive enumerability would be guaranteed here by the so-called step-by-step argument: if each step is recursive then f will be recursive; and three conditions: (i) each rule must be a recursive operation, (ii) the set of rules and axioms must be recursively enumerable, (iii) the relation between a positive integer and the expression which stands for it must be recursive. Conditions i-iii are Sieg's formulation of Church's conditions. See Sieg, [12], p. 165. For Gandy's formulation of the step-by-step argument, see [1], p. 77.

[12] See also Sieg's discussion of the "semi-circularity" of the step-by-step argument in his [12].

[13] [1], p. 71. Of course this shift presaged a much more dramatic shift of perspective in 1936, inaugurated by the work of Turing together with Post's earlier work.

the circularity problem in the writings of logicians working on computability at the time. However it seems plausible that the circularity problem was at least partly responsible for this shift away from logical systems, especially given the initial inconsistency of the λ-calculus.

2.1 The "Scope Problem": How General Are the Incompleteness Theorems?

We turn now to Gödel's role in these developments. Gödel was perhaps the first to suggest isolating the concept of effective computability.[14] His interest was driven, at least in part, by an important piece of unfinished business as far as the Incompleteness Theorems are concerned, in that it was not clear at the time to which formal systems the theorems apply, outside of the fragment of the type theory of *Principia Mathematica* he used.

In short, solving the scope problem depends on articulating a precise and adequate notion of effective computability, because the formal systems at issue in the Incompleteness Theorems, are to be given *effectively*.[15]

To this end, that is, with a wish to "make the incompleteness results less dependent on particular formalisms,"[17] Gödel introduced in 1934 the general recursive, or Herbrand-Gödel recursive functions. It is striking from the point of view of this note that he begins the section introducing those functions thus: *"Now we turn to some considerations which for the present have nothing to do with a formal system."*, and defines the notion of "formal system" in the next section as consisting of "symbols and *mechanical rules* relating to them."[18] Gödel seems here to have all the elements behind the Turing conception of computability in place. What is missing of course, is the model itself.

The Herbrand-Gödel calculus allows for forms of recursions that go beyond primitive recursion, however it was not clear to Gödel at the time, that the schema captured *all* recursions.[19]

Gödel gave, in the same presentation, a precise definition of the conditions a formal system must satisfy so that the incompleteness theorems apply to it. These included the restriction that:

---

[14] See Gandy's [1], p. 72.

[15] Gödel was careful not to claim complete generality for the Second Incompleteness Theorem in his 1931 paper:

> For this [formalist JK] viewpoint presupposes only the existence of a consistency proof in which nothing but finitary means of proof is used, and it is conceivable that there exist finitary proofs that cannot be expressed in the formalism of *P* (or of *M* and *A*).[16]

[16] [15], p. 195. *P* is a variant of *Principia Mathematica*.

[17] Sieg [16], p. 554.

[18] [15], p. 346 and 349, resp. Emphasis added.

[19] As Gödel would later write to Martin Davis, "...I was, at the time of these lectures, not at all convinced that my concept of recursion comprises all possible recursions." Quoted in [15], p. 341.

Supposing the symbols and formulas to be numbered in a manner similar to that used for the particular system considered above, then the class of axioms and the relation of immediate consequence shall be [primitive JK] recursive.[20]

By 1935, Gödel's reflections on computability in the higher order context began to point towards the possibility of a definitive notion of formal system. Nevertheless, an, in Gödel's terminology now, *absolute* definition of effective computability was still missing at that point.[21]

3   Turing's Analysis of Computability

In 1936, Turing gave a self-standing analysis of human effective computability and used it to solve the *Entscheidungsproblem*.

Rather than calculability in a logic, Turing analyzed effectivity informally but exactly, via the concept of a Turing Machine—a machine model of computability, consisting of a tape scanned by a reader, together with a set of simple instructions in the form of quadruples.[23]

We alluded to circularity in connection with approaches to computability that are centered on the idea of calculability in a logic.[24] The crucial point here is that Turing's analysis does not require the specification of a logic *at all*.

The reaction to Turing's work among the Princeton logicians was immediately positive. As Kleene would write in 1981, "Turing's computability is intrinsically persuasive but λ-definability is not intrinsically persuasive and general recursiveness scarcely so (its author Gödel being at the time not at all persuaded)."

For Gödel, as he would later explain to Hao Wang, Turing's model of human effective calculability is, in some sense, perfect:

---

[20] [15], p. 361.

[21] As Gödel wrote to Kreisel in 1965:

That my [incompleteness] results were valid for all possible formal systems began to be plausible for me (that is since 1935) only because of the Remark printed on p. 83 of 'The Undecidable' . . . But I was completely convinced only by Turing's paper.[22]

[22] Quoted in Sieg [17], in turn quoting from an unpublished manuscript of Odifreddi, p. 65.

[23] Or quintuples, in Turing's original presentation.

[24] As Sieg puts it [14], "The work of Gödel, Church, Kleene, and Hilbert and Bernays had intimate historical connections and is still of deep interest. It explicated calculability of functions by *exactly one core notion*, namely calculability of their values in logical calculi via (a finite number of) elementary steps. But no one gave convincing and non-circular reasons for the proposed rigorous restrictions on the steps that are permitted in calculations.".

The resulting definition of the concept of mechanical by the sharp concept of "performable by a Turing machine" is both correct and unique ...*Moreover it is absolutely impossible that anybody who understands the question and knows Turing's definition should decide for a different concept.*[25]

And Turing's analysis led to the complete solution of the scope problem:

In consequence of later advances, in particular of the fact that, due to A. M. Turing's work, a precise and unquestionably adequate definition of the general concept of formal system can now be given, the existence of undecidable arithmetical propositions and the non-demonstrability of the consistency of a system in the same system can now be proved rigorously for every consistent formal system containing a certain amount of finitary number theory.

Turing's analysis thus settled definitively the adequacy question for computability. For Gödel, and for the logicians of the time, the Turing Machine was not just another in the list of acceptable notions of computability—it is the *grounding* of all of them.

We suggested that the issue of circularity was in the background of the shift away from a logical orientation to the problem of isolating and precisifying the notion of intuitive computability. Once again this is the problem of how to spell out the concept of effectivity of a logical system that embeds one's notion of computability, without invoking that very notion. How did the Turing model of computation solve this problem for Gödel? We explain it thus: Turing's model of computation allows for an autonomous logical perspective, because it is itself logic free.

What was the role of confluence? *Church's Thesis* identified effective calculability with λ-calculability, and then with Herbrand-Gödel calculability, after the equivalence between the two was established. Prior to Turing's work, the available confluence was seen as justifying adequacy in a weak sense only. Once one has a grounding example in hand this changes—confluence now plays an epistemologically important evidentiary role.

It is striking that Emil Post, who called for the return to meaning and truth in the opening of his [19], and the downplaying of what he called "postulational thinking," aligns him both ideologically and mathematically with these developments. As it turns out, Post's recommendation to develop recursion theory mathematically, by stripping off the formalism with which the theory was encumbered, led to the formalism free development of recursion theory just along the lines he advocated. It also gave rise to the development of Post Systems, a model of computability very similar to Turing's.

The project of developing an autonomous logical perspective permeated Gödel's outlook from then on. Gödel alludes to it a number of times in his Princeton Bicentennial Lecture, in connection with finding absolute notions of

---

[25] Remark to Hao Wang, [18], p. 203. Emphasis added.

decidability and provability. One can also read the perspective into Gödel's over-arching goal of attaining decidability in set theory—for how else to achieve decidability in set theory, except by remaining, as we have called it, formalism free?

4   Kripke and the *Entscheidungsproblem*

Kripke advocates a, as he calls it, logical orientation to the problem of isolating the notion of effective computability.

> My main point is this: a computation is a special form of mathematical argument.[26]

He then asks, given that such an approach would have been very easily within reach of Gödel and other logicians working in the period immediately after the 1931 Incompleteness Theorems, why wasn't it noticed that a negative solution to the *Entscheidungsproblem* follows immediately?

> Suppose we had taken derivability by a computation expressible in a first-order language as one's basic definition of computability. Then given the Gödel Completeness Theorem, any conventional formalism for first-order logic will be sufficient to formalise such derivability. ... It will be a short and direct step to conclude the undecidability in this sense of the *Entschei-dungsproblem*.[27]

Kripke gives the argument, which necessarily includes the notion of *validity*. He concludes, very reasonably in our view, that Gödel would have been reluctant to invoke a notion of truth in the proof, had he arrived at that proof. Indeed it is well known that Gödel's initial proof of the Incompleteness Theorem followed from the undefinability of truth together with the observation that first order provability is not only definable, but r.e.[28]

We suggest in this talk that the evidence in the record regarding Gödel's response to the Turing model, together with the developments leading up to it (as recounted here), might also be taken into account in explaining Gödel's so-called blindness. It just might be that for Gödel, *grounding* the notion of computability required an altogether new *mathematical* idea, and not a logical one. We saw how close he came to Turing's conception in 1934, just before he turned his attention to the continuum problem in set theory.

References

1. Gandy, R.: The confluence of ideas in 1936. In: The universal turing machine: A half-century survey, pp. 55–111. Oxford Science Publications, Oxford University Press, New York (1988)

---

[26] [2], p. 80.

[27] ibid, p. 85.

[28] See Gödel's letter to van Heijenoort of February 22, 1964, in [20].

2. Kripke, S.: The church-turing "Thesis" as a special corollary of gödel's completeness theorem. In: Copeland, B.J., Posy, C.J., Shagrir, O. (eds.) Computability: Gödel, Church, and Beyond. MIT Press, Cambridge (2013)
3. Sieg, W.: Hilbert's Programs and Beyond. Oxford University Press, Oxford (2013)
4. Kennedy, J.: Turing, Gödel and the "Bright Abyss". In: Philosophical Explorations of the Legacy of Alan Turing, vol. 324 of Boston Studies in Philosophy (Springer)
5. Gödel, K.: Über formal unentscheidbare Sätze der Principia Mathematica und verwandter Systeme I. Monatsh. Math. Phys. 38, 173–198 (1931)
6. Church, A.: A set of postulates for the foundation of logic I, II. Ann. Math. 33(2), 346–366 (1932)
7. Kleene, S.C., Rosser, J.B.: The inconsistency of certain formal logics. Ann. Math. 36(2), 630–636 (1935)
8. Church, A.: The richard paradox. Amer. Math. Monthly 41, 356–361 (1934)
9. Kleene, S.C.: Origins of recursive function theory. Ann. Hist. Comput. 3, 52–67 (1981)
10. Kleene, S.C.: Recursive predicates and quantifiers. Trans. Am. Math. Soc. 53, 41–73 (1943)
11. Davis, M.: Why Gödel didn't have church's thesis. Inf. Control 54, 3–24 (1982)
12. Sieg, W.: Step by recursive step: Church's analysis of effective calculability. Bull. Symbolic Log. 3, 154–180 (1997)
13. Church, A.: A note on the Entscheidungsproblem. J. Symbolic Log. 1(1), 40–41 (1936). (Correction 1:101–102)
14. Sieg, W.: Church without dogma: Axioms for computability. In: Cooper, S.B., Löwe, B., Sorbi, A. (eds.) New Computational Paradigms, pp. 139–152. Springer, New York (2008)
15. Gödel, K.: Collected works. Vol. I. The University Press, New York (1986). Publications 1929–1936, Edited and with a preface by Solomon Feferman. Clarendon Press, Oxford
16. Sieg, W.: On computability. In: Philosophy of Mathematics. Handbook of the Philosophy of Science, pp. 535–630. Elsevier/North-Holland, Amsterdam (2009)
17. Sieg, W.: Gödel on computability. Philos. Math. 14(3), 189–207 (2006)
18. Wang, H.: A Logical Journey. Representation and Mind. MIT Press, Cambridge (1996)
19. Post, E.L.: Recursively enumerable sets of positive integers and their decision problems. Bull. Am. Math. Soc. 50, 284–316 (1944)
20. Gödel, K.: Collected Works. V: Correspondence H-Z. Oxford University Press, Oxford (2003). Feferman, S., et al. (eds.)

A Guided Tour to Computational Haplotyping

Gunnar W. Klau[1] and Tobias Marschall[2,3(✉)]

[1] Heinrich Heine University, Düsseldorf, Germany
[2] Center for Bioinformatics, Saarland University, Saarbrücken, Germany
[3] Max Planck Institute for Informatics, Saarbrücken, Germany
marschal@mpi-inf.mpg.de

Abstract. Human genomes come in pairs: every individual inherits one version of the genome from the mother and another version from the father. Hence, every chromosome exists in two similar yet distinct "copies", called haplotypes. The problem of determining the full sequences of both haplotypes is known as *phasing* or *haplotyping*. In this paper, we review different approaches for haplotyping and point out how they are formalized as optimization problems. We survey different technologies and, in this way, provide guidance on the characteristics of problem instances resulting from present day technologies. Furthermore, we highlight open algorithmic challenges.

1 Haplotyping

Humans and many other species are *diploid*. That is, every individual inherits two versions of each autosomal chromosome, one from its mother and one from its father. We refer to these two versions of a chromosome's DNA sequence as *haplotypes*. The two haplotypes inherited by mother and father are very similar but not identical, reflecting the genetic differences between the parents. Determining the two alleles present at a particular genetic locus is known as *genotyping*, and can be achieved using various technologies including microarrays and short-read sequencing.

In contrast to genotyping, *haplotyping* aims to reconstruct the full sequences of the two haplotypes (Fig. 1). Moving from (sequences of) genotypes to haplotypes is known as *phasing*. We use the terms *haplotyping* and *phasing* interchangeably in the following. The knowledge of haplotypes is important for fundamental and clinical research. On the one hand, haplotype information allows addressing questions in population genetics, for instance to study demographic history [1] and selection [2]. On the other hand, haplotype-specific phenotypes arise when two variants interact [3], e.g. when a disruptive variant in a coding region is combined with a silencing variant in an associated regulatory element. In particular, the interaction between variants in coding regions and enhancers is currently gaining attention [4]. All of this highlights the pressing need for techniques to reliably phase human genomes along whole chromosomes.

In this paper, we review different approaches to reconstruct haplotypes from various data sources. It is meant as an introduction to the topic for computer

© Springer International Publishing AG 2017
J. Kari et al. (Eds.): CiE 2017, LNCS 10307, pp. 50–63, 2017.
DOI: 10.1007/978-3-319-58741-7_6

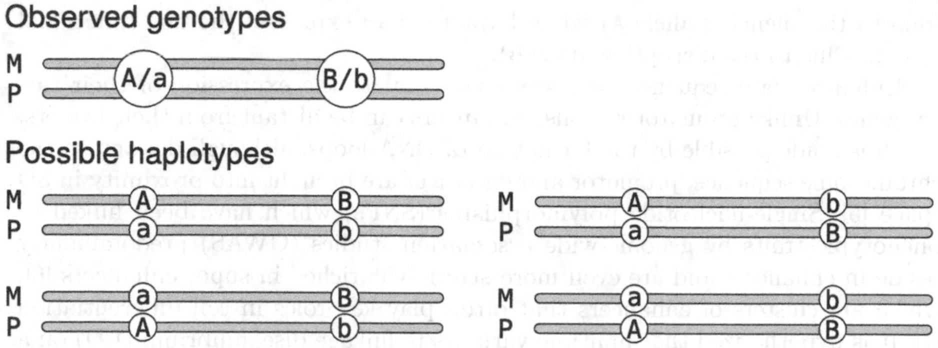

Observed genotypes

Possible haplotypes

Fig. 1. Difference between genotype and haplotype data. Gray bars represent the two homologous chromosomes, i.e. the two versions of a chromosome inherited by the mother (marked M) and the father (marked P). Genotype data (top) tells which pair of alleles is present at a site (white circles). Haplotype data (bottom) gives the full sequence of both chromosomal copies. All four possible haplotypes compatible with the genotypes above are shown.

scientists, who want to learn which problem formulations are relevant and how practical problem instances look like.

We start by motivating the relevance of haplotype phasing in more detail in Sect. 2. Then, in Sect. 3, we introduce three paradigms for haplotyping based on three different data sources: genotype data from populations, genotype data from families, and sequencing reads. From there on, we mainly focus on read-based phasing and introduce its most common formalization (the Minimum Error Correction Problem) in Sect. 4. Section 5 gives an overview of current technologies and hence provides insights into the problem instances encountered in practice. We briefly survey techniques to optimally solve the Minimum Error Correction problem in Sect. 6. Latest developments on using family information and read information in a unified way are presented in Sect. 7. We conclude the paper by pointing out some important open challenges in Sect. 8.

2   Relevance of Phasing

To illustrate the importance of phasing, consider two alleles "A" and "a" for a variant in a regulatory region, where the allele A strongly downregulates expression of a target gene; and two alleles "B" and "b" for a variant in the coding region of that gene, where the allele B is so disruptive that the resulting protein cannot perform its function. If a person is heterozygous in both variants, i.e. has genotypes A/a and B/b, four haplotype configurations are possible, as shown in Fig. 1. The two haplotypes can either be AB and ab or they are Ab and aB. For the former combination, sufficient amounts of functional protein might still be produced from haplotype ab. The latter combination can lead to strong phenotypic consequences because haplotype Ab hardly expresses the gene product (due to the silencing allele A) while haplotype aB expresses the non-functional protein (due to the disruptive allele B).

Enhancers are sequence elements that regulate the expression of their target genes. Unlike promotor regions, enhancers can be distant from their targets, which is made possible by the formation of DNA loops: although distant on the chromosome sequence, promotor and enhancer are brought into proximity in 3D space [5]. Single-nucleotide polymorphisms (SNPs) which have been linked to phenotypic traits by genome-wide association studies (GWAS) predominantly reside in enhancers and are even more strongly enriched in super enhancers [6], which are clusters of enhancers that often play key roles in cell differentiation [7]. It is hypothesized that multiple variants in linkage disequilibrium (LD) on a super enhancer often collectively bring forward the phenotype changes underlying GWAS hits [8]. That means, many variants located on the same haplotype are needed to explain the phenotype. Again, a likely explanation is physical interaction between different variant sites, and the proteins that bind to them, in three dimensional space.

Beyond that, latest findings show that combinatorial effects between (sometimes distant) regulatory variants that are *not in LD* with the respective GWAS hit SNP could substantially contribute to the observed phenotypes [4]. That is, variants that are not correlated with each other often interact to bring forward an effect, for instant when a change in an enhancer is reinforced by a change in the corresponding promotor. Such interactions have the potential to explain parts of the "missing heritability" [4,9], i.e. the gap between epidemiologically observed genetic contribution to a disease and known genetic variants explaining it. Note that all these short- and long-range interactions (within enhancers and between enhancer, promoter and coding sequence) take place on one copy of the two homologous chromosomes, that is, on single haplotypes. To study (common and rare) genomic diseases from this promising new angle, we hence *urgently need tools to obtain haplotype-resolved genomes*. We refer to [3,10] for further reading on the relevance of haplotype-resolved genomes, including many examples of haplotype-specific clinical conditions.

3   Phasing Approaches

There are three different paradigms for haplotyping, which we briefly review in the following.

3.1   Population-Based Haplotyping

Population-based haplotyping takes genotypes from many individuals as input and infers haplotypes by exploiting shared haplotype tracts that exist due to common ancestry and limited recombination [11]. Prominent examples of software systems for population-based phasing are Beagle [12], ShapeIt [13–15] and Eagle [16,17]. They have been successfully employed in population-scale sequencing efforts such as the 1000 Genomes [18] and the Genome of the Netherlands

[19, 20] projects. In regions of strong linkage disequilibrium, this approach yields highly accurate haplotypes. A main advantage of this technique lies in its applicability when only genotype data from a large cohort are available. This is often the case, as such data are comparatively cheap to obtain. Its main limitation lies in the fact that only variants that are common enough to allow for robust statistical inference can be reliably phased, which exclude low-frequency variants and *de novo* variants, which are private to one individual, although these types of variants are of particular medical interest.

Once a (large) set of haplotypes, known as *reference panel*, exists, additional samples can be phased with respect to this panel. To do this, one seeks to find two paths through the panel that explain the genotypes of the additional sample. This problem is known as *parsing genotypes with respect to a founder set* and admits a polynomial-time solution by dynamic programming [21].

3.2 Genetic Haplotyping

Genetic haplotyping takes genotype data for related individuals with a known pedigree as input and determines haplotypes based on the constraints imposed by Mendelian segregation. This approach is very successful when a large pedigree and high-quality genotypes are available. Examples of corresponding software tools include Merlin [22], HaploScribe [23], and Hapi [24]. However, genetic haplotyping has the intrinsic limitation of not being able to phase variants that are heterozygous in all individuals in the pedigree.

To illustrate the idea, consider a case where, at a given locus, the mother has a genotype of B/b, the father has genotype B/B and the child has B/b. The b allele in the child can only have been inherited from the mother and hence b is on the maternal haplotype. Likewise, the b allele in the mother is on the *transmitted* haplotype. By applying the same reasoning to all variants, one can construct chromosome-length haplotypes. That is, at every heterozygous site in the child the alleles are determined to be on the maternal or paternal haplotype and, for each parent, we determine whether an allele is on the transmitted haplotype or on the non-transmitted haplotype. Note, however, that recombination events, i.e. changes of which haplotype in a parent is transmitted, cannot be detected from trio data, but requires larger pedigrees.

3.3 Molecular Haplotyping

Molecular haplotyping experimentally observes individual haplotypes or fragments thereof in a single individual [10]. The most prominent technique to directly determine the sequence of haplotype fragments is *next-generation sequencing* (NGS). Each *read* generated by an NGS platform represents a piece of one of the two haplotypes. Stitching together these pieces (after mapping them to a reference) into longer haplotype fragments is known as *read-based phasing*. The ability to reconstruct haplotypes is limited by the length of the sequencing reads in comparison to the distance between heterozygous sites. Only reads that cover two or more heterozygous sites are helpful for read-based phasing and, to achieve a full-chromosome phasing, all pairs of variants must be (directly or indirectly) connected by sequencing reads. Therefore, data from third-generation technologies that deliver comparatively long reads, such as those marketed by PacBio or Oxford Nanopore, are better suited for this task than standard Illumina data. Most recently, the even more challenging task of de novo assembly (i.e., not using a reference genome) of both haplotypes of a diploid individual has been attempted, for instance using PacBio [25] or 10X Genomics [26] technology. Beyond next-generation sequencing, a number of specialized experimental protocols to determine haplotypes exist, as reviewed in [27].

3.4   Hybrid Approaches

Combinations of the above approaches have also been explored: Population-based haplotyping tools can make use of pedigree information [28,29] and sequencing data [30] and the combined use of sequencing data and pedigree information has been shown to yield markedly better result than sequencing data alone [31].

4   Minimum Error Correction (MEC)

The key challenge in molecular haplotyping is to distinguish true genetic variability from sequencing errors. Among the computational models addressing this task, the *Minimum Error Correction* problem (MEC) [32] is the most accepted and relevant model in practice. MEC asks for a minimum cost correction of the sequencing data to allow a conflict-free bipartition of the reads to the two chromosomal copies.

The input for MEC is a *fragment matrix* $F \in \{0, 1, `-`\}^{m \times n}$, where the rows $1 \leq i \leq m$ correspond to the fragments and the columns $1 \leq j \leq n$ correspond to the variant positions. A non-gap entry $F(i,j) \neq `-`$ specifies that read i covers variant j and gives evidence for the reference allele when $F(i,j) = 0$ or the alternative allele when $F(i,j) = 1$.

Two rows i_1 and i_2 of F are in *conflict* if there is a position j such that $F(i_1, j) \neq `-`$ and $F(i_2, j) \neq `-`$ but $F(i_1, j) \neq F(i_2, j)$. A set of rows is *conflict-free* if it does not contain conflicting row pairs.

A fragment matrix F is *feasible* if there exists a bipartition (I^0, I^1) of its rows such that both I^0 and I^1 are conflict-free. Such a bipartition determines the two haplotypes $h^0, h^1 \in \{0, 1, `-`\}^n$ in the following, natural way:

$$h^0(j) = \begin{cases} 0 & \text{if } F(i,j) = 0 \text{ for some } i \in I_0 \\ 1 & \text{if } F(i,j) = 1 \text{ for some } i \in I_0 \\ - & \text{otherwise,} \end{cases}$$

$$h^1(j) = \begin{cases} 0 & \text{if } F(i,j) = 0 \text{ for some } i \in I_1 \\ 1 & \text{if } F(i,j) = 1 \text{ for some } i \in I_1 \\ - & \text{otherwise.} \end{cases}$$

In practice, the entries in a fragment matrix are associated with phred-scaled base qualities $Q \in \mathbb{N}^{m \times n}$ that correspond to estimated probabilities of $10^{-Q(i,j)/10}$ that entry $F(i,j)$ has been wrongly sequenced. These phred scores serve as costs of flipping entries and allow less confident base calls to be corrected at lower cost compared to high confidence ones. The *error distance* of two fragment matrices F and F' is

$$d_Q(F, F') = \sum_{i=1}^{m} \sum_{j=1}^{n} \begin{cases} 0 & F(i,j) = F'(i,j) \\ Q(i,j) & F(i,j) \neq F'(i,j). \end{cases}$$

We can now state the weighted Minimum Error Correction problem formally.

Problem 1 (wMEC). Given fragment matrix F and quality matrix Q, find a conflict-free fragment matrix F' with minimum error distance $d_Q(F, F')$.

Solving the wMEC problem corresponds to minimizing the sum of phred-scaled probabilities and is therefore equivalent to finding a maximum likelihood bipartition of reads. MEC and its weighted variant wMEC are NP-hard [32], even if there are no internal gaps in the fragments [33,34], and do not admit a constant-factor approximation [35].

5   Relevant Technologies

We now briefly review the most relevant sources of haplotype information and explain their characteristics with respect to the MEC problem. To put read lengths into perspective, we note that, for human genomes, one observes around one heterozygous single-nucleotide variant (SNV) every 1 kbp on average. Read-based phasing requires reads that cover at least two heterozygous sites. Refer to Fig. 2 for an illustration of how the different data sources (surveyed in the following) give rise to rows in the fragment matrix used as input for MEC. This exemplifies the differences in number and distribution of non-dash characters between technologies.

5.1   Illumina Sequencing

Illumina instruments constitute the most common sequencing platform today. The standard mode of operation yields paired-end reads, where one DNA fragment (of usually 300–1000 bp) is sequenced from both ends. Each of these read ends typically have a length of 100–150 bp for HiSeq instruments and of 250–300 bp for MiSeq instruments. Given these characteristics, standard Illumina data are of limited use for read-based phasing since rather few read pairs cover two or more heterozygous variants.

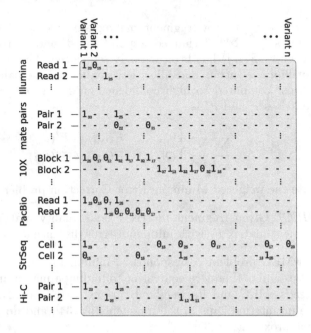

Fig. 2. Illustration of how different technologies contribute to the fragment matrix F which is the input to the MEC problem. For human chromsomes, the number of heterozygous variants n is usally on the order of 10^5. Weights are shown in subscripts (red). This toy example shows qualitative differences in terms of read length and span (i.e. distance from first to last non-dash variant in a row) between data types. (Color figure online)

5.2   Mate-Pair or Jumping Libraries

Mate-pair or jumping libraries are created by circularizing longer DNA fragments, followed by paired-end Illumina sequencing across the fusion site [36]. As a result, one obtains pairs of reads that are further apart than for regular paired-end sequencing. Typical insert sizes of mate pair libraries are between 3 kbp and 5 kbp. Mate pair reads do not cover more sequence than paired-end reads, but span longer chromosomal distances. Therefore the fraction of mate pairs usable for phasing is equally low (as for paired ends), but those pairs that do cover two heterozygous sites establish a phase connection over a longer distance. Combining several mate-pair libraries with different insert size increases the utility of these data for phasing.

5.3   Linked-Read Sequencing

10X Genomics markets a technology to perform linked-read sequencing of high-molecular weight DNA [37]. Long DNA fragments of up to 100 kbp are used as input and separated using micro-fluidics techniques. Each of these fragments is labeled with a barcode such that the probability of assigning the same barcode to different fragments is very low. Long fragments are broken into (barcoded) short fragments that are then subject to standard Illumina sequencing. Since short reads with the same barcode are highly likely to have originated from the same long fragment and hence from the same haplotype, this data type is by far more powerful for phasing than Illumina sequencing alone [37]. 10X data can be used in two different forms, either the mapped and barcoded reads can be used directly to form one row in the allele matrix per bar-code, or the phased blocks output by 10X' proprietary Long Ranger software can be taken as input rows. These blocks are typically several Mbp long.

5.4 Third-Generation Sequencing Technologies

The *PacBio and Oxford Nanopore* sequencing platforms both deliver long but error-prone reads. Although read lengths and error rates differ for the specific instrument and chemistry versions, error rates are typically around 10% and average read-lengths on the order of magnitude of 10 kbp. In contrast to Illumina sequencing, both platforms are prone also to indel errors. The resulting long reads usually contain multiple heterozygous sites and are hence phase informative, which is a key advantage for the purpose of reconstructing haplotypes. As an additional advantage, the read-mapping ambiguity in repetitive areas of the genome is reduced compared to short reads, because long reads are more likely to be longer than a repeat unit. This translates into the ability to resolve many difficult regions inaccessible to Illumina sequencing [38].

5.5 StrandSeq

StrandSeq is a specialized protocol for haplotype-resolved single-cell sequencing [39]. A single cell undergoes one cell division in the presence of bromodeoxyuridine (BrdU), which is incorporated instead of thymine during DNA replication, followed by selective degradation of the newly synthesized strands and library preparation. As a result, one can determine the haplotype of origin of each individual read based on its directionality. We refer to [39] for full details. Below 0.1-fold coverage is obtained from one single cell, so that many cells need to be processed to cover the full genome. While this technique is labor and cost intensive, it delivers chromosome-length resolution of haplotypes. Each single cell results in two rows (for two haplotypes) in our allele matrix. These two rows span the full chromosome but are rather sparse.

5.6 Chromsome Conformation Capturing

Hi-C is an experimental protocol to capture the three-dimensional conformation of a chromosome in the nucleus [40]. It is based on cross-linking DNA stretches that are in close proximity in space, which includes pieces of DNA that are far apart in terms of (one dimensional) chromosomal coordinates or are even located on different chromosomes. The cross-linked DNA fragments can then be sequenced, resulting in paired-end reads where each read-end comes from one of the fragments that have been in physical contact. Contacts between segments of the same copy of a chromosome are much more likely than between a chromosome and its homolog, because each chromosome is located in its own "territory". In other words, Hi-C read pairs that both map to the same chromosome are likely to have originated from the same haplotype. Hi-C data is hence valuable for phasing [40] and each read pair translates into one row in our allele matrix.

6 Solving the MEC Problem

Many exact and heuristic approaches exist to solve the MEC and wMEC problems. We refer to the review [41] for heuristic approaches and focus on exact methods, which are guaranteed to find an optimal solution, in the following.

The currently best exact approaches for MEC and wMEC are fixed-parameter tractable (FPT) algorithms and integer linear programming (ILP) based algorithms.

He et al. [42] proposed an exact dynamic programming algorithm with a runtime of $O(2^L mn)$, where m is the number of reads, n is the number of variants (as in Sect. 4), and L is the maximum number of variants covered by any read. Due to being exponential in L, a quantity proportional to the read length in practice, the algorithm is only suitable for short-read data. Deng et al. [43] suggested a different dynamic programming algorithm for MEC, which is exponential in the maximum coverage of a SNP, but is linear in the number of variants. Its runtime is $O(2^c nc)$, where c is the maximum coverage across all columns. In particular, the running time of this DP does not depend on the read length, which is beneficial for upcoming long-read sequencing data. In our previous approach WhatsHap [44,45], we independently arrived at a similar DP as [43] to solve wMEC (where [43] consider unweighted MEC), but achieve a better runtime of $O(2^c n)$ due to enumerating read bipartitions in Gray code order. The algorithm by Kuleshov [46] approached the weighted MEC problem in a message-passing framework and arrived at the same DP used in [43,45]. Pirola et al. [47] considered a restricted variant of MEC, in which up to k corrections are allowed per SNP position, and presented an FPT algorithm that runs in time $O(c^k Ln)$.

These DP-based algorithms work well for maximum coverage values up to 20 or 25. Instances with higher coverage cannot be solved to optimality in reasonable computing time with these methods. Note in this context that each row contributes to the "coverage" from its first to its last non-dash position, regardless of intermediate dash positions. That means that the presence of many sparse rows, such as generated by StrandSeq or Hi-C, lead to difficulties for these solvers and, in particular, instances arising from combining different technologies as outlined above can often not be solved anymore.

The first ILP formulation for MEC was given by Fouilhoux and Mahjoub [48] and is based on a reduction to the maximum bipartite induced subgraph problem. Chen et al. [49] introduced a linearization of a simple quadratic program for MEC, which they later adapted to wMEC [50]. This approach shows promising running times due to strong pre-processing rules, which are, however, only effective in the all-heterozygous case.

7 Minimum Error Correction on Pedigrees (PedMEC)

The information inherent to genotypes from a whole family and inherent to sequencing reads are orthogonal and complement each other. Hence, genetic phasing and read-based phasing can and should be treated in a unified framework. The Minimum Error Correction on Pedigrees (PedMEC) problem is such a generalization [31]. The input for PedMEC consists of multiple fragment matrices and genotypes for a whole pedigree of individuals. Consider three fragment/quality matrix pairs (F_m, Q_m), (F_f, Q_f), and (F_c, Q_c), for mother, father, and child, respectively, and per-site phred-scaled recombination costs $\mathcal{X} \in \mathbb{N}^n$. Solving the PedMEC problem for this mother-father-child trio requires to determine which entries to flip to make the fragment matrices feasible and to compute two transmission vectors $t_{m \to c}, t_{f \to c} \in \{0, 1\}^n$ such that the sum of phred-scaled flipping and transmission costs are minimal. The feasible fragment matrices give rise to haplotypes (h_m^0, h_m^1), (h_f^0, h_f^1), (h_c^0, h_c^1), for mother, father, and child, respectively.

The transmission vectors formalize Mendelian inheritance. The values $t_{m \to c}(j)$ and $t_{f \to c}(j)$ specify which allele at site j is transmitted from mother to child and from father to child, respectively, in the following way:

$$h_c^0(j) = \begin{cases} h_m^0(j) & t_{m \to c}(j) = 0 \\ h_m^1(j) & t_{m \to c}(j) = 1 \end{cases} \quad \text{and} \quad h_c^1(j) = \begin{cases} h_f^0(j) & t_{f \to c}(j) = 0 \\ h_f^1(j) & t_{f \to c}(j) = 1. \end{cases}$$

The notation can be extended in a straightforward manner to large pedigrees that contain many such relationships. We refer the reader to [31] for full details. It is demonstrated in [31] that a 2-fold sequencing coverage on each indvidual from a trio yields better results than 15-fold coverage when phasing each individual separately, highlighting the power of this unified approach.

8 Discussion and Future Challenges

In this paper, we review different paradigms for haplotyping and particularly emphasized read-based phasing and its formalization as the Minimum Error Correction (MEC) problem. We choose this emphasis because we argue that haplotyping based on direct evidence (such as long reads) will become the standard mode of operation once long read technologies have gained widespread adoption.

From our point of view, the most pressing challenges to be addressed by the bioinformatics community are:

1. Solvers for the MEC problem have predominantly been engineered to be efficient on consecutive reads; that is, on data where each row in the fragment

matrix contains exactly one block of consecutive non-dash characters. With the advent of technologies like Hi-C and StrandSeq, it is important to develop practically efficient solvers for these data types and, in particular, for fragment matrices constructed from a mix of different data sources.

2. Developing practically efficient algorithms for solving the PedMEC problem for higher coverages. The only available solver [31] is based on a dynamic programming algorithm that scales exponentially in the cumulative coverage across individuals. While this already allows to obtain high-quality haplotypes for small pedigrees, it does not scale well to larger pedigrees and higher coverages.

3. While the problem formulations for pedigree-based and read-based phasing have been unified in the PedMEC problem, a generalization that incorporates population-based phasing is missing. Such a generalization would allow to handle all available data sources in a unified way.

4. Phasing polyploid genomes is even more challenging in theory [35] and practice. Exact solvers for the k-ploid case do not exist yet, but are needed to resolve, for example, plant genomes.

References

1. Lawson, D.J., Hellenthal, G., Myers, S., Falush, D.: Inference of population structure using dense haplotype data. PLoS Genet. **8**(1), e1002453 (2012)
2. Sabeti, P.C., Varilly, P., Fry, B., et al.: Genome-wide detection and characterization of positive selection in human populations. Nature **449**(7164), 913–918 (2007)
3. Tewhey, R., Bansal, V., Torkamani, A., Topol, E.J., Schork, N.J.: The importance of phase information for human genomics. Nat. Rev. Genet. **12**(3), 215–223 (2011)
4. Corradin, O., Cohen, A.J., Luppino, J.M., Bayles, I.M., Schumacher, F.R., Scacheri, P.C.: Modeling disease risk through analysis of physical interactions between genetic variants within chromatin regulatory circuitry. Nat. Genet. **48**(11), 1313–1320 (2016)
5. Shlyueva, D., Stampfel, G., Stark, A.: Transcriptional enhancers: from properties to genome-wide predictions. Nat. Rev. Genet. **15**(4), 272–286 (2014)
6. Hnisz, D., Abraham, B.J., Lee, T.I., Lau, A., Saint-Andr, V., Sigova, A.A., Hoke, H.A., Young, R.A.: Super-enhancers in the control of cell identity and disease. Cell **155**(4), 934–947 (2013)
7. Whyte, W.A., Orlando, D.A., Hnisz, D., Abraham, B.J., Lin, C.Y., Kagey, M.H., Rahl, P.B., Lee, T.I., Young, R.A.: Master transcription factors and mediator establish super-enhancers at key cell identity genes. Cell **153**(2), 307–319 (2013)
8. Corradin, O., Saiakhova, A., Akhtar-Zaidi, B., Myeroff, L., Willis, J., Cowper-Sallari, R., Lupien, M., Markowitz, S., Scacheri, P.C.: Combinatorial effects of multiple enhancer variants in linkage disequilibrium dictate levels of gene expression to confer susceptibility to common traits. Genome Res. **24**(1), 1–13 (2014)
9. Eskin, E.: Discovering genes involved in disease and the mystery of missing heritability. Commun. ACM **58**(10), 80–87 (2015)
10. Glusman, G., Cox, H.C., Roach, J.C.: Whole-genome haplotyping approaches and genomic medicine. Genome Med. **6**(9), 73 (2014)
11. Browning, S.R., Browning, B.L.: Haplotype phasing: existing methods and new developments. Nat. Rev. Genet. **12**(10), 703–714 (2011)

12. Browning, S.R., Browning, B.L.: Rapid and accurate haplotype phasing and missing-data inference for whole-genome association studies by use of localized haplotype clustering. Am. J. Hum. Genet. **81**(5), 1084–1097 (2007)
13. Delaneau, O., Marchini, J., Zagury, J.F.: A linear complexity phasing method for thousands of genomes. Nat. Meth. **9**(2), 179–181 (2012)
14. Delaneau, O., Zagury, J.F., Marchini, J.: Improved whole-chromosome phasing for disease and population genetic studies. Nat. Meth. **10**(1), 5–6 (2013)
15. O'Connell, J., Sharp, K., Shrine, N., Wain, L., Hall, I., Tobin, M., Zagury, J.F., Delaneau, O., Marchini, J.: Haplotype estimation for biobank-scale data sets. Nat. Genet. **48**(7), 817–820 (2016)
16. Loh, P.R., Palamara, P.F., Price, A.L.: Fast and accurate long-range phasing in a UK Biobank cohort. Nat. Genet. **48**(7), 811–816 (2016)
17. Loh, P.R., Danecek, P., Palamara, P.F., Fuchsberger, C., Reshef, Y.A., Finucane, H.K., Schoenherr, S., Forer, L., McCarthy, S., Abecasis, G.R., Durbin, R., Price, A.L.: Reference-based phasing using the Haplotype Reference Consortium panel. Nat. Genet. **48**(11), 1443–1448 (2016)
18. The 1000 Genomes Project Consortium: A global reference for human genetic variation. Nature **526**(7571), 68–74 (2015)
19. The Genome of the Netherlands Consortium: Whole-genome sequence variation, population structure and demographic history of the dutch population. Nat. Genet. **46**, 818–825 (2014)
20. Hehir-Kwa, J.Y., Marschall, T., Kloosterman, W.P., et al.: A high-quality human reference panel reveals the complexity and distribution of genomic structural variants. Nat. Commun. **7**, 12989 (2016)
21. Rastas, P., Ukkonen, E.: Haplotype inference via hierarchical genotype parsing. In: Giancarlo, R., Hannenhalli, S. (eds.) WABI 2007. LNCS, vol. 4645, pp. 85–97. Springer, Heidelberg (2007). doi:10.1007/978-3-540-74126-8_9
22. Abecasis, G.R., Cherny, S.S., Cookson, W.O., Cardon, L.R.: Merlin–rapid analysis of dense genetic maps using sparse gene flow trees. Nat. Genet. **30**(1), 97–101 (2002)
23. Roach, J.C., Glusman, G., Hubley, R., Montsaroff, S.Z., Holloway, A.K., Mauldin, D.E., Srivastava, D., Garg, V., Pollard, K.S., Galas, D.J., Hood, L., Smit, A.F.A.: Chromosomal haplotypes by genetic phasing of human families. Am. J. Hum. Genet. **89**(3), 382–397 (2011)
24. Williams, A.L., Housman, D.E., Rinard, M.C., Gifford, D.K.: Rapid haplotype inference for nuclear families. Genome Biol. **11**, R108 (2010)
25. Chin, C.S., Peluso, P., Sedlazeck, F.J., Nattestad, M., Concepcion, G.T., Clum, A., Dunn, C., O'Malley, R., Figueroa-Balderas, R., Morales-Cruz, A., Cramer, G.R., Delledonne, M., Luo, C., Ecker, J.R., Cantu, D., Rank, D.R., Schatz, M.C.: Phased diploid genome assembly with single-molecule real-time sequencing. Nat. Meth. **13**(12), 1050–1054 (2016). Advance online publication
26. Weisenfeld, N.I., Kumar, V., Shah, P., Church, D., Jae, D.B.: Direct determination of diploid genome sequences. bioRxiv, 070425 (2016)
27. Snyder, M.W., Adey, A., Kitzman, J.O., Shendure, J.: Haplotype-resolved genome sequencing: experimental methods and applications. Nat. Rev. Genet. **16**(6), 344–358 (2015)
28. Marchini, J., Cutler, D., Patterson, N., Stephens, M., Eskin, E., Halperin, E., Lin, S., Qin, Z.S., Munro, H.M., Abecasis, G.R., Donnelly, P.: A comparison of phasing algorithms for trios and unrelated individuals. Am. J. Hum. Genet. **78**(3), 437–450 (2006)

29. Chen, W., Li, B., Zeng, Z., Sanna, S., Sidore, C., Busonero, F., Kang, H.M., Li, Y., Abecasis, G.R.: Genotype calling and haplotyping in parent-offspring trios. Genome Res. **23**(1), 142–151 (2013)
30. Delaneau, O., Howie, B., Cox, A.J., Zagury, J.F., Marchini, J.: Haplotype estimation using sequencing reads. Am. J. Hum. Genet. **93**(4), 687–696 (2013)
31. Garg, S., Martin, M., Marschall, T.: Read-based phasing of related individuals. Bioinformatics (Oxford, England) **32**(12), i234–i242 (2016)
32. Lippert, R., Schwartz, R., Lancia, G., Istrail, S.: Algorithmic strategies for the single nucleotide polymorphism haplotype assembly problem. Briefings Bioinform. **3**(1), 23–31 (2002)
33. Cilibrasi, R., Iersel, L., Kelk, S., Tromp, J.: On the complexity of several haplotyping problems. In: Casadio, R., Myers, G. (eds.) WABI 2005. LNCS, vol. 3692, pp. 128–139. Springer, Heidelberg (2005). doi:10.1007/11557067_11
34. Zhao, Y.Y., Wu, L.Y., Zhang, J.H., Wang, R.S., Zhang, X.S.: Haplotype assembly from aligned weighted SNP fragments. Comput. Biol. Chem. **29**(4), 281–287 (2005)
35. Bonizzoni, P., Dondi, R., Klau, G.W., Pirola, Y., Pisanti, N., Zaccaria, S.: On the minimum error correction problem for haplotype assembly in diploid and polyploid genomes. J. Comput. Biol. **23**(9), 718–736 (2016). A journal of computational molecular cell biology
36. Hanscom, C., Talkowski, M.: Design of large-insert jumping libraries for structural variant detection using illumina sequencing. Curr. Protoc. Hum. Genet. **80**, 7.22.1–7.22.9 (2014)
37. Zheng, G.X.Y., Lau, B.T., Schnall-Levin, M., et al.: Haplotyping germline and cancer genomes with high-throughput linked-read sequencing. Nat. Biotechnol. **34**(3), 303–311 (2016)
38. Chaisson, M.J.P., Huddleston, J., Dennis, M.Y., Sudmant, P.H., Malig, M., Hormozdiari, F., Antonacci, F., Surti, U., Sandstrom, R., Boitano, M., Landolin, J.M., Stamatoyannopoulos, J.A., Hunkapiller, M.W., Korlach, J., Eichler, E.E.: Resolving the complexity of the human genome using single-molecule sequencing. Nature **517**(7536), 608–611 (2015)
39. Porubský, D., Sanders, A.D., van Wietmarschen, N., Falconer, E., Hills, M., Spierings, D.C.J., Bevova, M.R., Guryev, V., Lansdorp, P.M.: Direct chromosome-length haplotyping by single-cell sequencing. Genome Res. **26**(11), 1565–1574 (2016)
40. Lieberman-Aiden, E., van Berkum, N.L., Williams, L., Imakaev, M., Ragoczy, T., Telling, A., Amit, I., Lajoie, B.R., Sabo, P.J., Dorschner, M.O., Sandstrom, R., Bernstein, B., Bender, M.A., Groudine, M., Gnirke, A., Stamatoyannopoulos, J., Mirny, L.A., Lander, E.S., Dekker, J.: Comprehensive mapping of long-range interactions reveals folding principles of the human genome. Science **326**(5950), 289–293 (2009)
41. Rhee, J.K., Li, H., Joung, J.G., Hwang, K.B., Zhang, B.T., Shin, S.Y.: Survey of computational haplotype determination methods for single individual. Genes Genomics **38**(1), 1–12 (2015)
42. He, D., Choi, A., Pipatsrisawat, K., Darwiche, A., Eskin, E.: Optimal algorithms for haplotype assembly from whole-genome sequence data. Bioinformatics **26**(12), i183–i190 (2010)
43. Deng, F., Cui, W., Wang, L.: A highly accurate heuristic algorithm for the haplotype assembly problem. BMC Genom. **14**(Suppl 2), S2 (2013)
44. Patterson, M., Marschall, T., Pisanti, N., Iersel, L., Stougie, L., Klau, G.W., Schönhuth, A.: WhatsHap: haplotype assembly for future-generation sequencing reads. In: Sharan, R. (ed.) RECOMB 2014. LNCS, vol. 8394, pp. 237–249. Springer, Cham (2014). doi:10.1007/978-3-319-05269-4_19

45. Patterson, M., Marschall, T., Pisanti, N., van Iersel, L., Stougie, L., Klau, G.W., Schönhuth, A.: WhatsHap: weighted haplotype assembly for future-generation sequencing reads. J. Comput. Biol. **22**(6), 498–509 (2015)
46. Kuleshov, V.: Probabilistic single-individual haplotyping. Bioinformatics (Oxford, England) **30**(17), i379–i385 (2014)
47. Pirola, Y., Zaccaria, S., Dondi, R., Klau, G.W., Pisanti, N., Bonizzoni, P.: HapCol: accurate and memory-efficient haplotype assembly from long reads. Bioinformatics **32**(11), 1610–1617 (2015)
48. Fouilhoux, P., Mahjoub, A.R.: Solving VLSI design and DNA sequencing problems using bipartization of graphs. Comput. Optim. Appl. **51**(2), 749–781 (2012)
49. Chen, Z.Z., Deng, F., Wang, L.: Exact algorithms for haplotype assembly from whole-genome sequence data. Bioinformatics (Oxford, England) **29**(16), 1938–1945 (2013)
50. Chen, Z.Z., Deng, F., Shen, C., Wang, Y., Wang, L.: Better ILP-based approaches to haplotype assembly. J. Comput. Biol. **23**(7), 537–552 (2016)

Outline of Partial Computability
in Computable Topology

Margarita Korovina[1(✉)] and Oleg Kudinov[2]

[1] A.P. Ershov Institute of Informatics Systems, SbRAS, Novosibirsk, Russia
rita.korovina@gmail.com
[2] Sobolev Institute of Mathematics, SbRAS, Novosibirsk, Russia
kud@math.nsc.ru

Abstract. In the framework of computable topology we investigate properties of partial computable functions, in particular complexity of various problems in computable analysis in terms of index sets, the effective Borel and Lusin hierarchies.

1 Introduction

Classical computability theory has a long term tradition to study partial computable functions [16,17]. While the class of effective topological spaces, in particular computable Polish spaces, is one of the main objects for investigation in the Effective Descriptive Set Theory (EDST) [3,14,19] the class of partial computable functions over effective topological spaces has not been deeply investigated yet. In this paper we address natural problems related to partial computable functions.

The paper is organised as follows. Sections 2 and 3 contain preliminaries and basic background. In Sect. 3.2 we introduce the class of effectively enumerable T_0–spaces with point recovering which contains computable Polish spaces among others and plays an important role in the description of the images of surjective partial computable functions. In Sect. 4 we recall the definition of a partial computable function in the settings of effectively enumerable spaces that is motivated by the following observations. It is well-known that in the domain–theoretic framework a partial computable real function is effectively continuous on its domain [2] and the domain is a $\Pi_2^0[\mathbb{R}]$ in the effective Borel hierarchy [13] (see also [6]). On the computable Polish spaces this definition agrees with several known approaches to partial computability [4,5,21]. We show that the class $\mathcal{PCF}$ of partial computable functions over effectively enumerable spaces is closed under composition. In Sect. 5 we work with computable Polish spaces. After showing the correspondence between the partial computable functions and the classical enumeration operators [16] we prove the existence of the principal computable numbering of $\mathcal{PCF}$ for computable Polish spaces. This allows us to

---

The research has been partially supported by the DFG grants CAVER BE 1267/14-1 and WERA MU 1801/5-1.

J. Kari et al. (Eds.): CiE 2017, LNCS 10307, pp. 64–76, 2017.
DOI: 10.1007/978-3-319-58741-7_7

study the complexity of index sets of important problems in computable analysis such as function equality and root verification. Then we turn to an investigation of images of partial computable functions. First we show the existence of a partial computable surjection between any computable Polish space and any effectively enumerable topological space with point recovering. Using this result we prove that for any computable Polish spaces $\mathcal{X}$ and $\mathcal{Y}$, the images of a partial computable functions $f : \mathcal{X} \to \mathcal{Y}$ are exactly Σ_1^1–subsets of Y.

2 Preliminaries

We refer the reader to [16,17] for basic definitions and fundamental concepts of recursion theory, to [3,7,14] for basic definitions and fundamental concepts of DST and EDST, in particular, for the definitions of the effective Borel and Lusin hierarchies. In the major part of our paper we work with the following notion of a computable Polish space. A computable Polish space is a complete separable metric space $\mathcal{X}$ without isolated points and with a metric d such that there is a countable dense set $\mathcal{B} = \{b_1, b_2, \dots\}$ called a *basis of X* that makes the following two relations $\{(n, m, i) \mid d(b_n, b_m) < q_i, \ q_i \in \mathbb{Q}\} \ \{(n, m, i) \mid d(b_n, b_m) > q_i, \ q_i \in \mathbb{Q}\}$ computably enumerable (c.f. [4,15,21]). The standard notations $B(x, y)$ and $\overline{B}(x, y)$ are used for open and closed balls with the center x and the radius y. We consider this concept in the framework of effectively enumerable spaces (see Sect. 3.1). We work with the Baire space $\mathcal{N} = (\omega^\omega, \alpha_\mathcal{N})$, the Cantor space $\mathcal{C} = (2^\omega, \alpha_\mathcal{C})$ with the standard topologies and numberings of the bases.

3 Basic Background

3.1 Effectively Enumerable Topological Spaces

Now we recall the notion of an effectively enumerable topological space. Let (X, τ, α) be a topological space, where X is a non-empty set, $B_\tau \subseteq 2^X$ is a base of the topology τ and $\alpha : \omega \to B_\tau$ is a numbering.

Definition 1 [11]. *A topological space (X, τ, α) is effectively enumerable if the following conditions hold.*

(1) There exists a computable function $g : \omega \times \omega \times \omega \to \omega$ such that

$$\alpha(i) \cap \alpha(j) = \bigcup_{n \in \omega} \alpha(g(i, j, n)).$$

(2) The set $\{i | \alpha(i) \neq \emptyset\}$ is computably enumerable.

For a computable Polish space $(X, \mathcal{B}, d)$ in a naturale way we define the numbering of the base of the standard topology as follows. First we fix a computable numbering $\alpha^* : \omega \setminus \{0\} \to (\omega \setminus \{0\}) \times \mathbb{Q}^+$. Then,

$$\alpha(0) = \emptyset,$$
$$\alpha(i) = B(b_n, r) \text{ if } i > 0 \text{ and } \alpha^*(i) = (n, r).$$

For $\alpha^*(i) = (n, r)$ later we use notation $n = u(i)$ and $r = r_i$.

It is easy to see that (X, τ, α) is an effectively enumerable topological space. Therefore we consider the computable Polish spaces as a proper subclass of the effectively enumerable topological spaces. For details we refer to [11]. In this paper for such effectively enumerable topological space (X, τ, α) we use the standard relations on the indices of basic balls defined as follows:

$$i \prec_X j \leftrightharpoons d(b_{u(i)}, b_{u(j)}) + r_i < r_j,$$
$$i \mid_X j \leftrightharpoons d(b_{u(i)}, b_{u(j)}) > r_i + r_j,$$

for details see, e.g., [18]. The relation $\prec_X$ is irreflexive and transitive and if $i \prec_X j$ then $\mathrm{cl}(\alpha(i)) \subseteq \alpha(j)$. It is easy to see that these relations are computably enumerable on the indices of basic balls. We recall the notion of an effectively open set.

Definition 2. *Let (X, τ, α) be an effectively enumerable topological space. A set $A \subseteq X$ is effectively open if there exists a computably enumerable set $V \subseteq \omega$ such that $A = \bigcup_{n \in V} \alpha(n)$.*

It is worth noting the set of all effectively open subsets of X is closed under intersection and union since the class of effectively enumerable sets is a lattice.

3.2 Effectively Enumerable T_0–spaces with Point Recovering

In this section we introduce effectively enumerable T_0–spaces with point recovering. Further on we will see that they play an important role in the description of images of surjective partial computable functions.

Definition 3. *Let $\mathcal{Y} = (Y, \lambda, \beta)$ be an effectively enumerable T_0–space. We say that $\mathcal{Y}$ admits point recovering if $\{A_x \mid x \in Y\}$ is a Σ_1^1-subset of $\mathcal{P}(\omega)$, where $A_x = \{n \mid x \in \beta(n)\}$. Here $\mathcal{P}(\omega)$ is considered as the Cantor space $\mathcal{C}$.*

Theorem 1. *Every computable Polish space $\mathcal{X} = (X, \tau, \alpha)$ admits point recovering. Moreover, $\{A_x \mid x \in X\}$ is a Π_2^0–subset of $\mathcal{C}$.*

Proof. For a computable Polish space $\mathcal{X} = (X, \tau, \alpha)$, to prove that the set $\{A_x \mid x \in X\}$ is a Π_2^0-set in the effective Borel hierarchy on $\mathcal{C}$ let us observe that, for $I \subseteq \omega$, $(\exists x \in X)\, I = A_x$ if and only if the following conditions hold.

Cond 1: $(\forall k \in \omega)(\exists m \in \omega)(\exists n \in \omega)(\exists r \in \mathbb{Q}^+)(\exists l \in \omega)(\exists r' \in \mathbb{Q}^+)\Big(k \in I \rightarrow$

$$\Big(\alpha^*(k) = (n, r) \wedge \alpha^*(m) = (l, r') \wedge r' < \frac{r}{2} \wedge m \prec_X k \wedge m \in I\Big)\Big)$$

where α^* is defined as on the page 2.

Cond 2: $(\forall k \in \omega)(\forall m \in \omega)\Big((k \in I \wedge m \in I) \rightarrow \alpha(k) \cap \alpha(m) \neq \emptyset\Big).$

Cond 3: $I \neq \emptyset$.

Cond 4: $(\forall k \in \omega)(\forall m \in \omega)\Big((k \in I \wedge k \prec_X m) \rightarrow m \in I\Big).$

Let us denote $\Psi(I) = \text{Cond } 1(I) \wedge \text{Cond } 2(I) \wedge \text{Cond } 3(I) \wedge \text{Cond } 4(I)$. By definition, Ψ is in Π_2^0-form.   □

Remark 1. It is easy to see that the conditions Cond $1(I)$ − Cond $4(I)$ in Theorem 1 can be rewritten in the special form

$$\forall \bar{k}\big(\eta(\bar{k}, I) \vee \Phi(\bar{k}, I)\big),$$

where η is a disjunction of formulas of the kind $k_i \notin I$ and Φ is a computable disjunction (possible infinite) of ∃–formulas with positive occurrences of I i.e. Φ does not contain formulas of the kind $k_i \notin I$. Indeed, for example, Cond $4(I)$ can be rewritten as follows:

$$(\forall k \in \omega)(\forall m \in \omega)\Big(k \notin I \vee m \in I \vee \neg k \prec_X m\Big).$$

Since $\neg k \prec_X m \leftrightharpoons (\forall l \in \omega)\, Q(m, k, l)$, where $Q(m, k, l)$ defines computable subset of ω^3, we have

$$\text{Cond } 4\,(I) \leftrightarrow (\forall k \in \omega)(\forall m \in \omega)(\forall l \in \omega)\Big(k \notin I \vee m \in I \vee Q(m, k, l)\Big).$$

Later we use this form in the proof of Theorem 2.

Proposition 1. *There exists effectively enumerable topological space that does not admit point recovering.*

Proof. Let us consider $\mathcal{C}$ as a subset of $\mathbb{R}$ and take $Y \subseteq \mathcal{C}$ such that $\{A_x \mid x \in Y\}$ is non-analytic. It is possible to do since the number of subsets $Y \subseteq \mathcal{C}$ such that $\{A_x \mid x \in Y\}$ is analytic is no more than continuum. Then put $X = \mathbb{R} \setminus Y$ and $\mathcal{X} = (X, \tau_X)$, where the topology τ_X is induced by $\tau_{\mathbb{R}}$.

It is clear that $\mathcal{X}$ is an effectively enumerable topological space since $\mathcal{C}$ is nowhere dense in $\mathbb{R}$. Taking into account that $\{A_x \mid x \in X\} = \{A_x \mid x \in \mathbb{R}\} \setminus \{A_x \mid x \in Y\}$ we conclude that $\mathcal{X}$ does not admit point recovering.   □

4 $\mathcal{PCF}$ over Effectively Enumerable Topological Spaces

In this section we recall the notion of a partial computable function $f : \mathcal{X} \to \mathcal{Y}$, where $\mathcal{X} = (X, \tau_X, \alpha)$ is an effectively enumerable topological space and $\mathcal{Y} = (Y, \tau_Y, \beta)$ is an effectively enumerable T_0–space.

Definition 4 [8]. *Let $\mathcal{X} = (X, \tau_X, \alpha)$ be an effectively enumerable topological space and $\mathcal{Y} = (Y, \tau_Y, \beta)$ be an effectively enumerable T_0–space. A function $f : \mathcal{X} \to \mathcal{Y}$ is called* partial computable *(pcf) if there exist a computable sequence of effectively open sets $\{O_n\}_{n \in \omega}$ and a computable function $H : \omega^2 \to \omega$ such that*

1. *$\text{dom}(f) = \bigcap_{n \in \omega} O_n$ and*
2. *$f^{-1}(\beta(m)) = \bigcup_{i \in \omega} \alpha(H(m, i)) \cap \text{dom}(f)$.*

In the following if a partial computable function f is everywhere defined we say f is a *total computable function*. For effectively enumerable topological spaces $\mathcal{X}$ and effectively enumerable T_0–space $\mathcal{Y}$ we denote the set of partial computable functions $f : \mathcal{X} \to \mathcal{Y}$ as $\mathcal{PCF}_{\mathcal{XY}}$.

Remark 2. It is worth noting that for computable Polish spaces, Definition 4 corresponds to the results in [5], where partial TTE–computable functions have been described via effective continuity and Π_2^0-domains. Moreover, for partial real functions, the classes of $\mathcal{PCF}_{\mathbb{RR}}$, the TTE–computable [20], the domain-computable [2] and the majorant–computable functions [9,12] coincide.

The following proposition is a straightforward corollary of Definition 4.

Proposition 2. *Let $\mathcal{X} = (X, \tau, \alpha)$ be an effectively enumerable topological space and $\mathcal{Y} = (Y, \lambda, \beta)$ be an effectively enumerable T_0-space.*

1. *If $f : \mathcal{X} \to \mathcal{Y}$ is a pcf, then f is continuous at every points of $\mathrm{dom}(f)$.*
2. *A total function $f : \mathcal{X} \to \mathcal{Y}$ is computable if and only if it is effectively continuous.*

Proposition 3 [8]. *Over effectively enumerable T_0–spaces, $\mathcal{PCF}$ is closed under composition.*

5 $\mathcal{PCF}$ over Computable Polish Spaces

In this section we consider partial computable functions over the subclass of effectively enumerable topological spaces which is the class of computable Polish spaces. We give a characterisation of partial computability in terms of classical enumeration operators (see e.g. [16]). Then based on this characterisation we show the existence of the principal computable numbering of $\mathcal{PCF}_{\mathcal{XY}}$ and compute the complexity of some problems such as function equality and root verification.

5.1   Characterisation of $\mathcal{PCF}_{\mathcal{XY}}$

Definition 5 [16]. *A function $\Gamma_e : \mathcal{P}(\omega) \to \mathcal{P}(\omega)$ is called an enumeration operator if*

$$\Gamma_e(A) = B \leftrightarrow B = \{j \mid \exists i\, c(i, j) \in W_e,\ D_i \subseteq A\},$$

where W_e is the e-th computably enumerable set, and D_i is the i-th finite set. A function $\Gamma_e : \mathcal{P}(\omega) \to \mathcal{P}(\omega)$ is called a reduced enumeration operator if

$$\Gamma_e(A) = B \leftrightarrow B = \{j \mid (\exists i \in A)\, c(i, j) \in W_e\},$$

where W_e is the e-th computably enumerable set.

Now we recall the notion of a computable function introduced in [11].

Definition 6 [11]. *Let $\mathcal{X} = (X, \tau, \alpha)$ be an effectively enumerable topological space and $\mathcal{Y} = (Y, \lambda, \beta)$ be an effectively enumerable T_0-space.*
A partial function $f : \mathcal{X} \to \mathcal{Y}$ is called computable if there exists an enumeration operator $\Gamma_e : \mathcal{P}(\omega) \to \mathcal{P}(\omega)$ such that, for every $x \in X$,

1. *If $x \in dom(f)$ then*

$$\Gamma_e(\{i \in \omega | x \in \alpha(i)\}) = \{j \in \omega \mid f(x) \in \beta(j)\}.$$

2. *If $x \notin dom(f)$ then, for all $y \in Y$,*

$$\bigcap_{j \in \omega} \{\beta(j) | j \in \Gamma_e(A_x)\} \neq \bigcap_{j \in \omega} \{\beta(j) | j \in B_y\},$$

where $A_x = \{i \in \omega | x \in \alpha(i)\}$, $B_y = \{j \in \omega | y \in \beta(j)\}$.

In this case we say that Γ_e completely defines the function f.

Remark 3. It is worth noting that if we work with effectively enumerable topological spaces then a function is computable if and only if there exists a reduced enumeration operator satisfying the requirements of Definition 6.

Proposition 4. *Let Γ_e be an enumeration operator, $\mathcal{X} = (X, \tau, \alpha)$ and $\mathcal{Y} = (Y, \lambda, \beta)$ be computable Polish spaces. Then $E = \{x \mid \Psi(\Gamma_e(A_x))\}$ is a Π_2^0-subset of X, where Ψ is a Π_2^0-condition from Theorem 1. Moreover, the function $f : \mathcal{X} \to \mathcal{Y}$ defined as follows: $dom(f) = E$ and, for $x \in dom(f)$, $f(x) = y \leftrightarrow \Gamma_e(A_x) = B_y$ is a partial computable function.*

Proof. Let us show that the condition $\Psi(\Gamma_e(A_x))$ defines a Π_2^0-subset of X. From Remark 1 it follows that $\Psi(I)$ is a conjunction of Π_2^0-formulas in the form

$$\forall \bar{k} \left(\bigvee_{i \in D} k_i \notin I \vee \Phi(\bar{k}, I) \right),$$

where D is a finite subset of the indices of $\bar{k}$ and Φ is a computable disjunction of $\exists$–formulas with positive occurrences of I. Therefore, for $i = 1, \ldots, 4$ every $\text{Cond } i(\Gamma_e(A_x))$ defines the set

$$\{x \mid \forall \bar{k} \, x \in A_{\bar{k}}^i\}, \text{ where}$$
$$A_{\bar{k}}^i = \{x \mid \bigvee_{j \in D} k_j \notin \Gamma_e(A_x) \vee \Phi_i(\bar{k}, \Gamma_e(A_x))\}$$
$$= \{x \mid \bigvee_{j \in D} k_j \notin \Gamma_e(A_x)\} \cup \{x \mid \Phi_i(\bar{k}, \Gamma_e(A_x))\}.$$

Let us make a close look at $A_{\bar{k}}^i$. The first element of the union is a Π_1^0-subset of $\mathcal{X}$ and the second one is a Σ_1^0-subset of $\mathcal{X}$ according to the following observation.

By the definition of the enumeration operator Γ_e,

$$\{x \mid k \in \Gamma_e(A_x)\} = \bigcup_{<k,j> \in W_e} \bigcap_{j \in D_k} \alpha(j).$$

Therefore it is effectively open and its complement is co-effectively closed. As corollary every set $\{x \mid \text{Cond } i(\Gamma_e(A_x))\}$ is a Π_2^0-subset of $\mathcal{X}$.

Let us show that f is a partial computable function. It is worth noting that $x \in \text{dom}(f) \leftrightarrow \Gamma_e(A_x) \in \{B_y \mid y \in Y\} \leftrightarrow \Psi(\Gamma_e(A_x))$. So $\text{dom}(f)$ is a Π_2^0-subset of X. For $x \in \text{dom}(f)$,

$$x \in f^{-1}(\beta(j)) \leftrightarrow f(x) \in \beta(j) \leftrightarrow \exists k \, (k \in \{i \mid x \in \alpha(i)\} \wedge c(k,j) \in W_e)$$
$$\leftrightarrow \bigvee_{c(k,j) \in W_e} x \in \alpha(k) \leftrightarrow x \in \bigcup_{m \in \omega} \alpha(H(j,m))$$

for a computable function $H : \omega \times \omega \to \omega$. Therefore f is a partial computable function. $\qquad \square$

Further on if Γ_e and f satisfy the conditions of Proposition 4 we say that Γ_e *defines* f.

Theorem 2. *Let $\mathcal{X} = (X, \tau, \alpha)$ and $\mathcal{Y} = (Y, \lambda, \beta)$ be computable Polish spaces. A function $f : \mathcal{X} \to \mathcal{Y}$ is computable if and only if it is partial computable.*

Proof. ($\to$)The claim follows from Proposition 4.

($\leftarrow$) Now suppose, $\text{dom}(f) = \bigcap_{n \in \omega} O_n$ and, for $x \in \text{dom}(f)$, $f(x) \in \beta(n) \leftrightarrow x \in \bigcup_{i \in \omega} \alpha(H(n,i))$, where $\{O_n\}_{n \in \omega}$ is a computable sequence of effectively open sets such that $O_{n+1} \subseteq O_n$ and $H : \omega^2 \to \omega$ is a computable function. It is worth noting that, for all $n \in \omega$ and $i \in \omega$, $O_n \cap \alpha(H(n,i))$ is an effectively open set. So, $O_n \cap \alpha(H(n,i)) = \bigcup_{t \in T_{ni}} \alpha(t)$, where $n \in \omega$, $i \in \omega$ and $\{T_{ni}\}_{n, i \in \omega}$ is a computable sequence of c.e. sets. Put

$$W_e = \{c(t,n) \mid (\exists i \in \omega) \, t \in T_{ni}\}.$$

Let Γ_e be a reduced enumeration operator that corresponds to W_e. By Proposition 4 this operator defines a function f_{Γ_e}. Let us show that $f = f_{\Gamma_e}$. We first prove that $\text{dom}(f) = \text{dom}(f_{\Gamma_e})$. If $x \in \bigcap_{n \in \omega} O_n$ then, by construction, $\Gamma_e(A_x) = B_{f(x)}$. Indeed, let $n \in \omega$ be such that $f(x) \in \beta(n)$. By definition, $(\exists i \in \omega) \, x \in \alpha(H(n,i))$. This means that $x \in \alpha(t)$ for $t \in T_{ni}$, therefore $n \in \Gamma_e(A_x)$. Conversely, if $n \in \Gamma_e(A_x)$ then $(\exists i \in \omega)(\exists t \in T_{ni}) \, f(x) \in \beta(n)$. So $\Gamma_e(A_x) = B_{f(x)}$ and $x \in \text{dom}(f_\Gamma)$. So $\text{dom}(f) \subseteq \text{dom}(f_{\Gamma_e})$.

If $x \notin \bigcap_{n \in \omega} O_n$ then there exists $k \in \omega$ such that $x \notin O_n$ for all $n \geq k$. In other words, $x \notin \bigcup_{t \in T_{ni}} \alpha(t)$ for all $n \geq k$. This means that, for all $n \geq k$, $\neg(\exists t \in T_{ni}) \, x \in \alpha(t)$, i.e., $n \notin \Gamma_e(A_x)$. Therefore $\Gamma_e(A_x)$ is finite and $B = \bigcap\{\beta(j) \mid j \in \Gamma_e(A_x)\}$ is a finite intersection of basic open balls. Since we consider spaces without isolated points, $B \neq \bigcap\{\beta(j) \mid j \in B_y\} = \{y\}$ for any $y \in Y$. In particular, $x \notin \text{dom}(f_{\Gamma_e})$.

Now if $x \in \text{dom}(f) = \text{dom}(f_{\Gamma_e})$ then, by the definitions of f and Γ,

$$\Gamma(A_x) = \{j | \exists s\, H(j, s) \in A_x\} = \{j | x \in f^{-1}(\beta(j))\} = \{j | f(x) \in \beta(j)\}.$$

Therefore $B_{f(x)} = B_{f_{\Gamma}(x)}$. Since any point y is uniquely defined by the set of basic neighborhoods, $f(x) = f_{\Gamma}(x)$. So Γ_e completely defines f. $\qquad\square$

5.2 Index Sets for $\mathcal{PCF}_{\mathcal{XY}}$

In this section we show that there exists a principal computable numbering of $\mathcal{PCF}_{\mathcal{XY}}$ for computable Polish spaces $\mathcal{X}$ and $\mathcal{Y}$. With respect to this principal computable numbering we investigate complexity of important problems such as totality and root verification. It turns out that for some problems the corresponding complexity does not depend on the choice of a computable Polish space while for other ones the corresponding choice plays a crucial role.

A function $\gamma : \omega \to \mathcal{PCF}_{\mathcal{XY}}$ is called a *numbering* of $\mathcal{PCF}_{\mathcal{XY}}$ if $\{\gamma(n) \mid n \in \omega\} = \mathcal{PCF}_{\mathcal{XY}}$. A numbering $\gamma : \omega \to \mathcal{PCF}_{\mathcal{XY}}$ is called *computable* if $\Gamma : \omega \times \mathcal{X} \to \mathcal{Y}$ such that $\Gamma(n, x) = \gamma(n)(x)$ is a partial computable function. The numbering γ is called *principal computable* if it is computable and every computable numbering ξ is computably reducible to γ. It is worth noting that these definitions agree with the notions of computable and principal computable numberings of $A \subseteq \mathcal{P}(\omega)$ [1]. The following proposition is a straightforward corollary of Proposition 4 and Theorem 2.

Proposition 5. *There exists the principal computable numbering γ of $\mathcal{PCF}_{\mathcal{XY}}$. In fact, $\gamma(e)$ is a pcf defined by Γ_e.*

Let $\perp_{\mathcal{XY}}$ denote the nowhere defined function.

Proposition 6. *The index set $Ix(\{(f \in \mathcal{PCF}_{\mathcal{XY}} \mid f \neq \perp_{\mathcal{XY}}\})$ is Σ_1^1-complete.*

Proof. Let us fix computable Polish spaces $\mathcal{X}$ and $\mathcal{Y}$. Our proof is based on the following lemmas.

Lemma 6.1 [8]. There exists an effective procedure which given any computable sequence $\{A_n\}_{n \in \omega}$ of effectively open subsets of $\mathcal{N}$ produces a computable sequence $\{E_n\}_{n \in \omega}$ of effectively open subsets of X such that $\bigcap_{n \in \omega} A_n = \emptyset \leftrightarrow \bigcap_{n \in \omega} E_n = \emptyset$.

Lemma 6.2. Let $\{E_n\}_{n \in \omega}$ be a computable sequence of effectively open subsets of lX. There is an algorithm producing a partial computable function $f : \mathcal{X} \to \mathcal{Y}$ such that $\text{dom}(f) = \bigcap_{n \in \omega} E_n$.

Lemma 6.3 [8]. Let us fix a Σ_1^1–complete $I \subseteq \omega$. Then there exists an effective procedure which generates a computable sequence $\{A_n^m\}_{m,n \in \omega}$ of effectively open subsets of $\mathcal{N}$ such that $\bigcap_{n \in \omega} A_n^m \neq \emptyset \leftrightarrow m \in I$.

In order to prove Proposition 6 first let us note that the standard Kleene-Addison algorithm [16] shows that $Ix(\{f | f \neq \perp\} \in \Sigma_1^1$. Our goal is to show that $I \leq_m Ix(\{f | f \neq \perp\})$. For that, we construct a computable sequence $\mathcal{F} = \{f_m\}_{m \in \omega}$ of partial computable functions such that $m \in I \leftrightarrow f_m \neq \perp$. Using Lemmas 6.3 and 6.1 we construct a computable sequence $\{A_n^m\}_{m, n \in \omega}$ of effectively open subsets of $\mathcal{N}$ and a computable sequence $\{E_n^m\}_{m, n \in \omega}$ of effectively open subsets of X such that $\bigcap_{n \in \omega} A_n^m \neq \emptyset \leftrightarrow m \in I$ and $\bigcap_{n \in \omega} A_n^m = \emptyset \leftrightarrow \bigcap_{n \in \omega} E_n^m = \emptyset$. Using Lemma 6.2 we can effectively construct a partial computable function f_m such that $\text{dom}(f_m) = \bigcap_{n \in \omega} E_n^m$. So, there exists a computable function $\chi : \omega \to \omega$ such that $f_m = \gamma(\chi(m))$. The function χ is a reduction, since $m \in I \leftrightarrow \gamma(\chi(m)) \neq \perp$. $\qquad \square$

Corollary 1. *For $\mathcal{PCF}_{\mathcal{XY}}$, the problem of function equality is Π_1^1-complete.*

Theorem 3 [10]. *For partial computable real functions $f : \mathbb{R} \to \mathbb{R}$, totality is Π_2^0-complete.*

Theorem 4 [8]. *For partial computable functions $f : \mathcal{N} \to \mathbb{R}$, totality is Π_1^1-complete.*

For $\mathcal{X} = \mathbb{R}$ (or $\mathcal{N}$) and $\mathcal{Y} = \mathbb{R}$, complexity of other important problems such as root verification can be found in [8].

Proposition 7. *Let $K \subseteq \mathcal{PCF}_{\mathcal{XY}}$. If $K \neq \emptyset$ and $\perp \notin K$ then $Ix(K)$ is Π_2^0-hard.*

Proof. The proof is based on the following observation. Suppose $A \in \Pi_2^0$, $f \in \mathcal{PCF}_{\mathcal{XY}}$ and $f \neq \perp_{\mathcal{XY}}$. Then there exists a computable function $g : \omega \to \omega$ such that

$$n \in A \to g(n) \in Ix(\{f\}) \text{ and } n \notin A \to g(n) \in Ix(\perp_{\mathcal{XY}}).$$

The existence of g for a computable Polish space $\mathcal{Y}$ can be proven in a similar way as for $\mathcal{Y} = \mathbb{R}$ (see [10]). $\qquad \square$

Corollary 2 (Generalised Rice's Theorem). *Let $K \subset \mathcal{PCF}_{\mathcal{XY}}$. Then $K \neq \emptyset$ if and only if $Ix(K) \notin \Delta_2^0$.*

5.3   Complexity of Images of Partial Computable Functions

First we propose a characterisation of effectively enumerable topological spaces that are images of partial computable surjections from computable Polish spaces.

Proposition 8. *Let $\mathcal{X} = (X, \tau, \alpha)$ be a computable Polish space and $\mathcal{Y} = (Y, \lambda, \beta)$ be an effectively enumerable T_0-space. Then the following assertions are equivalent.*

1. *There exists a partial computable surjection $f : \mathcal{X} \twoheadrightarrow \mathcal{Y}$.*
2. *The space $\mathcal{Y}$ admits point recovering.*

Proof. (1) → (2). Assume $f : \mathcal{X} \twoheadrightarrow \mathcal{Y}$ is a partial computable surjection. It means that $\mathrm{dom}(f) = \bigcap_{n \in \omega} O_n = \bigcap_{n \in \omega} \bigcup_{s \in \omega} \alpha(g(n,s))$ and for all $x \in \mathrm{dom}(f)$, $f(x) \in \beta(n) \leftrightarrow x \in \bigcup_{i \in \omega} \alpha(H(n,i))$, where $g : \omega^2 \to \omega$ and $H : \omega^2 \to \omega$ are computable functions. Recall that $A_x = \{n \mid x \in \alpha(n)\}$ and $B_y = \{m \mid y \in \beta(m)\}$. In order to show that $\mathcal{Y}$ admits recovering let us prove that $\{B_y \mid y \in Y\}$ is a Σ_1^1-subset of $\mathcal{P}(\omega)$ considered as the Cantor space. Since f is a surjection, for $I \subseteq \omega$, $(\exists y \in Y)\, I = B_y$ if and only if $(\exists x \in \mathrm{dom}(f))\, I = B_{f(x)}$. Let us make analysis. If $I = B_{f(x)}$ then

$$n \in I \leftrightarrow (\exists x \in \mathrm{dom}(f)) f(x) \in \beta(n) \leftrightarrow (\exists x \in \mathrm{dom}(f))\, x \in \bigcup_{i \in \omega} \beta(H(n,i))$$

$$\leftrightarrow (\exists x \in \mathrm{dom}(f))(\exists i \in \omega)\, H(n,i) \in A_x.$$

From Theorem 1 it follows that $J \in \{A_x \mid x \in X\} \leftrightarrow \Psi(J)$, where $\Psi(J)$ is a Π_2^0-subset of $\mathcal{C}$. It is easy to see that $x \in \mathrm{dom}(f) \leftrightarrow (\forall n \in \omega)\, x \in O_n \leftrightarrow (\forall n \in \omega)(\exists s \in \omega) g(n,s) \in A_x$. Finally, we have

$$(\exists y \in Y)I = B_y \leftrightarrow (\exists J \subseteq \omega)\ \Big(\Psi(J) \wedge (\forall n \in \omega)\Big(n \in I \leftrightarrow \big((\exists i \in \omega)\, H(n,i) \in J \wedge$$

$$(\forall m \in \omega)(\exists s \in \omega)\, g(n,s) \in J \big) \Big) \Big).$$

Now we can see that $\{B_y \mid y \in Y\}$ is a Σ_1^1-subset of $\mathcal{C}$.

(2) → (1). Let $\mathcal{Y}$ admit point recovering. We construct a required partial computable surjection in few steps: $\mathcal{X} \twoheadrightarrow \mathcal{N} \twoheadrightarrow \mathcal{C} \twoheadrightarrow \mathcal{C}^2 \twoheadrightarrow \mathcal{Y}$.

Step 1. It is known (see e.g. [9,14,19]) that there exists a homeomorphism $F : \mathcal{N} \to \mathcal{X}$ such that $F^{-1} : \mathcal{X} \to \mathcal{N}$ is a partial computable surjection.

Step 2. A partial computable surjection $g : \mathcal{N} \twoheadrightarrow \mathcal{C}$ is defined in a standard way $g(f) = \lambda n. f(n) \bmod 2$.

Step 3. A partial computable bijection $\lambda : \mathcal{C}^2 \twoheadrightarrow \mathcal{C}$ is defined in a standard way $\lambda(I,J) = \{2n \mid n \in I\} \cup \{2n+1 \mid n \in J\}$.

Step 4. Let us construct a partial computable surjection $h : \mathcal{C}^2 \twoheadrightarrow \mathcal{Y}$. Assume Θ is a Σ_1^1-condition that certifies point recovering of $\mathcal{Y}$ i.e. $I = B_y$ for some $y \in Y$ iff $\Theta(I) \leftrightharpoons (\exists J \subseteq \omega)\Phi(I,J)$, where $\Phi(I,J)$ is a Π_2^0-condition (see e.g. [14]). Put $D = \{(I,J) \mid \Phi(I,J)\} \subseteq \mathcal{C}^2$.

If $(I,J) \in D$ then $I = B_y$ for some $y \in Y$. Since Y is a T_0-space, this y is uniquely defined by I. Define $h(I,J) = y$. We have $(I,J) \in h^{-1}(\beta(n)) \leftrightarrow I = B_z$ for some $z \in \beta(n) \leftrightarrow \Phi(I,J) \wedge n \in I$. So, $\mathrm{dom}(h) = D$ is a Π_2^0-subset of $\mathcal{C}^2$ and $h^{-1}(\beta(n)) = D \cap (\{I \subseteq \omega \mid n \in I\} \times \mathcal{C})$. Therefore h is a partial computable surjection.

Step 5. A required partial computable surjection is the composition $f = h \circ \lambda^{-1} \circ g \circ F^{-1}$, i.e.,

$$\mathcal{X} \xrightarrow{F^{-1}} \mathcal{N} \xrightarrow{g} \mathcal{C} \xrightarrow{\lambda^{-1}} \mathcal{C}^2 \xrightarrow{h} \mathcal{Y}.$$

The construction is complete. □

In order to study the complexity of images of partial computable functions we use the effective Borel and Lusin hierarchies on computable Polish spaces [14,19]. In particular our proofs are based on the following properties of Borel and analytic subsets of a computable Polish space $\mathcal{X}$:

- A set B is a Π_2^0-set in the effective Borel hierarchy on $\mathcal{X}$ (a Π_2^0-subset of X) if and only if $B = \bigcap_{n \in \omega} A_n$ for a computable sequence of effectively open sets $\{A_n\}_{n \in \omega}$.
- A set $A \in$ is a Σ_1^1-set in the effective Lusin hierarchy on $\mathcal{X}$ (a Σ_1^1-subset of X) if and only if $A = \{y \mid (\exists x \in X)B(x, y)\}$, where B is a Π_2^0-subset of X.

Theorem 5. *Let $\mathcal{X}$ be a computable Polish space, $\mathcal{Y}$ be an effectively enumerable T_0-space and $Y_0 \subseteq Y$. Then the following assertions are equivalent.*

1. *Y_0 is the image of a partial computable function $f : \mathcal{X} \to \mathcal{Y}$.*
2. *$\{B_y \mid y \in Y_0\}$ is a Σ_1^1-subset of $\mathcal{C}$.*

Proof. 1) $\to$ 2). Assume Y_0 is the image of a partial computable function $f : \mathcal{X} \to \mathcal{Y}$. For $x \in \text{dom}(f)$ and $y = f(x) \in Y_0$, we have

$$n \in B_{f(x)} \leftrightarrow f(x) \in \beta(n) \leftrightarrow x \in \bigcup_{i \in \omega} \alpha(H(n, i)) \leftrightarrow \{H(n, i) \mid i \in \omega\} \cap A_x \neq \emptyset.$$

Then, for $I \subseteq \omega$,

$$I \in \{B_y \mid y \in Y_0\}$$
$$\leftrightarrow \exists J \in \{A_x \mid x \in \text{dom}(f)\}(\forall n \in \omega)\left(n \in I \leftrightarrow \{H(n, i) \mid i \in \omega\} \cap A_x \neq \emptyset\right).$$

This is a Σ_1^1-condition since

$$J \in \{A_x \mid x \in \text{dom}(f)\} \leftrightarrow (\forall m \in \omega)J \cap J_m \neq \emptyset,$$

where $\{J_m\}_{m \in \omega}$ is a computable sequence of c.e. sets such that $\text{dom}(f) = \bigcap_{m \in \omega} O_m$ and $O_m = \bigcup_{i \in J_m} \alpha(i)$.

2) $\to$ 1). Let $\{B_y \mid y \in Y_0\} \in \Sigma_1^1$. This means that $J \in \{B_y \mid y \in Y_0\} \leftrightarrow (\exists I \subseteq \omega) Q(I, J)$, where $Q(I, J)$ is a Π_2^0-condition on $\mathcal{C}$ (see e.g. [14]). Let us construct a partial computable function $h : \mathcal{C}^2 \to \mathcal{Y}$ such that $\text{dom}(h) = D$ and $\text{im}(h) = Y_0$. Put $D = \{(I, J) \mid Q(I, J)\} \subseteq \mathcal{C}^2$. If $Q(I, J)$ then $J = B_y$ for some $y \in Y$. Since $\mathcal{Y}$ is a T_0-space, y is uniquely defined by I. Define $h(I, J) = y$. We have $(I, J) \in h^{-1}(\beta(n)) \leftrightarrow J = B_z$ for some $z \in \beta(n) \leftrightarrow Q(I, J) \wedge n \in J$. So, $\text{dom}(h) = D$ is a Π_2^0-subset of $\mathcal{C}^2$ and $h^{-1}(\beta(n)) = D \cap (\{I \subseteq \omega \mid n \in I\} \times \mathcal{C})$. Therefore h is a partial computable function. Using Theorem 8 we construct the composition of partial computable surjections f, g and h as follows:

$$\mathcal{X} \overset{f}{\twoheadrightarrow} \mathcal{C} \overset{g}{\twoheadrightarrow} \mathcal{C}^2 \overset{h}{\to} \mathcal{Y}.$$

This is a required function.

Proposition 9. *Let $\mathcal{Y}$ be a computable Polish space, $Y_0 \subseteq Y$ and $\widetilde{Y}_0 = \{B_y \mid y \in Y_0\}$. Then Y_0 is a Σ_1^1-subset of $\mathcal{Y}$ if and only if $\widetilde{Y}_0$ is a Σ_1^1-subset of $\mathcal{C}$.*

Theorem 6. *Let $\mathcal{X}$ and $\mathcal{Y}$ be computable Polish spaces and $Y_0 \subseteq Y$. Then the following assertions are equivalent.*

1. Y_0 is the image of a partial computable function $f : \mathcal{X} \to \mathcal{Y}$.
2. Y_0 is a Σ_1^1-subset of Y.

Proof. The claim follows from Theorem 5 and Proposition 9. □

References

1. Ershov, Y.L.: Theory of numberings. In: Griffor, E.R. (ed.) Handbook of Computability Theory, pp. 473–503. Elsevier Science B.V., Amsterdam (1999)
2. Edalat, A.: Domains for computation in mathematics, physics and exact real arithmetic. Bull. Symbolic Log. **3**(4), 401–452 (1997)
3. Gao, S.: Invariant Descriptive Set Theory. CRC Press, New York (2009)
4. Gregoriades, V., Kispeter, T., Pauly, A.: A comparison of concepts from computable analysis and effective descriptive set theory. Math. Struct. Comput. Sci. 1–23(2016). https://doi.org/10.1017/S0960129516000128. (Published online: 23 June 2016)
5. Hemmerling, A.: Effective metric spaces and representations of the reals. Theor. Comput. Sci. **284**(2), 347–372 (2002)
6. Hemmerling, A.: On approximate and algebraic computability over the real numbers. Theor. Comput. Sci. **219**(1–2), 185–223 (1999)
7. Kechris, A.S.: Classical Descriptive Set Theory. Springer, New York (1995)
8. Korovina, M., Kudinov, O.: Complexity for partial computable functions over computable Polish spaces. Math. Struct. Comput. Sci. (2016). doi:10.1017/S0960129516000438. (Published online: 19 December 2016)
9. Korovina, M., Kudinov, O.: Computable elements and functions in effectively enumerable topological spaces. Mathematical structure in Computer Science (2016). doi:10.1017/S0960129516000141. (Published online: 23 June 2016)
10. Korovina, M., Kudinov, O.: Index sets as a measure of continuous constraint complexity. In: Voronkov, A., Virbitskaite, I. (eds.) PSI 2014. LNCS, vol. 8974, pp. 201–215. Springer, Heidelberg (2015). doi:10.1007/978-3-662-46823-4_17
11. Korovina, M., Kudinov, O.: Towards computability over effectively enumerable topological spaces. Electr. Notes Theor. Comput. Sci. **221**, 115–125 (2008)
12. Korovina, M., Kudinov, O.: Towards computability of higher type continuous data. In: Cooper, S.B., Löwe, B., Torenvliet, L. (eds.) CiE 2005. LNCS, vol. 3526, pp. 235–241. Springer, Heidelberg (2005). doi:10.1007/11494645_30
13. Korovina, M., Kudinov, O.: Characteristic properties of majorant-computability over the reals. In: Gottlob, G., Grandjean, E., Seyr, K. (eds.) CSL 1998. LNCS, vol. 1584, pp. 188–203. Springer, Heidelberg (1999). doi:10.1007/10703163_14
14. Moschovakis, Y.N.: Descriptive set theory. North-Holland, Amsterdam (2009)
15. Moschovakis, Y.N.: Recursive metric spaces. Fund. Math. **55**, 215–238 (1964)
16. Rogers, H.: Theory of Recursive Functions and Effective Computability. McGraw-Hill, New York (1967)

17. Soare, R.I.: Recursively Enumerable Sets and Degrees: A Study of Computable Functions and Computably Generated Sets. Springer Science and Business Media, Heidelberg (1987)
18. Spreen, D.: On effective topological spaces. J. Symb. Log. **63**(1), 185–221 (1998)
19. Selivanov, V.: Towards the effective descriptive set theory. In: Beckmann, A., Mitrana, V., Soskova, M. (eds.) CiE 2015. LNCS, vol. 9136, pp. 324–333. Springer, Cham (2015). doi:10.1007/978-3-319-20028-6_33
20. Weihrauch, K.: Computable Analysis. Springer, New York (2000)
21. Weihrauch, K.: Computability on computable metric spaces. Theor. Comput. Sci. **113**(1), 191–210 (1993)

Eliminating Unbounded Search
in Computable Algebra

Alexander G. Melnikov[✉]

Massey University, Auckland, New Zealand
alexander.g.melnikov@gmail.com

Abstract. Klaimullin, Melnikov and Ng [KMNa] have recently suggested a new systematic approach to algorithms in algebra which is
intermediate between computationally feasible algebra [CR91, KNRS07]
and abstract computable structure theory [AK00, EG00]. In this short
survey we discuss some of the key results and ideas of this new
topic [KMNa, KMNc, KMNb]. We also suggest several open problems.

1 Introduction

What does it mean for an infinite algebraic structure to be computable? Which
algebraic structures admit an algorithmic presentation? These questions are central to computable structure theory [AK00, EG00]. The main objects of computable structure theory are countably infinite cby a Turing machine:

Definition 1 (Mal'cev, Rabin). An algebraic structure $\mathcal{A}$ is *constructive* or
computable if its universe is the set of natural numbers $\mathbb{N}$, and the operations
and relations on $\mathcal{A}$ are (Turing) computable.

For example, every finitely generated group with algorithmically solvable
Word Problem [Hig61] has a computable isomorphic copy. In both combinatorial
group theory [Hig61] and computable structure theory [AK00, EG00] algorithms
are allowed to be computationally inefficient. It often does not make any difference, since one of the main aims of such studies is to show that some problems –
such as the Word Problem for a f.g. group – have no algorithmic solution at all,
let alone a computationally feasible solution [Nov55, Boo59, Hig61]. Also, Turing
computability can be viewed as a formalisation of one's constructive approach to
algebra [EG00]. Algorithmic investigations in algebra are related to other seemingly distant areas of pure mathematics such as topological group theory and
model theory (e.g., [MM, Mon13]).

In contrast with (Turing) computable structure theory, *automatic* and
polynomial-time algebra put resource bounds and other restrictions on algorithms representing algebraic structures. Automatic algebra studies algebraic
structures that are presented by (typically finite) automata [KN94, KN08,
KNRS07, ECH+92, NT08, BS11, NS07] Automatic structures have a number of
nice properties including quick computational characteristics, but automatic

© Springer International Publishing AG 2017
J. Kari et al. (Eds.): CiE 2017, LNCS 10307, pp. 77–87, 2017.
DOI: 10.1007/978-3-319-58741-7_8

structures tend to be rare (e.g.,[Tsa11]). A countably infinite algebraic structure is *polynomial-time* if the operations and relations polynomial time computable (in the length of the input, see survey [CR91]). As we will discuss later, in many cases one can show that a (Turing) computable algebraic structure has a polynomial copy (e.g., [CDRU09, CR, Gri90]), but this phenomenon is not yet understood.

One would expect that computable structure theory and computationally feasible algebra should have significant overlaps, but it is not quite the case. Kalimullin, Melnikov and Ng [KMNa, KMNc, KMNb] have initiated a systematic development of a theory the main purpose of which is to fill this gap. (Independently, Alaev [Ala] has suggested an alternate approach.) In this brief survey we explain the key ideas and results from these recent works [KMNa, KMNc, KMNb, Ala]. An expert working in computable structure theory may find some of the key results discussed below rather counter-intuitive. Although the new topic is in its infancy, it already has technical depth and offers seemingly challenging problems, some of which will be posed below.

2   Primitive Recursion and Computability Without Delay

We open this section with two elementary examples illustrating the power of unbounded search.

Example 2. Clearly, there exists a (Turing) computable 2-colouring of any computable infinite tree. Simply wait for a node v to get path-connected to the root. If we put any (computable and total) restriction on the waiting time, then we will need infinitely many colours in general.

Example 3. Suppose $(G, \cdot)$ is a (Turing) computable group. Then the operation $a \to a^{-1}$ is also Turing computable. We simply search through the domain of G until we find a g such that $a \cdot g = e$. Note this elementary fact no longer holds if we cannot use the unbounded search.

Eliminating all unbounded loops from Turing computability gives us the notion of a primitive recursive function. In several common algebraic classes we can show that every computable structure has a polynomial-time computable copy. As was noted in [KMNa], many known proofs of this sort (e.g., [CR91, CR92, CDRU09, Gri90]) are essentially focused on making the operations and relations on the structure primitive recursive, and then observing that the presentation that we obtain is polynomial-time. It appears that primitive recursion plays a rather important intermediate role in transforming (Turing) computable structures into feasible structures. This thesis is also supported by a number of negative results in the literature. Indeed, to illustrate that a structure has no polynomial time copy, it is sometimes easiest to argue that it does not even have a copy with primitive recursive operations, see e.g. [CR92]. It is thus natural to investigate into those structures that admit a presentation with primitive recursive operations:

Definition 4 ([KMNa]). A countable structure is *fully primitive recursive* (fpr) if its domain is $\mathbb{N}$ and the operations and predicates of the structure are (uniformly) primitive recursive[1].

One may ask whether the domain of a fpr A could be a primitive recursive subset of $\mathbb{N}$ as in [CR91], not the whole $\mathbb{N}$. This means that the structure could reveal its domain with an arbitrary delay. Such presentations upon a primitive recursive subset of $\mathbb{N}$ are called *primitive recursive* [CR92]. It follows from [Ala] (see [KMNa]) that there exist primitive recursive structures that have no fully primitive recursive presentation.

Our main goal is the complete elimination of all unbounded loops from algorithmic presentations of structures. In particular, we should not allow the domain to be revealed with an unbounded delay. Thus, a "truly primitive recursive" algebraic structure should (minimally) satisfy Definition 4 of a fpr structure. Note that one might impose further restrictions on fpr presentations. We will briefly discuss one such possible strengthening of Definition 4 in the next section.

The notion of a fpr structure is intermediate between computable structures and polynomial-time structures. It usually takes some work to produce a fpr copy of a computable structure (if it exists at all), and there are enough natural examples of computable structures that do not have a fpr copy. Similarly, not all fpr structures admit polynomial-time copies, but a good portion of our results can be extended to polynomial-time structures. The rest of the survey is focused on the systematic development of the theory of fpr structures.

3 Existence of a fpr Copy

Which computable structures admit fpr presentations?

Theorem 5. *In each of the following classes, every computable structure has a fully primitive recursive presentation:*

1. *Equivalence structures [CR91].*
2. *Linear orders [Gri90].*
3. *Torsion-free abelian groups [KMNa].*
4. *Boolean algebras [KMNa].*
5. *Abelian p-groups [KMNa].*

Proof (Proof idea). We use various structural properties specific to the class to predict the behaviour of the structure in certain locations even before the structure has revealed itself there. For example, in (3) we first produce a computable presentation of the group with a computable maximal linearly independent set [Dob, Nur74], and then we use the maximal free subgroup upon this set as a "safe" location. The case of Boolean algebras (4) is more interesting. The proof is not uniform and it goes through several cases. In the case of an atomic algebra we use a priority construction, the old theorem of Remmel [Rem81], and tree-presentations of BA's [Gon97] to produce a Π_3^0-isomorphic fpr copy.

---

[1] We also agree that all finite structures are fpr.

The reader perhaps thinks that most common classes will have the nice property from Theorem 5, but this is not the case.

Theorem 6. *In each of the following classes, there exists a computable structure that does not admit a fpr presentation*

1. *Torsion abelian groups [CR92].*
2. *Archimedean ordered abelian groups [KMNa].*
3. *Undirected graphs [KMNa].*

Proof (Proof idea). In (2) we produce an Archimedean group of rank 2 that (essentially) encodes a computable real which does not have a primitive recursive rapid approximation. The proof uses that every such group is computably categorical, so we need to diagonalise only against computable isomorphisms. The proofs of (1) and (3) are brute-force diagonalisation arguments, but (3) is a lot more subtle. In the proof we need to monitor all potential fpr copies at once and force them to reveal themselves at certain locations used for diagonalisation.

Parts (1) and (2) of Theorem 6 contrast with (3) and (5) of Theorem 5, and (3) of Theorem 6 refutes the (natural) conjecture that every relational computable structure has a fpr copy. Note that Theorems 5 and 6 (combined) confirm the intermediate nature of fpr structures. Although there are some observable patterns in the proofs of Theorems 6 and 5, it is not clear whether there is any meaningful and general enough sufficient condition for a computable structure to have (or not have) a fpr copy. We leave open:

Question 7. Fix the listing $M_0, M_1, \ldots$ of all partial computable structures. What is the complexity of the index set $\{e : M_e \text{ has a fpr copy}\}$?

3.1   Strongly fpr Structures

In this subsection we briefly discuss one possible strengthening of the notion of a fpr structure. We say that a fpr structure is *strongly primitive recursive* of it possesses a primitive recursive Skolem function [KMNa].

We note that this approach resembles the earlier notion of a *honest witness* due to Cenzer and Remmel [CR91]. The following structures have strongly primitive recursive copies:

– The additive groups $\mathbb{Z}$ and $\mathbb{Q}$ and their direct sums.
– The countable atomless Boolean algebra.
– The order-type ω.

There exist 1-decidable fpr structures that are not strongly fpr, see [KMNa] for a proof. (See also [Ala] for a similar approach.)

4 Uniqueness of a fpr Copy

Recall that a structure is *computably categorical* or *autostable* if it has a unique computable copy up to computable isomorphism [AK00, EG00]. There has been a lot of work on computably categorical structures [KS99, EG00, Gon80, Smi81, LaR77, Nur74, HKS03]. A function $f : \omega \rightarrow^{onto} \omega$ is *fully primitive recursive* (fpr) if f and f^{-1} are both primitive recursive.

Definition 8. ([KMNa]). A fully primitive recursive structure $\mathcal{A}$ is *fpr categorical* if it has a unique fully primitive recursive presentation up to fully primitive recursive isomorphism.

Example 9. [KMNa]

1. The additive group $\mathbb{V}_p \cong \bigoplus_{i \in \omega} \mathbb{Z}_p$ is fpr-categorical
2. The dense linear order $(\mathbb{Q}, <)$ without end points is *not* fpr-categorical
3. The successor structure $\mathcal{S} = (\omega, S)$, where $S(x) = x + 1$, is *not* fpr-categorical

Even though Theorem 5 typically produces the most "boring" fpr presentations in each class, Theorem 10 below says that almost all structures in these classes have complex ("irregular", "unpredictable") fpr presentations.

Theorem 10. ([KMNa]).

1. *An equivalence structure S is fpr-categorical iff it is either of the form $F \cup E$, where F is finite and E has only classes of size 1, or S has finitely many classes at most one of which is infinite.*
2. *A linear order is fpr-categorical iff it is finite.*
3. *A Boolean algebra is fpr-categorical iff it is finite.*
4. *An abelian p-group is fpr-categorical iff it has the form $F \oplus \mathbb{V}$, where $p\mathbb{V} = \mathbf{0}$ and F is finite.*
5. *A torsion-free abelian group is fpr-categorical iff it is the trivial group $\mathbf{0}$.*

Proof (Informal Discussion). In some simple cases we can appeal to Theorem 5 and combine it with known facts from computable structure theory. For example, suppose a Boolean algebra $\mathcal{B}$ has an atomless element and is not computably categorical. Although Theorem 5 is not uniform in general, it is uniformly computable for such BAs. It follows that we can produce two fpr copies of $\mathcal{B}$ that are not even computably isomorphic. Nonetheless, some cases require a direct proof. For instance, in the case of an atomic BA we cannot appeal to Theorem 5 directly since the isomorphism between $\mathcal{B}$ and its fpr copy is not even $0''$ in general. Instead, in this case we need to simultaneously build two fpr copies of $\mathcal{B}$ and diagonalise against all potential fpr-isomorphisms. Although the proof is not hard, it does require some care.

Note that Theorem 10 resembles the following result of Khoussainov and Nerode [KN94]: A structure is automatically categorical iff it is finite.

According to Definition 8, every fpr-categorical structure must have a fully primitive recursive (thus, computable) copy. Theorem 10 suggests that fpr-categorical structures are necessarily computably categorical. Surprisingly, this is not the case:

Theorem 11 (*[KMNa]*). *There exists a fpr-categorical structure which is not computably categorical.*

Proof (Informal Discussion). The proof of Theorem 11 is quite combinatorially involved. We build the structure $\mathcal{A}$ carefully and force any fpr isomorphic copy to be fpr-isomorphic to it. The rigidity helps to make the inverse of each such isomorphism primitive recursive. Also, for this purpose we introduce a "local coordinate system" that will allow us to recognise a local part of the structure without looking through the whole structure. Finally, we combine these strategies with a diagonalisation strategy. For that, we produce a computable copy $\mathcal{B}$ of $\mathcal{A}$ and kill off all $\phi_e : \mathcal{A} \cong \mathcal{B}$. We heavily rely on the fact that we *can* delay the computation in $\mathcal{B}$.

Question 12. Which of the common algebraic classes (such that groups, fields, integral domains, ...) contain examples of fpr-categorical but not computably categorical structures?

5   The fpr-Degrees of a Structure

Note that the inverse of a primitive recursive function does not have to be primitive recursive. It leads to a reduction [KMNc]. Let $\mathbf{FPR}(\mathcal{A})$ be the collection of all fpr presentations of a countably infinite structure $\mathcal{A}$. For $\mathcal{A}_1, \mathcal{A}_2 \in \mathbf{FPR}(\mathcal{A})$, write $\mathcal{A}_1 \leq_{pr} \mathcal{A}_2$ if there exists a primitive recursive isomorphism from $\mathcal{A}_1$ onto $\mathcal{A}_2$. Clearly, $\leq_{pr}$ is reflexive and transitive. We write $A_1 \simeq_{pr} A_2$ if $A_1 \leq_{pr} A_2$ and $A_2 \leq_{pr} A_1$. In particular, we can look at the fpr-degrees of a given countably infinite structure $\mathcal{A}$.

Definition 13 ([KMNc]). The fully primitive recursive degrees of a countably infinite algebraic structure $\mathcal{A}$ is the quotient structure $\mathbf{FPR}(\mathcal{A}) = (\mathbf{FPR}(\mathcal{A}), \leq_{pr})/ \simeq_{pr}$.

The fpr-degrees $\mathbf{FPR}(\mathcal{A})$ is a computability-theoretic invariant of $\mathcal{A}$ that encodes/reflects the non-primitive recursive content of the isomorphism type of $\mathcal{A}^2$.

Question 14. Does $|\mathbf{FPR}(\mathcal{A})| = 1$ imply that $\mathcal{A}$ is fpr-categorical?

As strange as it may sound, we still don't know the answer to the question above. It takes quite a bit of effort to prove the rather satisfying:

Theorem 15 (M. and Ng, to appear). *For every undirected graph $\mathcal{G}$, $|\mathbf{FPR}(\mathcal{G})| = 1$ implies that $\mathcal{G}$ is fpr-categorical.*

---

[2] If $\mathcal{A}$ is fpr-categorical then $\mathbf{FPR}(\mathcal{A})$ contains a unique degree. Nonetheless, $A \simeq_{pr} B$ does not necessarily imply that there exists a fpr isomorphism $A \to B$ (it is easy to construct a counter-example).

Proof (Informal discussion). In fact, we have proved more. TFAE:

(1) $|\mathbf{FPR}(\mathcal{G})| = 1$.

(2) $\mathcal{G}$ is fpr-categorical.

(3) Given any two f.p.r. copies $\mathcal{A} \cong \mathcal{B}$ of $\mathcal{G}$, there exist primitive recursive isomorphisms $f : \mathcal{A} \mapsto \mathcal{B}$ and $g : \mathcal{B} \mapsto \mathcal{A}$, and a primitive recursive function $t : \mathbb{N} \mapsto \mathbb{N}$ such that given $a \in \mathcal{A}$, either $Orb(a) = \{(gf)^n(a) : n \in \omega\}$ has size at most $t(a)$, or every permutation u of $Orb(a)$ can be extended to an automorphism of $\mathcal{G}$.

Note that given (3) we can run a primitive recursive back-and-forth construction to produce a fpr isomorphism between two fpr copies. First, check whether $Orb(a)$ has size $\leq t(a)$. If "yes" then match $Orb(a)$ with $Orb(f(a))$. Otherwise, if $Orb(a)$ has not yet closed after $t(a)$ steps, then do the back-and-forth on $Orb(a)$ and $Orb(f(a))$ essentially ignoring the rest of the structure. Unfortunately, the implication (1) $\rightarrow$ (2) is quite non-trivial.

Question 16. Is there a structure $\mathcal{A}$ such that $1 < |\mathbf{FPR}(\mathcal{A})| < \infty$?

6   A Sub-hierarchy of Computable Categoricity

In this section we discuss fpr-degrees of computably categorical structures (see Definition 13).

A total computable function is *honest* if its graph is primitive recursive [Kri96] This means that we can see whether $f(x) = y$ for a given (x, y) without any unbounded delay. Thus, is some sense this is a non-deterministic version of primitive recursion. Clearly, if f is honest then so is f^{-1}, but the composition of two honest functions does not have to be honest.

In Theorem 10, all computably categorical fpr structures satisfy at least one of the properties defined below. (Let $\mathcal{A}$ be a fpr structure.)

– $\mathcal{A}$ is **bottom-categorical** if $\mathbf{FPR}(\mathcal{A})$ has the $\leq_{pr}$-least element.
– $\mathcal{A}$ is **top-categorical** if $\mathbf{FPR}(\mathcal{A})$ has the $\leq_{pr}$-greatest element.
– $\mathcal{A}$ is **honestly categorical** if for each $\mathcal{A}_1, \mathcal{A}_2 \in \mathbf{FPR}(\mathcal{A})$ there exists a honest isomorphism from $\mathcal{A}_1$ onto $\mathcal{A}_2$.
– $\mathcal{A}$ is **relatively fpr categorical** if there exists a pair of (oracle) primitive recursive schemata $\mathcal{P}_+$ and $\mathcal{P}_-$ such that for any isomorphic copy $\mathcal{B}$ (upon $\mathbb{N}$) of the fpr structure $\mathcal{A}$ the maps $\mathcal{P}_+^{\mathcal{B}} : \mathcal{A} \rightarrow \mathcal{B}$ and $\mathcal{P}_-^{\mathcal{B}} : \mathcal{B} \rightarrow \mathcal{A}$ are isomorphisms that are inverses of each other.

For example, every finitely generated structure in a finite functional language is bottom-categorical. Every computably categorical linear order or Boolean algebra is top-categorical. In many standard classes (including e.g. linear orders) honest categoricity is equivalent to the usual computable categoricity. All algebraically natural examples of fpr categorical structures are relatively fpr-categorical [KMNa] (see also (1) of Example 9). We summarise all these notions in the diagram below.

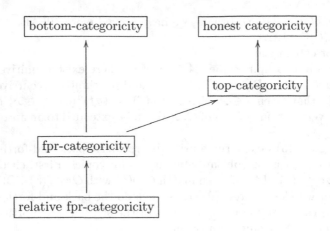

Theorem 17 ([KMNc]). *The diagram above is proper[3].*

Proof (Informal Discussion). All implications are straightforward. Note that every relatively fpr-categorical structure is (relatively) computably categorical. It follows from Theorem 11 that fpr-categoricity does not imply relative fpr-categoricity. The failures of the other missing implications are witnessed by priority constructions.

An alternate way to compare computably categorical fpr structures uses relative computation and the primitive recursive jump $0'_{PR}$[4]. Then a structure $\mathcal{A}$ is $0^{(n)}_{PR}$-categorical if it has a unique fpr-presentation, up to fully $0^{(n)}_{PR}$-isomorphism.

Theorem 18 ([KMNc]). *For every $n > 0$ there exists a fully primitive recursive structure which is fully $0^{(n)}_{PR}$-categorical but not fully $0^{(n-1)}_{PR}$-categorical.*

Proof (Informal Discussion). In [KMNc], Kalimllin, M. and Ng have proved more than is stated in Theorem 18. The PR-degree any honest function can be realized as the fpr-degree of categoricity of some structure (we omit formal definitions). Remarkably, for any n, the PR-degree of $0^{(n)}_{PR}$ is honest.

It is unclear whether there is any interesting relationship between full $0^{(n)}_{PR}$-categoricity and any of the properties from Theorem 17.

---

[3] This means that all implications that are shown at the diagram above are proper. Furthermore, these implications (and their transitive closures) are the only implications that hold.

[4] A total function g is primitively recursively reducible to a function f ($g \leq_{PR} f$) if $g = \Phi^f$ for some f-primitive recursive schema Φ^f. This leads to the definitions of $g \equiv_{PR} f$ and $g <_{PR} f$ as well as the notion of primitive recursive (PR-) degree. For a total function f let $\{\Phi^f_n\}$ be the Gödel numbering of all f-primitive recursive schemata for functions with one variable. Define the primitive recursive jump f'_{PR} to be the function $f'_{PR}(n, x) = \Phi^f_n(x)$. It is easy to check that $f \leq_{PR} g \implies f'_{PR} \leq_{PR} g'_{PR}$ and $f <_{PR} f'_{PR}$.

7 Cantor's Back-and-Forth Method Revisited

As we know from Example 9 (2), $\eta = (\mathbb{Q}, <)$ has two fpr copies that are not fpr isomorphic. This means that Cantor's back-and-forth proof is no longer "effective" in absence of unbounded search.

Recall that a structure $\mathcal{X}$ is homogeneous if every isomorphism $f : F_1 \to F_2$ between any two finitely generated substructures $F_1, F_2 \subseteq \mathcal{X}$ is extendable to an automorphism of $\mathcal{X}$. See [Mac11] for a survey on homogeneous structures. The following structures are homogeneous:

- η, the dense linear order without end-points.
- $\mathcal{R}$, the Random Graph.
- $\mathcal{P} \cong \bigoplus_{i \in \omega} \mathbb{Z}_p^\infty$, the universal divisible abelian p-group.
- $\mathcal{B} \cong I(\eta)$, the countable atomless Boolean algebra.

Apart from homogeneity, there is little in common between the three homogeneous structures $\eta, \mathcal{R}, \mathcal{P}$. Nonetheless, they do share essentially the same back-and-forth proof of their uniqueness up to isomorphism. The (Turing) computable construction of an isomorphism for these structures has only one potentially unbounded search at every substage. Recently Klamullin, M. and Ng have announced the following rather counterintuitive result.

Theorem 19 ([KMNb]). *The fpr-degree structures of the dense linear order η, the random graph $\mathcal{R}$, and the universal divisible abelian p-group $\mathcal{P}$ are pairwise non-isomorphic.*

Proof (Proof idea). It follows that both $\mathbf{FPR}(\mathcal{R})$ and $\mathbf{FPR}(\mathcal{P})$ have no greatest element, while $\mathbf{FPR}(\eta)$ does. Also, $\mathbf{FPR}(\mathcal{P})$ has no maximal elements, while $\mathbf{FPR}(\mathcal{R})$ does. Some these facts require a non-trivial proof.

Let $\mathcal{B}$ be the atomless Boolean algebra. Recall that $\mathcal{B} \cong I(\eta)$, the interval Boolean algebra of η. The reader might find the question below a bit strange:

Question 20. Is $\mathbf{FPR}(\mathcal{B}) \cong \mathbf{FPR}(\eta)$?

In fact, we suspect that $\mathbf{FPR}(\mathcal{B}) \not\cong \mathbf{FPR}(\eta)$. In spite of the differences in their fpr-degree structure, there is at least one fundamental property that is common for $\mathbf{FPR}(\eta)$, $\mathbf{FPR}(\mathcal{B})$, $\mathbf{FPR}(\mathcal{P})$, and $\mathbf{FPR}(\mathcal{R})$. Kalimullin, M. and Ng have announced:

Theorem 21 ([KMNb]). *For each of the structures $\eta, \mathcal{B}, \mathcal{P}, \mathcal{R}$, the fpr-degrees are not linearly ordered under $\leq_{pr}$.*

Although the proof of the theorem above requires some work, the result is somewhat unsatisfying since we don't know the answer to the following:

Question 22. Is there a structure $\mathcal{A}$ for which $\mathbf{FPR}(\mathcal{A})$ is linearly ordered and $|\mathbf{FPR}(\mathcal{A})| > 1$? Can $\mathcal{A}$ be chosen homogeneous?

We also conjecture that the fpr-degrees of the above structures enjoy some density properties, but this is still work in progress.

References

[AK00]   Ash, C., Knight, J.: Computable Structures and the Hyperarithmetical Hierarchy. Studies in Logic and the Foundations of Mathematics, vol. 144. North-Holland Publishing Co., Amsterdam (2000)

[Ala]   Alaev, P.E.: Existence and uniqueness of structures computable in polynomial time. Algebra Log. **55**(1), 72–76 (2016)

[Boo59]   Boone, W.: The word problem. Ann. Math **70**, 207–265 (1959)

[BS11]   Braun, G., Strüngmann, L.: Breaking up finite automata presentable torsion-free abelian groups. Internat. J. Algebra Comput. **21**(8), 1463–1472 (2011)

[CDRU09] Cenzer, D., Downey, R.G., Remmel, J.B., Uddin, Z.: Space complexity of abelian groups. Arch. Math. Log. **48**(1), 115–140 (2009)

[CR]   Cenzer, D., Remmel, J.B.: Polynomial time versus computable boolean algebras. In: Arslanov, M., Lempp, S. (eds.) Recursion Theory and Complexity, Proceedings 1997 Kazan Workshop, pp. 15–53. de Gruyter, Berlin (1999)

[CR91]   Cenzer, D.A., Remmel, J.B.: Polynomial-time versus recursive models. Ann. Pure Appl. Logic **54**(1), 17–58 (1991)

[CR92]   Cenzer, D.A., Remmel, J.B.: Polynomial-time abelian groups. Ann. Pure Appl. Logic **56**(1–3), 313–363 (1992)

[Dob]   Dobritsa, V.: Some constructivizations of abelian groups. Siberian J. Math. **793**(24), 167–173 (1983). (in Russian)

[ECH+92] Epstein, D.B.A., Cannon, J.W., Holt, D.F., Levy, S.V.F., Paterson, M.S., Thurston, W.P.: Word Processing in Groups. Jones and Bartlett Publishers, Boston (1992)

[EG00]   Ershov, Y., Goncharov, S.: Constructive Models. Siberian School of Algebra and Logic. Consultants Bureau, New York (2000)

[Gon80]   Goncharov, S.: The problem of the number of non auto equivalent construc tivizations. Algebra i Logika **19**(6), 621–639, 745 (1980)

[Gon97]   Goncharov, S.: Countable Boolean Algebras and Decidability. Siberian School of Algebra and Logic. Consultants Bureau, New York (1997)

[Gri90]   Grigorieff, S.: Every recursive linear ordering has a copy in dtime-space(n, log(n)). J. Symb. Log. **55**(1), 260–276 (1990)

[Hig61]   Higman, G.: Subgroups of finitely presented groups. Proc. Roy. Soc. Ser. A **262**, 455–475 (1961)

[HKS03]   Hirschfeldt, D.R., Khoussainov, B., Shore, R.A.: A computably categorical structure whose expansion by a constant has infinite computable dimension. J. Symbolic Log. **68**(4), 1199–1241 (2003)

[KMNa]   Kalimullin, I., Melnikov, A., Ng, K.M.: Algebraic structures computable without delay (Submitted)

[KMNb]   Kalimullin, I., Melnikov, A., Ng, K.M.: Cantor's back-and-forth method and computability without delay (to appear)

[KMNc]   Kalimullin, I., Melnikov, A., Ng, K.M.: The diversity of categoricity without delay. Algebra i Logika (to appear)

[KN94]   Khoussainov, B., Nerode, A.: Automatic presentations of structures. In: Leivant, D. (ed.) LCC 1994. LNCS, vol. 960, pp. 367–392. Springer, Heidelberg (1995). doi:10.1007/3-540-60178-3_93

[KN08]   Khoussainov, B., Nerode, A.: Open questions in the theory of automatic structures. Bull. EATCS **94**, 181–204 (2008)

[KNRS07] Khoussainov, B., Nies, A., Rubin, S., Stephan, F.: Automatic structures: richness and limitations. Log. Methods Comput. Sci. **3**(2:2), 1–18 (2007)

[Kri96] Kristiansen, L.: Papers on Subrecursion Theory, Dr Scient Thesis, Research report 217. Ph.D. thesis, University of Oslo (1996)

[KS99] Khoussainov, B., Shore, R.A.: Effective model theory: the number of models and their complexity. In: Cooper, S.B., Truss, J.K. (eds.) Models and Computability (Leeds, 1997). London Mathematical Society Lecture Notes, vol. 259, pp. 193–239. Cambridge University Press, Cambridge (1999)

[LaR77] LaRoche, P.: Recursively presented boolean algebras. Notices AMS **24**, 552–553 (1977)

[Mac11] Macpherson, D.: A survey of homogeneous structures. Discrete Math. **311**(15), 1599–1634 (2011)

[MM] Melnikov, A., Montalban, A.: Computable polish group actions (to appear)

[Mon13] Montalbán, A.: A computability theoretic equivalent to Vaught's conjecture. Adv. Math. **235**, 56–73 (2013)

[Nov55] Novikov, P.: On the algorithmic unsolvability of the word problem in group theory. Trudy Mat. Inst. Steklov **44**, 1–143 (1955)

[NS07] Nies, A., Semukhin, P.: Finite automata presentable abelian groups. In: Artemov, S.N., Nerode, A. (eds.) LFCS 2007. LNCS, vol. 4514, pp. 422–436. Springer, Heidelberg (2007). doi:10.1007/978-3-540-72734-7_29

[NT08] Nies, A., Thomas, R.M.: FA-presentable groups and rings. J. Algebra **320**(2), 569–585 (2008)

[Nur74] Nurtazin, A.: Computable classes and algebraic criteria of autostability. Math. Inst. SB USSRAS, Novosibirsk, Summary of Scientific Schools (1974)

[Rem81] Remmel, J.B.: Recursive boolean algebras with recursive atoms. J. Symb. Log. **46**(3), 595–616 (1981)

[Smi81] Smith, R.L.: Two theorems on autostability in p-Groups. In: Lerman, M., Schmerl, J.H., Soare, R.I. (eds.) Logic Year 1979–80. LNM, vol. 859, pp. 302–311. Springer, Heidelberg (1981). doi:10.1007/BFb0090954

[Tsa11] Tsankov, T.: The additive group of the rationals does not have an automatic presentation. J. Symbolic Log. **76**(4), 1341–1351 (2011)

Computable Transformations of Structures

Russell Miller[1,2]([⊠])

[1] Queens College – C.U.N.Y., 65-30 Kissena Blvd., Queens, NY 11367, USA
Russell.Miller@qc.cuny.edu
[2] Graduate Center of C.U.N.Y., 365 Fifth Avenue, New York, NY 10016, USA
http://qcpages.qc.cuny.edu/~rmiller

Abstract. The *isomorphism problem*, for a class of structures, is the set of pairs of structures within that class which are isomorphic to each other. Isomorphism problems have been well studied for many classes of computable structures. Here we consider isomorphism problems for broader classes of countable structures, using Turing functionals and applying the notions of finitary and countable computable reductions which have been developed for equivalence relations more generally.

1 Introduction

In much of mathematics, two first-order structures which are isomorphic to each other are treated as being exactly the same for all purposes: the objects of study are really the equivalence classes under isomorphism, rather than the structures themselves. Computable structure theory addresses this situation at a deeper level. It is well known that two isomorphic structures may have substantially different algorithmic properties, and therefore, when we consider questions of computability for first-order structures, isomorphism is far too coarse an equivalence relation to be ignored. The fundamental equivalence relation in this discipline is *computable isomorphism*: two structures (both countable, with domain ω) are computably isomorphic if some Turing-computable function on ω is in fact an isomorphism between them. In this case, essentially all known computability-theoretic properties transfer from either structure to the other. In theoretical computer science, complexity theory would go deeper yet, but we will not focus on those questions here.

Of course, the question of whether two structures are isomorphic remains extremely important in computable structure theory. Instead of being so low-level as to be ignored (as in much of model theory), it becomes an object of serious study. The statement that structures $\mathcal{A}$ and $\mathcal{B}$ are isomorphic is on its face a Σ_1^1 sentence about $\mathcal{A}$ and $\mathcal{B}$. (For the rest of this article, all structures are countable with domain ω.) In specific cases, however, it may be not be a Σ_1^1-complete question, but may lie at various levels in the hyperarithmetical hierarchy instead. For example, for algebraically closed fields, isomorphism

---

The author was supported by Grant # DMS – 1362206 from the N.S.F., and by grants from the PSC-CUNY Research Award Program and the Queens College Research Enhancement Fund.

J. Kari et al. (Eds.): CiE 2017, LNCS 10307, pp. 88–97, 2017.
DOI: 10.1007/978-3-319-58741-7_9

depends solely on the characteristic of the field and its transcendence degree over its prime subfield, both of which can be expressed with just a few first-order quantifiers. Vector spaces over $\mathbb{Q}$ are quite similar: for both, the question of whether two computable models of the given theory are isomorphic is a Π_3^0-complete question. The study of isomorphism problems often turns into a search for *invariants*, such as the characterstic and the transcendence degree, which determine isomorphism.

On the other hand, it is known that for computable graphs (which simply means computable symmetric irreflexive subsets of ω^2), the isomorphism problem is Σ_1^1-complete. In [5], Friedman and Stanley created a framework for showing other isomorphism problems to be equally difficult. They showed, for example, that given any two computable graphs $\mathcal{G}_0$ and $\mathcal{G}_1$, one can produce computable linear orders $\mathcal{L}_0$ and $\mathcal{L}_1$ such that

$$\mathcal{G}_0 \cong \mathcal{G}_1 \iff \mathcal{L}_0 \cong \mathcal{L}_1.$$

The "production" of these linear orders is a hyperarithmetic procedure – indeed, a computable procedure – and therefore the isomorphism problem for computable linear orders must also be Σ_1^1-complete.

However, if one knows that the isomorphism problem for computable algebraically closed fields is Π_3^0-complete and wishes to show the same for computable rational vector spaces, a hyperarithmetic procedure in general is insufficient. For classification at these levels, effective procedures are required, and have been examined in [2–4,8], by Calvert, Cummins, Knight, S. Miller, and Vanden Boom, in various combinations. The results in [2, Sect. 4] yield a computable function f which accepts as input the indices (e_0, e_1) of any two computable algebraically closed fields and computes the indices $(i_0, i_1) = f(e_0, e_1)$ of two computable rational vector spaces which are isomorphic if and only if the original two algebraically closed fields were. (By an *index* e for a computable structure $\mathcal{A}$, we mean a number such that the e-th partial computable function φ_e is the characteristic function of the atomic diagram of $\mathcal{A}$, under a fixed coding into ω of atomic sentences in the language of $\mathcal{A}$ with constants from ω.)

The method here works well for computable structures, but the results are sometimes surprising. For example, the isomorphism problem for computable algebraic fields of characteristic 0 (that is, subfields of the algebraic closure $\overline{\mathbb{Q}}$) turns out to be only Π_2^0-complete. It does reduce to the isomorphism problem for algebraically closed fields, but not vice versa – which is puzzling, since $\mathbf{ACF}_0$ has straightforward invariants determining the isomorphism type, while no such invariants are known for the class of algebraic fields.

2   Equivalence Relations on Cantor Space

Our purpose here is to bring to this situation methods from the study of Borel reductions on equivalence relations. We begin by introducing that topic, which has been well studied in descriptive set theory with a focus on equivalence relations on Cantor space 2^ω, and more recently has been extended by many authors to the context of equivalence relations on ω itself.

Let E and F be equivalence relations on domains S and T, respectively. A *reduction* of E to F is a function $g : S \to T$ such that:

$$(\forall x_0, x_1 \in S) \ [x_0 \ E \ x_1 \iff g(x_0) \ F \ g(x_1)].$$

If this holds, then by computing g and deciding the relation F, one can decide E as well. Thus E is "no harder to decide" than F, at least modulo the difficulty of computing g.

The next definition was given in [10].

Definition 1. *Let E and F be equivalence relations on subsets $\mathfrak{C}$ and $\mathfrak{D}$ of 2^ω, respectively. A* computable reduction *of E to F is a reduction $g : \mathfrak{C} \to \mathfrak{D}$ given by a computable function Φ (that is, an oracle Turing functional) on the reals involved:*

$$(\forall A \in \mathfrak{C})(\forall x \in \omega) \ \chi_{g(A)}(x) = \Phi^A(x).$$

If such a reduction exists, then E is computably reducible to F, *denoted $E \leq_0 F$.*

Descriptive set theorists usually eschew this definition in favor of the more general concept of a *Borel reduction*, which is to say, a reduction that happens to be a Borel function. This is the context in which Friedman and Stanley developed their work. More recently, computability theorists have taken to considering computable functions (from ω to ω) as reductions, in the context of equivalence relations on ω. The term "computable reduction" therefore often refers to that context, but we will use it here as well for the reductions described in Definition 1, trusting the reader to distinguish the two concepts based on the equivalence relations in question.

Another refinement of reducibilities on equivalence relations was introduced by Ng and the author in [11]. Studying equivalence relations on ω, they defined *finitary reducibilities*. In the context of Cantor space, it is natural to extend their notion to all cardinals $\mu < 2^\omega$ (as indeed was suggested in their article), yielding the following definitions, which also appeared in [10].

Definition 2. *For equivalence relations E and F on domains S and T, and for any cardinal $\mu < |S|$, we say that a function $g : S^\mu \to T^\mu$ is a μ-ary reduction of E to F if, for every $\boldsymbol{x} = (x_\alpha)_{\alpha \in \mu} \in S^\mu$, we have*

$$(\forall \alpha < \beta < \mu) \ [x_\alpha \ E \ x_\beta \iff g_\alpha(\boldsymbol{x}) \ F \ g_\beta(\boldsymbol{x})],$$

where $g_\alpha : S^\mu \to T$ are the component functions of $g = (g_\alpha)_{\alpha < \mu}$. For limit cardinals μ, a related notion applies with $<\mu$ in place of μ: a function $g : S^{<\mu} \to T^{<\mu}$ which restricts to a ν-ary reduction of E to F for every cardinal $\nu < \mu$ is called a $(<\mu)$-ary reduction. (For $\mu = \omega$, an ω-ary reduction is a countable *reduction, and a $(<\omega)$-ary reduction is a* finitary *reduction.)*

When $S \subseteq 2^\omega$ and $T \subseteq 2^\omega$ and the μ-ary reduction g is computable, we write $E \leq_0^\mu F$, with the natural adaptation $E \leq_\alpha^\mu F$ for α-jump μ-ary reductions. Likewise, when a $(<\mu)$-ary reduction g is α-jump computable, we write $E \leq_\alpha^{<\mu} F$, When $\alpha > 0$, it is important to note that $\Phi^{((\boldsymbol{x})^{(\alpha)})}$ is required to equal $g(\boldsymbol{x})$; this allows more information in the oracle than it would if we had required $\Phi^{((x_0^{(\alpha)} \oplus x_1^{(\alpha)} \oplus \cdots))} = g(\boldsymbol{x})$, with the jumps of the individual inputs taken separately.

In our context for applying these notions, the domains S and T will be subsets of Cantor space, defined by

$$S = \{A \subseteq \omega : A \text{ codes the atomic diagram of a structure in } \mathfrak{C}\},$$

for some class $\mathfrak{C}$ of countable structures with domain ω, with T likewise defined by $\mathfrak{D}$. For us the equivalence relation on each of these domains will be isomorphism on the structures coded. One could explore further, of course, using elementary equivalence of those structures, or bi-embeddability, or other equivalence relations on structures.

3   Early Examples

To begin with, we consider the situation described in the introduction. The models of $\mathbf{ACF}_0$ form a particularly simple class of structures, with isomorphism equivalent to having the same transcendence degree (since we have restricted here to characteristic 0; similar remarks apply to any other fixed characteristic). Isomorphism between algebraic fields of characteristic 0 – that is, the subfields of $\overline{\mathbb{Q}}$ – seems a more challenging problem. However, analysis of computable models in these classes yields the opposite conclusion: isomorphism of computable models of $\mathbf{ACF}_0$ is Π_3^0-complete, whereas isomorphism of computable algebraic fields is only Π_2^0 (and is complete at this level). The latter remark follows from a lemma which appears as [13, Corollary 3.9].

Lemma 1. *Two algebraic field extensions E and F of $\mathbb{Q}$ are isomorphic if and only if every finitely generated subfield of each one embeds into the other.*   □

By the Primitive Element Theorem, the condition here can be expressed by saying that, for every irreducible polynomial $q \in \mathbb{Q}[X]$, E possesses a root of q if and only if F does. For computable fields E and F, this is clearly a Π_2^0 condition.

When we broaden our analysis to the classes $\mathfrak{C}$ of all models of $\mathbf{ACF}_0$ with domain ω and $\mathfrak{D}$ of all algebraic field extensions of $\mathbb{Q}$ with domain ω, we gain a richer view of the situation. Write ACF and Alg for the sets of atomic diagrams of elements of $\mathfrak{C}$ and $\mathfrak{D}$, respectively, and $\cong_{ACF}$ and $\cong_{Alg}$ for the isomorphism relations on these sets of reals. First of all, it is clear that $\cong_{Alg} \not\leq_0 \cong_{ACF}$, as a full computable reduction would require every one of the continuum-many isomorphism classes in Alg to map to a distinct isomorphism class in ACF, and ACF has only countably many isomorphism classes in all. (To see that Alg has uncountably many, write p_n for the n-th prime and notice that for every $A \neq B \subseteq \omega$, the fields $\mathbb{Q}[\sqrt{p_n} : n \in A]$ and $\mathbb{Q}[\sqrt{p_n} : n \in B]$ cannot be isomorphic, as no finite set of square roots of primes generates the square root of any other prime.)

On the other hand, it is not difficult to give a binary computable reduction of Alg to ACF. Such a reduction is simply a Turing functional which, given the atomic diagrams of two algebraic fields F_0 and F_1, computes the diagrams of algebraically closed fields K_0 and K_1 as follows. Fix an enumeration $q_0, q_1, \ldots$ of all irreducible polynomials in $\mathbb{Q}[X]$. (We use here the fact that $\mathbb{Q}$ has a splitting algorithm, which was proven by Kronecker in [9].) At stage 0 we start with $\mathbb{Q}$ as K_0 and $\mathbb{Q}(t)$ as K_1 (with t transcendental).

Now for each s, compute the greatest number $n_s \leq s$ such that

$$(\forall n \leq n_s)(\forall i < \deg(q_n)) \; [(\exists \text{ roots } x_0 < \ldots < x_i \leq s \text{ of } q_n(X) \text{ in } F_0)$$
$$\Longleftrightarrow (\exists \text{ roots } y_0 < \ldots < y_i \leq s \text{ of } q_n(X) \text{ in } F_1)].$$

(Here $x_0 < \ldots < x_i \leq s$ refers to the order of the x_j in ω, not in F_0, which is not an ordered field. Hence this statement is decidable from the atomic diagrams of F_0 and F_1.) At stage $s + 1$, if $(\forall t \leq s) \; n_t < n_{s+1}$, we adjoin a new element to each of K_0 and K_1, independent over all previous elements. If not, we adjoin no new independent elements. In either case, we also take one more step towards making K_0 and K_1 into models of **ACF$_0$**.

At the end of this process, K_0 and K_1 will be models of **ACF$_0$**. If $F_0 \cong F_1$, then new transcendentals were adjoined to each at infinitely many stages, so both have infinite transcendence degree, yielding $K_0 \cong K_1$ as desired. Otherwise, Lemma 1 yields some (least) n for which q_n has more roots in F_0 than in F_1 (without loss of generality). In this case, once all the roots in F_0 have appeared, n_s will never exceed n, and so no further transcendentals will ever again be added to either field. But at every finite stage, K_1 has larger transcendence degree than K_0, since we started that way at stage 0, and so $K_0 \not\cong K_1$ as desired.

The process above can be converted into a countable reduction, yielding the next result.

Proposition 1. *Alg $\leq_0^\omega$ ACF.*

We sketch the construction of a countable computable reduction. Let d_n be the degree of the polynomial q_n. For each F, define the path $p_F \in \omega^\omega$ by

$$p_F(n) = |\{x \in F : q_n(x) = 0\}|.$$

Each such path is confined to the *possible nodes* satisfying $p_F(n) \leq d_n$ for all n, which form a finite-branching subtree. We assign numbers to these nodes: those at level 1 are numbered $0, \ldots, d_0$; those at the next level are numbered $d_0 + 1, d_0 + 2, \ldots, d_0 + (d_0 + 1)(d_1 + 1)$, and so on. The only important aspect of this computable numbering is that each node has a label greater than its predecessor's label.

Given the atomic diagrams of algebraic fields $F_0, F_1, \ldots$, we construct models $K_0, K_1, \ldots$ of **ACF$_0$** to satisfy, for each i:

- If there exists $j < i$ such that $p_{F_i} = p_{F_j}$, then K_i has the same transcendence degree (over $\mathbb{Q}$) as K_j.
- Otherwise, there exists some least n such that $(\forall j < i) \; p_{F_i} \restriction n \neq p_{F_j} \restriction n$. Let d be the label of the node $p_F \restriction n$. Then K_i will have transcendence degree d.

Clearly, satisfying these conditions will ensure that we have a countable reduction from *Alg* to *ACF*. To satisfy them, we guess effectively at the path

p_{F_i} for each i, at each stage s, with the guesses converging to the actual path p_{F_i}. If our guesses produce an n as described in the second item, then K_i at this stage has transcendence degree n; if not, then it is isomorphic to K_j at this stage.

This is not really a generalization of the binary reduction constructed above. Here we use the fact that transcendentals can be destroyed as well as created in the construction of a computable field: we build only finitely much of K_i at any stage, and therefore any element previously considered transcendental in K_i can consistently be turned into a large rational number at the next stage. (The 1-type of a transcendental element is a nonprincipal type.) For a given K_i, once the guesses stabilize on the true value n (if one exists), the transcendentals in K_i at that stage remain independent forever, and any more transcendentals subsequently added to K_i are later destroyed this way. On the other hand, if $F_i \cong F_j$ for some smaller j, then for the n belonging to the least such j, there is some stage after which we always have the same guess $p_{F_i} \restriction n = p_{F_j} \restriction n$ for both paths. From then on, the independent elements in K_i corresponding to those in F_j will stay independent forever, and any subsequent ones will later be destroyed. (Notice that this means that every K_i will have finite transcendence degree. So we have actually given a countable reduction of Alg to a slightly smaller class than ACF.)

On the other hand, there is no computable reduction, not even a binary reduction, from ACF to Alg. This follows from the Π^0_3-completeness of the isomorphism problem for the set of indices for computable algebraically closed fields of characteristic 0: any such reduction would show this Π^0_3-complete set to be Π^0_2, by Lemma 1. Thus the non-reducibility result for computable structures carries over to the general case.

4 ACF_0 and Equivalence Structures

For further insights about $\mathbf{ACF_0}$, we consider another class of structures: the class $\mathfrak{E}$ of countable equivalence structures with no infinite equivalence classes. (An *equivalence structure* consists of a single equivalence relation R on the domain, with equality also in the language.) For computable members of $\mathfrak{E}$, the isomorphism problem is Π^0_3-complete, the same level of complexity as for $\mathbf{ACF_0}$. (If we had allowed infinite equivalence classes, the complexity level would be Π^0_4 instead.) However, there are 2^ω-many nonisomorphic structures in $\mathfrak{E}$. Our goal is to distinguish these two classes using the new notions of this article. First, we show that the class $\mathfrak{C}$ of countable models of $\mathbf{ACF_0}$ is no harder than $\mathfrak{E}$.

Proposition 2. *$ACF \leq_0 Eq$, where Eq is the isomorphism relation on the reals in the class $\mathfrak{E}$.*

Proof. Given an algebraically closed field K as oracle, our reduction Φ builds an equivalence relation R, beginning by creating a single R-equivalence class of size $(2n-1)$ and infinitely many R-equivalence classes of size $2n$ for each $n > 0$.

Then it begins to guess (separately for each $n > 0$) whether K has transcendence degree $\geq n$.

At each stage $s + 1$, for each $n \leq s$, we find the least n-tuple $\boldsymbol{x} \in K$ with all $x_i \leq s$ such that, for all $i \leq s$, $q_{n,i}(\boldsymbol{x}) \neq 0$ in K. (Here we use a fixed ordering of ω^n and a fixed list $\{q_{n,i}\}_{i \in \omega}$ of $\mathbb{Q}[X_1, \ldots, X_n]$.) If this is the same tuple as at stage s, we do nothing. If it is a new tuple, then we take the unique equivalence class of size $2n - 1$ currently in R, add one more element to it, and create a new R-equivalence class of size $(2n - 1)$ to replace it. This is the entire construction.

Now if K has transcendence degree $< n$, then every R-class of size $(2n - 1)$ ever created will eventually become a class of size $2n$. On the other hand, if K has transcendence degree $\geq n$, then eventually an independent n-tuple will be found in K, and from that stage on, the unique R-class of size $(2n - 1)$ will never have another element added to it. Thus R has exactly one class of size $(2n - 1)$ for each $n \leq$ the transcendence degree of K, along with infinitely many classes of each even size. Thus we have a computable full reduction from ACF to Eq. $\qquad\square$

The next proposition is also not surprising. In fact, given that isomorphism on computable models of $\mathbf{ACF_0}$ is Π_3^0-complete as a set (and that isomorphism on computable sturctures in $\mathfrak{E}$ is Π_3^0, the proposition holds immediately for computable structures. This is because a computable binary reduction from E to F (where these are equivalence relations on ω) is in fact simply a many-one reduction from the set E to the set F. Since we wish to establish it for all of $\mathfrak{E}$ and $\mathfrak{C}$, rather than just for computable structures, we give the entire proof.

Proposition 3. $Eq \leq_0^2 ACF$. That is, there is a computable binary reduction from Eq to ACF.

Proof. Given two equivalence relations R_0 and R_1, we must produce algebraically closed fields K_0 and K_1, isomorphic if and only if R_0 and R_1 are. To do so, we wish to test, for each k and n, whether

$$R_0 \text{ has } \geq k \text{ classes of size exactly } n \iff R_1 \text{ has } \geq k \text{ classes of size exactly } n.$$

A pair $\langle k, n \rangle$ for which this may fail will be assigned a number $t_{n,k}$. If R_0 turns out to have k classes of size n while R_1 does not, then K_0 will have transcendence degree $\geq t_{k,n}$ and K_1 will not. If no such k and n exist, then both fields will have infinite transcendence degree. To determine what to do, we will search first for a finite subset $X_{k,n}$ forming k-many R_0-classes of size n, and then for a corresponding subset $\tilde{Y}_{k,n}$ of R_1-classes. If the R_1-classes appear first, then they will form $Y_{k,n}$ and we will then search for $\tilde{X}_{k,n}$ instead.

At stage 0 we begin with $\mathbb{Q}$ as both K_0 and K_1. At each stage, finitely many steps are taken so that each field will be algebraically closed at the end of the construction, but at no stage will $\overline{\mathbb{Q}}$ yet be a subfield of either K_0 or K_1. Thus, putative transcendentals can always be "destroyed," by being turned into elements algebraic over $\mathbb{Q}$. We write $R_{i,s}$ for the restriction of R_i to $\{0, \ldots, s\}$.

At stage $s+1$ we first address that pair $\langle k, n \rangle \leq s$ for which $t_{k,n,s}$ is smallest; then that for which $t_{k,n,s}$ is second-smallest, and so on. If none of these steps ends the stage, we will subsequently address those $\langle k, n \rangle$ with $t_{k,n,s}$ undefined.

If $t_{k,n,s}$ is defined, then so is one (but not both) of the finite subsets $X_{k,n,s}$ ($\subseteq R_{0,s}$) or $Y_{k,n,s}$ ($\subseteq R_{1,s}$). The instructions are symmetric; we give them here with $X_{k,n,s}$ defined, in which case its elements formed k distinct R_0-classes of size n when it was first chosen.

1. If any of these k R_0-classes contains more than n elements in $R_{0,s+1}$, then $X_{k,n,s+1}$ and $t_{k,n,s+1}$ become undefined, and we destroy enough transcendentals in both K_0 and K_1 to ensure that both have transcendence degree $t_{k',n',s+1}$, for that pair $\langle k', n' \rangle$ with the greatest $t_{k',n',s+1} < t_{k,n,s}$ (If there is no such $\langle k', n' \rangle$, then K_0 and K_1 both get transcendence degree 0.) The stage ends here, with all remaining values becoming undefined as well. Otherwise, $X_{k,n,s+1} = X_{k,n,s}$ and K_0 remains the same, and we consider (2)–(5) below.
2. Otherwise, if $\tilde{Y}_{k,n,s}$ was undefined, and $R_{1,s+1}$ contains k distinct classes of size exactly n, then the elements of the first k of these classes are defined to form $\tilde{Y}_{k,n,s+1}$, and we add just as many transcendentals to K_1 as needed to make its transcendence degree $\geq t_{k,n,s}$.
3. If $\tilde{Y}_{k,n,s}$ was undefined, but (2) does not apply, then nothing changes.
4. If $\tilde{Y}_{k,n,s}$ was defined, then its elements formed k distinct R_1-classes. If all these classes still have size exactly n in $R_{1,s+1}$, then nothing changes.
5. Otherwise $\tilde{Y}_{k,n,s}$ was defined, but one of its R_1-classes now has size $> n$. In this case $\tilde{Y}_{k,n,s+1}$ is undefined, and we destroy just enough transcendentals in K_1 to make its transcendence degree $< t_{k,n,s+1} = t_{k,n,s}$. ($K_0$ still has trancendence degree $\geq t_{k,n,s}$.)

If either (1), (2), (3), or (5) applies, then the stage ends here. If (4) applies, then we continue to the pair $\langle k, n \rangle$ with the next-smallest $t_{k,n,s}$. If no more pairs have $t_{k,n,s}$ defined, then we now go in order through those pairs $\langle k, n \rangle \leq s$ for which $t_{k,n,s}$ is undefined. For the least pair $\langle k, n \rangle$ (if any) among these such that either $R_{0,s+1}$ or $R_{1,s+1}$ contains at least k distinct classes of size exactly n, we define $t_{k,n,s+1} = s + 1$, and either

- let $X_{k,n,s+1}$ contain the (kn) elements of $R_{0,s+1}$ forming those R_0-classes, and add transcendentals to K_0 so that it has transcendence degree $t_{k,n,s+1}$; or else
- let $Y_{k,n,s+1}$ contain the (kn) elements of $R_{1,s+1}$ forming those R_1-classes, and add transcendentals to K_1 so that it has transcendence degree $t_{k,n,s+1}$.

This completes the stage.

If $R_0 \not\cong R_1$, then fix the least stage s_0 at which, for some $\langle k, n \rangle$, we have found k classes truly of size n in R_{0,s_0} (WLOG) and set them to equal X_{k,n,s_0}, and R_1 does not possess k classes of this this size, and for all $\langle k', n' \rangle$ with $t_{k',n',s_0} < t_{k,n,s_0}$, X_{k',n',s_0} and $\tilde{Y}_{k',n',s_0}$ are defined and have stabilized. Such an s_0 must exist, since some $\langle k, n \rangle$ do exist. At this stage, K_0 will be given transcendence degree t_{k,n,s_0}, which will equal $t_{k,n} = \lim_s t_{k,n,s}$, while K_1 will have lesser transcendence degree at that stage. Moreover, every $\tilde{Y}_{k,n,s}$ ever subsequently found will later become undefined, with the transcendence degree of K_1 threrefore dropping back below $t_{k,n}$ infinitely often, and so $K_1 \not\cong K_0$.

However, if $R_0 \cong R_1$, then for every $\langle k, n \rangle$ for which $t_{k,n,s}$ stabilizes, both K_0 and K_1 will have transcendence degree $\geq \lim_s t_{k,n,s}$. Moreover, there will be infinitely many such pairs $\langle k, n \rangle$, since every element of R_0 lies in a finite R_0-class, and likewise for R_1. Therefore, both K_0 and K_1 will have infinite transcendence degree, leaving them isomorphic. □

Proposition 4. $Eq \leq_0^3 ACF$, but $Eq \not\leq_0^4 ACF$. That is, there is a computable ternary reduction from Eq to ACF, but no 4-ary computable reduction.

The details of the proof are too extensive to present here; they will be described in the author's talk at the C.i.E. meeting.

In light of the Π_3^0-completeness of isomorphism for computable algebraic fields, this proposition seems like a surprise. Section 4.2 of [11] makes it more plausible. The discussion there centers on the fact that, since isomorphism on models of **ACF$_0$** is given by transcendence degree, it is essentially just a matter (for a computable model K) of counting the elements in the following Σ_2^0 basis for K:

$$\{x \in K : (\forall \text{ nonzero } h \in \mathbb{Z}[X_0, \ldots, X_x]) \, h(0, 1, \ldots, x) \neq 0 \text{ in } K\}.$$

It is shown in [11] that, if $E_{\text{card}}^{\emptyset'}$ is the relation (on indices e of Σ_2^0 sets $W_e^{\emptyset'}$) of having the same cardinality, then $E_{\text{card}}^{\emptyset'}$ is complete under ternary reducibility among Π_3^0 equivalence relations on ω, but not complete among them under 4-ary reducibility. Since the relation (for indices of computable fields in general) of having the same transcendence degree over the prime subfield is computably reducible to $E_{\text{card}}^{\emptyset'}$, it is not so surprising that isomorphism on ACF in general loses its power at the same specific finitary level of reduction.

The arguments in [11] do prove the following, using the notion of *jump-reduction* from [10], with a functional whose oracle is the jump of the inputs.

Lemma 2. $Eq \leq_1^3 ACF$. That is, there is a Turing functional Γ such that $\Gamma^{(E_0 \oplus E_1 \oplus E_2)'} = K_0 \oplus K_1 \oplus K_2$ is a ternary reduction from Eq to ACF.

In light of Proposition 4, it is natural to enquire into other theories admitting similar notions of dimension. Baldwin and Lachlan showed in [1] that, if T is an ω_1-categorical theory that is not ω-categorical (such as **ACF$_0$**), the countable models of T form an $(\omega+1)$-sequence under elementary embedding. One suspects, therefore, that the class of such models might be complete among Π_α^0-definable equivalence relations on 2^ω under ternary computable reducibility but not under 4-ary computable reducibility, just as holds for models of **ACF$_0$** with $\alpha = 3$.

5 Transformations and Functors

As a final remark, we note that these computable transformations, as in Definition 1, form part of the larger concept of a *computable functor*. These were defined by Poonen, Schoutens, Shlapentokh, and the author in [12], and subsequently, in [6,7], he and Harrison-Trainor, Melnikov, and Montalbán broadened their applicability. We give their definition here.

Definition 3. *Let $\mathfrak{C}$ and $\mathfrak{D}$ be categories of structures with domain ω, for which the morphisms from $\mathcal{S}$ to $\mathcal{T}$ are maps from the domain ω of $\mathcal{S}$ to the domain ω of $\mathcal{T}$. A computable functor is a functor $\mathcal{F} : \mathfrak{C} \to \mathfrak{D}$ for which there exist Turing functionals Φ and Φ_* such that*

- *for every $\mathcal{S} \in \mathfrak{C}$, the function $\Phi^{\mathcal{S}}$ computes (the atomic diagram of) the structure $\mathcal{F}(\mathcal{S})$; and*
- *for every morphism $g : \mathcal{S} \to \mathcal{T}$ in $\mathfrak{C}$, we have $\Phi_*^{\mathcal{S} \oplus g \oplus \mathcal{T}} = \mathcal{F}(g)$ in $\mathfrak{D}$.*

It would be natural to examine how close the computable transformations defined earlier in this article come to being computable as functors. The functional Φ_* computing the functor on morphisms is likely to require one or more jumps of $\mathcal{S}$ and $\mathcal{T}$ as oracle, but if not, then the conclusions in [6,12] about computable functors would all apply here as well.

References

1. Baldwin, J.T., Lachlan, A.H.: On strongly minimal sets. J. Symb. Log. **36**(1), 79–96 (1971)
2. Calvert, W.: The isomorphism problem for classes of computable fields. Arch. Math. Log. **43**, 327–336 (2004)
3. Calvert, W., Cummins, D., Knight, J.F., Miller, S.: Comparing classes of finite structures. Algebra Log. **43**, 365–373 (2004)
4. Calvert, W., Knight, J.F.: Classification from a computable viewpoint. Bull. Symb. Log. **12**, 191–218 (2006)
5. Friedman, H., Stanley, L.: A Borel reducibility for classes of countable structures. J. Symb. Log. **54**, 894–914 (1989)
6. Harrison-Trainor, M., Melnikov, A., Miller, R., Montalbán, A.: Computable functors and effective interpretability. J. Symb. Log. **82**(1), 77–97 (2017)
7. Harrison-Trainor, M., Miller, R., Montalbán, A.: Borel functors and infinitary interpretations (submitted for publication)
8. Knight, J.F., Miller, S., Vanden Boom, M.: Turing computable embeddings. J. Symb. Log. **72**(3), 901–918 (2007)
9. Kronecker, L.: Grundzüge einer arithmetischen Theorie der algebraischen Größen. J. Math. **92**, 1–122 (1882)
10. Miller, R.: Computable reducibility for Cantor space (submitted for publication)
11. Miller, R., Ng, K.M.: Finitary reducibility on equivalence relations. J. Symb. Log. **81**(4), 1225–1254 (2016)
12. Miller, R., Poonen, B., Schoutens, H., Shlapentokh, A.: A computable functor from graphs to fields (submitted for publication)
13. Miller, R., Shlapentokh, A.: Computable categoricity for algebraic fields with splitting algorithms. Trans. Amer. Math. Soc. **367**(6), 3981–4017 (2015)

Formulas with Reversal

Narad Rampersad[(✉)]

Department of Mathematics and Statistics, University of Winnipeg,
515 Portage Avenue, Winnipeg, MB R3B 2E9, Canada
n.rampersad@uwinnipeg.ca

Abstract. We introduce the new concept of "formula with reversal". We show some simple formulas with reversal that have high index and we give a partial characterization of unavoidable formulas with reversal.

One of the fundamental objects of study in combinatorics on words is the pattern. A *pattern* is a word over an alphabet of *variables* and is meant to describe some kind of repetitive structure. For instance the pattern XX over the single variable X is meant to describe the repetition of the same word twice in succession. A word x over an alphabet Σ *encounters* a pattern p over an alphabet Δ if there is a non-erasing morphism $h : \Delta^* \to \Sigma^*$ such that $h(p)$ is a factor of x. For example, the word waverers encounters the pattern XX via the map $X \to$ er. If the word x does not encounter p then it *avoids p*.

The first systematic analysis of the avoidability of patterns was done by Bean, Ehrenfeucht, and McNulty [2], and, independently, by Zimin [14]. One of their major results was a characterization of which patterns are avoidable on a finite alphabet. The least k such that a pattern p is avoidable on a k-letter alphabet is called the *index* of p. Baker, McNulty, and Taylor [1] produced an example of a pattern with index 4 and left it as an open question whether or not there are patterns of higher index. Clark [4] produced an example of a pattern of index 5 and at present no pattern is known to have index higher than 5.

Cassaigne [3] introduced the related concept of a *formula*. A formula is a set of patterns $\{p_1, \ldots, p_s\}$, usually written with "dots" as $p_1 \cdot p_2 \cdots p_s$. The p_i are referred to as *fragments* of the formula. The formula is avoided by a word x if there is no non-erasing morphism $h : \Delta^* \to \Sigma*$ such that $h(p_i)$ is a factor of x for every $i = 1, \ldots, s$. Clark used this notion of formula to produce his example of a pattern of index 5.

Recently, there has been some interest in studying patterns with reversal, such as XXX^R. A word x encounters XXX^R if there is a morphism $h : \{X\}^* \to \Sigma^*$ such that $h(X)h(X)(h(X))^R$ is a factor of x, where $(h(x))^R$ denotes the reversal of the word $h(x)$. Du, Mousavi, Rowland, Schaeffer, and Shallit [10] and Currie and Rampersad [8,9] examined the patterns XXX^R and XX^RX. Currie and Lafrance [5] determined the index of every binary pattern with reversal. For some recent algorithmic results, see [11–13].

Of course, one can generalize patterns with reversal to formulas with reversal in the same way one generalizes patterns to formulas. The work we will present in this talk is that of Currie, Mol, and Rampersad [6,7] on formulas with reversal.

© Springer International Publishing AG 2017
J. Kari et al. (Eds.): CiE 2017, LNCS 10307, pp. 98–100, 2017.
DOI: 10.1007/978-3-319-58741-7_10

The first results [6] are a construction of some simple formulas with reversal that have index 5. Clark's example of a formula of index 5, which does not use reversal, is quite complicated; with reversals, our examples are quite simple. For each $k \geq 1$, we define the formula

$$\psi_i = XY_1Y_2\cdots Y_kX \cdot Y_1^R \cdot Y_2^R \cdots \cdot Y_k^R.$$

Theorem 1. *The formula ψ_1 has index 4.*

Theorem 2. *The formula ψ_2 has index 5.*

Theorem 3. *For $k \geq 1$, the formula ψ_k has index ≥ 4.*

Our next result is an attempt to generalize the so-called Zimin characterization of unavoidable formulas. Let $X_1, X_2, \ldots$ be variables. Define the *Zimin words* recursively by $Z_0 = \epsilon$ and for $n \geq 1$ we have $Z_n = Z_{n-1}X_nZ_{n-1}$. Zimin showed that a formula (without reversal) on n variables is unavoidable if and only if it is encountered by Z_n.

We generalize this to formulas with reversal as follows. Let $X_1, X_2, \ldots$ and $Y_1, Y_2, \ldots$ be variables. For any variable X define $X^\sharp = \{X, X^R\}$. For $m \geq 0$ and $n \geq 0$ we define the *Zimin formula with reversal $Z_{m,n}$* by

$$Z_{m,0} = X_1^\sharp \cdots X_m^\sharp$$

and

$$Z_{m,n} = Z_{m,n-1}Y_nZ_{m,n-1}.$$

Given a formula with reversals ϕ, we say that a variable X is *two-way* in ϕ if both X and X^R occur in ϕ. Otherwise, we say that X is *one-way*. Note that in $Z_{m,n}$ the X_i are two-way and the Y_i are one-way. We have the following partial result [7].

Theorem 4. *Let ϕ be a formula with reversals with $m \geq 0$ two-way variables and $n \leq 2$ one-way variables. Then ϕ is unavoidable if and only if it is encountered by $Z_{m,n}$.*

In this abstract we have been somewhat informal with our definitions. For full details and proofs, please see the papers [6, 7].

References

1. Baker, K., McNulty, G., Taylor, W.: Growth problems for avoidable words. Theoret. Comput. Sci. **69**, 319–345 (1989)
2. Bean, D.R., Ehrenfeucht, A., McNulty, G.F.: Avoidable patterns in strings of symbols. Pacific J. Math. **85**, 261–294 (1979)
3. Cassaigne, J.: Motifs évitables et régularité dans les mots, Ph.D. thesis, Université Paris VI (1994)
4. Clark, R.J.: Avoidable formulas in combinatorics on words, Ph.D. thesis, University of California, Los Angeles (2001)

5. Currie, J.D., Lafrance, P.: Avoidability index for binary patterns with reversal. Electron. J. Combin. **23**(1) (2016). Paper #P1.36
6. Currie, J., Mol, L., Rampersad, N.: A family of formulas with reversal of high avoidability index (Submitted)
7. Currie, J., Mol, L., Rampersad, N.: On avoidability of formulas with reversal (Submitted)
8. Currie, J., Rampersad, N.: Binary words avoiding xx^Rx and strongly unimodal sequences. J. Integer Seq. **15**, Article 15.10.3 (2015)
9. Currie, J., Rampersad, N.: Growth rate of binary words avoiding xxx^R. Theoret. Comput. Sci. **609**, 456–468 (2016)
10. Du, C.F., Mousavi, H., Rowland, E., Schaeffer, L., Shallit, J.: Decision algorithms for Fibonacci-automatic words, II: related sequences and avoidability. Theoret. Comput. Sci. **657**, 146–162 (2017)
11. Gawrychowski, P., I, T., Inenaga, S., Köppl, D., Manea, F.: Efficiently finding all maximal alpha-gapped repeats. In: Proceedings of the STACS, pp. 39:1–39:14 (2016)
12. Gawrychowski, P., Manea, F., Nowotka, D.: Testing generalised freeness of words. In: Proceedings of the STACS, pp. 337–349 (2014)
13. Kosolobov, D., Manea, F., Nowotka, D.: Detecting unary patterns. ArXiv preprint: https://arxiv.org/abs/1604.00054
14. Zimin, A.I.: Blocking sets of terms. Math. USSR Sbornik **47**(2), 353–364 (1984). (English translation)

Compressibility and Probabilistic Proofs

Alexander Shen[✉]

LIRMM CNRS & University of Montpellier, Montpellier, France
alexander.shen@lirmm.fr

Abstract. We consider several examples of probabilistic existence proofs using compressibility arguments, including some results that involve Lovász local lemma.

1 Probabilistic Proofs: A Toy Example

There are many well known probabilistic proofs that objects with some properties exist. Such a proof estimates the probability for a random object to violate the requirements and shows that it is small (or at least strictly less than 1). Let us look at a toy example.

Consider a $n \times n$ Boolean matrix and its $k \times k$ minor (the intersection of k rows and k columns chosen arbitrarily). We say that the minor is *monochromatic* if all its elements are equal (either all zeros or all ones).

Proposition 1. *For large enough n and for $k = O(\log n)$, there exists a $(n \times n)$-matrix that does not contain a monochromatic $(k \times k)$-minor.*

Proof. We repeat the same simple proof three times, in three different languages. (Probabilistic language) Let us choose matrix elements using independent tosses of a fair coin. For a given k columns and k rows, the probability of getting a monochromatic minor at their intersection in 2^{-k^2+1}. (Both zero-minor and one-minor have probability 2^{-k^2}.) There are at most n^k choices for columns and the same number for rows, so by the union bound the probability of getting at least one monochromatic minor is bounded by

$$n^k \times n^k \times 2^{-k^2+1} = 2^{2k\log n - k^2 + 1} = 2^{k(2\log n - k) + 1}$$

and the last expression is less then 1 if, say, $k = 3\log n$ and n is sufficiently large. (Combinatorial language) Let us count the number of bad matrices. For a given choice of columns and rows we have 2 possibilities for the minor and $2^{n^2-k^2}$ possibilities for the rest, and there is at most n^k choices for raws and columns, so the total number of matrices with monochromatic minor is

_____

A. Shen—On leave from IITP RAS, Moscow, Russia.
Supported by ANR-15-CE40-0016-01 RaCAF grant.

J. Kari et al. (Eds.): CiE 2017, LNCS 10307, pp. 101–111, 2017.
DOI: 10.1007/978-3-319-58741-7_11

$$n^k \times n^k \times 2 \times 2^{n^2-k^2} = 2^{n^2+2k\log n-k^2+1} = 2^{n^2+k(2\log n-k)+1},$$

and this is less than 2^{n^2}, the total number of Boolean $(n \times n)$-matrices. (Compression language) To specify the matrix that has a monochromatic minor, it is enough to specify $2k$ numbers between 1 and n (rows and column numbers), the color of the monochromatic minor (0 or 1) and the remaining $n^2 - k^2$ bits in the matrix (their positions are already known). So we save k^2 bits (compared to the straightforward list of all n^2 bits) using $2k\log n + 1$ bits instead (each number in the range $1 \ldots n$ requires $\log n$ bits; to be exact, we may use $\lceil \log n \rceil$), so we can compress the matrix with a monochromatic minor if $2k\log n + 1 \ll k^2$, and not all matrices are compressible.

Of course, these three arguments are the same: in the second one we multiply probabilities by 2^{n^2}, and in the third one we take logarithms. However, the compression language provides some new viewpoint that may help our intuition.

2   A Bit More Interesting Example

In this example we want to put bits (zeros and ones) around the circle in a "essentially asymmetric" way: each rotation of the circle should change at least a fixed percentage of bits. More precisely, we are interested in the following statement (Fig. 1):

Proposition 2. *There exists $\varepsilon > 0$ such for every sufficiently large n there exists a sequence $x_0 x_1 \ldots x_{n-1}$ of bits such that for every $k = 1, 2, \ldots, n-1$ the cyclic shift by k positions produces a sequence*

$$y_0 = x_k, y_1 = x_{k+1}, \ldots, y_{n-1} = x_{k-1},$$

that differs from x in at least εn positions (the Hamming distance between x and y is at least εn).

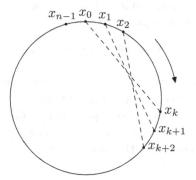

Fig. 1. A string $x_0 \ldots x_{n-1}$ is bad if most of the dotted lines connect equal bits

Proof. Assume that some rotation (cyclic shift by k positions) transforms x into a string y that coincides almost everywhere with x. We may assume that $k \leq n/2$: the cyclic shift by k positions changes as many bits as the cyclic shift by $n - k$ (the inverse one). Imagine that we dictate the string x from left to right. First k bits we dictate normally. But then the bits start to repeat (mostly) the previous ones (k positions before), so we can just say "the same" or "not the same", and if ε is small, we know that most of the time we say "the same". Technically, we have εn different bits, and at least $n - k \geq n/2$ bits to dictate after the first k, so the fraction of "not the same" signals is at most 2ε. It is well known that strings of symbols where some symbols appear more often than others can be encoded efficiently. Shannon tells us that a string with two symbols with frequencies p and q (so $p + q = 1$) can be encoded using

$$H(p,q) = p \log \frac{1}{p} + q \log \frac{1}{q}$$

bits per symbol and that $H(p,q) = 1$ only when $p = q = 1/2$. In our case, for small ε, one of the frequencies is close to 0 (at most 2ε), and the other one is close to 1, so $H(p,q)$ is significantly less than 1. So we get a significant compression for every string that is bad for the theorem, therefore most string are good (so good string do exist).

More precisely, every string $x_0 \ldots x_{n-1}$ that does not satisfy the requirements, can be described by

- k [$\log n$ bits]
- $x_0, \ldots, x_{k-1}$ [k bits]
- $x_k \oplus x_0, x_{k+1} \oplus x_1, \ldots, x_{n-1} \oplus x_{n-k-1}$ [$n - k$ bits where the fraction of 1s is at most 2ε, compressed to $(n - k)H(2\varepsilon, 1 - 2\varepsilon)$ bits]

For $\varepsilon < 1/4$ and for large enough n the economy in the third part (compared to $n - k$) is more important than $\log n$ in the first part.

Of course, this is essentially a counting argument: the number of strings of length $(n - k)$ where the fraction of 1s is at most 2ε, is bounded by $2^{H(2\varepsilon, 1-2\varepsilon)(n-k)}$ and we show that the bound for the number of bad strings,

$$\sum_{k=1}^{n/2} 2^k 2^{H(2\varepsilon, 1-2\varepsilon)(n-k)}$$

is less than the total number of strings (2^n). Still the compression metaphor makes the proof more intuitive, at least for some readers.

3   Lovász Local Lemma and Moser–Tardos Algorithm

In our examples of probabilistic proofs we proved the existence of objects that have some property by showing that *most* objects have this property (in other words, that the probability of this property to be true is close to 1 under some natural distribution). Not all probabilistic proofs go like that. One of the exceptions is the famous Lovász local lemma (see, e.g., [1]). It can be used in the situations where the union bound does not work: we have too many bad events, and the sum of their probabilities exceeds 1 even if probability of each one is very small. Still Lovász local lemma shows that these bad events do not cover the probability space entirely, assuming that the bad events are "mainly independent". The probability of avoiding these bad events is exponentially small, still Lovász local lemma provides a positive lower bound for it.

This means, in particular, that we cannot hope to construct an object satisfying the requirements by random trials, so the bound provided by Lovász local lemma does not give us a randomized algorithm that constructs the object with required properties with probability close to 1. Much later Moser and Tardos [4,5] suggested such an algorithm — in fact a very simple one. In other terms, they suggested a different distribution under which good objects form a majority.

We do not discuss the statement of Lovász local lemma and Moser–Tardos algorithm in general (see [8]). Instead, we provide two examples when they can be used, and the compression-language proofs that can be considered as ad hoc versions of Moser–Tardos argument. They are (1) satisfiability of formulas in conjunctive normal form (CNF) and (2) strings without forbidden factors.

4   Satisfiable CNF

A CNF (*conjunctive normal form*) is a propositional formula that is a conjuction of *clauses*. Each clause is a disjunction of *literals*; a literal is a propositional variable or its negation. For example, CNF

$$(\neg p_1 \lor p_2 \lor p_4) \land (\neg p_2 \lor p_3 \lor \neg p_4)$$

consists of two clauses. First one prohibits the case when $p_1 = \text{TRUE}$, $p_2 = \text{FALSE}$, $p_4 = \text{FALSE}$; the second one prohibits the case when $p_2 = \text{TRUE}$, $p_3 = \text{FALSE}$, $p_4 = \text{TRUE}$. A CNF is *satisfiable* if it has a *satisfying assigment* (that makes all clauses true, avoiding the prohibited combinations). In our example there are many satisfying assignments. For example, if $p_1 = \text{FALSE}$ and $p_3 = \text{TRUE}$, all values of other variables are OK.

We will consider CNF where all clauses include n literals with n different variables (from some pool of variables that may contain much more than n variables). For a random assignment (each variable is obtained by an independent tossing of a fair coin) the probability to violate a clause of this type is 2^{-n} (one of 2^n combinations of values for n variables is forbidden). Therefore, *if the number of clauses of this type is less than 2^n, then the formula is satisfiable*. This is a tight bound: using 2^n clauses with the same variables, we can forbid all the combinations and get an unsatisfiable CNF.

The following result says that we can guarantee the satisfiability for formuli with much more clauses. In fact, the total number of clauses may be arbitrary (but still we consider finite formulas, of course). The only thing we need is the "limited dependence" of clauses. Let us say that two clauses are *neighbors* if they have a common variable (or several common variables). The clauses that are not neighbors correspond to independent events (for a random assignment). The following statement says that if the number of neighbors of each clause is bounded, then CNF is guaranteed to be satisfiable.

Proposition 3. *Assume that each clause in some CNF contains n literals with different variables and has at most 2^{n-3} neighbor clauses. Then the CNF is satisfiable.*

Note that 2^{n-3} is a rather tight bound: to forbid all the combinations for some n variables, we need only 2^n clauses.

Proof. It is convenient to present a proof using the compression language, as suggested by Lance Fortnow. Consider the following procedure $\text{FIX}(C)$ whose argument is a clause (from our CNF).

> { C is false }
> $\text{FIX}(C)$:
> $\text{RESAMPLE}(C)$
>   **for** all C' that are neighbors of C:
>   **if** C' is false **then** $\text{FIX}(C')$
> { C is true; other clauses that were true remain true }

Here $\text{RESAMPLE}(C)$ is the procedure that assigns fresh random values to all variables in C. The pre-condition (the first line) says that the procedure is called only in the situation where C is false. The post-condition (the last line) says that *if the procedure terminates*, then C is true after termination, and, moreover, all other clauses of our CNF that were true before the call remain true. (The ones that were false may be true or false.)

Note that up to now we do not say anything about the termination: note that the procedure is randomized and it may happen that it does not terminate (for example, if all RESAMPLE calls are unlucky to choose the same old bad values).

Simple observation: if we have such a procedure, we may apply it to all clauses one by one and after all calls (assuming they terminate and the procedure works according to the specification) we get a satisfying assignment.

Another simple observation: it is easy to prove the "conditional correctness" of the procedure $\text{FIX}(C)$. In other words, it achieves its goal assuming that (1) it terminates; (2) all the recursive calls $\text{FIX}(C')$ achieve their goals. It is almost obvious: the $\text{RESAMPLE}(C)$ call may destroy (=make false) only clauses that are neighbors to C, and all these clauses are FIX-ed after that. Note that C is its own neighbor, so the **for**-loop includes also a recursive call $\text{FIX}(C)$, so after all these calls (that terminate and satisfy the post-condition by assumption) the clause C and all its neighbors are true and no other clause is damaged.

Note that the last argument remains valid even if we delete the only line that really changes something, i.e., the line RESAMPLE(C). In this case the procedure never changes anything but still is conditionally correct; it just does not terminate if one of the clauses is false.

It remains to prove that the call FIX(C) terminates with high probability. In fact, it terminates with probability 1 if there are no time limits and with probability exponentially close to 1 in polynomial time. To prove this, one may use a compression argument: we show that *if the procedure works for a long time without terminating, then the sequence of random bits used for resampling is compressible.* We assume that each call of RESAMPLE() uses n fresh bits from the sequence. Finally, we note that this compressibility may happen only with exponentially small probability.

Imagine that FIX(C) is called and during its recursive execution performs many calls

$$\text{RESAMPLE}(C_1), \ldots, \text{RESAMPLE}(C_N)$$

(in this order) but does not terminate (yet). We stop it at some moment and examine the values of all the variables.

Lemma 1. *Knowing the values of the variables after these calls and the sequence $C_1, \ldots, C_N$, we can reconstruct all the Nn random bits used for resampling.*

Proof. Let us go backwards. By assumption we know the values of all variables after the calls. The procedure RESAMPLE(C_N) is called only when C_N is false, and there is only one n-tuple of values that makes C_N false. Therefore we know the values of all variables before the last call, and also know the random bits used for the last resampling (since we know the values of variables after resampling).

The same argument shows that we can reconstruct the values of variables before the preceding call RESAMPLE(C_{N-1}), and random bits used for the resampling in this call, etc.

Now we need to show that the sequence of clauses $C_1, \ldots, C_N$ used for resampling can be described by less bits than nN (the number of random bits used). Here we use the assumption saying each clause has at most 2^{n-3} neighbors and that the clauses C' for which FIX(C') is called from FIX(C), are neighbors of C.

One could try to say that since C_{i+1} is a neighbor of C_i, we need only $n-3$ bits to specify it (there are at most 2^{n-3} neighbors by assumption), so we save 3 bits per clause (compared to n random bits used by resampling). But this argument is wrong: C_{i+1} is not always the neighbor of C_i, since we may return from a recursive call that causes resampling of C_i and then make a new recursive call that resamples C_{i+1}.

To get a correct argument, we should look more closely at the tree of recursive calls generated by one call FIX(C) (Fig. 2). In this tree the sons of each vertex correspond to neighbor clauses of the father-clause. The sequence of calls is determined by a walk in this tree, but we go up and down, not only up (as we assumed in the wrong argument). How many bits we need to encode this walk (and therefore the sequence of calls)? We use one bit to distinguish between steps

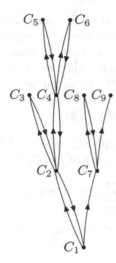

Fig. 2. The tree of recursive calls for $\text{Fix}(C_1)$ (up to some moment)

up and down. If we are going down, no other information is needed. If we are going up (and resample a new clause), we need one bit to say that we are going up, and $n - 3$ bits for the number of neighbor we are going to. For accounting purposes we combine these bits with a bit needed to encode the step back (this may happen later or not happen at all), and we see that in total we need at most $(n - 3) + 1 + 1 = n - 1$ bits per each resampling. This is still less than n, so we save one bit for each resampling. If N is much bigger than the number of variables, we indeed compress the sequence of random bits used for resampling, and this happens with exponentially small probability.

This argument finishes the proof.

5 Tetris and Forbidden Factors

The next example is taken from word combinatorics. Assume that a list of binary strings $F_1, \ldots, F_k$ is given. These F_i are considered as "forbidden factors": this means that we want to construct a (long) string X that does not have any of F_i as a factor (i.e., none of F_i is a substring of X). This may be possible or not depending on the list. For example, if we consider two strings $0, 11$ as forbidden factors, every string of length 2 or more has a forbidden factor (we cannot use zeros at all, and two ones are forbidden).

The more forbidden factors we have, the more chances that they block the growth in the sense that every sufficiently long string has a forbidden factor. Of course, not only the number of factors matters: e.g., if we consider $0, 00$ as forbidden factors, then we have long strings of ones without forbidden factors. However, now we are interested in quantitative results of the following type: *if the number of forbidden factors of length j is a_j, and the numbers a_j are "not too big", then there exists an arbitrarily long string without forbidden factors.*

This question can be analyzed with many different tools, including Lovász local lemma (see [9]) and Kolmogorov complexity. Using a complexity argument, Levin proved that if $a_j = 2^{\alpha j}$ for some constant $\alpha < 1$, then there exists a constant M and an infinite sequence that does not contain forbidden factors of length smaller than M. (See [10, Sect. 8.5] for Levin's argument and other related results.) A nice sufficient condition was suggested by Miller [3]: we formulate the statement for the arbitrary alphabet size.

Proposition 4. *Consider an alphabet with m letters. Assume that for each $j \geq 2$ we have a_j "forbidden" strings of length j. Assume that there exist some constant $x > 0$ such that*

$$\sum_{j \geq 2} a_j x^j < mx - 1$$

Then there exist arbitrarily long strings that do not contain forbidden substrings.

Remarks. 1. We do not consider $j = 1$, since this means that some letters are deleted from the alphabet.

2. By compactness the statement implies that there exists an infinite sequence with no forbidden factors.

3. The constant x should be at least $1/m$, otherwise the right hand side is negative. This means that a_j/m^j should be small, and this corresponds to our intuition (a_j should be significantly less than m^j, the total number of strings of length j).

The original proof from [3] uses some ingenious potential function defined on strings: Miller shows that if its value is less than 1, then one can add some letter preserving this property. It turned out (rather mysteriously) that exactly the same condition can be obtained by a completely different compression argument (following [2,6]), so probably the inequality is more fundamental than it may seem! (It appears once more in Golod – Shafarevich theorem that provides yet another proof for the same statement; we do not go into details here, see, e.g., [7].)

Proof. Here is the idea of the compression argument. We start with an empty string and add randomly chosen letters to its right end. If some forbidden string appears as a suffix, it is immediately deleted. So forbidden strings may appear only as suffixes, and only for a short time. After this "backtracking" we continue adding new letters. (This resembles the famous "tetris game" when blocks fall down and then disappear under some conditions.)

We want to show that if this process is unsuccessful in the sense that after many steps we still have a short string, then the sequence of added random letters is compressible, so this cannot happen always, and therefore a long string without forbidden factors exists. Let us consider a "record" (log file) for this process that is a sequence of symbols "+" and "+⟨deleted string⟩" (for each forbidden string we have a symbol, plus one more symbol without a string). If a letter was added and no forbidden string appears, we just add '+' to the record. If we have to delete some forbidden string s after a letter was added, we write this string in brackets after the $+$ sign. Note that we do *not* record the added letters, only the deleted substrings. (It may happen that several forbidden suffixes appear; in this case we may choose any of them.)

Lemma 2. *At every stage the current string and the record uniquely determine the sequence of random letters used.*

Proof. Having this information, we can reconstruct the configuration going backwards. This reversed process has steps where a forbidden string is added (and we know which one, since it is written in brackets in the record), and also steps when a letter is deleted (and we know which letter is deleted, i.e., which random letter was added when moving forwards).

If after many (say, T) steps we still have a short current string, then the sequence of random letters can be described by the record (due to the Lemma; we ignore the current string part since it is short). As we will see, the record can be encoded with less bits than it should have been (i.e., less than $T \log m$ bits). Let us describe this encoding and show that it is efficient (assuming the inequality $\sum a_j x^j < mx - 1$).

We use arithmetic encoding for the lengths. Arithmetic encoding for M symbols starts by choosing positive reals $q_1, \ldots, q_M$ such that $q_1 + \ldots + q_M = 1$. Then we split the interval $[0, 1]$ into parts of length $q_1, \ldots, q_M$ that correspond to these M symbols. Adding a new symbol corresponds to splitting the current interval in the same proportion and choosing the right subinterval. For example, the sequence (a, b) corresponds to bth subinterval of ath interval; this interval has length $q_a q_b$. The sequence $(a, b, \ldots, c)$ corresponds to interval of length $q_a q_b \ldots q_c$ and can be reconstructed given any point of this interval (assuming $q_1, \ldots, q_M$ are fixed); to specify some binary fraction in this interval we need at most $-\log(q_a q_b \ldots q_c) + O(1)$ bits, i.e., $-\log q_a - \log q_b - \ldots - \log q_c + O(1)$ bits.

Now let us apply this technique to our situation. For $+$ without brackets we use $\log(1/p_0)$ bits, and for $+\langle s \rangle$ where s is of length j, we use $\log(1/p_j) + \log a_j$ bits. Here p_j are some positive reals to be chosen later; we need $p_0 + \sum p_j = 1$. Indeed, we may split p_j into a_j equal parts (of size p_j/a_j) and use these parts as q_s in the description of arithmetical coding above; splitting adds $\log a_j$ to the code length for strings of length j.

To bound the total number of bits used for encoding the record, we perform amortised accounting and show that the average number of bits per letter is less than $\log m$. Note that the number of letters is equal to the number of $+$ signs in the record. Each $+$ without brackets increases the length of the string by one letter, and we want to use less that $\log m - c$ bits for its encoding, where $c > 0$ is some constant saying how much is saved as a reserve for amortized analysis. And $+\langle s \rangle$ for a string s of length j decreases the length by $j - 1$, so we want to use less than $\log m + c(j - 1)$ bits (using the reserve).

So we need:

$$\log(1/p_0) < \log m - c;$$
$$\log(1/p_j) + \log a_j < \log m + c(j - 1)$$

together with

$$p_0 + \sum_{j \geq 2} p_j = 1.$$

Technically is it easier to use non-strict inequalities in the first two cases and a strict one in the last case (and then increase p_i a bit):

$$\log(1/p_0) \leq \log m - c; \quad \log(1/p_j) + \log a_j \leq \log m + c(j-1); \quad p_0 + \sum_{j \geq 2} p_j < 1.$$

Then for a given c we take minimal possible p_i:

$$p_0 = \frac{1}{m2^{-c}}$$

$$p_j = \frac{a_j(2^{-c})^j}{m2^{-c}}$$

and it remains to show that the sum is less than 1 for a suitable choice of c. Let $x = 2^{-c}$, then the inequality can be rewritten as

$$\frac{1}{mx} + \sum_{j \geq 2} \frac{a_j x^j}{mx} < 1, \quad \text{or} \quad \sum_{j \geq 2} a_j x^j < mx - 1,$$

and this is our assumption.

Now we see the role of this mystical x in the condition: it is just a parameter that determines the constant used for the amortised analysis.

Acknowledgement. Author thanks his LIRMM colleagues, in particular Pascal Ochem and Daniel Gonçalves, as well as the participants of Kolmogorov seminar in Moscow.

References

1. Alon, N., Spencer, J.H.: The Probabilistic Method. Wiley, New York (2004)
2. Gonçalves, D., Montassier, M., Pinlou, A.: Entropy compression method applied to graph colorings. https://arxiv.org/pdf/1406.4380.pdf
3. Miller, J.: Two notes on subshifts. Proc. AMS **140**, 1617–1622 (2012)
4. Moser, R.: A constructive proof of the Lovász local lemma. https://arxiv.org/abs/0810.4812
5. Moser, R., Tardos, G.: A constructive proof of the general Lovász local lemma. J. ACM **57**(2), 11.1–11.15 (2010)
6. Ochem, P., Pinlou, A.: Application of entropy compression in pattern avoidance. Electron. J. Comb. **21**(2), paper P2.7 (2014)
7. Rampersad, N.: Further applications of a power series method for pattern avoidance. https://arxiv.org/pdf/0907.4667.pdf
8. Rumyantsev, A., Shen, A.: Probabilistic constructions of computable objects and a computable version of Lovász local lemma. Fundam. Informaticae **132**, 1–14 (2013). https://arxiv.org/abs/1305.1535

9. Rumyantsev, A.Y., Ushakov, M.A.: Forbidden substrings, Kolmogorov complexity and almost periodic sequences. In: Durand, B., Thomas, W. (eds.) STACS 2006. LNCS, vol. 3884, pp. 396–407. Springer, Heidelberg (2006). doi:10.1007/11672142_32

10. Shen, A., Uspensky, V.A., Vereshchagin, N.: Kolmogorov complexity and algorithmic randomness, to be published by the AMS (2013). www.lirmm.fr/~ashen/kolmbook-eng.pdf. Russian version published by MCCME (Moscow)

Delayed-Input Cryptographic Protocols

Ivan Visconti[✉]

DIEM, University of Salerno, Fisciano, Italy
visconti@unisa.it

Abstract. The *delayed-input* witness-indistinguishable proof of knowledge of Lapidot and Shamir (LS) [CRYPTO 1989] is a powerful tool for designing round-efficient cryptographic protocols. Since LS was designed for the language of Hamiltonian graphs, when used as subprotocol it usually requires expensive NP reductions.

We first overview how LS works, how it can be used to obtain round-efficient protocols as shown by Ostrovsky and Visconti [ECCC 2012] and why it suffers of intrinsic efficiency limitations.

Then we will overview some recent advances on delayed-input cryptographic protocols and their applications. We will in particular consider the efficient witness-indistinguishable proofs of knowledge of Ciampi, Persiano, Scafuro, Siniscalchi and Visconti [TCC 2016a, Eurocrypt 2016], and the round-efficient non-malleable commitments of Ciampi, Ostrovsky, Siniscalchi and Visconti [Crypto 2016, Eprint 2016].

1 Introduction

Cryptographic protocols are designed to securely implement useful functionalities with the goal of preserving data privacy against adversarial behaviors.

The traditional setting assumes that inputs that are relevant for the computation are already known to players when a cryptographic protocol starts. As a consequence, external protocols require more communication rounds since the task of obtaining data and the task of using the same data in a subprotocol are not parallelizable.

Delayed-input cryptographic protocols. In contrast to the traditional paradigm, Lapidot and Shamir (LS) [14] proposed a 3-round witness-indistinguishable proof of knowledge (WIPoK) where the input is not needed until the last round is played. Their approach works for the language of Hamiltonian Graphs and therefore can be used for any NP language at the cost of expensive NP reductions. The security of LS is maintained even in case the adversarial verifier (resp., prover) chooses the inputs adaptively in the first (resp., second) round.

2 Applications of LS

The delayed-input feature of LS has been used multiple times in literature. Notable examples are [10,13] that were among the first papers making crucial

© Springer International Publishing AG 2017
J. Kari et al. (Eds.): CiE 2017, LNCS 10307, pp. 112–115, 2017.
DOI: 10.1007/978-3-319-58741-7_12

use of the delayed-input property of LS to obtain some round-optimal construc-tions. Whenever round efficiency was desired, LS has been critically used several times in subsequent work (e.g., [2–4,9,11,12,15,17,18]).

A detailed description of LS is available in [16] where the use of LS allowed to reduce to 6 the number of rounds of the public-coin non-black-box zero-knowledge argument system of Barak [1], then obtaining a 7-round resettable witness-indistinguishable argument system.

Round-efficient non-malleable commitments. Recently, Ciampi et al. in [4] showed how to use LS to obtain a 3-round concurrent non-malleable commit-ment scheme. The key idea in their construction consists of executing a regular (i.e., potentially malleable) non-interactive commitment c and a 3-round (non-concurrent) non-malleable commitment c'. Moreover the receiver sends a puzzle in the second round. Then, LS is used to prove knowledge of either the message committed in c or of a solution of the puzzle that is committed in c'. Notice that since all the above protocols are played in parallel, c and c' are well defined only during the 3rd round. As such, the delayed-input property of LS is cru-cial in order to give the above proof of knowledge without penalizing the round complexity.

While the above result relies on the use of superpolynomial hardness assump-tions, more recently, Ciampi et al. have shown in [3] how to rely on standard one-way functions only, at the price of adding one more round. LS is used again to guarantee knowledge of inputs used in some other subprotocols that are played in parallel. Therefore the delayed-input property is again essential.

3   Limits of LS and Recent Improvements

A major limit in the use of LS is that an NP reduction to the language of Hamiltonian graphs is required in order to run LS. As such, the use of LS is limited to feasibility results and has limited impact on practical constructions.

The limits of LS have been recently bypassed by new efficient constructions of 3-round WIPoKs for languages useful in practice. The first construction was given by Ciampi et al. in [5] and guarantees unconditionally some partial delayed-input property. Instead, a more recent construction of Ciampi et al. [6] allows to obtain 3-round delayed-input WIPoKs for languages useful in practice, at the cost of relying on the standard Decisional Diffie-Hellman (DDH) assumption. Both constructions make use of some special properties of Σ-protocols [8] and propose some interesting new techniques that extend the applicability of the well known OR-composition of Cramer et al. [7].

The unconditional construction of [5]. Ciampi et al. in [5] showed how to compose two Σ-protocols for two inputs x_0 and x_1 so that the resulting Σ-protocol can be run without requiring knowledge of both x_0 and x_1 in advance (knowledge of one of them suffices to start the protocol). The key idea of their construction consists in the following two main observations: (1) the commonly used Σ-protocols do not need the instance when the protocol starts; (2) one can get a instance-dependent trapdoor commitment from a Σ-protocol. By carefully combining these two observations they show (unconditionally) a Σ-protocol that is an OR-composition of two Σ-protocols such that only one out of x_0 and x_1 is required when the protocol starts. This property is not enjoyed by the well known OR-composition techniques of Cramer et al. [7].

The DDH-based construction of [6]. Ciampi et al. in [6] showed delayed-input efficient WIPoKs by combining Σ-protocols with a specific instance-dependent trapdoor commitment for the language of Diffie-Hellman tuples. The proposed construction is extremely powerful since no input is required when the protocol starts, and the prover can prove knowledge of witnesses for at least k instances out of n instances. This result therefore matches in large part the composition power of [7] giving however the delayed-input property, at the price of relying on the standard DDH assumption.

Acknowledgments. I thank my coauthors Michele Ciampi, Rafail Ostrovsky, Giuseppe Persiano, Alessandra Scafuro and Luisa Siniscalchi for the good time that we have spent together working on delayed-input cryptographic protocols. I also thank Helger Lipmaa for valuable comments on a preliminary version of this paper. This work has been supported in part by "GNCS - INdAM", in part by University of Salerno through grants FARB-2014/2015 and in part by the EU COST Action IC1306.

References

1. Barak, B.: How to go beyond the black-box simulation barrier. In: 42nd Annual Symposium on Foundations of Computer Science, FOCS 2001, 14–17 October, Las Vegas, Nevada, USA, pp. 106–115. IEEE Computer Society (2001)
2. Chung, K.-M., Ostrovsky, R., Pass, R., Venkitasubramaniam, M., Visconti, I.: 4-round resettably-sound zero knowledge. In: Lindell, Y. (ed.) TCC 2014. LNCS, vol. 8349, pp. 192–216. Springer, Heidelberg (2014). doi:10.1007/978-3-642-54242-8_9
3. Ciampi, M., Ostrovsky, R., Siniscalchi, L., Visconti, I.: 4-round concurrent non-malleable commitments from one-way functions. IACR Cryptology ePrint Archive 2016, 621 (2016). http://eprint.iacr.org/2016/621
4. Ciampi, M., Ostrovsky, R., Siniscalchi, L., Visconti, I.: Concurrent non-malleable commitments (and more) in 3 rounds. In: Robshaw, M., Katz, J. (eds.) CRYPTO 2016. LNCS, vol. 9816, pp. 270–299. Springer, Heidelberg (2016). doi:10.1007/978-3-662-53015-3_10
5. Ciampi, M., Persiano, G., Scafuro, A., Siniscalchi, L., Visconti, I.: Improved OR-composition of sigma-protocols. In: Kushilevitz, E., Malkin, T. (eds.) TCC 2016. LNCS, vol. 9563, pp. 112–141. Springer, Heidelberg (2016). doi:10.1007/978-3-662-49099-0_5
6. Ciampi, M., Persiano, G., Scafuro, A., Siniscalchi, L., Visconti, I.: Online/offline OR composition of sigma protocols. In: Fischlin, M., Coron, J.-S. (eds.) EUROCRYPT 2016. LNCS, vol. 9666, pp. 63–92. Springer, Heidelberg (2016). doi:10.1007/978-3-662-49896-5_3

7. Cramer, R., Damgård, I., Schoenmakers, B.: Proofs of partial knowledge and simplified design of witness hiding protocols. In: Desmedt, Y.G. (ed.) CRYPTO 1994. LNCS, vol. 839, pp. 174–187. Springer, Heidelberg (1994). doi:10.1007/3-540-48658-5_19

8. Damgård, I.: On Σ-protocols (2010). http://www.cs.au.dk/~ivan/Sigma.pdf

9. Crescenzo, G., Persiano, G., Visconti, I.: Constant-round resettable zero knowledge with concurrent soundness in the bare public-key model. In: Franklin, M. (ed.) CRYPTO 2004. LNCS, vol. 3152, pp. 237–253. Springer, Heidelberg (2004). doi:10.1007/978-3-540-28628-8_15

10. Crescenzo, G., Persiano, G., Visconti, I.: Improved setup assumptions for 3-round resettable zero knowledge. In: Lee, P.J. (ed.) ASIACRYPT 2004. LNCS, vol. 3329, pp. 530–544. Springer, Heidelberg (2004). doi:10.1007/978-3-540-30539-2_37

11. Goyal, V., Richelson, S., Rosen, A., Vald, M.: An algebraic approach to non-malleability. In: 55th IEEE Annual Symposium on Foundations of Computer Science, FOCS 2014, Philadelphia, PA, USA, 18–21 October, pp. 41–50 (2014)

12. Hazay, C., Venkitasubramaniam, M.: On the power of secure two-party computation. In: Robshaw, M., Katz, J. (eds.) CRYPTO 2016. LNCS, vol. 9815, pp. 397–429. Springer, Heidelberg (2016). doi:10.1007/978-3-662-53008-5_14

13. Katz, J., Ostrovsky, R.: Round-optimal secure two-party computation. In: Franklin, M. (ed.) CRYPTO 2004. LNCS, vol. 3152, pp. 335–354. Springer, Heidelberg (2004). doi:10.1007/978-3-540-28628-8_21

14. Lapidot, D., Shamir, A.: Publicly verifiable non-interactive zero-knowledge proofs. In: Menezes, A.J., Vanstone, S.A. (eds.) CRYPTO 1990. LNCS, vol. 537, pp. 353–365. Springer, Heidelberg (1991). doi:10.1007/3-540-38424-3_26

15. Mittelbach, A., Venturi, D.: Fiat–shamir for highly sound protocols is instantiable. In: Zikas, V., Prisco, R. (eds.) SCN 2016. LNCS, vol. 9841, pp. 198–215. Springer, Cham (2016). doi:10.1007/978-3-319-44618-9_11

16. Ostrovsky, R., Visconti, I.: Simultaneous resettability from collision resistance. Electronic Colloquium on Computational Complexity (ECCC) 19, 164 (2012)

17. Wee, H.: Black-box, round-efficient secure computation via non-malleability amplification. In: 51th Annual IEEE Symposium on Foundations of Computer Science, FOCS 2010, 23–26 October, Las Vegas, Nevada, USA, pp. 531–540. IEEE Computer Society (2010)

18. Yung, M., Zhao, Y.: Generic and practical resettable zero-knowledge in the bare public-key model. In: Naor, M. (ed.) EUROCRYPT 2007. LNCS, vol. 4515, pp. 129–147. Springer, Heidelberg (2007). doi:10.1007/978-3-540-72540-4_8

Contributed Papers

A Deterministic Algorithm for Testing the Equivalence of Read-Once Branching Programs with Small Discrepancy

Stefan Arnold and Jacobo Torán[✉]

Institute of Theoretical Computer Science, University of Ulm, Ulm, Germany
{stefan.arnold,jacobo.toran}@uni-ulm.de

Abstract. The problem to test the equivalence of two given read-once branching programs is a well-known problem in the class BPP that is not known to be solvable in deterministic polynomial time. The standard probabilistic algorithm to solve the problem reduces it to an instance of Polynomial Identity Testing and then applies the Schwartz-Zippel Lemma to test the equivalence. This method needs $O(n \log n)$ random bits, where n is the number of variables in the branching programs. We provide a new method for testing the equivalence of read-once branching programs that uses $O(\log n + \log |D|)$ random bits, where D is the set of assignments for which the two branching programs compute different results. This means $O(n)$ random bits in the worst case and a deterministic polynomial time algorithm when the discrepancy set D is at most polynomial.

We also show that the equivalence test can be extended to the more powerful model of deterministic, decomposable negation normal forms (d-DNNFs).

1 Introduction

Branching programs (also known as binary decision diagrams, BDDs) are a well-known model of computation used for the representation of Boolean functions. A branching program consists of a directed acyclic graph with one start vertex and two output vertices labeled by 0 and 1. All the inner vertices are labeled by variables x_i and have two outgoing edges, one for the case $x_i = 0$ and the other one for $x_i = 1$. A branching program computes a Boolean function in a natural way by starting at the start vertex, querying the variables that label the vertices and following the edges according to the values of the queried variables until the output-0 or the output-1 vertex is reached. Branching programs have many applications in areas like computer-aided design, model checking, or formal verification. This is due to the fact that their expressive power lies between those of formulas and Boolean circuits, which allows on one hand compact representations for many Boolean functions and on the other hand effective algorithms for several logical operations (see e.g. [19]). Certain fundamental problems on this model, like testing the equivalence of two programs, are however intractable.

© Springer International Publishing AG 2017
J. Kari et al. (Eds.): CiE 2017, LNCS 10307, pp. 119–128, 2017.
DOI: 10.1007/978-3-319-58741-7_13

Because of this fact more restricted models like read-once branching programs (ROBPs; also called free BDDs) [13] and ordered binary decision diagrams (OBDDs) [3] have been introduced. A ROBP is a branching program with the additional property that in every path from the start vertex to an output vertex, each variable is queried at most once. An OBDD is a restricted type of ROBP in which for some permutation of the variables, in every path from start to output, the variables are queried in a relative order consistent with the permutation. There are explicit examples of Boolean functions for which the size of the ROBP model is exponentially larger than that of a general branching program, or the size of an OBDD is exponentially larger than that of a ROBP (see [19]). On the other hand, the equivalence of two OBDDs [3] (and even the equivalence of an OBDD and a ROBP [10]) can be tested deterministically in polynomial time, while there is a randomized polynomial time algorithm for testing whether two ROBPs are equivalent [2]. In fact, equivalence testing for read-once branching programs is one of the few natural problems solvable in BPP (bounded error probabilistic polynomial time) but not known to be in P. As such a rare example it is often mentioned in complexity theory courses and appears in introductory books in the area [16,18]. Wegener conjectured that the problem is not in P ([19] pg. 144).

The probabilistic algorithm for the equivalence of ROBPs from Blum, Chandra, and Wegman [2] arithmetizes the branching programs, transforming them into multilinear polynomials that coincide on all 0–1 inputs if and only if the programs are equivalent. Since the polynomials are multilinear, evaluated over any field (not only over $\mathbb{F}_2^n$) they coincide for any possible input exactly when the branching programs compute the same function. To test equivalence one just has to evaluate both polynomials over randomly chosen points from a large enough field, as shown by the Schwartz-Zippel Lemma [17,20]. In order to guarantee a constant error probability the algorithm needs $O(n \log n)$ random bits, where n is the number of variables of the branching programs.

In this paper we introduce a different probabilistic algorithm for testing the equivalence of ROBPs that uses less random bits than the algorithm in [2]. Our algorithm can be seen as a partial derandomization. Let f and $g \colon \{0,1\}^n \to \{0,1\}$ be the functions computed by the two given branching programs, and let D be the discrepancy of the programs: $D = \{x \in \{0,1\}^n \mid f(x) \neq g(x)\}$. Observe that $|D|$ can take any value between 0 and 2^n. Our algorithm uses $O(\log n + \log |D|)$ random bits (if $|D|$ is given). In the worst case, the number of random bits needed is $O(n)$. This also implies a deterministic polynomial equivalence test when both branching programs compute functions that cannot be too far from each other (at most polynomial size discrepancy). Similar derandomization results for situations in which the search space is small have been obtained for other problems. For example, for the perfect matching problem, which can be computed in randomized NC but is not known to be in deterministic NC, there are deterministic NC algorithms that solve the problem when the number of perfect matchings is bounded by a polynomial [1,11].

Our algorithm has two basic ingredients. First, it uses the fact that in the ROBP model it is possible to compute efficiently the number of assignments that lead to a certain vertex in the program. The second component needed to reduce the number of probabilistic choices is the randomness-efficient Isolating Lemma from [4]. The algorithm randomly assigns weights in a restricted range to the Boolean variables of a read-once branching program. It transforms this program into a different one which accepts only the assignments that are accepted by the first program and have a certain weight. If two input programs are equivalent, then, for any weight function and for every possible weight, they must have the same number of accepting assignments with such a weight. Using the Isolating Lemma we show that with high probability there will be a unique element of minimum weight in the discrepancy set D. This implies that the number of accepting assignments having this weight in both programs has to be different if $D \neq \emptyset$ (the number of assignments with this weight that are not in D is the same for both programs). This fact is then checked using the counting properties of the ROBP model.

It has been observed in the literature that the equivalence test for ROBPs from [2] can be extended to the more powerful computation model of deterministic decomposable negation normal forms (d-DNNFs) [9]. We show in Sect. 4 that this is also true for our algorithm, by proving that the same bounds for the number of random bits can be achieved for testing the equivalence of d-DNNFs. For this we show that as in the ROBP model, it is possible to count in polynomial time the number of satisfying assignments having a certain weight for a given d-DNNF.

2 Preliminaries

We denote by $B = \{0, 1\}$ the set of Boolean values. For an arbitrary set S we write $\mathcal{P}(S)$ for the power set of S. By $\delta_{x,y}$ we denote the Kronecker delta for arbitrary numbers x and y, that is, we set $\delta_{x,y} = 1$ if $x = y$ and $\delta_{x,y} = 0$ otherwise.

Definition 1 *(Branching program [14, 19]). A branching program (BP) on the variable set $X_n = \{x_1, \ldots, x_n\}$ consists of a directed acyclic graph whose inner nodes have outdegree 2, and a labeling of the nodes and edges. This graph has a distinguished node, called the source, which has indegree 0, and two sinks with outdegree 0, each one labeled with an element from B. The inner nodes get labels from X_n. For each inner node, one of the outgoing edges gets the label 0 and the other one gets the label 1.*

A branching program computes a function $f : B^n \to B$ by following for each input $a = a_1 a_2 \ldots a_n \in B^n$ a path from the source to a sink, taking at every node labeled by x_i the outgoing edge labeled by a_i. The function value $f(a)$ is the label of the sink that is finally reached.

Input strings $a \in B^n$ will also be called assignments throughout this paper. For a branching program A and an assignment a, $A(a)$ denotes the result of applying A on a. $|A|$ is the number of nodes in program A.

In the next section, we will consider an easy extension of the branching program model that can be used for computing functions $f\colon B^n \to R$ for any finite range R, by allowing a sink node for every element in R.

Read-once BPs are a restricted form of BPs introduced by Masek [13].

Definition 2 *(Read-once BP). A read-once BP (ROBP) is a BP where, for each i, each path from the input vertex to any of the output vertices contains at most one node labeled by x_i.*

In a ROBP it is impossible to construct an inconsistent path from source to sink, i.e., a path according to which a variable would get values 0 and 1 simultaneously. This fact makes it possible to test efficiently some properties in the ROBP model. For example, the question of whether there is some assignment for which the function computed by the program evaluates to 1 (satisfiability test) is just a reachability question between the source and the 1-labeled sink in the graph of the ROBP. Also it is possible to count in polynomial time the number of assignments for which the program evaluates to 1. This is a well-known fact. For self-containedness and because this result plays an important role in our algorithm, we sketch a proof of it.

Lemma 3. *There is a polynomial time algorithm that on input a ROBP for a function $f\colon B^n \to B$ computes the number $|\{a \in B^n \mid f(a) = 1\}|$ of assignments for which the function evaluates to 1.*

Proof. Observe that it is not sufficient to count the number of paths leading to the output-1 vertex since a path does not have to include all the variables and can therefore correspond to more than one assignment.

We describe an algorithm that computes for each vertex v of the ROBP the number $n(v)$ of assignments that lead to it. Note that for any inner node v of the ROBP, exactly half of the assignments leading from the source to v will follow the outgoing edge of v labeled by 0, and the remaining half will follow the outgoing edge labeled by 1. The algorithm first finds a topological ordering $v_1, \ldots, v_n$ of the vertices in the graph. Clearly for the source v_1, $n(v_1) = 2^n$. The algorithms visits the remaining nodes according to the topological ordering, such that a node v_i is visited only after each of its parents has been visited. Suppose that v_i has k incoming edges and these edges come from nodes $v_{i_1}, \ldots, v_{i_k}$. Then $n(v_i)$ is the sum of numbers assigned to $v_{i_1}, \ldots, v_{i_k}$, divided by two: $n(v_i) = (n(v_{i_1}) + \cdots + n(v_{i_k}))/2$. This is because for each predecessor v_{i_j} of v_i, only half of the assignments reaching v_{i_j} follow each edge leaving v_{i_j}. (If a predecessor has two edges leading to v_i, then it is automatically counted twice in the sum). The algorithm returns $n(s_1)$ for the sink vertex s_1 labeled by 1.

It follows by induction that the algorithm correctly computes the number of assignments leading to any node in the ROBP and consequently returns the correct result. A topological ordering of the nodes can be found in time linear in the number of nodes and edges. Note that the total number of operations to compute all the numbers $n(v)$ is also linear in the number of nodes and edges. □

3   An Algorithm for Testing the Equivalence of ROBPs

The algorithm is based on the following well-known Isolation Lemma from [4], which uses less randomness than the original one from [15]. Let $S = \{x_1, \ldots, x_n\}$ be a finite set. A weight function w for S is a function that assigns nonnegative integer weights in a certain range to the elements of S. Such a function can be extended to any subset $I \subseteq S$ by defining $w(I) = \sum_{x_i \in I} w(x_i)$. Let $\mathcal{F} \subseteq \mathcal{P}(S)$ be a family of subsets of S. We call the pair $(S, \mathcal{F})$ a set system. The lemma states that for any set system $(S, \mathcal{F})$ there is a polynomial-time method for choosing the weight function w randomness-efficiently such that with high probability there is a unique subset of S in $\mathcal{F}$ with minimum weight under w.

Lemma 4. *(Generalized Isolating Lemma [4]) There is a randomized scheme that for a set system $(S, \mathcal{F})$, with $|S| = n$ and $|\mathcal{F}|$ bounded from above by t, computes a weight function $w \colon S \to [0, n^7]$ such that with probability at least $1/4$ there is a unique minimum weight set under w in $\mathcal{F}$. The input for the scheme consists of the two numbers n and t. The scheme needs $O(\log n + \log t)$ random bits and runs in time polynomial in n and $\log t$.*

We next show that the following problem can efficiently be solved: Given a ROBP computing a function f and a weight function assigning values within a polynomial range to the variables of the program, count the number of assignments a that have a certain weight and fulfill $f(a) = 1$. For this we consider a variation on the ROBP model in which in addition to the 0 and 1 sink vertices, more output vertices (taking care of the different possible weights) are allowed. For simplicity we will refer to this model also as a ROBP.

Lemma 5. *There is an algorithm that on input a ROBP A with variables $X_n = \{x_1, \ldots, x_n\}$, computing a function $f \colon B^n \to B$, and a weight function $w \colon X_n \to \{0, \ldots, m\}$ transforms A into a new ROBP A^w with output vertices 0 and 1_k for $k \in \{0, \ldots, mn\}$ such that for any k, the set of assignments reaching vertex 1_k in A^w is exactly the set of assignments x with $f(x) = 1$ and $w(x) = k$. The running time of this algorithm is polynomial in $|A|$ and m.*

Proof. Observe that the set of possible weights under w that an assignment can get is a subset of $\{0, \ldots, mn\}$. The ROBP A^w is constructed from A in the following way: For the start vertex s in A we provide one start vertex s' in A^w. For every vertex $v \neq s$ in A we provide $mn + 1$ vertices $v_0, \ldots, v_{mn}$, one for each possible weight. Later, some of these vertices might not be reachable and can be deleted. For any inner node v in A labeled with a variable x_i and any edge $e = (v, u)$, we include in A^w the set of edges $\{(v_k, u_k) \mid 0 \leq k \leq mn\}$ if e is labeled by 0; if it is labeled by 1, we include $\{(v_k, u_{k+j}) \mid 0 \leq k \leq mn, j = w(x_i)$ and $k + j \leq mn\}$. In this way, the branching program A^w mirrors the behavior of A but additionally the set of paths reaching a vertex v_k in A^w corresponds to partial assignments having weight exactly k under w. Observe that in the definition of the edges in A^w we do not need to include any edge leading to a vertex with weight more than mn because no assignment can have such a weight. Finally, for any $k \in \{0, \ldots, mn\}$ the copy 1_k in A^w of the output-1 node in A corresponds to the set of assignments $a \in B^n$ with $f(a) = 1$ and $w(a) = k$.

The number of vertices of the new branching program A^w is bounded by $mn|A|$. By the same arguments as in Lemma 3, for any weight k it is possible to count efficiently the number of assignments that lead from the start vertex in A^w to the output vertex 1_k. □

Let A_1 and A_2 be two ROBPs computing functions $f, g \colon B^n \to B$ with $f \neq g$ and let D be the set of assignments that distinguish A_1 from A_2, that is, $D = \{x \in B^n \mid f(x) \neq g(x)\}$. Suppose that there is a weight function $w \colon X_n \to \{0, \ldots, m\}$ under which there is a unique minimum weight element $a \in D$, and let $k = w(a)$. Consider the two ROBPs A_1^w and A_2^w transformed according to the previous lemma from A_1 and A_2 with respect to the weight function w. The numbers of assignments leading to the output nodes 1_k in A_1^w and in A_2^w are not the same. This is because a is the only assignment in D that reaches either the vertex 1_k in A_1^w or the vertex 1_k A_2^w. All the other assignments in D have weight greater than k and therefore cannot reach vertex 1_k. Moreover, the set of assignments in $B^n \setminus D$ that reach 1_k in A_1^w coincides with those that reach 1_k in A_2^w because A_1 and A_2 agree on $B^n \setminus D$. Therefore by counting the number of assignmets reaching the vertices 1_k for all $k \leq mn$ in the two branching programs A_1^w and A_2^w, an algorithm detects that A_1 and A_2 are not equivalent. Observe that for a fixed weight function $w \colon X_n \to \{0, \ldots, m\}$, the described algorithm is completely deterministic and has a running time polynomially bounded in $|A_1|, |A_2|$, and m. Putting these thoughts together with Lemma 4, we get our main result in this section.

Theorem 6. *There is a randomized algorithm that, given as input two read-once branching programs A_1 and A_2 over a set of variables X_n and an upper bound $t \geq |\{x \in B^n \mid A_1(x) \neq A_2(x)\}|$, decides with constant error probability whether A_1 and A_2 are equivalent. The algorithm uses $O(\log n + \log t)$ random bits. Its running time is polynomial in $|A_1|, |A_2|$, and $\log t$.*

By simulating the random bits deterministically we obtain:

Corollary 7. *There is a deterministic algorithm that, given as input two read-once branching programs A_1 and A_2 over a set of variables X_n and an upper bound $t \geq |\{x \in B^n \mid A_1(x) \neq A_2(x)\}|$, decides whether A_1 and A_2 are equivalent. The running time of the algorithm is polynomial in $|A_1|, |A_2|$, and t.*

When t can be polynomially bounded in n, this means a deterministic polynomial time algorithm.

4   An Extension to a Stronger Model

Our randomized equality test for ROBDs can also be extended to the more powerful model of so-called deterministic decomposable negation normal form (d-DNNF) sentences. The decomposable normal form (DNNF) was introduced by Darwiche [5,6]. Several restricted versions of DNNF have since been defined and studied (see, e.g., [8] for an overview).

Definition 8 *(Negation normal form [8]). A sentence in negation normal form (NNF) on the variable set $X_n = \{x_1, \ldots, x_n\}$ is a rooted, directed acyclic graph where each leaf node is labeled with 0, 1, or by some literal (x_i or $\neg x_i$) from X_n; and each internal node is labeled with $\wedge$ or $\vee$ and can have arbitrary many children. The* size *of a sentence in NNF is the number of its edges.*

The model defined is just that of unbounded fanin Boolean circuits, with negations only allowed at the inputs. We call nodes labeled by $\wedge$ conjunctions or and-nodes; nodes labeled by $\vee$ will be called disjunctions or or-nodes. A NNF A on X_n calculates a function $f: B^n \to B$ in the natural way: The function computed at a leaf node is just the value of its label, the function computed at an and-node or an or-node is the conjunction or disjunction, respectively, of the functions computed at its child nodes, and the function f computed by A is the function computed at the root node of A.

Two of the restricting properties considered by Darwiche are decomposability [5,6] and determinisim [7].

A NNF A is called *decomposable* if for each conjunction C in A, the conjuncts of C do not share variables. That is, for any two children C_i and C_j of and-node C, the set of variables appearing in the subtree rooted at C_i and the set of variables appearing in the subtree rooted at C_j are disjoint.

A NNF A is called *deterministic* if for each disjunction C in A, each two disjuncts of C are logically contradictory. That is, for any two children C_i and C_j of or-node C, we have $C_i \wedge C_j \vDash$ false.

Decomposable NNFs are called DNNF for short, whereas the abbreviation d-DNNF denotes deterministic decomposable NNFs. It is known that every ROBP can be transformed in polynomial time into a d-DNNF computing the same function (see eg. [8]). When determinism is replaced in d-DNNF by an even stronger property called *decision*, one obtains a model that is actually equivalent to ROBPs. There are d-DNNF sentences that cannot be transformed to additionally satisfy the decision property with only a polynomial increase in size [6]. In this sense, d-DNNF is a strict generalization of ROBPs.

In order to generalize our algorithm for the equivalence of ROBPs in order to test equivalence in the d-DNNF model we need to show that it is possible to count in polynomial time the number of satisfying assignments a of a d-DNNF with a particular weight $w(a) = w_0$. We describe an algorithm fulfilling this.

Lemma 9. *There is a polynomial time algorithm that on input a d-DNNF A over variable set X_n, a weight function $w: X_n \to \{0, \ldots, m\}$, and a desired weight value w_0 outputs the number of assignments a that satisfy A and have value w_0 under w.*

Proof. Let v be a node in A. We denote by Vars(v) the set of variables appearing (negated or not) in the node labels of the subgraph rooted at v. Furthermore, for

$r \in \{0, \ldots, mn\}$ we denote by $C_v[r]$ the number of assignments a to variables in $\mathrm{Vars}(v)$ such that $w(a) = r$ and a is satisfying for the function computed at v.

Our algorithm will visit each node v in A and determine $\mathrm{Vars}(v)$ as well as the table entries $C_v[0], \ldots, C_v[mn]$. Before visiting the nodes in A, the algorithm determines a topological ordering of the nodes. Using this ordering, it will make sure that node v is visited only after each child of v has been visited.

The operations performed while visiting node v depend on the type of v. There are five possibilities:

1. v is labeled by a constant (either 0 or 1). In this case, the algorithm sets $\mathrm{Vars}(v) = \emptyset$ and $C_v[r] = \delta_{r,0}$ for $r \in \{0, \ldots, mn\}$ if v is labeled by 1, otherwise $C_v[r] = 0$ for $r \in \{0, \ldots, mn\}$.
2. v is labeled by a variable x_i. In this case, the algorithm sets $\mathrm{Vars}(v) = \{x_i\}$ and $C_v[r] = \delta_{r,w(x_i)}$ for $r \in \{0, \ldots, mn\}$.
3. v is labeled by a negated variable $\neg x_i$. In this case, the algorithm sets $\mathrm{Vars}(v) = \{x_i\}$ and $C_v[r] = \delta_{r,0}$ for $r \in \{0, \ldots, mn\}$.
4. v is labeled by $\wedge$. Let $v_1, \ldots, v_k$ be the children of v. Note that $\mathrm{Vars}(v_i)$ and the table C_{v_i} have already been computed for $i \in [k]$. The algorithm sets

$$\mathrm{Vars}(v) = \bigcup_{i \in [k]} \mathrm{Vars}(v_i).$$

For the computation of C_v, our algorithm will take into account the children of v only one after the other, similar as if the and-node v with fan-in k had first been replaced by a binary tree of $k - 1$ and-nodes with fan-in 2 each. To this end, the algorithm determines the intermediate tables $C_{v\langle 0 \rangle}, \ldots, C_{v\langle k \rangle}$ where $v\langle i \rangle$ is an imaginary and-node with children $v_1, \ldots, v_i$. Note that v is equivalent to $v\langle k \rangle$. The computation starts with $C_{v\langle 0 \rangle}[r] = \delta_{r,0}$ for $r \in \{0, \ldots, mn\}$ and finally reaches $C_{v\langle k \rangle}[r] = C_v[r]$. Let s be a number from $\{0, \ldots, r\}$. Since d-DNNF A satisfies the decomposability property, the combination of any satisfying assignment of weight s for $v\langle i - 1 \rangle$ with any satisfying assignment of weight $r - s$ for v_i yields a satisfying assignment of weight r for $v\langle i \rangle$. Thus, $C_{v\langle i \rangle}$ follows from $C_{v\langle i-1 \rangle}$ and C_{v_i} by the convolution formula

$$C_{v\langle i \rangle}[r] = \sum_{s=0}^{r} C_{v\langle i-1 \rangle}[s] \, C_{v_i}[r - s].$$

Using these steps, the algorithm obtains C_v in time polynomial in $|A|$ and m.
5. v is labeled by $\vee$. Let $v_1, \ldots, v_k$ be the children of v. The algorithm sets

$$\mathrm{Vars}(v) = \bigcup_{i \in [k]} \mathrm{Vars}(v_i).$$

Note that each combination of a satisfying assignment for the function represented by v_i with an arbitrary assignment to the variables in $\mathrm{Vars}(v) \setminus \mathrm{Vars}(v_i)$ yields a satisfying assignment for v. Our algorithm uses this fact in the computation of C_v. Let $X \subseteq X_n$ be a subset of the variables and denote by $E_X[r]$

the number of all assignments a to variables in X with $w(a) = r$. Suppose for a moment that the table E_X would be available to our algorithm. The number of assignments to the variables in $\text{Vars}(v)$ that satisfy the function computed at child v_i of v and have weight r is

$$\sum_{s=0}^{r} C_{v_i}[s] \, E_{\text{Vars}(v) \setminus \text{Vars}(v_i)}[r - s].$$

Since d-DNNF A satisfies the determinism property, we obtain $C_v[r]$ by summation over all children v_i:

$$C_v[r] = \sum_{i=1}^{k} \sum_{s=0}^{r} C_{v_i}[s] \, E_{\text{Vars}(v) \setminus \text{Vars}(v_i)}[r - s].$$

It remains to show that $E_X[r]$ can also be efficiently calculated for any set X of variables. Again, we use dynamic programming to solve this problem. We clearly have $E_\emptyset[r] = \delta_{r,0}$. Now assume that $E_Y[r]$ was already known for any set Y of variables and let x_i be a variable not in Y. Considering the two possible cases $x_i = 0$ and $x_i = 1$, we can easily compute $E_{Y \cup \{x_i\}}[r]$:

$$E_{Y \cup \{x_i\}}[r] = E_Y[r] + E_Y[r - w(x_i)].$$

(If $r < w(x_i)$, we simply have $E_{Y \cup \{x_i\}}[r] = E_Y[r]$.) Thus, starting with $Y = \emptyset$, we can take the variables into account one after the other, until we obtain the table E_X.

By induction, the algorithm correctly computes C_v for all nodes v of A in polynomial time. It returns $C_{\tilde{v}}[w_0]$ from the table for the root node $\tilde{v}$.   □

The described algorithm can in fact compute for the output vertex v of the d-DNNF A, a weight frequency tuple $C_v[0], \ldots, C_v[mn]$ with the number of satisfying assignments with each possible weight between 0 and mn. If two given d-DNNFs are not equivalent, and w is a weight function having a unique minimum weight element in the discrepancy set of both forms, the above algorithm can check that the weight frequency tuples corresponding to both programs are not identical. As in the case of ROBPs, the randomnes needed for the equivalence test is only needed for choosing the weight function, and we obtain:

Theorem 10. *There is a randomized algorithm that, given as input two deterministic decomposable normal forms A_1 and A_2 over a set of variables X_n and an upper bound $t \geq |\{x \in B^n \mid A_1(x) \neq A_2(x)\}|$, decides with constant error probability whether A_1 and A_2 are equivalent. The algorithm uses $O(\log n + \log t)$ random bits. Its running time is polynomial in $|A_1|$, $|A_2|$, and $\log t$.*

References

1. Agrawal, M., Hoang, T.M., Thierauf, T.: The polynomially bounded perfect matching problem is in NC^2. In: Thomas, W., Weil, P. (eds.) STACS 2007. LNCS, vol. 4393, pp. 489–499. Springer, Heidelberg (2007). doi:10.1007/978-3-540-70918-3_42
2. Blum, M., Chandra, A.K., Wegman, M.N.: Equivalence of free boolean graphs can be decided probabilistically in polynomial time. Inf. Process. Lett. 10, 80–82 (1980)
3. Randal, E.: Bryant: graph based algorithms for Boolean function manipulation. IEEE Trans. Comput. 35, 677–691 (1986)
4. Chari, S., Rohatgi, P., Srinivasan, A.: Randomness-optimal unique element isolation with applications to perfect matching and related problems. SIAM J. Comput. 24, 1036–1050 (1995)
5. Darwiche, A.: Compiling knowledge into decomposable negation normal form. In: Proceedings of the 16th International Joint Conference on Artifical Intelligence (IJCAI 1999), pp. 284–289 (1999)
6. Darwiche, A.: Decomposable negation normal form. J. ACM 48, 608–647 (2001)
7. Darwiche, A.: On the tractable counting of theory models and its application to truth maintenance and belief revision. J. Appl. Non-Class. Logics 11, 11–34 (2001)
8. Darwiche, A., Marquis, P.: A knowledge compilation map. J. Artifi. Intell. Res. 17, 229–264 (2002)
9. Darwiche, A., Huang, J.: Testing equivalence probabilistically. Technical report D-123 Computer Science Department, UCLA (2002)
10. Fortune, S., Hopcroft, J., Schmidt, E.M.: The complexity of equivalence and containment for free single variable program schemes. In: Ausiello, G., Böhm, C. (eds.) ICALP 1978. LNCS, vol. 62, pp. 227–240. Springer, Heidelberg (1978). doi:10.1007/3-540-08860-1_17
11. Grigoriev, D., Karpinski, M.: The matching problem for bipartite graphs with polynomially bounded permanents is in NC. In: 28th Annual Symposium on Foundations of Computer Science (FOCS), pp. 166–172 (1987)
12. Lee, C.Y.: Representation of switching circuits by binary-decision programs. Bell Syst. Tech. J. 38, 985–999 (1959)
13. Masek, W.J.: A fast algorithm for the string editing problem and decision graph complexity. M.Sc. thesis, MIT, Cambridge MA (1976)
14. Meinel, C.: Modified Branching Programs and Their Computational Power. LNCS, vol. 370. Springer, Heidelberg (1989)
15. Mulmuley, K., Vazirani, U.V., Vazirani, V.V.: Matching is as easy as matrix inversion. Combinatorica 7, 105–113 (1987)
16. Schöning, U., Pruim, R.: Gems of Theoretical Computer Science. Springer, Heidelberg (1998)
17. Jacob, T.: Schwartz: fast probabilistic algorithms for verification of polynomial identities. J. ACM 27, 701–717 (1980)
18. Sipser, M.: Introduction to the Theory of Computation. PWS Publishing Company, Boston (1997)
19. Wegener, I.: Branching Programs and Binary Decision Diagrams. SIAM, Philadelphia (2000)
20. Zippel, R.: Probabilistic algorithms for sparse polynomials. In: Ng, E.W. (ed.) Symbolic and Algebraic Computation. LNCS, vol. 72, pp. 216–226. Springer, Heidelberg (1979). doi:10.1007/3-540-09519-5_73

Counting Substrate Cycles in Topologically Restricted Metabolic Networks

Robert D. Barish$^{(\boxtimes)}$ and Akira Suyama

Graduate School of Arts and Sciences, University of Tokyo,
Meguro-ku Komaba 3-8-1, Tokyo 153-8902, Japan
rbarish@genta.c.u-tokyo.ac.jp

Abstract. Substrate cycles in metabolic networks play a role in various forms of homeostatic regulation, ranging from thermogenesis to the buffering and redistribution of steady-state populations of metabolites. While the general problem of enumerating these cycles is $\#P$-hard, it is unclear if this result holds for realistic networks where e.g. pathological vertex degree distributions or minors may not exist. We attempt to address this gap by showing that the problem of counting directed substrate cycles ($\#DirectedCycle$) remains $\#P$-complete (implying $\#P$-hardness for enumeration) for any superclass of cubic weakly-3-connected bipartite planar digraphs, and at the limit where all reactions are reversible, that the problem of counting undirected substrate cycles ($\#UndirectedCycle$) is $\#P$-complete for any superclass of cubic 3-connected bipartite planar graphs where the problem of counting Hamiltonian cycles is $\#P$-complete. Lastly, we show that unless $NP = RP$, no $FPRAS$ can exist for either counting problem whenever the Hamiltonian cycle decision problem is NP-complete.

1 Introduction

In graph theoretic terms, substrate cycles can be thought of as cycles in *reaction-centric graphs* of metabolic networks, where vertices correspond to enzymes and directed (resp. undirected) edges correspond to irreversible (resp. reversible) flows of metabolite species between enzymes. Cycles in reaction-centric graphs are also known as cyclical *Elementary Fundamental Modes* (EFMs) [1], and may correspond to special cases of these objects denoted *extreme pathways* [2], where the former are minimal sets of reactions or enzymes that can maintain a particular steady-state reaction and the latter are the extreme rays of a *flux cone* for a given biochemical or metabolic network at steady state.

Substrate cycles have been linked to thermogenesis in the flight muscles of bumble bees [3] as well as in the brown adipose tissue of mammals [4,5], and have also been shown to have an important role in buffering steady-state populations [6], and in sensitizing regulatory mechanisms related to metabolism

---

The original version of this chapter was revised: Incorrect capitalization has been corrected. The erratum to this chapter is available at 10.1007/978-3-319-58741-7_37

© Springer International Publishing AG 2017
J. Kari et al. (Eds.): CiE 2017, LNCS 10307, pp. 129–140, 2017.
DOI: 10.1007/978-3-319-58741-7_14

[5,7]. However, despite their apparent importance to biochemists, the only hardness results we are aware of for enumerating substrate cycles are an indirect consequence of proofs regarding the NP-hardness of counting all simple (not necessarily induced) cycles in digraphs (see e.g. ("Theorem 17.4"; pp. 343–344) of Arora and Barak [8]) via reduction from the NP-complete Hamiltonian cycle decision problem [9], where digraphs can correspond to reaction-centric graphs where all reactions are irreversible, and the recent proof due to Yamamoto [10] that the problem of counting simple (not necessarily induced) cycles in arbitrary undirected graphs is $\#P$-complete as well as polynomial time inapproximable unless $NP = RP$, where undirected graphs can correspond to reaction-centric graphs where all reactions are reversible. Moreover, the only complexity theoretic results we are aware of for enumerating EFMs is that counting EFMs in substrate-centric graphs, where metabolites are represented as vertices and reactions as edges, is $\#P$-complete [11] via reduction from the $\#P$-complete problem of counting perfect matchings [12], and that no *polynomial total-time* algorithm (see Johnson et al. [13] for a definition of this term) can exist for enumerating EFMs containing a specified reaction unless $P = NP$ [14].

However, a question arises as to the practical relevance of these hardness results which were proven for general graphs. Consider substrate-centric graphs, where both metabolites are encoded as vertices which are connected by edges corresponding to enzymes. We can note that these graphs have been shown to have e.g. bounded diameters and to exhibit $P(k) \propto k^{-\gamma}$ power law scaling for their vertex degree distributions in a range of archaea, bacteria, and eukaryotic organisms [15]. It could therefore arguably be the case that the aforementioned hardness results are simply the consequence of pathological families or classes of graphs that have little relevance to actual metabolic networks.

Motivated by these concerns, we attempt to sharpen known hardness results for enumerating substrate cycles to apply to some family of hypothetical reaction-centric graphs, F, that is simultaneously more constrained than any realistic large scale metabolic network and "physical" in the sense that one would expect the constraints defining F to be satisfied individually by subgraphs composed of some reasonable fraction of enzymes in any given metabolic pathway. After examining the metabolic pathways in the Kyoto Encyclopedia of Genes and Genomes (KEGG) database [16], we decided that some form of vertex degree uniformity, vertex or edge connectivity, bipartiteness, and planarity constraints would reasonably meet these dual criteria (see e.g. Figure 1). We then proceeded to prove that the problem of counting substrate cycles remains $\#P$-complete ($\implies$ enumeration is $\#P$-hard) on reaction-centric graphs belonging to any superclass of the highly restricted class of cubic weakly-3-connected bipartite planar digraphs (Theorem 1), and on any superclass of cubic 3-connected bipartite planar undirected graphs where the problem of counting Hamiltonian cycles is $\#P$-complete (Theorem 2).

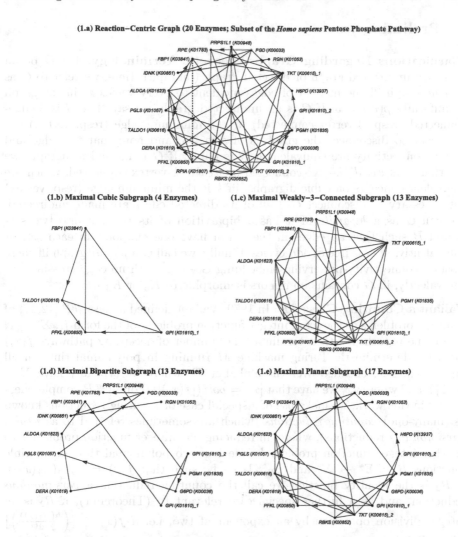

Fig. 1. Reaction-centric graph (1.a), simplified by not considering {ATP, NADP$^+$, NADPH} or other cofactors, for the largest set of *Homo sapiens* enzymes in the Kyoto Encyclopedia of Genes and Genomes (KEGG) database [16] (Release 81.0, January 1st, 2017) assigned to the pentose phosphate pathway, carrying out distinct reactions, and allowing for a weakly-connected reaction-centric graph. Enzyme's names are followed by their KEGG Orthology (KO) numbers, and may have a '_1' or '_2' postfix to indicate reactions involving distinct metabolite inputs. Edges that are directed (undirected) correspond to irreversible (reversible) flows of metabolites. Undirected edges constitute minimal substate cycles, e.g., the (dashed) undirected edge in (1.a) corresponds to a substrate cycle for a pair of enzymes that, if unchecked and provided a pool of ATP for to reduce to ADP, will continuously interconvert fructose-6-phosphate and fructose-1, 6-bisphosphate, releasing heat in the process. Maximal subgraphs of the (1.a) reaction-centric graph (in terms of vertex/enzyme counts) are shown, which are: (1.b) cubic; (1.c) weakly-3-connected; (1.d) bipartite; and (1.e) planar.

2   Preliminaries

Clarifications Regarding Graph Theoretic Terminology. Let G be an arbitrary undirected graph or directed graph (digraph). Here, we refer to G as a *cubic* graph iff the number of edges incident to every vertex v_i in the graph is uniformly $\rho(v_i) = 3$. If G is an undirected graph, we say that G is 3-edge-connected (resp. 3-vertex-connected) iff the minimum edge (resp. vertex) cut necessary to disconnect the graph is of size ≥ 3, and note that the edge and vertex connectivity are equivalent if G is cubic. If G is instead a digraph, we say that G is *weakly-k-edge-connected* or *weakly-k-vertex-connected*, which are equivalent concepts on cubic digraphs, iff k is the minimum edge (resp. vertex) cut necessary to disconnect G if all of its directed edges are made undirected. We call G as a *bipartite* iff it has a bipartition of its vertices into two sets A and B such that all edges in the graph have one endpoint in each set, or equivalently, iff G is free of odd cycles. Finally, we call G a *planar* graph iff there exists a connectivity-preserving embedding $G \rightarrow \mathbb{R}^2$ with no edge crossings, or equivalently, iff G contains no minors isomorphic to K_5 or $K_{3,3}$.

Valiant's Counting Class $\#P$. In 1979, Valiant defined a class $\#P$ [12,17] of counting problems as the set of integer function problems of the form $f : \Sigma^* \rightarrow \mathbb{N}$ where one is tasked with determining the number of accepting pathways $f(x_i)$ for a nondeterministic Turing machine M running in polynomial time on all input strings x_i encoded over the alphabet Σ (where we typically have $\Sigma = \{0,1\}$), and where $\forall x_i$ we have that $|x_i| = poly(|f(x_i)|)$. For all $\#P$-completeness results in this work, we make use of a special case of 1-Turing reductions known as (many-one) *counting reductions* (which are sometimes referred to as *weakly parsimonious* reductions), where, in reducing one integer function problem f to another integer function problem h, one has two polynomial time compatible functions $R_1 : \Sigma^* \rightarrow \Sigma^*$ and $R_2 : \mathbb{N} \rightarrow \mathbb{N}$, such that $f(x) = R_2(h(R_1(x)))$. If R_2 is the identity function we call the counting reduction a *parsimonious* reduction, and as a further special case relevant to (Theorem 1), if R_2 is an integer division operation by an exponent of two, i.e. if $f(x) = \left\lfloor \left(\frac{h(R_1(x))}{2^{R_3(x)}} \right) \right\rfloor$ where $R_3(x) : \Sigma^* \rightarrow \mathbb{N}_{>0}$, then we refer to the counting reduction as a *right-bit-shift* reduction [18].

Randomized approximation schemes and Approximation Preserving reductions (*AP*-reductions). Let $f : \Sigma^* \rightarrow \mathbb{N}$ be a counting problem in the complexity class $\#P$, let $x \in \Sigma^*$ be some appropriate input for f, and let $f(x) = N$. Following Karp and Luby [19], we define a Randomized Approximation Scheme (RAS) as a randomized algorithm which takes f and x as input and outputs some value $\hat{f}_{(\epsilon,\delta)}$ such that $Pr\left[\left(\frac{|\hat{f}_{(\epsilon,\delta)} - f(x)|}{f(x)} \right) > \epsilon \right] < \delta$, where we have some *error rate* parameter $0 < \epsilon < 1$ and some *accuracy parameter* $0 < \delta < 1$. A Fully Polynomial-time Randomized Approximation Scheme (FPRAS) is simply a RAS that has a running time polynomially bounded by $|x|$, ϵ^{-1}, and δ^{-1}. Now let $\{f, h\} : \Sigma^* \rightarrow \mathbb{N}$ be two counting problems in the complexity class $\#P$.

An *Approximation Preserving reduction* (AP-reduction) from f to h (denoted $f \leq_{AP} h$), as originally defined by Dyer et al. [20], is a probabilistic oracle Turing machine M, taking as input a string $x \in \Sigma^*$ and error parameter $0 < \epsilon < 1$, and satisfying the following three conditions: (1) letting x be an instance of h and $0 < \delta < 1$ (where δ^{-1} is polynomially bounded by $|x|$ and ϵ^{-1}), we have that all calls to M specify an input of the form $\{x, \epsilon\}$; (2) we have that M is a RAS for f if it is a RAS for g; (3) the time complexity for M is polynomially bounded by $|x|$ and ϵ^{-1}. Here, if $f \leq_{AP} h$ and $h \leq_{AP} f$, we call f and h *AP-interreducible* and write $f \equiv_{AP} h$.

3   Hardness Results for Counting Cycles on Undirected Graphs and Digraphs

Theorem 1. The problem of counting directed simple (not necessarily induced) cycles on cubic weakly-3-connected bipartite planar digraphs having vertex in-degree and out-degree at most two, $\#DirectedCycle(C3BP::In_2Out_2)$, is $\#P$-complete.

We note that Theorem 1 is equivalent to the statement that counting substrate cycles in reaction-centric graphs corresponding to cubic 3-connected bipartite planar digraphs ($\implies$ all reactions are irreversible) is $\#P$-complete.

Claim 1.1. $\#DirectedCycle(C3BP::In_2Out_2) \in \#P$.

Consider a nondeterministic Turing machine that accepts an input string encoding a cubic 3-connected bipartite planar digraph G and a directed edge-wise (or vertex-wise) path P in G, and accepts this input if and only if P is a simple directed cycle in G (i.e. a directed cycle which visits any given vertex at most once). Here, we define $\#DirectedCycle(C3BP::In_2Out_2)$ as an integer function problem $f : \Sigma^* \to \mathbb{N}$, where one is tasked with determining the number of accepting branches $f(x)$ for such a nondeterministic Turing machine $\implies$ $\#DirectedCycle(C3BP::In_2Out_2) \in \#P$.

Claim 1.2. $\#DirectedCycle(C3BP::In_2Out_2)$ is $\#P$-hard.

We proceed with a strategy of creating some digraph H in polynomial time from an arbitrary simple cubic 3-connected bipartite planar digraph having in-degree and out-degree at most two, G, of order n, where via a gadget substitution scheme we amplify the number of simple length L cycles in the graph by a factor $(2^m)^L$ allowing us to calculate the number of Hamiltonian cycles via a simple integer division operation. However, in lieu of substituting this gadget in place of the edges of G in the manner of the proof for ("Theorem 17.4"; pp. 343–344) in Arora and Barak [8] and in the manner of Yamamoto [10], which precludes H from being 3-connected, we instead replace each of the n vertices in G with the gadget in such a manner as to ensure that H is a cubic weakly-3-connected bipartite planar digraph having in-degree and out-degree at most two iff G is cubic weakly-3-connected bipartite planar digraph having in-degree and out-degree at most two. To do so we define a surgery for cubic graphs which we

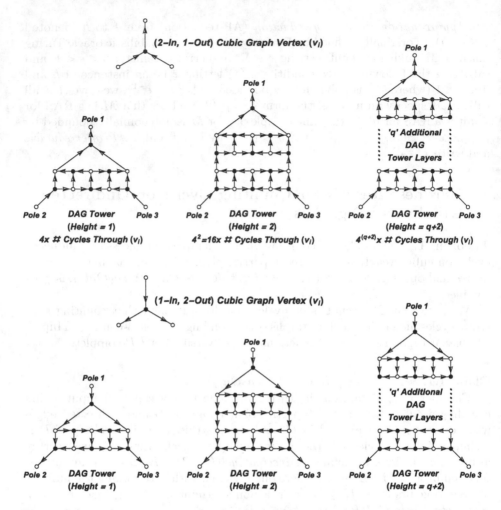

Fig. 2. Scheme for generating a family of cubic 3-connected bipartite planar Directed Acyclic Graph (DAG) *vertex substitution gadgets* where the *height* q of the gadget can be specified as desired. If the appropriate (in terms of matching in-degree and out-degree counts) height q gadget is substituted in place of all vertices in some cubic 3-connected bipartite planar graph G, creating a new cubic 3-connected bipartite planar graph H, this will multiply the number of length L cycles in G by the factor $(4^q)^L$.

denote *tripole substitution* wherein a selected vertex v_i in a cubic graph G is substituted with a target graph H having three degree $\rho(v_i) = 1$ *pole vertices* that are identified (via some bijection if necessary) as the vertices connected to the v_i vertex in G via an edge of identical orientation.

In Fig. 2 we illustrate the scheme for our *height* (using this term to satisfy the visual metaphor of a tower) q gadget used for tripole substitution at all vertices of G to create H. We can now make the following two observations: (1)

that the gadget is a Directed Acyclic Graph (DAG), which implies that there will be no gadget internal cycles; (2) that the substitution of a height q gadget at every vertex in G to create H will amplify the number of simple length L cycles in G by the factor $(4^q)^L$. We set $q = \lceil n * log_4(n) \rceil$, which implies a rate of growth for q of $\mathcal{O}(n * ln(n))$, and which correspondingly implies that we can construct H in time polynomial in the number of vertices in G. Next, we note there can be at most $n^{(n-1)}$ directed cycles of length at most $(n-1)$ in a digraph as this term corresponds to the number of distinct length $(n-1)$ strings that can be created from an alphabet Σ of size $|\Sigma| = n$. This then implies that if the graph G is non-Hamiltonian we have an upperbound of $(4^q)^{(n-1)} * n^{(n-1)} = (4^{\lceil (n*log_4(n)) \rceil})^{(n-1)} * n^{(n-1)}$ directed cycles. However, if the graph G has at least one Hamiltonian cycle, H has at least $(4^q)^n = (4^{\lceil (n*log_4(n)) \rceil})^n$ directed cycles. Here, if we remove the ceiling function in the expression for q, the lowerbound number of cycles for Hamiltonian G is exactly a factor of n larger than the upperbound R for the number of cycles for non-Hamiltonian G, and with the ceiling function, at least a factor of n larger since $\forall \{n \in \mathbb{R}_{>0}, q \in \mathbb{R}_{\geq 0}\}$ we have that $sgn \left[\frac{\partial}{\partial q} \left(\frac{(4^q)^n}{(4^q)^{(n-1)} * n^{(n-1)}} \right) = (4^q * n^{(1-n)} * ln(4)) \right] = (+1)$.

Putting everything together, we have that the number of Hamiltonian cycles in G is equal to $\left\lfloor \left(\frac{(\# \ Directed \ Cycles \ in \ H)}{(4^{\lceil (n*log_4(n)) \rceil})^n} \right) \right\rfloor = \left\lfloor \left(\frac{(\# \ Directed \ Cycles \ in \ H)}{2^{(2n*\lceil (n*log_4(n)) \rceil)}} \right) \right\rfloor$. As this closed-form expression is a polynomial time computable integer division operation consistent with the requirements for a right-bit-shift reduction [18], and as the problem of counting Hamiltonian cycles on cubic 3-connected bipartite planar digraphs is $\#P$-complete under parsimonious reductions (though unremarked upon by Plesník, his construction in [21] is odd-cycle-free, and moreover becomes a parsimonious counting reduction from a variant of $\#3SAT$, and transitively $\#SAT$, if one reorients one edge in his "OR" gadget (Barish and Suyama; in preparation)), we have that $\#DirectedCycle(C3BP :: In_2Out_2)$ is $\#P$-hard.

Claim 1.3. $\#DirectedCycle(C3BP :: In_2Out_2)$ is $\#P$-complete under right-bit-shift reductions.

As we have that the counting problem is in $\#P$ (Claim 1.1), is $\#P$-hard (Claim 1.2), and because the reduction described in the proof argument for (Claim 1.2) is a right-bit-shift reduction [18], by definition we have that the counting problem is $\#P$-complete under right-bit-shift reductions.

Claim 1.4. Unless we have that $NP = RP$, there does not exist a $FPRAS$ for $\#DirectedCycle(C3BP :: In_2Out_2)$.

Let G be an arbitrary cubic 3-connected bipartite planar digraph of order n, let H be the gadget substituted graph constructed as described in (Claim 1.2), and let R represent an upperbound for the fraction of directed cycles in H not corresponding to Hamiltonian cycles. Proceeding now in much the same manner as Yamamoto [10], from the argument in (Claim 1.2) we have that

$R = \left(\frac{(4^{\lceil (n*log_4(n)) \rceil})^{(n-1)} * n^{(n-1)}}{(4^{\lceil (n*log_4(n)) \rceil})^n} \right) \leq \frac{1}{n} \implies \left(\frac{(\# \ Directed \ Cycles \ in \ H)}{(4^{\lceil (n*log_4(n)) \rceil})^n} \right) \leq$

$\left\lfloor \left(\frac{(\# \ Directed \ Cycles \ in \ H)}{(4^{\lceil (n*log_4(n)) \rceil})^n} \right) \right\rfloor + \frac{1}{4}$ for $n \geq 4$. As the Hamiltonian cycle decision problem is NP-complete for cubic 3-connected bipartite planar digraphs (see again

(Claim 1.2) regarding a discussion of this point), by the argument at the end of "Theorem 3" in Dyer et al. [20], we have that $\#DirectedCycle(C3BP::In_2Out_2) \equiv_{AP} \#SAT$. Finally, we note that in Zuckerman [22] it is proven that no $FPRAS$ can exist for approximating $\#SAT$ unless $NP = RP$, and that this holds for all counting problems in $\#P$ that are AP-interreducible in polynomial time. Therefore, there can be no $FPRAS$ for $\#DirectedCycle(C3BP::In_2Out_2)$ unless we have that $NP = RP$.

Theorem 2. The problem of counting undirected simple (not necessarily induced) cycles on any superclass (superset) of cubic 3-connected bipartite planar graphs where the problem of counting Hamiltonian cycles is $\#P$-complete, $\#UndirectedCycle(Superclass\ C3BP::\#HC = \#P)$, is $\#P$-complete.

We note that Theorem 2 is equivalent to the statement that counting substrate cycles in reaction-centric graphs corresponding to any superclass of cubic 3-connected bipartite planar undirected graphs ($\implies$ all reactions are reversible) is $\#P$-complete whenever the problem of counting all Hamiltonian cycles is $\#P$-complete.

Claim 2.1. $\#UndirectedCycle(Superclass\ C3BP::\#HC = \#P) \in \#P$.
    This follows straightforwardly from the method of argument in (Claim 1.1).

Claim 2.2. $\#UndirectedCycle(Superclass\ C3BP::\#HC = \#P)$ is $\#P$-hard.
    Let G be an arbitrary simple undirected cubic 3-connected bipartite planar graph of order n. In Fig. 3 we illustrate the scheme for our *depth q* gadget used for tripole substitution at all vertices of G to create H. We can now make the following three observations: (1) that H is an undirected cubic graph that is 3-connected iff G is 3-connected, bipartite iff G is bipartite, and planar iff G is planar; (2) that $n = |V_G|$ depth q gadgets will imply the existence of some number of undirected cycles internal to the set of gadgets; (3) that the substitution of a height q gadget at every vertex in G to create H will amplify the number of simple length L cycles in G by some *gadget amplification factor* denoted Ψ. Specifically, for cycles traversing a depth q gadget, we have the recurrence relation $a_{(q+1)} = 2 + 5 * a_q$ where $a_{(1)} = 7 \implies a_q = \Psi = \frac{1}{2}(3 * 5^q - 1)$, and for the number of cycles internal to a depth q gadget, we have the recurrence relation $b_{(q+1)} = (1 + 3 * 2 * \Psi + b_q) = (1 + 6 * \frac{1}{2}(3 * 5^q - 1) + b_q)$ where $b_{(1)} = 7 \implies b_q = \frac{1}{4}(9 * 5^q - 8q - 9)$. We set $q = \lceil log_5(\frac{1}{3}(1 + 2 * n^n)) \rceil$, which implies a rate of growth for q of $\mathcal{O}(ln(n^n)) = \mathcal{O}(n * ln(n))$, and which correspondingly implies that we can construct H in time polynomial in the order of G. Noting as in (Claim 1.3) that there can be at most $n^{(n-1)}$ undirected cycles of length at most $(n-1)$ in a graph, we have that if the graph G is non-Hamiltonian, post-amplification there are at most $((\frac{1}{2}(3 * 5^q - 1))^{(n-1)} * n^{(n-1)} + n * (\frac{1}{4}(9 * 5^q - 8q - 9)))$ undirected cycles. However, if the graph G has at least one Hamiltonian cycle, H has at least $(\frac{1}{2}(3 * 5^q - 1))^n + n * (\frac{1}{4}(9 * 5^q - 8q - 9))$ undirected cycles. Here, if we remove the ceiling function in the expression for q, the lowerbound number of cycles for Hamiltonian G is exactly a factor of n larger than the upperbound R for the number of cycles for non-Hamiltonian G, and with the ceiling function,

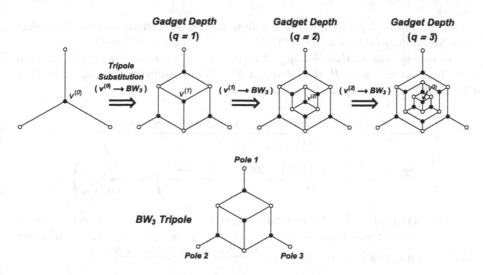

Fig. 3. Let a "BW_3 tripole" be a BW_3 graph modified to have pole vertices {Pole 1, Pole 2, Pole 3} joined by an edge to the degree $\rho(v_i) = 2$ vertices in the original BW_3 graph. Here we show a recursive BW_3 tripole substitution scheme for generating a family of undirected cubic 3-connected bipartite planar *vertex substitution gadgets* where the *depth* q of the gadget can be specified as desired. If a depth q gadget is substituted in place of all vertices in some undirected cubic 3-connected bipartite planar graph G, creating a new cubic 3-connected bipartite planar graph H, this will multiply the number of length L cycles in G by $\Psi^L = (\frac{1}{2}(3 * 5^q - 1))^L$ and generate $(n * (\frac{1}{4}(9 * 5^q - 8q - 9)))$ total gadget internal undirected cycles.

at least a factor of n larger since $\forall\{n \in \mathbb{R}_{>0},\, q \in \mathbb{R}_{\geq 0}\}$ we have that:

$$sgn\left[\frac{\partial}{\partial q}\left(\frac{(\frac{1}{2}(3 * 5^q - 1))^n + n * (\frac{1}{4}(9 * 5^q - 8q - 9))}{(\frac{1}{2}(3 * 5^q - 1))^{(n-1)} * n^{(n-1)} + n * (\frac{1}{4}(9 * 5^q - 8q - 9))}\right)\right]$$

$$= sgn\left[\frac{\partial}{\partial q}\left(\frac{(\frac{1}{2}(3 * 5^q - 1))^n}{(\frac{1}{2}(3 * 5^q - 1))^{(n-1)} * n^{(n-1)}}\right) = \left(\frac{3}{2}(5^q * n^{(1-n)} * ln(5))\right)\right] = (+1)$$

$$(1)$$

Now, setting $q \geq log_5(\frac{1}{3}(1 + 2 * n^n))$, letting $\varphi = n * (\frac{1}{4}(9 * 5^q - 8q - 9))$ be the number of cycles internal to the vertex substitution gadgets, and letting $\Omega = (\frac{1}{2}(3 * 5^q - 1))^n$ be the number of undirected cycles in H per Hamiltonian cycle in G, we can write the relation:

$$(\#\ Hamiltonian\ Cycles\ in\ G) = \left\lfloor \left(\frac{(\#\ Undirected\ Cycles\ in\ H) - \varphi}{\Omega}\right)\right\rfloor$$

$$= \left\lfloor \left(\frac{(\#\ Undirected\ Cycles\ in\ H) - (\frac{3n(n^n - 1)}{2}) + 2n * log_5(\frac{2n^n + 1}{3})}{(n^n)^n}\right)\right\rfloor \quad (2)$$

As this closed-form expression is polynomial time computable, we have that $\#UndirectedCycle(Superclass\ C3BP::\#HC = \#P)$ is $\#P$-hard.

We briefly remark that, letting R represent an upperbound for the fraction of undirected cycles in H that are not Hamiltonian cycles, and again setting $q \geq log_5(\frac{1}{3}(1 + 2 * n^n))$, we have the bound:

$$
R = \left(\frac{(\frac{1}{2}(3 * 5^q - 1))^{(n-1)} * n^{(n-1)} + n * (\frac{1}{4}(9 * 5^q - 8q - 9))}{(\frac{1}{2}(3 * 5^q - 1))^n} \right)
$$
$$
= \left(\frac{(\frac{(n^n)^n}{n}) + (\frac{3n^{(n+1)}}{2}) - (\frac{3n}{2}) - 2n * log_5(\frac{2n^n+1}{3})}{(n^n)^n} \right) \approx \left(\frac{(\frac{(n^n)^n}{n})}{(n^n)^n} \right) = \frac{1}{n}
$$
(3)

Thus, letting $\|x\|$ be the nearest-integer function, for $n > 2$ (which holds without the approximation shown in (Expr. 3)) we have that the number of Hamiltonian cycles in G is equal to $\left\| \left(\frac{(\#\ Undirected\ Cycles\ in\ H)}{(n^n)^n} \right) \right\|$.

Claim 2.3. $\#UndirectedCycle(Superclass\ C3BP::\#HC = \#P)$ is $\#P$-complete under weakly parsimonious reductions.

As we have that the counting problem is in $\#P$ (Claim 2.1) and is $\#P$-hard (Claim 2.2), by definition we have that the counting problem is $\#P$-complete.

Claim 2.4. Unless we have that $NP = RP$, there does not exist a $FPRAS$ for $\#UndirectedCycle(Superclass\ C3BP::\#HC = \#P)$ whenever we have that the Hamiltonian cycle decision problem is NP-complete.

This claim follows straightforwardly from (Expr. 3) and the method of argument in (Claim 1.4).

Corollary 2.1. The problem of counting all undirected cycles on cubic 2-connected planar graphs, $\#UndirectedCycle(C2P)$, is $\#P$-complete.

This corollary follows from the proof argument for (Theorem 2) and the proof by Liskiewicz, Ogihara, and Toda [18] that the problem of counting all Hamiltonian cycles in an input graph is $\#P$-complete under right-bit-shift reductions on cubic 2-connected planar graphs.

Corollary 2.2. The problem of counting all undirected cycles on cubic 3-connected planar graphs ($\#UndirectedCycle(C3P)$), and on cubic 2-connected bipartite planar graphs ($\#UndirectedCycle(C2BP)$), is NP-hard.

This corollary follows from the proof argument for (Theorem 2) and the NP-completeness of the Hamiltonian cycle decision problem on cubic 3-connected planar graphs [23], cubic 3-connected bipartite graphs [24], and cubic 2-connected bipartite planar graphs [24].

4   Concluding Remarks

We note that substrate cycles in reaction-centric graphs corresponding to closed systems (or, more appropriately, approximations thereof) such as individual cells, organelles like the mitochondria, or assuming some appropriate generalization to arbitrary chemical reaction networks, the upper atmospheres of gas giants, are equivalent to EFMs (see for example the proof argument for "Theorem 6" in Acuna et al. [11]). Furthermore, if all reactions are irreversible, there is also a one-to-one between substrate cycles and extreme pathways. Therefore, we have that our hardness results concerning substrate cycles have the same implications for counting and enumerating EFMs and, at the limit where all reactions are irreversible, extreme pathways in metabolic networks with the specified topological restrictions.

References

1. Schuster, S., Hilgetag, C.: On elementary flux modes in biochemical reaction systems at steady state. J. Biol. Syst. **2**, 165–182 (1994)
2. Schilling, C.H., Letscher, D., Palsson, B.O.: Theory for the systemic definition of metabolic pathways and their use in interpreting metabolic function from a pathway-oriented perspective. J. Theor. Biol. **203**, 229–248 (2000)
3. Clark, M.G., Bloxham, D.P., Holland, P.C., Lardy, H.A.: Estimation of the fructose diphosphatase-phos-phofructokinase substrate cycle in the flight muscle of Bombus affinis. Biochem. J. **134**, 589–597 (1973)
4. Kazak, L., et al.: A Creatine-driven substrate cycle enhances energy expenditure and thermogenesis in beige fat. Cell **163**, 643–655 (2015)
5. Newsholme, E.A., Crabtree, B.: Substrate cycles in metabolic regulation and in heat generation. Biochem. Soc. Symp. **41**, 61–110 (1976)
6. Hervagault, J.F., Canu, S.: Bistability and irreversible transitions in a simple substrate cycle. J. Theor. Biol. **127**, 439–449 (1987)
7. Adolfsen, K.J., Brynildsen, M.P.: Futile cycling increases sensitivity toward oxidative stress in Escherichia coli. Metab. Eng. **29**, 26–35 (2015)
8. Arora, S., Barak, B.: Computational Complexity: A Modern Approach. Cambridge University Press, New York (2009)
9. Karp, R.M.: Reducibility among combinatorial problems. In: Miller, R.E., Thatcher, J.W. (eds.) Complexity of Computer Computations, pp. 85–103 (1972)
10. Yamamoto, M.: Approximately counting paths and cycles in a graph. Discrete Appl. Math. **217**, 381–387 (2017)
11. Acuna, V., et al.: Modes and cuts in metabolic networks: complexity and algorithms. BioSystems **95**, 51–60 (2009)
12. Valiant, L.G.: The complexity of computing the permanent. Theor. Comput. Sci. **8**, 189–201 (1979)
13. Johnson, D.S., Yannakakis, M., Papadimitriou, C.H.: On generating all maximal independent sets. Inf. Process. Lett. **27**, 119–123 (1988)
14. Acuna, V., et al.: A note on the complexity of finding and enumerating elementary modes. BioSystems **99**, 210–214 (2010)
15. Jeong, H., Tombor, B., Albert, R., Oltvai, Z.N., Barabasi, A.L.: The large-scale organization of metabolic networks. Nature **407**, 651–654 (2000)
16. Kanehisa, M., Goto, S.: KEGG: Kyoto encyclopedia of genes and genomes. Nucleic Acids Res. **28**, 27–30 (2000)
17. Valiant, L.G.: The complexity of enumeration and reliability problems. SIAM J. Comput. **8**, 410–421 (1979)

18. Liskiewicz, M., Ogihara, M., Toda, S.: The complexity of counting self-avoiding walks in subgraphs of two-dimensional grids and hypercubes. Theor. Comput. Sci. **304**, 129–156 (2003)
19. Karp, R.M., Luby, M.: Monte-Carlo algorithms for enumeration and reliability problems. In: Proceedings of the 24th Annual Symposium on Foundations of Computer Science (FOCS), pp. 56–64 (1983)
20. Dyer, M., Greenhill, C., Goldberg, L.A., Jerrum, M.: On the relative complexity of approximate counting problems. Algorithmica **38**, 471–500 (2004)
21. Plesnik, J.: The NP-completeness of the Hamiltonian cycle problem in planar digraphs with degree bound two. Inf. Process. Lett. **8**, 199–201 (1979)
22. Zuckerman, D.: On unapproximable versions of NP-complete problems. SIAM J. Comput. **25**, 1293–1304 (1996)
23. Garey, M.R., Johnson, D.S., Tarjan, R.E.: The planar Hamiltonian circuit problem is NP-complete. SIAM J. Comput. **5**, 704–714 (1976)
24. Akiyama, T., Nishizeki, T., Saito, N.: NP-completeness of the Hamiltonian cycle problem for bipartite graphs. J. Inf. Process. **3**, 73–76 (1980)

Turing Computable Embeddings, Computable Infinitary Equivalence, and Linear Orders

Nikolay Bazhenov[1,2](✉) (iD)

[1] Sobolev Institute of Mathematics, Novosibirsk, Russia
bazhenov@math.nsc.ru
[2] Novosibirsk State University, Novosibirsk, Russia

Abstract. We study Turing computable embeddings for various classes of linear orders. The concept of a Turing computable embedding (or *tc*-embedding for short) was developed by Calvert, Cummins, Knight, and Miller as an effective counterpart for Borel embeddings. We are focused on *tc*-embeddings for classes equipped with *computable infinitary Σ_α equivalence*, denoted by $\sim_\alpha^c$. In this paper, we isolate a natural subclass of linear orders, denoted by WMB, such that $(WMB, \cong)$ is not universal under *tc*-embeddings, but for any computable ordinal $\alpha \geq 5$, $(WMB, \sim_\alpha^c)$ is universal under *tc*-embeddings. Informally speaking, WMB is not *tc*-universal, but it becomes *tc*-universal if one imposes some natural restrictions on the effective complexity of the syntax. We also give a complete syntactic characterization for classes $(K, \cong)$ that are Turing computably embeddable into some specific classes $(\mathcal{C}, \cong)$ of well-orders. This extends the similar result of Knight, Miller, and Vanden Boom for the class of all finite linear orders $\mathcal{C}_{fin}$.

Keywords: Turing computable embedding · Linear order · Ordinal · Computable infinitary equivalence · Computable structure

1 Introduction

In this paper we study computability-theoretic complexity for linear orders. One approach to comparing different classes of structures is to investigate certain effective properties that can be realized by a structure in the class. For example, the *categoricity spectrum* of a computable structure $\mathcal{S}$ is the collection of all Turing degrees capable of computing isomorphisms among arbitrary computable copies of $\mathcal{S}$. It is well-known that computable ordinals cannot realize every possible categoricity spectrum [1]. On the other hand, any categoricity spectrum can be realized by a computable partial order [2]. Thus, one can say that partial orders are computationally "harder" (with respect to categoricity spectra) than well-orders. Nevertheless, there are several ways to compare computability-theoretic complexity of two classes of structures, see, e.g., [2–5].

Friedman and Stanley [6] introduced the notion of *Borel embedding* to compare complexity of the classification problems for classes of countable structures.

© Springer International Publishing AG 2017
J. Kari et al. (Eds.): CiE 2017, LNCS 10307, pp. 141–151, 2017.
DOI: 10.1007/978-3-319-58741-7_15

Calvert, Cummins, Knight, and Miller [3] developed *Turing computable embeddings* as an effective counterpart of Borel embeddings. The formal definition of a Turing computable embedding (or *tc-embedding* for short) is given in Sect. 2.1. TC-embeddings for familiar classes of structures have been further studied in [7–11]. Note that these works study *tc*-embeddings only for classes equipped with the relation of isomorphism.

VanDenDriessche [12] and Wright [13] began the investigation of Turing computable embeddings for classes with equivalence relations other than isomorphism. Wright [13] considered computable isomorphism and some special relations on trees. VanDenDriessche [12] studied classes of abelian p-groups equipped with *computable infinitary Σ_α equivalence*.

Assume that $\mathcal{A}$ and $\mathcal{B}$ are structures in the same language, and α is a computable ordinal. We say that $\mathcal{A}$ and $\mathcal{B}$ are *computably infinitarily Σ_α equivalent* (Σ_α^c *equivalent*), denoted by $\mathcal{A} \sim_\alpha^c \mathcal{B}$, if $\mathcal{A}$ and $\mathcal{B}$ satisfy the same computable infinitary Σ_α sentences. Recall that the *computable dimension* of a computable structure $\mathcal{S}$ is the number of computable copies of $\mathcal{S}$, up to computable isomorphism.

In [12], it was shown that for any computable $\alpha \geq 2$, the class of abelian p-groups is universal for $\sim_\alpha^c$ under *tc*-embeddings. In contrast, the class of abelian p-groups is not universal for isomorphism under *tc*-embeddings (see, e.g., [9, pp. 852–853]). Moreover, Goncharov [14] proved that for a computable abelian group G, the computable dimension of G is either 1 or ω. Thus, the class of abelian p-groups is not universal in the sense of Hirschfeldt, Shore, Khoussainov, and Slinko (see [2, Definition 1.21]).

The results above motivate the following:

Problem. Find classes of structures K such that:

– K is not universal for $\cong$ under *tc*-embeddings, and
– there is $\beta_0 < \omega_1^{CK}$ such that for any computable $\alpha \geq \beta_0$, K is universal for $\sim_\alpha^c$ under *tc*-embeddings.

We call such classes *quasi-tc-universal*. Informally speaking, a quasi-*tc*-universal class is non-*tc*-universal, but it becomes *tc*-universal if we impose some natural restrictions on the effective complexity of our syntactic descriptions. For the exact definition of a quasi-*tc*-universal class, see Sect. 2.1. The results of [12] show that the class of abelian p-groups is quasi-*tc*-universal.

Here we focus on subclasses of linear orders. Friedman and Stanley [6] proved that the class of all linear orders is universal for isomorphism under *tc*-embeddings. In this paper, we isolate a natural subclass of linear orders, denoted by WMB, which is quasi-*tc*-universal.

Recall that for a linear order $\mathcal{L}$, the *block relation* on $\mathcal{L}$ is the set

$$B(\mathcal{L}) = \big\{(a,b) : a,b \in \mathcal{L} \text{ and there are only finitely many } x$$
$$\text{with } (a \leq_\mathcal{L} x \leq_\mathcal{L} b) \vee (b \leq_\mathcal{L} x \leq_\mathcal{L} a)\big\}.$$

The relation $B(\mathcal{L})$ is a congruence on $\mathcal{L}$. We define the class WMB ("well-orders modulo block relation") as follows:

$$WMB = \{\mathcal{L} : \mathcal{L} \text{ is a linear order, and } \mathcal{L}/B(\mathcal{L}) \text{ is a well-order}\}.$$

The outline of the paper is as follows. Section 2 contains necessary preliminaries. In Sect. 3, we prove that WMB is a quasi-tc-universal class. We leave open whether the class of well-orders is quasi-tc-universal. In Sect. 4, we study tc-embeddings for subclasses of computable ordinals, equipped with isomorphism. In particular, for a computable ordinal α, we obtain a necessary and sufficient condition for a class $(K, \cong)$ to be Turing computably embeddable into $(\mathcal{C}^\alpha, \cong) := (\{\omega^\alpha \cdot (t+1) : t \in \omega\}, \cong)$. This extends the result of Knight, Miller, and Vanden Boom [7, Theorem 4.1], which provides the similar condition for the class of all finite linear orders $\mathcal{C}^0$. Section 5 discusses some open questions and consequences of our results.

2   Preliminaries

We consider only computable languages, and structures with universe contained in ω. We assume that any considered class of structures K is closed under isomorphism, modulo the restriction on the universes. In addition, we assume that all the structures from K have the same language. For a structure $\mathcal{S}$, $D(\mathcal{S})$ denotes the atomic diagram of $\mathcal{S}$. We identify formulas with their Gödel numbers.

We treat linear orders as structures in the language $\{\leq^2\}$. If $\mathcal{L}$ is a linear order and $a \leq_{\mathcal{L}} b$, then $[a, b[$ is the interval on the universe $\{x : a \leq_{\mathcal{L}} x <_{\mathcal{L}} b\}$.

For a language L, *infinitary formulas* of L are formulas of the logic $L_{\omega_1 \omega}$. For a countable ordinal α, infinitary Σ_α and Π_α formulas are defined in a standard way (see, e.g., [15, Chap. 6]). We give a short informal description for the class of *computable infinitary formulas* of L. These formulas allow disjunctions and conjunctions over computably enumerable (c.e.) sets of formulas. Let α be a non-zero computable ordinal.

1. Computable Σ_0 and Π_0 formulas are quantifier-free first-order L-formulas.
2. A computable Σ_α formula (Σ_α^c formula) is a c.e. disjunction $\bigvee_i \exists \bar{u}_i \psi_i(\bar{x}, \bar{u}_i)$, where ψ_i is a computable Π_{β_i} formula for some $\beta_i < \alpha$.
3. A computable Π_α formula (Π_α^c formula) is a c.e. conjunction $\bigwedge_i \forall \bar{u}_i \psi_i(\bar{x}, \bar{u}_i)$, where ψ_i is a computable Σ_{β_i} formula for some $\beta_i < \alpha$.

For the exact definition of computable infinitary formulas and their properties, the reader is referred to Chap. 7 in [15].

For a computable non-zero ordinal α and a Turing degree $\mathbf{d}$, let

$$\mathbf{d}_{(\alpha)} = \begin{cases} \mathbf{d}^{(\alpha-1)}, & \text{if } \alpha < \omega, \\ \mathbf{d}^{(\alpha)}, & \text{if } \alpha \geq \omega. \end{cases}$$

This notation is very convenient: e.g., a set $X \subseteq \omega$ is Σ_α^0 if and only if X is c.e. relative to $\mathbf{0}_{(\alpha)}$.

A total function $G \colon \omega \to \omega$ is $\mathbf{d}$-*limitwise monotonic* if there exists a $\mathbf{d}$-computable function $g(x, s)$ with the following properties:

1. $g(x, s) \leq g(x, s + 1)$ for all x and s, and
2. $G(x) = \lim_s g(x, s)$ for all x.

2.1   Turing Computable Embeddings

Definition 2.1 ([12, Definition 1], **after** [3, Definition 2] **and** [7, Definition 1.3]). *Suppose that K_1 and K_2 are classes of countable structures, and for $i \in \{1, 2\}$, E_i is an equivalence relation on K_i. A Turing computable embedding (tc-embedding) of (K_1, E_1) into (K_2, E_2) is an operator $\Phi = \varphi_e$ with the following properties:*

1. *for every $\mathcal{A} \in K_1$, there exists $\mathcal{B} \in K_2$ such that the characteristic function $\chi_{D(\mathcal{B})}$ of the set $D(\mathcal{B})$ is equal to $\varphi_e^{D(\mathcal{A})}$ (such a structure $\mathcal{B}$ is denoted by $\Phi(\mathcal{A})$),*
2. *for any $\mathcal{A}, \mathcal{A}' \in K_1$, we have $\mathcal{A} E_1 \mathcal{A}'$ iff $\Phi(\mathcal{A}) E_2 \Phi(\mathcal{A}')$.*

We write $(K_1, E_1) \leq_{tc} (K_2, E_2)$ if there is a tc-embedding from (K_1, E_1) into (K_2, E_2). If $(K_1, E_1) \leq_{tc} (K_2, E_2)$ and $(K_2, E_2) \leq_{tc} (K_1, E_1)$, then we write $(K_1, E_1) \equiv_{tc} (K_2, E_2)$. Note that the relation $\leq_{tc}$ is a preorder.

One of important results in the theory of tc-embeddings is the following:

Pullback Theorem *(Knight, Miller, and Vanden Boom* [7]). *Suppose that $(K_1, \cong) \leq_{tc} (K_2, \cong)$ via Φ. Then for any computable infinitary sentence ψ in the language of K_2, we can effectively find a computable infinitary sentence ψ^* in the language of K_1 such that for all $\mathcal{A} \in K_1$, we have $\mathcal{A} \models \psi^*$ if and only if $\Phi(\mathcal{A}) \models \psi$. Moreover, for a non-zero $\alpha < \omega_1^{CK}$, if ψ is a Σ_α^c (Π_α^c) formula, then so is ψ^*.*

Fokina, Knight, Melnikov, Quinn, and Safransky [9] obtained a sufficient condition for not being universal for $\cong$ under tc-embeddings:

Theorem 2.1 ([9, Theorem 3.1]). *Let K_1 and K_2 be classes of structures. Suppose that K_1 contains a pair of non-isomorphic structures $\mathcal{A}_1$ and $\mathcal{A}_2$ such that $\omega_1^{D(\mathcal{A}_1)} = \omega_1^{D(\mathcal{A}_2)} = \omega_1^{CK}$, and $\mathcal{A}_1, \mathcal{A}_2$ satisfy the same computable infinitary sentences. If K_2 contains no such pair, then $(K_1, \cong) \not\leq_{tc} (K_2, \cong)$.*

We now give a formal definition of a quasi-tc-universal class:

Definition 2.2 *We say that a class of structures K is quasi-tc-universal if it satisfies the following:*

(i) *there is a class of structures K^* such that $(K^*, \cong) \not\leq_{tc} (K, \cong)$, and*
(ii) *there is $\beta_0 < \omega_1^{CK}$ such that for any computable $\alpha \geq \beta_0$ and any class K', we have $(K', \sim_\alpha^c) \leq_{tc} (K, \sim_\alpha^c)$.*

For more results on tc-embeddings, we refer the reader to [7, 12, 16].

2.2   Pairs of Computable Structures

The proof of our results uses the technique of pairs of computable structures developed by Ash and Knight [15,17]. Here we give necessary preliminaries on this technique.

Suppose that L is a language, $\mathcal{A}$ and $\mathcal{B}$ are L-structures, and α is a countable ordinal. We say that $\mathcal{A} \leq_\alpha \mathcal{B}$ if every infinitary Π_α sentence true in $\mathcal{A}$ is true in $\mathcal{B}$. The relations $\leq_\alpha$ are called *standard back-and-forth relations*.

Let α be a computable ordinal. A family $K = \{\mathcal{A}_i : i \in I\}$ of L-structures is called α-*friendly* if the structures $\mathcal{A}_i$ are uniformly computable in $i \in I$, and the relations

$$B_\beta = \left\{ (i, \bar{a}, j, \bar{b}) : \; i, j \in I, \; \bar{a} \text{ is from } \mathcal{A}_i, \; \bar{b} \text{ is from } \mathcal{A}_j, \; (\mathcal{A}_i, \bar{a}) \leq_\beta (\mathcal{A}_j, \bar{b}) \right\}$$

are computably enumerable uniformly in $\beta < \alpha$.

Theorem 2.2 (Ash and Knight [17], see also [15, Theorem 18.6]). *Let α be a non-zero computable ordinal. Suppose that $\mathcal{A}$ and $\mathcal{B}$ are computable L-structures such that $\mathcal{B} \leq_\alpha \mathcal{A}$ and the family $\{\mathcal{A}, \mathcal{B}\}$ is α-friendly. Then for any Π_α^0 set S, there is a uniformly computable sequence of structures $\{\mathcal{C}_n\}_{n \in \omega}$ such that*

$$\mathcal{C}_n \cong \begin{cases} \mathcal{A}, & \text{if } n \in S, \\ \mathcal{B}, & \text{otherwise.} \end{cases}$$

The key feature of Theorem 2.2 is its uniformity: given a Kleene's notation of an ordinal α, computable indices of structures $\mathcal{A}, \mathcal{B}$, and a Π_α^0 index for S, one can effectively find a computable index for the sequence $\{\mathcal{C}_n\}_{n \in \omega}$. This uniformity is very helpful: it was used, e.g., in the proof of [15, Theorem 18.9].

2.3   Back-and-Forth Relations for Ordinals

In order to use the technique of pairs of computable structures, we give a brief overview on the standard back-and-forth relations for the class of ordinals. First, we recall some infinitary formulas from [15]. For a computable ordinal α, we assume that $\mu_\alpha^=(x, y)$ is the infinitary formula from [15, Proposition 7.2] which satisfies the following:

- for any linear order $\mathcal{L}$ and any $a, b \in \mathcal{L}$, we have $\mathcal{L} \models \mu_\alpha^=(a, b)$ iff $a \leq_{\mathcal{L}} b$ and the interval $[a, b[$ is isomorphic to ω^α; and
- $\mu_\alpha^=$ is a $\Pi_{2\alpha+1}^c$ formula.

Using the proof of [15, Proposition 7.2], it is not difficult to construct a formula $\mu_\alpha^<(x, y)$ such that:

- for any linear order $\mathcal{L}$ and any $a, b \in \mathcal{L}$, we have $\mathcal{L} \models \mu_\alpha^<(a, b)$ iff $a \leq_{\mathcal{L}} b$ and $[a, b[$ is isomorphic to some ordinal $\beta < \omega^\alpha$; and
- $\mu_\alpha^<$ is a $\Sigma_{2\alpha}^c$ formula.

Note that for an order $\mathcal{L}$, the block relation $B(\mathcal{L})$ is definable by the formula $B(x,y) := \mu_1^<(x,y) \vee \mu_1^<(y,x)$.

Ash [1] obtained a complete description of the standard back-and-forth relations for well-orders. Here we give only an excerpt from the description.

Lemma 2.1 (Ash [1], see also [15, Sect. 15.3.3]). *For a countable ordinal β and a non-zero $t \in \omega$, we have $\omega^\beta \cdot (t+1) \leq_{2\beta+1} \omega^\beta \cdot t$ and $\omega^{\beta+1} + \omega^\beta + 1 \leq_{2\beta+2} \omega^{\beta+1} + 1$.*

It is well-known that ordinals behave nicely with respect to α-friendliness:

Lemma 2.2 ([15, Proposition 15.11]). *Suppose that $\alpha, \beta_0, \beta_1, \ldots, \beta_n$ are computable ordinals. Then there exists an α-friendly family $\{\mathcal{L}_0, \mathcal{L}_1, \ldots, \mathcal{L}_n\}$ such that for $i \leq n$, $\mathcal{L}_i$ is isomorphic to β_i.*

3 Quasi-TC-Universality of WMB

Theorem 3.1. *The class WMB is quasi-tc-universal.*

Proof. First, we show that WMB is not universal for isomorphism under tc-embeddings. In order to prove this, we need the following fact:

Proposition 3.1. *Assume that $\mathcal{A} \in WMB$ and $\omega_1^{D(\mathcal{A})} = \omega_1^{CK}$. Then there is a computable infinitary sentence ξ such that $\mathcal{A} \models \xi$, and for any linear order $\mathcal{B}$, $\mathcal{B} \models \xi$ implies $\mathcal{B} \cong \mathcal{A}$.*

Proof. Assume that $\mathcal{A}_1 = \mathcal{A}/B(\mathcal{A})$. Since the block relation is definable by a Σ_2^c formula, the well-order $\mathcal{A}_1$ has a $D(\mathcal{A})^{(2)}$-computable copy. In particular, $\mathcal{A}_1$ is hyperarithmetical relative to $D(\mathcal{A})$. Therefore, $\mathcal{A}_1$ is isomorphic to some ordinal $\beta < \omega_1^{D(\mathcal{A})} = \omega_1^{CK}$.

For an element $a \in \mathcal{A}$, let the *block of a* be the set $[a]_B = \{x : (x,a) \in B(\mathcal{A})\}$. In other words, a block in $\mathcal{A}$ is just an element of the quotient structure $\mathcal{A}_1$. Notice that every block in $\mathcal{A}$ is isomorphic to one of the following orders: ω, ω^*, ζ, or some non-zero $k < \omega$.

Now we are ready to define the sentence ξ. Here we give only an informal description of ξ. The omitted details can be easily reconstructed from [15]. We build ξ such that:

(a) ξ says that the quotient of an order (modulo the block relation) is isomorphic to β, and

(b) for every block $b \in \mathcal{A}_1$, ξ describes the isomorphism type of b (which is one of ω, ω^*, ζ, or k).

Since β is computable, one can use the formulas $\mu_\alpha^=$ and $\mu_\alpha^<$, $\alpha < \omega_1^{CK}$, to ensure that ξ is a computable infinitary sentence. Moreover, the definition of ξ guarantees that for an order $\mathcal{B}$, we have $\mathcal{B} \models \xi$ iff $\mathcal{B}$ is isomorphic to $\mathcal{A}$. $\square$

Corollary 3.1. *Let UG be the class of symmetric irreflexive graphs. Then we have $(UG, \cong) \not\leq_{tc} (WMB, \cong)$.*

Proof. In [9, Theorem 3.17], it was proved that there is a family of graphs $\{G_\beta : \beta < 2^{\aleph_0}\}$ such that all of the graphs satisfy the same computable infinitary sentences, $\omega_1^{D(G_\beta)} = \omega_1^{CK}$, and $G_\beta \not\cong G_\gamma$ for all $\beta < \gamma$.

By Proposition 3.1, the class WMB contains no pair described in Theorem 2.1. Thus, Theorem 2.1 implies that there is no tc-embedding from $(UG, \cong)$ into $(WMB, \cong)$. □

The next two propositions prove that for any computable $\alpha \geq 5$, WMB is universal for Σ_α^c equivalence under tc-embeddings.

Proposition 3.2. *Let α be a computable ordinal such that $\alpha \geq 4$. For any class K, we have $(K, \sim_{\alpha+1}^c) \leq_{tc} (WMB, \sim_{\alpha+1}^c)$.*

Proof. We fix an effective enumeration $\{\psi_n\}_{n \in \omega}$ of all $\Sigma_{\alpha+1}^c$ sentences in the language of K. Suppose that $\mathcal{A}$ is a structure from K.

We give the detailed proof for the case $\alpha = 2\beta$. By Lemma 2.2, one can choose a $(2\beta + 1)$-friendly family $\{\mathcal{L}_1, \mathcal{L}_2\}$ such that $\mathcal{L}_1 \cong \omega^\beta$ and $\mathcal{L}_2 \cong \omega^\beta \cdot 2$. Recall that by Lemma 2.1, we have $\omega^\beta \cdot 2 \leq_{2\beta+1} \omega^\beta$. Therefore, one can apply the relativized version of Theorem 2.2 and produce a $D(\mathcal{A})$-computable sequence $\{\mathcal{C}_n^{\mathcal{A}}\}_{n \in \omega}$ such that

$$\mathcal{C}_n^{\mathcal{A}} \cong \begin{cases} \omega^\beta, & \text{if } \mathcal{A} \not\models \psi_n, \\ \omega^\beta \cdot 2, & \text{if } \mathcal{A} \models \psi_n. \end{cases}$$

Let ζ denote the order type of the integers. We define the order $\Phi(\mathcal{A}) \in WMB$ as follows:

$$\Phi(\mathcal{A}) = \sum_{n \in \omega} (\zeta + \mathcal{C}_n^{\mathcal{A}}).$$

Note that here we slightly abuse the notation and identify $\Phi(\mathcal{A})$ and its natural $D(\mathcal{A})$-computable copy.

We now show that Φ is a tc-embedding from $(K, \sim_{\alpha+1}^c)$ into $(WMB, \sim_{\alpha+1}^c)$. Let $\mathcal{A}_1, \mathcal{A}_2 \in K$. Note that $\mathcal{A}_1 \sim_{\alpha+1}^c \mathcal{A}_2$ implies $\Phi(\mathcal{A}_1) \cong \Phi(\mathcal{A}_2)$. We define two auxiliary formulas:

1. the Π_3^c formula $Z(x)$ saying that the block of x is isomorphic to ζ:

$$Z(x) := \forall y[B(x, y) \rightarrow [\exists u(u < y \,\&\, B(x, u)) \,\&\, \exists v(y < v \,\&\, B(x, v))]];$$

2. for $n \in \omega$, the Σ_5^c formula $Z(x; n)$ saying that x belongs to the n-th ζ-block:

$$Z(x; n) := \exists z_0 \exists z_1 \ldots \exists z_n [B(z_n, x) \,\&\, \&_{i \leq n} Z(z_i) \,\&\, \&_{i < n} (z_i < z_{i+1}) \,\&$$
$$\&_{i \neq j} \neg B(z_i, z_j) \,\&\, \forall u((u \leq z_n \,\&\, Z(u)) \rightarrow \vee_{i \leq n} B(z_i, u))].$$

Assume that for some $n \in \omega$, we have $\mathcal{A}_1 \models \psi_n$ and $\mathcal{A}_2 \not\models \psi_n$. Recall that $2\beta + 1 = \alpha + 1 \geq 5$. Consider the following $\Sigma^c_{2\beta+1}$ sentence:

$$\xi_n := \exists x \exists y \exists u \exists v [Z(u;n) \,\&\, Z(v;n+1) \,\&\, (u < x < y < v) \,\&\, \neg B(u,x) \,\&$$
$$\neg B(y,v) \,\&\, \neg \mu^<_\beta(x,y)].$$

Since $\mathcal{C}^{\mathcal{A}_1}_n \cong \omega^\beta \cdot 2$ and $\mathcal{C}^{\mathcal{A}_2}_n \cong \omega^\beta$, we obtain that $\Phi(\mathcal{A}_1) \models \xi_n$ and $\Phi(\mathcal{A}_2) \not\models \xi_n$. Thus, $\Phi(\mathcal{A}_1) \not\sim^c_{2\beta+1} \Phi(\mathcal{A}_2)$.

Using the uniformity of Theorem 2.2, it is not difficult to show that Φ is a computable operator (i.e., $\Phi = \varphi_e$ for some e). Therefore, Φ is a tc-embedding from $(K, \sim^c_{\alpha+1})$ into $(WMB, \sim^c_{\alpha+1})$.

The proof for the case $\alpha = 2\beta+1$ is essentially the same, modulo the following modifications:

- we use the orders $(\omega^{\beta+1} + 1)$ and $(\omega^{\beta+1} + \omega^\beta + 1)$ in place of ω^β and $\omega^\beta \cdot 2$, respectively;
- the subformula $\neg \mu^<_\beta(x,y)$ in the sentence ξ_n should be replaced with the $\Pi^c_{2\beta+1}$ formula $\forall w(y < w < v \to B(v,w)) \,\&\, \mu^=_\beta(x,y)$. $\qquad\square$

Proposition 3.3. *Let δ be a computable limit ordinal. For any class K, we have $(K, \sim^c_\delta) \leq_{tc} (WMB, \sim^c_\delta)$.*

Proof. First, we find a computable sequence $\{\alpha_n\}_{n\in\omega}$ of computable ordinals with limit δ. Without loss of generality, we may assume that for any n, we have $\alpha_n = 2\beta_n + 1 \geq 5$. Fix an effective enumeration $\{\psi_{n,k}\}_{n,k\in\omega}$ such that for any n, $\{\psi_{n,k}\}_{k\in\omega}$ is an enumeration of all $\Sigma^c_{\alpha_n}$ sentences in the language of K.

Using essentially the same argument as in Proposition 3.2, one can build a $D(\mathcal{A})$-computable sequence $\{\mathcal{C}^\mathcal{A}_{n,k}\}_{n,k\in\omega}$ such that

$$\mathcal{C}^\mathcal{A}_{n,k} \cong \begin{cases} \omega^{\beta_n}, & \text{if } \mathcal{A} \not\models \psi_{n,k}, \\ \omega^{\beta_n} \cdot 2, & \text{if } \mathcal{A} \models \psi_{n,k}. \end{cases}$$

The order $\Phi(\mathcal{A})$ is defined as follows:

$$\Phi(\mathcal{A}) = \sum_{\langle n,k \rangle \in \omega} (\zeta + \mathcal{C}^\mathcal{A}_{n,k}),$$

where $\langle n, k \rangle$ is the Gödel number of the pair (n, k).

Assume that $\mathcal{A}_1, \mathcal{A}_2 \in K$. Notice that $\mathcal{A}_1 \sim^c_\delta \mathcal{A}_2$ iff $\mathcal{A}_1 \sim^c_{\alpha_n} \mathcal{A}_2$ for all n. Therefore, one can follow the lines of the proof of Proposition 3.2 and show that $\mathcal{A}_1 \sim^c_\delta \mathcal{A}_2$ iff $\Phi(\mathcal{A}_1) \sim^c_\delta \Phi(\mathcal{A}_2)$. The computability of Φ follows from the uniformity of Theorem 2.2. We note here that the more formal version of the proof can be obtained by using [15, Theorem 18.9] in place of Theorem 2.2. This concludes the proof of Proposition 3.3 and Theorem 3.1. $\qquad\square$

4 TC-Embeddings for Ordinals

For a computable ordinal β, we consider the following classes:

$$\mathbb{K}^{2\beta+1} = \{\mathcal{L} : \mathcal{L} \cong \omega^\beta \cdot (t+1) \text{ for some } t \in \omega\},$$
$$\mathbb{K}^{2\beta+2} = \{\mathcal{L} : \mathcal{L} \cong \omega^{\beta+1} + \omega^\beta \cdot (t+1) \text{ for some } t \in \omega\}.$$

Notice that $\mathbb{K}^1$ is the class of all finite linear orders. In this section, we sketch the proof of the following:

Theorem 4.1. *Suppose that α is a computable successor ordinal. For a class K, we have $(K, \cong) \leq_{tc} (\mathbb{K}^\alpha, \cong)$ if and only if there exists a computable sequence $\{\psi_n\}_{0<n<\omega}$ of Σ_α^c sentences in the language of K such that*

1. *for $\mathcal{A} \in K$ and $m < n$, if $\mathcal{A} \models \psi_n$, then $\mathcal{A} \models \psi_m$,*
2. *for $\mathcal{A}, \mathcal{B} \in K$, if $\mathcal{A} \not\cong \mathcal{B}$, then there is some n such that ψ_n is true in only one of $\mathcal{A}, \mathcal{B}$, and*
3. *for each $\mathcal{A} \in K$, there exists some n such that $\mathcal{A} \not\models \psi_n$.*

Proof. We note here that for $\alpha = 1$, this result was proved by Knight, Miller, and Vanden Boom [7, Theorem 4.4]. For reasons of space, we give the proof only for $\alpha = 2\beta + 1$.

($\Rightarrow$). Assume that for a non-zero $n \in \omega$, ξ_n is the $\Sigma_{2\beta+1}^c$ sentence:

$$\xi_n := \exists x_1 \exists x_2 \ldots \exists x_{n+1} [x_1 < x_2 < \ldots < x_{n+1} \ \& \ \&_{i \leq n} \neg \mu_\beta^{\leq}(x_i, x_{i+1})].$$

Suppose that $(K, \cong) \leq_{tc} (\mathbb{K}^{2\beta+1}, \cong)$ via Φ. We apply Pullback Theorem and obtain computable sequence $\{\xi_n^*\}_{n \in \omega}$ of $\Sigma_{2\beta+1}^c$ sentences in the language of K. It is straightforward to show that the sequence $\{\xi_n^*\}_{n \in \omega}$ satisfies the three properties above.

($\Leftarrow$). First, we obtain the following generalization of Theorem 2.2:

Theorem 4.2. *Suppose that α is a non-zero computable ordinal and $\{\mathcal{A}_k : k \in \omega\}$ is an α-friendly family of L-structures such that $\mathcal{A}_{k+1} \leq_\alpha \mathcal{A}_k$ for all k. Then for any $\mathbf{0}_{(\alpha)}$-limitwise monotonic function $G(x)$, there is a uniformly c.e. sequence of structures $\{\mathcal{C}_n\}_{n \in \omega}$ such that for each n, we have $\mathcal{C}_n \cong \mathcal{A}_{G(n)}$.*

For reasons of space, the proof of Theorem 4.2 is omitted. We emphasize that this proof is uniform. In addition, in Theorem 4.1 one essentially does not need general limitwise monotonic functions $G(x)$: every function G that is used here is constant. Nevertheless, Theorem 4.2 may be useful for more intricate constructions.

Suppose that $\mathcal{A} \in K$ and $\mathbf{d} = \deg_T(D(\mathcal{A}))$. We define the $\mathbf{d}_{(2\beta+1)}$-limitwise monotonic function $G^\mathcal{A}$: for all x, we have $G^\mathcal{A}(x) = 1 + card(\{n : \mathcal{A} \models \psi_n\})$.

Proposition 15.11 from [15] implies that there is a $(2\beta + 1)$-friendly family $\{\mathcal{L}_k : 0 < k < \omega\}$ such that $\mathcal{L}_t \cong \omega^\beta \cdot t$ for all t. In addition, by Lemma 2.1, we have $\omega^\beta \cdot (t+1) \leq_{2\beta+1} \omega^\beta \cdot t$. Recall that any c.e. linear order is computable.

Thus, we apply the relativized Theorem 4.2 to the function $G^{\mathcal{A}}$ and obtain a $D(\mathcal{A})$-computable structure $\Phi(\mathcal{A}) \cong \omega^\beta \cdot G^{\mathcal{A}}(0)$.

It is not difficult to show that for $\mathcal{A}_1, \mathcal{A}_2 \in K$, we have $\mathcal{A}_1 \cong \mathcal{A}_2$ iff $\Phi(\mathcal{A}_1) \cong \Phi(\mathcal{A}_2)$. As in Theorem 3.1, the computability of operator Φ follows from the uniformity of Theorem 4.2. This concludes the proof of Theorem 4.1.   □

Corollary 4.1. *Suppose that α and β are computable successor ordinals. Then we have $(\mathbb{K}^\alpha, \cong) \leq_{tc} (\mathbb{K}^\beta, \cong)$ iff $\alpha \leq \beta$.*

5   Further Discussion

Our results on linear orders are connected with other classes of structures:

Corollary 5.1

1. *Let VS be the class of $\mathbb{Q}$-vector spaces, and let FVS be the class of finite-dimensional $\mathbb{Q}$-vector spaces. Then we have $(\mathbb{K}^2, \cong) \equiv_{tc} (FVS, \cong) \lesssim_{tc} (VS, \cong)$.*
2. *Let E be the class of equivalence structures, each containing exactly one class of each finite size and any number of infinite classes. Then $(\mathbb{K}^3, \cong) \lesssim_{tc} (E, \cong)$.*

Proof. This is a consequence of Theorem 4.1, and Theorems 4.7, 4.13, Proposition 4.9 from [7].   □

Following the lines of Theorem 4.1, it is not difficult to prove:

Corollary 5.2. *For a computable ordinal β, let SBA^β be the class of all super-atomic Boolean algebras with the Fréchet rank β. Then we have $(SBA^{\beta+1}, \cong) \equiv_{tc} (\mathbb{K}^{2\beta+1}, \cong)$.*

In conclusion, we formulate two open questions:

Problem 5.1. Is the class of well-orders quasi-tc-universal?

Problem 5.2. For a computable limit ordinal α, find natural classes $\mathbb{K}^\alpha$ that satisfy Theorem 4.1.

Acknowledgements. The author is grateful to Sergey Goncharov for fruitful discussions on the subject. The author also thanks the anonymous reviewers for their helpful suggestions. The reported study was funded by RFBR, according to the research project No. 16-31-60058 mol_a_dk.

References

1. Ash, C.J.: Recursive labelling systems and stability of recursive structures in hyper-arithmetical degrees. Trans. Am. Math. Soc. **298**(2), 497–514 (1986). doi:10.1090/S0002-9947-1986-0860377-7
2. Hirschfeldt, D.R., Khoussainov, B., Shore, R.A., Slinko, A.M.: Degree spectra and computable dimensions in algebraic structures. Ann. Pure Appl. Logic **115**(1–3), 71–113 (2002). doi:10.1016/S0168-0072(01)00087-2
3. Calvert, W., Cummins, D., Knight, J.F., Miller, S.: Comparing classes of finite structures. Algebra Logic **43**(6), 374–392 (2004). doi:10.1023/B:ALLO.0000048827.30718.2c
4. Fokina, E.B., Friedman, S.-D.: On Σ_1^1 equivalence relations over the natural numbers. Math. Log. Q. **58**(1–2), 113–124 (2012). doi:10.1002/malq.201020063
5. Harrison-Trainor, M., Melnikov, A., Miller, R., Montálban, A.: Computable functors and effective interpretability. J. Symbolic Logic (to appear). doi:10.1017/jsl.2016.12
6. Friedman, H., Stanley, L.: A Borel reducibility theory for classes of countable structures. J. Symbolic Logic **54**(3), 894–914 (1989). doi:10.2307/2274750
7. Knight, J.F., Miller, S., Vanden Boom, M.: Turing computable embeddings. J. Symbolic Logic **72**(3), 901–918 (2007). doi:10.2178/jsl/1191333847
8. Chisholm, J., Knight, J.F., Miller, S.: Computable embeddings and strongly minimal theories. J. Symbolic Logic **72**(3), 1031–1040 (2007). doi:10.2178/jsl/1191333854
9. Fokina, E., Knight, J.F., Melnikov, A., Quinn, S.M., Safranski, C.: Classes of Ulm type and coding rank-homogeneous trees in other structures. J. Symbolic Logic **76**(3), 846–869 (2011). doi:10.2178/jsl/1309952523
10. Ocasio-González, V.A.: Turing computable embeddings and coding families of sets. In: Cooper, S.B., Dawar, A., Löwe, B. (eds.) CiE 2012. LNCS, vol. 7318, pp. 539–548. Springer, Heidelberg (2012). doi:10.1007/978-3-642-30870-3_54
11. Andrews, U., Dushenin, D.I., Hill, C., Knight, J.F., Melnikov, A.G.: Comparing classes of finite sums. Algebra Logic **54**(6), 489–501 (2016). doi:10.1007/s10469-016-9368-7
12. VanDenDriessche, S.M.: Embedding computable infinitary equivalence into p-groups. Ph.D. thesis, University of Notre Dame (2013)
13. Wright, M.: Turing computable embeddings of equivalences other than isomorphism. Proc. Am. Math. Soc. **142**, 1795–1811 (2014). doi:10.1090/S0002-9939-2014-11878-8
14. Goncharov, S.S.: Groups with a finite number of constructivizations. Sov. Math. Dokl. **23**, 58–61 (1981)
15. Ash, C.J., Knight, J.F.: Computable structures and the hyperarithmetical hierarchy. In: Studies in Logic and the Foundations of Mathematics, vol. 144. Elsevier Science B.V., Amsterdam (2000)
16. Knight, J.F.: Using computability to measure complexity of algebraic structures and classes of structures. Lobachevskii J. Math. **35**(4), 304–312 (2014). doi:10.1134/S1995080214040192
17. Ash, C.J., Knight, J.F.: Pairs of recursive structures. Ann. Pure Appl. Logic **46**(3), 211–234 (1990). doi:10.1016/0168-0072(90)90004-L

Degrees of Categoricity of Rigid Structures

Nikolay A. Bazhenov[1,2]([⊠]) [ID] and Mars M. Yamaleev[3]([⊠]) [ID]

[1] Sobolev Institute of Mathematics, Novosibirsk, Russia
[2] Novosibirsk State University, Novosibirsk, Russia
bazhenov@math.nsc.ru
[3] Kazan (Volga Region) Federal University, Kazan, Russia
Mars.Yamaleev@kpfu.ru

Abstract. We prove that there exists a properly 2-c.e. Turing degree **d** which cannot be a degree of categoricity of a rigid structure.

Keywords: Categoricity spectrum · Strong degree of categoricity · Rigid structure · 2-c.e. Turing degrees

1 Introduction

The study of effective categoricity for computable structures goes back to the works of Fröhlich and Shepherdson [1], and Mal'tsev [2,3]. In recent years, the focus of the research in the area is on computable categoricity relative to Turing degrees.

Definition 1. *Let* **d** *be a Turing degree. A computable structure* $\mathcal{A}$ *is* **d**-*computably categorical if for every computable copy* $\mathcal{B}$ *of* $\mathcal{A}$, *there is a* **d**-*computable isomorphism from* $\mathcal{A}$ *onto* $\mathcal{B}$. *The* categoricity spectrum *of* $\mathcal{A}$ *is the set*

$$\mathrm{CatSpec}(\mathcal{A}) = \{\mathbf{d} : \mathcal{A} \text{ is } \mathbf{d}\text{-computably categorical}\}.$$

A Turing degree **d** *is the* degree of categoricity *of* $\mathcal{A}$ *if* **d** *is the least degree in the spectrum* $\mathrm{CatSpec}(\mathcal{A})$.

Categoricity spectra and degrees of categoricity were introduced in [4]. Suppose that n is a natural number and α is an infinite computable ordinal. Fokina, Kalimullin, and Miller [4] proved that each Turing degree **d** that is 2-c.e. in and above $\mathbf{0}^{(n)}$ is the degree of categoricity for a computable structure. Csima, Franklin, and Shore [5] extended this result to hyperarithmetical degrees. They

N.A.Bazhenov—Supported by RFBR project No. 16-31-60058 mol_a_dk.

M.M.Yamaleev—Supported by RFBR projects No. 15-01-08252, 16-31-50048, and by the subsidy allocated to Kazan Federal University for the state assignment in the sphere of scientific activity (No. 1.1515.2017/PCh). The work is performed according to the Russian Government Program of Competitive Growth of Kazan Federal University and by the research grant of Kazan Federal University.

J. Kari et al. (Eds.): CiE 2017, LNCS 10307, pp. 152–161, 2017.
DOI: 10.1007/978-3-319-58741-7_16

proved that each degree **d** that is 2-c.e. in and above $\mathbf{0}^{(\alpha+1)}$ is a degree of categoricity. Anderson and Csima [6] investigated Turing degrees that cannot be degrees of categoricity. In particular, they constructed a Σ_2^0-set whose degree is not a degree of categoricity. Miller [7] built the first example of a computable structure without a degree of categoricity. Much work has been done of late in investigating categoricity spectra for structures in familiar classes: algebraic fields [7,8], Boolean algebras [9–12], linear orderings [13,14], rigid structures [15], decidable structures [16,17], etc.

Definition 2 ([4]). *Assume that $\mathcal{A}$ is a computable structure. A Turing degree* **d** *is the* strong degree of categoricity *of $\mathcal{A}$ if* **d** *is the degree of categoricity for $\mathcal{A}$, and there exist computable copies $\mathcal{B}$ and $\mathcal{C}$ of $\mathcal{A}$ such that for each isomorphism f from $\mathcal{B}$ onto $\mathcal{C}$, we have $\deg_T(f) \geq$* **d**.

Fokina, Frolov, and Kalimullin [15] proved that for each non-zero c.e. Turing degree **d**, there is a **d**-computably categorical rigid structure with no degree of categoricity. They also posed the following question.

Problem 1 ([15, Problem 3.2]). Can a properly 2-c.e. degree be a degree of categoricity of a rigid structure?

In this work, we make the first step towards the solution of Problem 1. Namely, we prove the following result.

Theorem 1. *There exists a properly 2-c.e. degree* **d** *which cannot be a degree of categoricity of a rigid structure.*

In the next section, we give some special definitions which will be involved in the proof. The third section will be devoted to the proof of the main result.

2 Preliminaries

By reducibility we always mean Turing reducibility, and we consider only Turing degrees.

Now we recall the notion of spectral dimension. This notion allows to make a distinction between "degree of categoricity" and "strong degree of categoricity" using a quantitative characteristic.

Assume that $\mathcal{A}$ and $\mathcal{B}$ are isomorphic computable structures. The *isomorphism spectrum* of the pair $(\mathcal{A}, \mathcal{B})$ is the set

$$\mathrm{IsoSpec}(\mathcal{A}, \mathcal{B}) = \{\mathbf{d} : \text{there is a } \mathbf{d}\text{-computable isomorphism from } \mathcal{A} \text{ onto } \mathcal{B}\}.$$

Note that we always have $\mathrm{CatSpec}(\mathcal{A}) \subseteq \mathrm{IsoSpec}(\mathcal{A}, \mathcal{B})$.

Definition 3 ([18]). *Let S be a computable structure. The* spectral dimension *of S is the least ordinal $k \leq \omega$ such that there exists a sequence of pairs of computable structures $\{(\mathcal{A}_i, \mathcal{B}_i)\}_{0 \leq i < k}$ with the following properties: $\mathcal{A}_i \cong \mathcal{B}_i \cong S$ for all $i < k$, and*

$$\mathrm{CatSpec}(S) = \bigcap_{i<k} \mathrm{IsoSpec}(\mathcal{A}_i, \mathcal{B}_i).$$

Let $\mathrm{SpecDim}(S)$ denote the spectral dimension of S.

Suppose that $\mathbf{d}$ is the degree of categoricity of a structure $\mathcal{S}$. It is easy to see that $\mathbf{d}$ is the strong degree of categoricity of $\mathcal{S}$ if and only if $\mathcal{S}$ has spectral dimension one. Furthermore, if we consider degrees of categoricity for rigid structures (recall that a rigid structure has a single element in its group of automorphisms), then we can work only with finite spectral dimensions (see Theorem 2 below).

Theorem 2 ([18, Theorem 3.1]). *Suppose that $\mathcal{S}$ is a computable structure with finite automorphism group, and $\mathbf{d}$ is the degree of categoricity for $\mathcal{S}$. Then* $\mathrm{SpecDim}(\mathcal{S}) < \omega$.

3   Main Result

The next theorem shows that there is a properly 2-c.e. degree which cannot be a strong degree of categoricity for a rigid structure. In the course of the proof we will add comments in italicized parentheses. They are not properly part of the formal proof and are intended to make the intuition clearer.

Theorem 3. *There exists a 2-c.e. degree $\deg(D)$ that satisfies the following condition: for any computable structures $\mathcal{A}, \mathcal{B}$ and any Δ_2^0 isomorphism g from $\mathcal{A}$ onto $\mathcal{B}$, the condition $\deg(g) = \deg(D)$ implies that there exist a computable structure $\mathcal{N}$ and an isomorphism f from $\mathcal{A}$ onto $\mathcal{N}$ with $\deg(f) \not\leq \deg(D)$.*

Note that if a computable structure $\mathcal{A}$ is rigid and $\deg(D)$ is its strong degree of categoricity then the degrees of the isomorphisms between any computable copies of $\mathcal{A}$ must be below $\deg(D)$. The left-hand side of this implication contradicts Theorem 3 and we obtain the desired $\deg(D)$. In fact, Theorem 3 can be easily generalized to the case of any finite spectral dimension. Now we present the simplest version (for the case when the spectral dimension is equal to one) since it describes ideas in a neat way. Then we will show how these ideas can be transferred to the case of a finite spectral dimension. Finally, together with Theorem 2 we obtain that there is a 2-c.e. degree which is not a degree of categoricity for a rigid structure. Recall that Fokina, Kalimullin, and Miller [4, Theorem 2.1] proved that any c.e. degree is a strong degree of categoricity for some rigid structure. Thus, the degree of D from Theorem 3 is automatically a properly 2-c.e. degree.

Proof. First, we give some preliminary information regarding structures. During the construction we need to effectively enumerate all pairs of computable structures which are Δ_2^0-isomorphic. Thus, we fix a strongly computable sequence of finite objects $\{\mathcal{A}_e[s], \mathcal{B}_e[s], g_e[s]\}_{e,s\in\omega}$ such that the following conditions hold:

(A) For all e and s, we have:
  (1) $\mathcal{A}_e[s]$ and $\mathcal{B}_e[s]$ are finite graphs,
  (2) $\mathcal{A}_e[s] \subseteq \mathcal{A}_e[s+1]$,
  (3) $\mathcal{B}_e[s] \subseteq \mathcal{B}_e[s+1]$,

(4) $g_e[s]: \mathcal{A}_e[s] \cong \mathcal{B}_e[s]$.

(Note that $g_e[s]$ can be taken as $\Phi_e^K[s]$ in order to cover all Δ_2^0 isomorphisms. However, some indices e can correspond to Σ_2^0-functions).

(B) If $\mathcal{A} \cong \mathcal{B}$ are infinite computable graphs and $g: \mathcal{A} \cong \mathcal{B}$ is Δ_2^0, then there is an index e such that:

(1) $\mathcal{A} = \cup_s \mathcal{A}_e[s]$,
(2) $\mathcal{B} = \cup_s \mathcal{B}_e[s]$,
(3) $g = \lim_s g_e[s]$.

(Obviously, it is enough to consider only graphs, since the class of graphs is universal relative to categoricity spectra [4, Proposition 4.1].)

Thus, in the course of the construction we enumerate all pairs of (partial) computable graphs. If we suspect that the corresponding isomorphism is Turing equivalent to D, then we build $\mathcal{N}$ and f for that particular tuple. Since our suspicions will arise only at some kind of expansionary stages, we need to define $\mathcal{N}$ and f only at those stages. Before the construction, we give some information on stages.

Assume that $t < s$ are two consecutive expansionary stages. At stage s we have $\mathcal{N}_e[t]$, $f_e[t]$, $\mathcal{A}_e[t]$, $\mathcal{B}_e[t]$, $g_e[t]$, $\mathcal{A}_e[s]$, $\mathcal{B}_e[s]$, and $g_e[s]$. Thus, we need to define $\mathcal{N}_e[s]$ and $f_e[s]$. Consider the following two functions: $h_g[s] = g_e^{-1}[s] \circ id_{\mathcal{B}_e}[t] \circ g_e[t] \circ f_e^{-1}[t]$ and $h_{id}[s] = id_{\mathcal{A}_e}[t] \circ f_e^{-1}[t]$. Each of the functions is an isomorphic embedding of $\mathcal{N}_e[t]$ into $\mathcal{A}_e[s]$. Clearly, we can extend the function $h_{id}[s]$ (the function $h_g[s]$, respectively) to the isomorphism H_{id} (H_g, respectively) from $\mathcal{N}_e[s]$ onto $\mathcal{A}_e[s]$, where $N_e[s]$ is an initial segment of ω (this can be done if $N_e[t]$ is also an initial segment of ω). Thus, we have two ways to define $f_e[s]$:

(1) If $f_e[s] = H_{id}^{-1}$, then we say that $f_e[s]$ is the *id-extension* of $f_e[t]$.
(2) If $f_e[s] = H_g^{-1}$, then we say that $f_e[s]$ is the *g-extension* of $f_e[t]$.

Note that if $f_e[s]$ is a g-extension and $g_e(y_0)[t] \neq g_e(y_0)[s]$ for some $y_0 \in A_e[t]$ then $f_e(y_0)[s] = h_g^{-1}(y_0)[s] \neq f_e(y_0)[t]$.

The purpose of the priority construction is to choose one of these two extensions when we really need to extend the function (e.g., when a stage is expansionary for the corresponding tuple).

We make some standard assumptions for the priority argument: e.g., we assume that each functional is nondecreasing in each of its arguments, and the standard notation $P[s]$ means that all parameters in P use their approximations at stage s. When we talk about a Turing reduction to a function g, we mean the reduction to the graph $Gr_g = \{\langle x, y \rangle : g(x) = y\}$. Recall that at stage s we have $g_e[s]$ which is an isomorphism from $\mathcal{A}_e[s]$ onto $\mathcal{B}_e[s]$. Thus, if we want to see that this isomorphism grew, then we can wait until $\{0, 1, \ldots, x\} \subseteq A_e[s]$; in particular, this means that $g_e(0), \ldots, g_e(x)$ are defined.

Requirements. We need to satisfy the following series of requirements:

$$\mathcal{R}_e : (g_e: \mathcal{A}_e \cong \mathcal{B}_e) \wedge (g_e \text{ is } \Delta_2^0) \wedge (D = \Phi_e^{g_e}) \wedge (g_e = \Theta_e^D) \Rightarrow$$
$$\exists\, f_e, \mathcal{N}_e \text{ such that } f_e: \mathcal{A}_e \cong \mathcal{N}_e \text{ and } f_e \not\leq_T D,$$

where the list $\{\mathcal{A}_e, \mathcal{B}_e, g_e, \Phi_e, \Theta_e\}_{e\in\omega}$ effectively enumerates all mentioned above triples $(\mathcal{A}_e[s], \mathcal{B}_e[s], g_e[s])$ and p.c. functionals Φ_e and Θ_e. The subrequirement $f_e \not\leq_T D$ is transformed into an infinite series (where $\{\Psi_i\}_{i\in\omega}$ is an enumeration of all p.c. functionals):

$$\mathcal{P}_{e,i} : f_e \neq \Psi_i^D.$$

For the sake of convenience and organizing the strategies, we will construct a single copy of f_e and $\mathcal{N}_e$ for each $e \in \omega$. Thus, it will be a global requirement. However, if the condition $(g_e : \mathcal{A}_e \cong \mathcal{B}_e) \wedge (g_e \text{ is } \Delta_2^0) \wedge (D = \Phi_e^{g_e}) \wedge (g_e = \Theta_e^D)$ is not true, then we do not need to ensure that f_e and $\mathcal{N}_e$ are correct. We construct them according to the highest priority strategy $\mathcal{S}_{e,i}$ with index e which requires attention.

(Also note that we can alternatively use $\mathcal{R}_e$ as the mother strategy with outcomes ∞ and fin on a tree of strategies, where below the ∞-outcome we arrange its child strategies $\mathcal{P}_{e,i}$ and construct f_e and $\mathcal{N}_e$. However, in our particular case, we can use a finite injury argument, and we do not need the power of the tree construction.)

Henceforth, we will build a 2-c.e. set D and a sequence of pairs $\{f_e, \mathcal{N}_e\}_{e\in\omega}$. We satisfy the infinite series of the following requirements:

$$\mathcal{S}_{e,i} : (g_e : \mathcal{A}_e \cong \mathcal{B}_e) \wedge (g_e \text{ is } \Delta_2^0) \wedge (D = \Phi_e^{g_e}) \wedge (g_e = \Theta_e^D) \ \Rightarrow \ f_e \neq \Psi_i^D.$$

Intuition for a single strategy. In a standard manner, by lower-case Greek letters we denote use-functions of the corresponding p.c. functionals. W.l.o.g, we assume the following: if $\Phi^X(x) \downarrow$, then $\Phi^X(y) \downarrow$ for all $y < x$ and $\varphi(y) < \varphi(x)$. We define the *length agreement functions* as follows.

The *major length agreement function* of $\mathcal{S}_{e,i}$:

$$L(e)[s] = \max\left\{x : \ (\forall y \leq x)\Big[\big(\Phi_e^{g_e}(y)[s] = D(y)[s]\big)\wedge\right.$$

$$\left.\big(\forall z < \varphi_e(y)[s]\big)\big(\Theta_e^D(z)[s] = g_e(z)[s]\big) \wedge \big(\{0,1,\ldots,\varphi_e(y)[s]\} \subseteq A_e[s]\big)\Big]\right\}.$$

The *minor length agreement function* of $\mathcal{S}_{e,i}$:

$$l(e,i)[s] = \max\big\{x : (\forall y \leq x)\big(f_e(y)[s] = \Psi_i^D(y)[s]\big)\big\}.$$

Note that the major length agreement function of $\mathcal{S}_{e,i}$ serves us similarly to the usual length agreement function of the mother strategy $\mathcal{R}_e$ and it plays a global role for strategies $\mathcal{S}_{e,i}$. Thus, we proceed to considering the minor length agreement function of $\mathcal{S}_{e,i}$ only when we successfully pass its major length agreement function. The minor length agreement function of $\mathcal{S}_{e,i}$ plays a local role and directly helps to satisfy $\mathcal{S}_{e,i}$.

Our intuition for the strategy $\mathcal{S}_{e,i}$ in isolation is as follows. By a "big" number we mean an integer which is greater than all numbers mentioned in the construction so far (in particular, it is greater than all use-functions which we currently want to preserve). We omit indices if they are clear from the context.

(1) Assign a "big" witness $x = x_{e,i}$.

(2) Wait for a stage s_0 such that $x < L(c)[s_0]$.

(Stage s_0 can be considered as an analogue of usual expansionary stages. However, it is a stage when we must define an extension of the current f_e.)

(3) Define $f_e[s_0]$ as the id-extension of $f_e[0]$.

(After seeing that g_e grew enough, at least up to $\varphi_e(x)[s_0]$, we must extend f_e too.)

(4) Wait for a stage $s_1 > s_0$ such that $\varphi_e(x)[s_1] = \varphi_e(x)[s_0] < l(e,i)[s_1]$.

(Now we want to witness the equalities for $f_e[s_0]$ in the parts where we extended f_e, note that $x < L(e)[s_1]$. In fact, $\varphi_e(x)[s_1] < l(e,i)[s_1]$ is a pretty rough bound, but it is enough. The same roughness holds for functions $g_e[s]$: we need only $g_e(z)[s] = \Theta_e^D(z)[s]$ for all z such that $\langle z, g_e(z) \rangle < \varphi_e(x)[s]$ (which is in $Gr_{g_e}[s]$) instead of all $z < \varphi_e(x)[s]$. Anyway, if $Gr_{g_e}[s]$ changes below $\varphi_e(x)[s]$, then $g_e[s]$ also changes at some point below $\varphi_e(x)[s]$.)

(5) Enumerate x into D.

(Thus, either the strategy is diagonalized forever or Gr_{g_e} changes at a later stage at some number below $\varphi_e(x)[s_0]$. Therefore, if Gr_{g_e} changes below $\varphi_e(x)[s_0]$, then g_e will also change at some $y_0 < \varphi_e(x)[s_0]$, and this will allow us to modify $f_e(y_0)$ by defining f_e as the g-extension of $f_e[s_0]$. Note that it is not necessary to extend f_e here.)

(6) Wait for a stage $s_2 > s_1$ such that $x < L(e)[s_2]$.

(Similarly to item (2), at stage s_2 it is time to define $f_e[s_2]$. We define it as the g-extension of $f_e[s_0]$. This gives a new value for some $y_0 < \varphi_e(x)[s_0]$. Moreover, this means that $\Psi_i^D(y_0)$ could change and become equal to $f_e(y_0)[s_1]$ again. Thus, we are forced to return D to its old state by extracting x from D.)

(7) Extract x from D and define $f_e[s_2]$ as the g-extension of $f_e[s_0]$.

(Note that this extraction may allow $g_e(y_0)$ to change back. Thus, the next g-extension of f_e can change it back too. However, in this case we define $f_e[s_3]$ as the id-extension of $f_e[s_2]$, unless one of the higher priority strategies does something different. Note that $f_e(y_0)$ can be changed only because of g-extension and it can happen only due to the higher priority strategies.)

Notice that the strategy is pretty similar to the construction of a properly 2-c.e. degree, but it has some additional features. We use the finite injury argument and put all the $\mathcal{S}$-requirements into a priority list. We say that $\mathcal{S}_{e,i}$ has higher priority than $\mathcal{S}_{e',i'}$ if $\langle e', i' \rangle > \langle e, i \rangle$.

When we *initialize* a strategy, we cancel its declaration about satisfaction and we also cancel its witness. For each $e \in \omega$, we build only one copy of $\mathcal{N}_e$ and f_e. The auxiliary parameter s_e^- denotes the previous expansionary stage (namely, $s_e^-[s] = s'$ is the previous stage at which we defined $\mathcal{N}_e[s']$ and $f_e[s']$). If we do not redefine s_e^- explicitly, then its value is not changed. Recall that we define $f_e[s]$ either as id-extension of $f_e[s']$ or as g-extension of $f_e[s']$, meanwhile $\mathcal{N}_e[s]$ is defined automatically as the corresponding initial segment.

We say that $\mathcal{S}_{e,i}$ *requires attention* at stage $s+1$ if it is not satisfied and one of the following conditions holds (we choose first case which applies):

(σ1) $x[s]$ is not defined.

(σ2) $x[s] \notin D[s]$, $x[s] < L(e)[s]$, and $f_e(y)[s_e^-[s]]$ is undefined for some $y < \varphi_e(x)[s]$.

(σ3) $x[s] \notin D[s]$, $x[s] < L(e)[s]$, and $\varphi_e(x)[s] < l(e,i)[s]$.

(σ4) $x[s] \in D[s]$ and $x[s] < L(e)[s]$.

(Note that we need to define f_e and $\mathcal{N}_e$ only in cases (σ2) and (σ4).)

Construction. *Stage* 0. We define $D[0] = \emptyset$, $s_e^-[0] = -1$, $\mathcal{N}_e[-1] = \emptyset$, and $f_e[-1] = \emptyset$, for all $e \in \omega$. We initialize all strategies.

Stage $s+1$. Find the least $\langle e, i \rangle$ such that $\mathcal{S}_{e,i}$ requires attention at stage $s+1$. Initialize all lower priority strategies. Consider cases in the following ordering, and act according to first case which applies, then proceed to the next stage $s + 2$.

(1) If $\mathcal{S}_{e,i}$ requires attention due to case (σ1), then define witness $x[s + 1]$ as a "big" number.

(2) If $\mathcal{S}_{e,i}$ requires attention due to case (σ2), then define $\mathcal{N}_e[s]$ and $f_e[s]$ in such a way that $f_e[s]$ is the *id*-extension of $f_e[s^-]$, and $f_e[s]: \mathcal{A}_e[s] \cong \mathcal{N}_e[s]$. Define $s_e^-[s + 1] = s$.
   (Here $s^- = s_e^-[s]$, and $s^- + 1$ is the previous stage such that a strategy $\mathcal{S}_{e,i_0}$ (for some i_0) required attention due to (σ2) or (σ4).)

(3) If $\mathcal{S}_{e,i}$ requires attention due to case (σ3), then put x into D (in other words, define $D(x)[s + 1] = 1$).

(4) If $\mathcal{S}_{e,i}$ requires attention due to case (σ4), then extract x from D (in other words, define $D(x)[s + 1] = 0$). Define $\mathcal{N}_e[s]$ and $f_e[s]$ in such a way that $f_e[s]$ is the g_e-extension of $f_e[s^-]$, and $f_e[s]: \mathcal{A}_e[s] \cong \mathcal{N}_e[s]$. Set $s_e^-[s+1] = s$ and declare that $\mathcal{S}_{e,i}$ is satisfied.

(Again, here $s^- = s_e^-[s]$, and $s^- + 1$ is the previous stage at which $\mathcal{S}_{e,i_0}$ (for some i_0) required attention due to (σ2) or (σ4).)

Verification. It is easy to see that D is a 2-c.e. set by construction. We prove by induction that all $\mathcal{S}$-requirements are satisfied, then we show that all $\mathcal{R}$-requirements are satisfied too.

Since the base step of the induction is a simpler case of an inductive step, we only prove the latter. Assume that requirements $\mathcal{S}_{e',i'}$ are satisfied for all $\langle e', i' \rangle < \langle e, i \rangle$. We will prove that $\mathcal{S}_{e,i}$ is also satisfied. By inductive assumption, fix a stage s_0 such that all higher priority strategies are declared to be satisfied. Thus, $\mathcal{S}_{e,i}$ is not initialized after stage s_0. So, assume that the final witness $x = x_{e,i}[s_1]$ was chosen at stage $s_1 > s_0$.

If $\mathcal{S}_{e,i}$ will not require attention after s_1, then it is satisfied vacuously: either g_e is not a Δ_2^0 isomorphism from $\mathcal{A}_e$ onto $\mathcal{B}_e$, or $D \neq \Phi_e^{g_e}$, or $g_e \neq \Theta_e^D$.

(Note also that g_e can be Σ_2^0: it is possible that for some z, $g_e(z)[s]$ changes infinitely often. This does not allow us to see the true state of the oracle of computation $\Phi_e^{g_e}$; however, in this case, the requirements $\mathcal{S}_{e,j}, j \in \omega$, are obsolete and they do not need to be satisfied.)

Assume that the left-hand side of $S_{e,i}$ is true. Then the strategy can require attention at most three times. If $S_{e,i}$ requires the next attention after s_1, then this must be due to case $(\sigma 2)$, say at stage $s_2 + 1 > s_1$. Indeed, this is a corollary of the following properties: $x \notin D[s_2]$ and f_e cannot be defined up to $\varphi_e(x)[s_2]$ (recall that for each stage t at which a higher priority strategy required attention, x was chosen as a "big" number after stage t; in particular, x is greater than every number from the range of $f_e[s_0]$). Thus, at the stage $s_2 + 1$ we extend $f_e[s_2]$ as an id-extension. Stages of this kind (for $S_{e,j}$, $j \in \omega$) ensure that $\mathcal{N}_e$ is an infinite structure and f_e is an isomorphism from $\mathcal{A}_e$ onto $\mathcal{N}_e$ (if the left-hand side of the requirement $\mathcal{R}_e$ is true). Moreover, the range of $f_e[s_2 + 1]$ became bigger than $\varphi_e(x)[s_2]$.

The next attention (at some stage $s_3 + 1 > s_2 + 1$) can appear if $f_e(y)[s_3] = \Psi_i^D(y)[s_3]$ for all $y \leq \varphi_e(x)[s_2]$. If this attention does not happen, then $S_{e,i}$ is satisfied by easy diagonalization. Thus, x goes into D at stage $s_3 + 1$ by case (3) of construction.

Since at stages $s_2 + 1$ and $s_3 + 1$ we initialized all lower priority strategies, no number less than $\theta_e(\varphi_e(x))[s_2]$ can enter D or exit D after stage s_2 (except x; also recall that now the higher priority strategies do not act and all the lower priority strategies have witnesses $> \theta_e(\varphi_e(x))[s_2]$). Now assume that the last attention happens at stage $s_4 + 1 > s_3 + 1$. Thus, we declare that our requirement is satisfied and henceforth, $S_{e,i}$ does not initialize other strategies. Moreover, the following situation is impossible: for some $j > i$, the strategy $S_{e,j}$ requires attention due to one of cases $(\sigma 2)$ or $(\sigma 4)$ after stage $s_3 + 1$ and before stage $s_4 + 1$ (since the truth of $(\sigma 2)$ for $S_{e,j}$ implies the truth of $(\sigma 4)$ for $S_{e,i}$). Thus, we have $g_e[s_2] \upharpoonright \varphi_e(x)[s_2] \neq g_e[s_4] \upharpoonright \varphi_e(x)[s_2]$ (otherwise, $0 = D(x)[s_2] = D(x)[s_4] = 1$). This allows us to define the g_e-extension of $f_e[s_2]$; thus, $f_e[s_2] \upharpoonright \varphi_e(x)[s_2] \neq f_e[s_4] \upharpoonright \varphi_e(x)[s_2]$. Also, by case (4) of the construction, we define $D(x)[s_4 + 1] = 0$.

However, another important point is that

$$D \upharpoonright (\theta_e(\varphi_e(x))[s_2]) = D[s_4 + 1] \upharpoonright \theta_e(\varphi_e(x))[s_2] = D[s_2] \upharpoonright \theta_e(\varphi_e(x))[s_2],$$

hence $g_e[s_e^-[s]] \upharpoonright \varphi_e(x)[s_2] = g_e[s_4] \upharpoonright \varphi_e(x)[s_2]$ for all $s > s_4 + 1$ and so

$$f_e[s_e^-[s]] \upharpoonright \varphi_e(x)[s_2] = f_e[s_4] \upharpoonright \varphi_e(x)[s_2].$$

Therefore,

$$f_e \upharpoonright (\varphi_e(x)[s_2]) = f_e[s_4] \upharpoonright \varphi_e(x)[s_2] \neq f_e[s_2] \upharpoonright \varphi_e(x)[s_2] =$$
$$\Psi_i^D[s_2] \upharpoonright \varphi_e(x)[s_2] = \Psi_i^D \upharpoonright (\varphi_e(x)[s_2]),$$

since $D[s_4 + 1] \upharpoonright \psi_i(\varphi_e(x))[s_2] = D[s_2] \upharpoonright \psi_i(\varphi_e(x))[s_2]$ will not change after $s_4 + 1$ (unless our $S_{e,i}$ is initialized). Thus, $S_{e,i}$ is satisfied.

(Recall also that all new witnesses will be greater than $\psi_i(\varphi_e(x))[s_2]$. Hence, the value of $f_e[s_4] \upharpoonright \varphi_e(x)[s_2]$ will not change after stage $s_4 + 1$, since $g_e[s_4] \upharpoonright \varphi_e(x)[s_2]$ will not change after stage $s_4 + 1$ since all new witnesses for D will be greater than $\theta_e(\varphi_e(x))[s_2]$ too.)

We proceed to the $\mathcal{R}$-requirements. Consider a requirement $\mathcal{R}_e$. Assume that g_e is an isomorphism from $\mathcal{A}_e$ onto $\mathcal{B}_e$, and the left-hand side of $\mathcal{R}_e$ is true (otherwise, $\mathcal{R}_e$ is satisfied trivially). Since $\mathcal{S}_{e,i}$ are satisfied for all $i \in \omega$, we have $f_e \not\leq_T D$. Moreover, there is an infinite series of stages $s_e^-[s] = s$ at which we have $f_e[s] : \mathcal{A}_e[s] \cong \mathcal{N}_e[s]$, by construction. Also, if g_e is Δ_2^0, then $f_e = \lim_s f_e[s]$. Therefore, if $g_e \colon \mathcal{A}_e \cong \mathcal{B}_e$, then $f_e \colon \mathcal{A}_e \cong \mathcal{N}_e$. Hence, all $\mathcal{R}$-requirements are satisfied. This finishes the verification and the proof of Theorem 3.

Remark. Now we can sketch how to prove the result for any finite spectral dimension. For each non-zero n, we consider n pairs of (partial) computable graphs: $(\mathcal{A}_e^0, \mathcal{B}_e^0)$, $(\mathcal{A}_e^1, \mathcal{B}_e^1)$, ..., $(\mathcal{A}_e^n, \mathcal{B}_e^n)$. We also consider $g_e = g_e^0 \oplus g_e^1 \oplus \ldots \oplus g_e^n$, where g_e^i is a Δ_2^0 isomorphism from $\mathcal{A}_e^i$ onto $\mathcal{B}_e^i$. We construct $f_e = f_e^0 \oplus f_e^1 \oplus \ldots \oplus f_e^n$. For each $\mathcal{A}_e^i$, we reserve its own $\mathcal{N}_e^i$. This means that whenever g_e^i changes, we have an opportunity to change f_e^i. Thus, eventually we will ensure that $f_e^0 \oplus f_e^1 \oplus \ldots \oplus f_e^n \not\leq_T D$. From this we deduce that $f_e^{i_0} \not\leq_T D$ for some $i_0 \leq n$.

Assume that $\mathcal{S}$ is a rigid structure and $\deg(D)$ is the degree of categoricity of $\mathcal{S}$. By Theorem 2, one can find a non-zero natural number N and a sequence of computable structures $(\mathcal{A}_i, \mathcal{B}_i)_{0 \leq i < N}$ with the following properties:

1. $\mathcal{A}_i \cong \mathcal{B}_i \cong \mathcal{S}$, and
2. if g_i is the (unique) isomorphism from $\mathcal{A}_i$ onto $\mathcal{B}_i$, then we have $\mathbf{d} = \deg(g_0 \oplus g_1 \oplus \ldots \oplus g_n)$.

On the other hand, the construction sketched above guarantees that there is a computable copy $\mathcal{N}^i$ such that the isomorphism from $\mathcal{A}_i$ onto $\mathcal{N}^i$ is not D-computable; a contradiction. Therefore, $\deg(D)$ cannot be a degree of categoricity of a rigid structure.

References

1. Fröhlich, A., Shepherdson, J.C.: Effective procedures in field theory. Phil. Trans. R. Soc. Lond. A **248**(950), 407–432 (1956)
2. Mal'tsev, A.I.: Constructive algebras. I. Russ. Math. Surv. **16**(3), 77–129 (1961)
3. Mal'tsev, A.I.: On recursive abelian groups. Sov. Math. Dokl. **32**, 1431–1434 (1962)
4. Fokina, E.B., Kalimullin, I., Miller, R.: Degrees of categoricity of computable structures. Arch. Math. Logic **49**(1), 51–67 (2010)
5. Csima, B.F., Franklin, J.N.Y., Shore, R.A.: Degrees of categoricity and the hyperarithmetic hierarchy. Notre Dame J. Form. Logic **54**(2), 215–231 (2013)
6. Anderson, B.A., Csima, B.F.: Degrees that are not degrees of categoricity. Notre Dame J. Form. Logic **57**(3), 389–398 (2016)
7. Miller, R.: $\mathbf{d}$-computable categoricity for algebraic fields. J. Symb. Log. **74**(4), 1325–1351 (2009)
8. Miller, R., Shlapentokh, A.: Computable categoricity for algebraic fields with splitting algorithms. Trans. Amer. Math. Soc. **367**(6), 3955–3980 (2015)
9. Bazhenov, N.A.: Degrees of categoricity for superatomic Boolean algebras. Algebra Logic **52**(3), 179–187 (2013)

10. Bazhenov, N.A.: Δ_2^0-categoricity of Boolean algebras. J. Math. Sci. **203**(4), 444–454 (2014)
11. Bazhenov, N.A.: Autostability spectra for Boolean algebras. Algebra Logic **53**(6), 502–505 (2015)
12. Bazhenov, N.A.: Degrees of autostability relative to strong constructivizations for Boolean algebras. Algebra Logic **55**(2), 87–102 (2016)
13. Frolov, A.N.: Effective categoricity of computable linear orderings. Algebra Logic **54**(5), 415–417 (2015)
14. Bazhenov, N.A.: Degrees of autostability for linear orders and linearly ordered abelian groups. Algebra Logic **55**(4), 257–273 (2016)
15. Fokina, E., Frolov, A., Kalimullin, I.: Categoricity spectra for rigid structures. Notre Dame J. Form. Logic. **57**(1), 45–57 (2016)
16. Goncharov, S.S.: Degrees of autostability relative to strong constructivizations. Proc. Steklov Inst. Math. **274**, 105–115 (2011)
17. Bazhenov, N.A.: Autostability spectra for decidable structures. Math. Struct. Comput. Sci. (to appear). doi:10.1017/S096012951600030X.
18. Bazhenov, N.A., Kalimullin, I., Yamaleev, M.M.: Degrees of categoricity vs. strong degrees of categoricity. Algebra Logic **55**(2), 173–177 (2016)

Flexible Indexing of Repetitive Collections

Djamal Belazzougui[1], Fabio Cunial[2(✉)], Travis Gagie[3], Nicola Prezza[4],
and Mathieu Raffinot[5]

[1] DTISI-CERIST, Algiers, Algeria
dbelazzougui@cerist.dz
[2] MPI-CBG, Dresden, Germany
cunial@mpi-cbg.de
[3] UDP and CeBiB, Santiago, Chile
travis.gagie@mail.udp.cl
[4] DTU, Copenhagen, Denmark
npre@dtu.dk
[5] CNRS, Bordeaux, France
mathieu.raffinot@u-bordeaux.fr

Abstract. Highly repetitive strings are increasingly being amassed by
genome sequencing experiments, and by versioned archives of source code
and webpages. We describe practical data structures that support counting and locating all the exact occurrences of a pattern in a repetitive
text, by combining the run-length encoded Burrows-Wheeler transform
(RLBWT) with the boundaries of Lempel-Ziv 77 factors. One such variant
uses an amount of space comparable to LZ77 indexes, but it answers count
queries between two and four orders of magnitude faster than all LZ77 and
hybrid index implementations, at the cost of slower locate queries. Combining the RLBWT with the compact directed acyclic word graph answers
locate queries for short patterns between four and ten times faster than a
version of the run-length compressed suffix array (RLCSA) that uses comparable memory, and with very short patterns our index achieves speedups
even greater than ten with respect to RLCSA.

1 Introduction

Locating and counting all the exact occurrences of a pattern in a massive,
highly repetitive collection of similar texts is a fundamental primitive in the
post-genome era, in which genomes from multiple related species, from multiple
strains of the same species, or from multiple individuals, are being sequenced
at an increasing pace. Most data structures designed for such repetitive collections take space proportional to a specific measure of repetition, for example
the number z of factors in a Lempel-Ziv parsing [1, 15], or the number r of runs
in a Burrows-Wheeler transform [17]. In previous work we achieved competitive
theoretical tradeoffs between space and time in locate queries, by combining
data structures that depend on multiple measures of repetition that all grow
sublinearly in the length of a repetitive string [3]. Specifically, we described
a data structure that takes approximately $O(z + r)$ words of space, and that

© Springer International Publishing AG 2017
J. Kari et al. (Eds.): CiE 2017, LNCS 10307, pp. 162–174, 2017.
DOI: 10.1007/978-3-319-58741-7_17

reports all the occurrences of a pattern of length m in a text of length n in $O(m(\log\log n + \log z) + \mathsf{pocc} \cdot \log^{\epsilon} z + \mathsf{socc} \cdot \log\log n)$ time, where pocc and socc are the number of primary and of secondary occurrences, respectively (defined in Sect. 2). This compares favorably to the reporting time of Lempel-Ziv 77 (LZ77) indexes [15], and to the space of solutions based on the run-length encoded Burrows-Wheeler transform (RLBWT) and on suffix array samples [17]. We also introduced a data structure whose size depends on the number of right-extensions of maximal repeats, and that reports all the occ occurrences of a pattern in $O(m \log\log n + \mathsf{occ})$ time. The main component of our constructions is the RLBWT, which we use for counting the number of occurrences of a pattern, and which we combine with the compact directed acyclic word graph, and with data structures from LZ indexes, rather than with suffix array samples, for answering locate queries. In this paper we describe and implement a range of practical variants of such theoretical approaches, and we compare their space-time tradeoffs to a representative set of state-of-the-art indexes for repetitive collections.

2 Preliminaries

Let $\Sigma = [1..\sigma]$ be an integer alphabet, let $\# = 0 \notin \Sigma$ be a separator, and let $T \in [1..\sigma]^{n-1}$ be a string. We denote by $\overline{T}$ the reverse of T, and by $\mathcal{P}_{T\#}(W)$ the set of all starting positions of a string $W \in [0..\sigma]^{+}$ in the circular version of $T\#$. We set $\Sigma^r_{T\#}(W) = \{a \in [0..\sigma] : |\mathcal{P}_{T\#}(Wa)| > 0\}$ and $\Sigma^{\ell}_{T\#}(W) = \{a \in [0..\sigma] : |\mathcal{P}_{T\#}(aW)| > 0\}$. A *repeat* $W \in \Sigma^{+}$ is a string with $|\mathcal{P}_{T\#}(W)| > 1$. A repeat W is *right-maximal* (respectively, *left-maximal*) iff $|\Sigma^r_{T\#}(W)| > 1$ (respectively, iff $|\Sigma^{\ell}_{T\#}(W)| > 1$). A *maximal repeat* is a repeat that is both left- and right-maximal. We say that a maximal repeat W is *rightmost* (respectively, *leftmost*) if no string WV with $V \in [0..\sigma]^{+}$ is left-maximal (respectively, if no string VW with $V \in [0..\sigma]^{+}$ is right-maximal).

For reasons of space we assume the reader to be familiar with the notion of *suffix tree* $\mathsf{ST}_{T\#} = (V, E)$ of $T\#$, i.e. the compact trie of all suffixes of $T\#$ (see e.g. [12] for an introduction). We denote by $\ell(\gamma)$, or equivalently by $\ell(u, v)$, the label of edge $\gamma = (u, v) \in E$, and we denote by $\ell(v)$ the concatenation of all edge labels in the path from the root to node $v \in V$. It is well known that a string W is right-maximal (respectively, left-maximal) in $T\#$ iff $W = \ell(v)$ for some internal node v of $\mathsf{ST}_{T\#}$ (respectively, iff $W = \overline{\ell(v)}$ for some internal node v of $\mathsf{ST}_{\overline{T}\#}$). Since left-maximality is closed under prefix operation, there is a bijection between the set of all maximal repeats of $T\#$ and the set of all nodes of the suffix tree of $T\#$ that lie on paths that start from the root and that end at nodes labelled by rightmost maximal repeats (a symmetrical observation holds for the suffix tree of $\overline{T}\#$).

The *compact directed acyclic word graph* of $T\#$ (denoted by $\mathsf{CDAWG}_{T\#}$ in what follows) is the minimal compact automaton that recognizes the set of suffixes of $T\#$ [4,7]. It can be seen as a minimization of $\mathsf{ST}_{T\#}$ in which all leaves are merged to the same node (the sink) that represents $T\#$ itself, and in which all nodes except the sink are in one-to-one correspondence with the maximal repeats of $T\#$ [20] (the source corresponds to the empty string). As in the suffix tree, transitions are labelled by substrings of $T\#$, and the subgraph of $\mathsf{ST}_{T\#}$ induced by maximal repeats is isomorphic to a spanning tree of $\mathsf{CDAWG}_{T\#}$.

For reasons of space we assume the reader to be familiar with the notion and uses of the Burrows-Wheeler transform (BWT) of T and of the FM index, including the C array, LF mapping, and backward search (see e.g. [9]). In this paper we use $\mathsf{BWT}_{T\#}$ to denote the BWT of $T\#$, and we use $\mathrm{range}(W) = [\mathrm{sp}(W)..\mathrm{ep}(W)]$ to denote the lexicographic interval of a string W in a BWT that is implicit from the context. We say that $\mathsf{BWT}_{T\#}[i..j]$ is a *run* iff $\mathsf{BWT}_{T\#}[k] = c \in [0..\sigma]$ for all $k \in [i..j]$, and moreover if every substring $\mathsf{BWT}_{T\#}[i'..j']$ such that $i' \leq i$, $j' \geq j$, and either $i' \neq i$ or $j' \neq j$, contains at least two distinct characters. We denote by $r_{T\#}$ the number of runs in $\mathsf{BWT}_{T\#}$, and we call *run-length encoded BWT* (denoted by $\mathsf{RLBWT}_{T\#}$) any representation of $\mathsf{BWT}_{T\#}$ that takes $O(r_{T\#})$ words of space, and that supports rank and select operations (see e.g. [16,17,21]). Since the difference between $r_{T\#}$ and $r_{\overline{T}\#}$ is negligible in practice, we denote both of them by r when T is implicit from the context.

Repetition-aware string indexes. The *run-length compressed suffix array* of $T\#$, denoted by $\mathsf{RLCSA}_{T\#}$ in what follows, consists of a run-length compressed rank data structure for $\mathsf{BWT}_{T\#}$, and of a sampled suffix array, denoted by $\mathsf{SSA}_{T\#}$ [17]. The average time for locating an occurrence is inversely proportional to the size of $\mathsf{SSA}_{T\#}$, and fast locating needs a large SSA regardless of the compressibility of the dataset. Mäkinen et al. suggested ways to reduce the size of the SSA [17], but they did not perform well enough in real repetitive datasets for the authors to include them in the software they released.

The *Lempel-Ziv 77 factorization* of T [24], abbreviated with LZ77 in what follows, is the greedy decomposition of T into *phrases* or *factors* $T_1 T_2 \cdots T_z$ defined as follows. Assume that T is virtually preceded by the set of distinct characters in its alphabet, and assume that $T_1 T_2 \cdots T_i$ has already been computed for some prefix of length k of T: then, T_{i+1} is the longest prefix of $T[k+1..n]$ such that there is a $j \leq k$ that satisfies $T[j..j + |T_{i+1}| - 1] = T_{i+1}$. For reasons of space we assume the reader to be familiar with LZ77 indexes: see e.g. [10,13]. Here we just recall that a *primary occurrence* of a pattern P in T is one that crosses or ends at a phrase boundary in the LZ77 factorization $T_1 T_2 \cdots T_z$ of T. All other occurrences are called *secondary*. Once we have computed primary occurrences, locating all socc secondary occurrences reduces to two-sided range reporting, and it takes $O(\mathrm{socc} \cdot \log \log n)$ time with a data structure of $O(z)$ words of space [13]. To locate primary occurrences, we use a data structure for four-sided range reporting on a $z \times z$ grid, with a marker at (x, y) if the x-th LZ factor in lexicographic order is preceded in the text by the lexicographically y-th reversed prefix ending at a phrase boundary. This data structure takes $O(z)$ words of space, and it returns all the phrase boundaries that are immediately followed by a factor in the specified range, and immediately preceded by a reversed prefix in the specified range, in $O((1 + k) \log^\epsilon z)$ time, where k is the number of phrase boundaries reported [5]. Kärkkäinen and Ukkonen used two PATRICIA trees

[18], one for the factors and the other for the reversed prefixes ending at phrase boundaries [13]. Their approach takes $O(m^2)$ total time if T is not compressed. Replacing the uncompressed text by an augmented compressed representation, we can store T in $O(z \log n)$ space such that later, given P, we can find all occ occurrences of P in $O(m \log m + \text{occ} \cdot \log \log n)$ time [10].

Alternatively, if all queried patterns are of length at most M, we could store in a FM index the substrings of T that consist of characters within distance M from the closest phrase boundary, and use that to find primary occurrences (see e.g. [22] and references therein). This approach is known as *hybrid indexing*.

Composite repetition-aware string indexes. Combining $\text{RLBWT}_{T\#}$ with the set of all starting positions $p_1, p_2, \ldots, p_z$ of the LZ factors of T, yields a data structure that takes $O(z + r)$ words of space, and that reports all the pocc primary occurrences of a pattern $P \in [1..\sigma]^m$ in $O(m(\log \log n + \log z) + \text{pocc} \cdot \log^\epsilon z)$ time [3]. Since such data structure is at the core of this paper, we summarize it in what follows. The same primary occurrence of P in T can cover up to m factor boundaries. Thus, we consider every possible way of placing, inside P, the rightmost boundary between two factors, i.e. every possible split of P in two parts $P[1..k-1]$ and $P[k..m]$ for $k \in [2..m]$, such that $P[k..m]$ is either a factor or a proper prefix of a factor. For every such k, we use four-sided range reporting queries to list all the occurrences of P in T that conform to the split, as described before. We encode the sequence $p_1, p_2, \ldots, p_z$ implicitly, as follows: we use a bitvector $\text{last}[1..n]$ such that $\text{last}[i] = 1$ iff $\text{SA}_{\overline{T}\#}[i] = n - p_j + 2$ for some $j \in [1..z]$, i.e. iff $\text{SA}_{\overline{T}\#}[i]$ is the last position of a factor. We represent such bitvector as a predecessor data structure with partial ranks, using $O(z)$ words of space [23]. Let $\text{ST}_{T\#} = (V, E)$ be the suffix tree of $T\#$, and let $V' = \{v_1, v_2, \ldots, v_z\} \subseteq V$ be the set of loci in $\text{ST}_{T\#}$ of all the LZ factors of T. Consider the list of node labels $L = \ell(v_1), \ell(v_2), \ldots, \ell(v_z)$, sorted in lexicographic order. It is easy to build a data structure that takes $O(z)$ words of space, and that implements in $O(\log z)$ time function $\mathbb{I}(W, V')$, which returns the (possibly empty) interval of W in L (see e.g. [3]). Together with last, $\text{RLBWT}_{T\#}$ and $\text{RLBWT}_{\overline{T}\#}$, this data structure is the output of our construction.

Given P, we first perform a backward search in $\text{RLBWT}_{T\#}$ to determine the number of occurrences of P in $T\#$: if this number is zero, we stop. During backward search, we store in a table the interval $[i_k..j_k]$ of $P[k..m]$ in $\text{BWT}_{T\#}$ for every $k \in [2..m]$. Then, we compute the interval $[i'_{k-1}..j'_{k-1}]$ of $\overline{P[1..k-1]}$ in $\text{BWT}_{\overline{T}\#}$ for every $k \in [2..m]$, using backward search in $\text{RLBWT}_{\overline{T}\#}$: if $\text{rank}_1(\text{last}, j'_{k-1}) - \text{rank}_1(\text{last}, i'_{k-1} - 1) = 0$, then $P[1..k-1]$ never ends at the last position of a factor, and we can discard this value of k. Otherwise, we convert $[i'_{k-1}..j'_{k-1}]$ to the interval $[\text{rank}_1(\text{last}, i'_{k-1} - 1) + 1..\text{rank}_1(\text{last}, j'_{k-1})]$ of all the reversed prefixes of T that end at the last position of a factor. Rank operations on last can be implemented in $O(\log \log n)$ time using predecessor queries. We get the lexicographic interval of $P[k..m]$ in the list of all distinct factors of T, in $O(\log z)$ time, using operation $\mathbb{I}(P[k..m], V')$. We use such intervals to query the four-sided range reporting data structure.

It is also possible to combine $\text{RLBWT}_{T\#}$ with $\text{CDAWG}_{T\#}$, building a data structure that takes $O(e_{T\#})$ words of space, and that reports all the occ occurrences of P in $O(m \log \log n + \texttt{occ})$ time, where $e_{T\#}$ is the number of right-extensions of maximal repeats of $T\#$ [3]. Specifically, for every node v in the CDAWG, we store $|\ell(v)|$ in a variable $v.\texttt{length}$. Recall that an arc (v, w) in the CDAWG means that maximal repeat $\ell(w)$ can be obtained by extending maximal repeat $\ell(v)$ to the right *and to the left*. Thus, for every arc $\gamma = (v, w)$ of the CDAWG, we store the first character of $\ell(\gamma)$ in a variable $\gamma.\texttt{char}$, and we store the length of the right extension implied by γ in a variable $\gamma.\texttt{right}$. The length $\gamma.\texttt{left}$ of the left extension implied by γ can be computed by $w.\texttt{length} - v.\texttt{length} - \gamma.\texttt{right}$. For every arc of the CDAWG that connects a maximal repeat W to the sink, we store just $\gamma.\texttt{char}$ and the starting position $\gamma.\texttt{pos}$ of string $W \cdot \gamma.\texttt{char}$ in T. The total space used by the CDAWG is $O(e_{T\#})$ words, and the number of runs in $\mathsf{BWT}_{T\#}$ can be shown to be $O(e_{T\#})$ as well [3] (an alternative construction could use $\mathsf{CDAWG}_{\overline{T}\#}$ and $\mathsf{RLBWT}_{\overline{T}\#}$).

Once again, we use the RLBWT to count the number of occurrences of P in T in $O(m \log \log n)$ time: if this number is not zero, we use the CDAWG to report all the $\texttt{occ}$ occurrences of P in $O(\texttt{occ})$ time, using a technique already sketched in [6]. Specifically, since we know that P occurs in T, we perform a blind search for P in the CDAWG, as is typically done with PATRICIA trees. We keep a variable i, initialized to zero, that stores the length of the prefix of P that we have matched so far, and we keep a variable j, initialized to one, that stores the starting position of P inside the last maximal repeat encountered during the search. For every node v in the CDAWG, we choose the arc γ such that $\gamma.\texttt{char} = P[i + 1]$ in constant time using hashing, we increment i by $\gamma.\texttt{right}$, and we increment j by $\gamma.\texttt{left}$. If the search leads to the sink by an arc γ, we report $\gamma.\texttt{pos} + j$ and we stop. If the search ends at a node v that is associated with a maximal repeat W, we determine all the occurrences of W in T by performing a depth-first traversal of all nodes reachable from v in the CDAWG, updating variables i and j as described before, and reporting $\gamma.\texttt{pos} + j$ for every arc γ that leads to the sink. The total number of nodes and arcs reachable from v is $O(\texttt{occ})$.

3   Combining RLBWT and LZ Factors in Practice

In this paper we implement[1] a range of practical variants of the combination of RLBWT and LZ factorization described in Sect. 2. Specifically, in addition to the version described in Sect. 2 (which we call *full* in what follows), we design a variant in which we drop $\mathsf{RLBWT}_{\overline{T}\#}$, simulating it with a bidirectional index, in order to save space (we call this *bidirectional* in what follows); a variant in which we drop $\mathsf{RLBWT}_{\overline{T}\#}$, the four-sided range reporting data structure, and the subset of suffix tree nodes, in order to save even more space (we call this variant *light* in what follows); and another variant in which, to reduce space even further, we use a *sparse* version of the LZ parsing, i.e. we skip a fixed number of characters after each factor (we call this index *sparse* in what follows). In addition, we design a number of optimizations to speed up locate queries in practice: we will describe them in the full version of the paper.

---

[1] Our source code is available at https://github.com/nicolaprezza/lz-rlbwt and https://github.com/nicolaprezza/slz-rlbwt and it is based on SDSL [11].

Our representation of the RLBWT is based on the one described in [21], which we summarize here for completeness, but is more space-efficient. The authors of [21] store one character per run in a string $H \in \Sigma^r$, they mark with a one the beginning of each run in a bitvector $V_{all}[0..n-1]$, and for every $c \in \Sigma$ they store the lengths of all runs of character c consecutively in a bit-vector V_c: specifically, every c-run of length k is represented in V_c as 10^{k-1}. This representation allows one to map rank and access queries on $\mathsf{BWT}_{T\#}$ to rank, select and access queries on H, V_{all}, and V_c. By gap-encoding the bitvectors, this representation takes $r(2\log(n/r) + \log\sigma)(1 + o(1))$ bits of space. We reduce the multiplicative factor of the term $\log(n/r)$ by storing in V_{all} just one out of $1/\epsilon$ ones, where $0 < \epsilon \le 1$ is a given constant (we set $\epsilon = 1/8$ in all our experiments). Note that we are still able to answer all queries on the RLBWT, by using the V_c vectors to reconstruct the positions of the missing ones in V_{all}. However, query time gets multiplied by $1/\epsilon$. We represent H as a Huffman-encoded string (wt_huff<> in SDSL), and we gap-encode bitvectors with Elias-Fano (sd_vector<> in SDSL).

Full index. Our first variant is an engineered version of the data structure described in Sect. 2. We store both $\mathsf{RLBWT}_{T\#}$ and $\mathsf{RLBWT}_{\overline{T}\#}$. A gap-encoded bitvector end$[0..n-1]$ of $z\log(n/z)(1+o(1))$ bits marks the rank, among all the suffixes of $\overline{T}\#$, of every suffix $\overline{T}[i..n-1]\#$ such that $n-i-2$ is the last position of an LZ factor of T. Symmetrically, a gap-encoded bitvector begin$[0..n-1]$ of $z\log(n/z)(1+o(1))$ bits marks the rank, among all the suffixes of $T\#$, of every suffix $T[i..n-1]\#$ such that i is the first position of an LZ factor of T.

Geometric range data structures are implemented with wavelet trees (wt_int in SDSL). We manage to fit the 4-sided data structure in $2z\log z(1+o(1))$ bits, and the 2-sided data structure in $z(2\log n + 1)(1+o(1))$ bits: we will detail such implementations in the full version of the paper. Finally, we need a way to compute the lexicographic range of a string among all the LZ factors of T. We implement a simpler and more space-efficient strategy than the one proposed in [3], which we will describe in the full version of the paper. In summary, the full index takes $(6z\log n + 2(1+\epsilon)r\log(n/r) + 2r\log\sigma) \cdot (1 + o(1))$ bits of space, and it supports count queries in $O(m \cdot (\log(n/r) + \log\sigma))$ time and locate queries in $O((m + \mathsf{occ}) \cdot \log n)$ time.

Bidirectional index. To save space we can drop $\mathsf{RLBWT}_{\overline{T}\#}$ and simulate it using just $\mathsf{RLBWT}_{T\#}$, by applying the synchronization step performed in bidirectional BWT indexes (see e.g. [2] and references therein). This strategy penalizes the time complexity of locate queries, which becomes quadratic in the length of the pattern. Moreover, since in our implementation we store run-lengths separately for each character, a synchronization step requires σ rank queries to find the number of characters smaller than a given character inside a BWT interval. This operation could be performed in $O(\log\sigma)$ time if the string were represented as a wavelet tree. In summary, the bidirectional variant of the index takes $(6z\log n + (1+\epsilon)r\log(n/r) + r\log\sigma) \cdot (1 + o(1))$ bits of space, it supports count queries in $O(m \cdot (\log(n/r) + \log\sigma))$ time, and it supports locate queries in $O(m^2\sigma\log(n/r) + (m + \mathsf{occ}) \cdot \log n)$ time.

Light index. Once we have computed the interval of the pattern in $\mathsf{BWT}_{T\#}$, we can locate all its primary occurrences by just forward-extracting at most m characters for each occurrence inside the range: this is because every primary occurrence of the pattern overlaps with the last position of an LZ factor. We implement forward extraction by using select queries on $\mathsf{RLBWT}_{T\#}$. This approach requires just $\mathsf{RLBWT}_{T\#}$, the 2-sided range data structure, a gap-encoded bitvector end_T that marks the last position of every LZ factor in the text, a gap-encoded bitvector end_{BWT} that marks the last position of every LZ factor in $\mathsf{BWT}_{T\#}$, and z integers of $\log z$ bits each, connecting corresponding ones in end_{BWT} and in end_T: this array plays the role of the sparse suffix array sampling in RLCSA.

Sparse index. We can reduce the size of the index even further by *sparsifying the LZ factorization*. Intuitively, the factorization of a highly-repetitive collection of strings $T = T_1 T_2 \cdots T_k$, where $T_2, \ldots, T_k$ are similar to T_1, is much denser inside T_1 than it is inside $T_2 \cdots T_k$. Thus, excluding long enough contiguous regions from the factorization (i.e. not outputting factors inside such regions) could reduce the number of factors in dense regions. Formally, let $d > 0$, and consider the following generalization of LZ77, denoted here by LZ77-d: we factor T as $X_1 Y_1 X_2 Y_2 \cdots X_{z_d} Y_{z_d}$, where z_d is the size of the factorization, $Y_i \in \Sigma^d$ for all $i \in [1..z_d]$, and X_i is the longest prefix of $X_i Y_i \cdots X_{z_d} Y_{z_d}$ that starts at least once inside the range of positions $[1..|X_1 Y_1 \cdots X_{i-1} Y_{i-1}|]$. To make the light index work with LZ77-d, we need to sample the suffix array of $T\#$ at the lexicographic ranks that correspond to the last position of every X_i, and we need to redefine primary occurrences as those that are not fully contained inside an X factor. To answer a locate query, we also need to extract d additional characters before each occurrence of the pattern, in order to detect primary occurrences that start inside a Y factor. Finally, the 2-sided range data structure needs to be built on the sources of the X factors. The sparse index takes $\left(z_d(3\log n + \log(n/z_d)) + (1+\epsilon)r\log(n/r) \right) \cdot (1+o(1))$ bits of space, it answers locate queries in $O((\mathsf{occ}+1) \cdot (m+d) \cdot \log n)$ time, and count queries in $O(m(\log(n/r) + \log \sigma))$ time. Setting d large enough makes z_d up to three times smaller than the number of LZ factors in realistic highly-repetitive collections.

4   Combining RLBWT and CDAWG in Practice

In this paper we also engineer[2] the combination of RLBWT and CDAWG described in Sect. 2, and in particular we study the effects of two representations of the CDAWG. In the first one, the graph is encoded as a sequence of variable-length integers: every integer is represented as a sequence of bytes, in which the seven least significant bits of every byte are used to encode the integer, and the most significant bit flags the last byte of the integer. Nodes are stored in the sequence according to their topological order in the graph obtained from the CDAWG by inverting the direction of all arcs: to encode a pointer from a

---

[2] Our source code is available at https://github.com/mathieuraffinot/locate-cdawg.

node v to its successor w in the CDAWG, we store the difference between the first byte of v and the first byte of w in the sequence. If w is the sink, such difference is replaced by a shorter code. We choose to store the length of the maximal repeat that corresponds to each node, rather than the offset of $\ell(v)$ inside $\ell(w)$ for every arc (v, w), since such lengths are short and their number is smaller than the number of arcs in practice.

In the second encoding we exploit the fact that the subgraph of the suffix tree of $T\#$ induced by maximal repeats is a spanning tree of $\text{CDAWG}_{T\#}$. Specifically, we encode such spanning tree with the balanced parenthesis scheme described in [19], and we resolve the arcs of the CDAWG that belong to the tree using corresponding tree operations. Such operations work on node identifiers, thus we need to convert a node identifier to the corresponding first byte in the byte sequence of the CDAWG, and vice versa. We implement such translation by encoding the monotone sequence of the first byte of every node with the quasi-succinct representation by Elias and Fano, which uses at most $2 + \log(N/n)$ bits per starting position, where N is the number of bytes in the byte sequence and n is the number of nodes [8].

5   Experimental Results

We test our implementations on five DNA datasets from the Pizza&Chili repetitive corpus[3], which include the whole genomes of approximately 36 strains of the same eukaryotic species, a collection of 23 and approximately 78 thousand substrings of the genome of the same bacterium, and an artificially repetitive string obtained by concatenating 100 mutated copies of the same substring of the human genome. We compare our results to the FM index implementation in SDSL [11] with sampling rate 2^i for $i \in [5..10]$, to an implementation of RLCSA[4] with the same sampling rates, to the five variants in the implementation of the LZ77 index described in [14], and to a recent implementation of the compressed hybrid index [22]. The FM index uses RRR bitvectors in its wavelet tree. For brevity, we call LZ1 the implementation of the LZ77 index that uses the suffix trie and the reverse trie. For each process, and for each pattern length 2^i for $i \in [3..10]$, we measure the maximum resident set size and the number of CPU seconds that the process spends in user mode[5], both for locate and for count queries, discarding the time for loading the indexes and averaging our measurements over one thousand patterns[6]. We experiment with skipping 2^i characters before opening a new phrase in the sparse index, where $i \in [5..10]$.

---

[3] http://pizzachili.dcc.uchile.cl/repcorpus.html.
[4] We compile the sequential version of https://github.com/adamnovak/rlcsa with **PSI_FLAGS** and **SA_FLAGS** turned off (in other words, we use a gap-encoded bitvector rather than a succinct bitvector to mark sampled positions in the suffix array). The block size of psi vectors (**RLCSA_BLOCK_SIZE**) is 32 bytes.
[5] We perform all experiments on a single core of a 6-core, 2.50 GHz, Intel Xeon E5-2640 processor, with access to 128 GiB of RAM and running CentOS 6.3. We measure resources with GNU Time 1.7, and we compile with GCC 5.3.0.
[6] We use as patterns random substrings of each dataset, containing just DNA bases, generated with the **genpatterns** tool from the Pizza&Chili repetitive corpus.

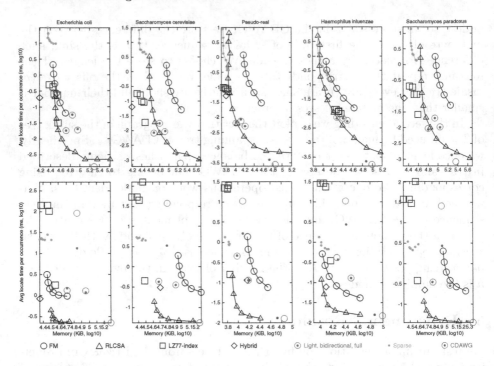

Fig. 1. Locate queries: space-time tradeoffs of our indexes (color) and of the state of the art (black). Top row: patterns of length 16. Bottom row: patterns of length 512. The full, bidirectional, and light indexes are shown with (red dots) and without (red circles) speed optimizations. The CDAWG is shown in succinct (blue dots) and non-succinct (blue circles) version. (Color figure online)

The first key result of our experiments is that, in highly-repetitive strings, the sparse index takes an amount of space that is comparable to LZ indexes, and thus typically smaller than the space taken by RLCSA and by the FM index, while supporting count operations that are approximately as fast as RLCSA and as the FM index, and thus typically faster than LZ indexes. This new tradeoff comes at the cost of slower locate queries.

Specifically, the gap between sparse index and LZ variants in the running time of count queries is large for short patterns: the sparse index is between two and four orders of magnitude faster than all variants of the LZ index, with the largest difference achieved by patterns of length 8 (Fig. 2, bottom). The difference between the sparse index and variant LZ1 shrinks as pattern length increases. Locate queries are between one and three orders of magnitude slower in the sparse index than in LZ indexes, and comparable to RLCSA with sampling rates equal to or greater than 2048 (Fig. 1, top). However, for patterns of length approximately 64 or larger, the sparse index becomes between one and two orders of magnitude *faster* than all variants of the LZ index, except LZ1. As a function of pattern length, the running time per occurrence of the sparse index grows

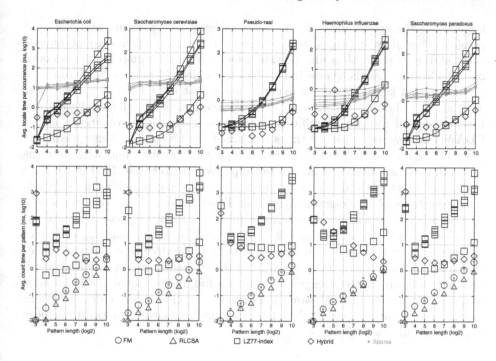

Fig. 2. Locate time per occurrence (top) and count time per pattern (bottom), as a function of pattern length, for the sparse index with skip rate 2^i, $i \in [5..10]$, the LZ77 index, and the hybrid index. Count plots show also the FM index and RLCSA.

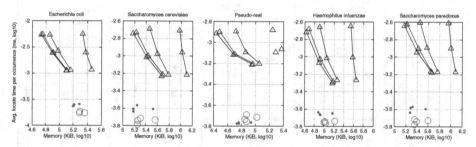

Fig. 3. Space-time traeoffs of the CDAWG (blue) compared to RLCSA (triangles) with sampling rate 2^i, $i \in [3..5]$. Patterns of length 8, 6, 4, 2 (from left to right). The CDAWG is shown in succinct (blue dots) and non-succinct (blue circles) version. (Color figure online)

more slowly than the running time of LZ1, suggesting that the sparse index might even approach LZ1 for patterns of length between 1024 and 2048 (Fig. 2, top). Compared to the hybrid index, the sparse index is again orders of magnitude faster in count queries, especially for short patterns (Fig. 2, bottom). As with LZ1, the difference shrinks as pattern length increases, but since the size of the hybrid index depends on maximum pattern length, the hybrid index becomes larger than the sparse index for patterns of length between 64 and 128, and possibly even shorter (Fig. 4, top). As with LZ indexes, faster count queries come at the expense of locate queries, which are approximately 1.5 orders of magnitude slower in the sparse index than in the hybrid index (Fig. 2, top).

The second key result of our experiments is that the CDAWG is efficient at locating very short patterns, and in this regime it achieves the smallest query time among all indexes. Specifically, the running time per occurrence of the CDAWG is between 4 and 10 times smaller than the running time per occurrence of a version of RLCSA that uses comparable memory, and with patterns of length two the CDAWG achieves speedups even greater than 10 (Fig. 3). Note that short exact patterns are a frequent use case when searching large repetitive collections of versioned source code. The CDAWG does not achieve any new useful tradeoff with long patterns. Using the succinct representation of the CDAWG saves between 20% and 30% of the disk size and resident set size of the non-succinct representation, but using the non-succinct representation saves between 20% and 80% of the query time of the succinct representation, depending on dataset and pattern length. Finally, our full, bidirectional and light index implementations exhibit the same performance as the sparse index for count queries, but it turns out that they take too much space in practice to achieve any new useful tradeoff (Fig. 1).

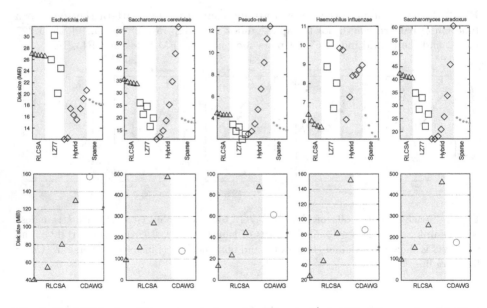

Fig. 4. Top) Disk size of the sparse index with skip rate 2^i, $i \in [10..15]$, compared to the hybrid index with maximum pattern length 2^i, $i \in [3..10]$, the LZ77 index, and RLCSA with sampling rate 2^i, $i \in [10..15]$. (Bottom) Disk size of the CDAWG compared to RLCSA with sampling rate 2^i, $i \in [2..5]$. The CDAWG is shown in succinct (blue dots) and non-succinct (blue circles) version. (Color figure online)

References

1. Arroyuelo, D., Navarro, G., Sadakane, K.: Stronger Lempel-Ziv based compressed text indexing. Algorithmica **62**, 54–101 (2012)
2. Belazzougui, D.: Linear time construction of compressed text indices in compact space. In: Proceedings of the STOC, pp. 148–193 (2014)
3. Belazzougui, D., Cunial, F., Gagie, T., Prezza, N., Raffinot, M.: Composite repetition-aware data structures. In: Cicalese, F., Porat, E., Vaccaro, U. (eds.) CPM 2015. LNCS, vol. 9133, pp. 26–39. Springer, Cham (2015). doi:10.1007/978-3-319-19929-0_3
4. Blumer, A., et al.: Complete inverted files for efficient text retrieval and analysis. JACM **34**, 578–595 (1987)
5. Chan, T.M., Larsen, K.G., Pătraşcu, M.: Orthogonal range searching on the RAM, revisited. In: Proceediings of the SoCG, pp. 1–10 (2011)
6. Crochemore, M., Hancart, C.: Automata for matching patterns. In: Rozenberg, G., et al. (eds.) Handbook of Formal Languages, pp. 399–462. Springer, Heidelberg (1997)
7. Crochemore, M., Vérin, R.: Direct construction of compact directed acyclic word graphs. In: Apostolico, A., Hein, J. (eds.) CPM 1997. LNCS, vol. 1264, pp. 116–129. Springer, Heidelberg (1997). doi:10.1007/3-540-63220-4_55
8. Elias, P., Flower, R.A.: The complexity of some simple retrieval problems. JACM **22**, 367–379 (1975)
9. Ferragina, P., Manzini, G.: Indexing compressed texts. JACM **52**(4), 552–581 (2005)
10. Gagie, T., Gawrychowski, P., Kärkkäinen, J., Nekrich, Y., Puglisi, S.J.: LZ77-based self-indexing with faster pattern matching. In: Pardo, A., Viola, A. (eds.) LATIN 2014. LNCS, vol. 8392, pp. 731–742. Springer, Heidelberg (2014). doi:10.1007/978-3-642-54423-1_63
11. Gog, S., Beller, T., Moffat, A., Petri, M.: From theory to practice: plug and play with succinct data structures. In: Gudmundsson, J., Katajainen, J. (eds.) SEA 2014. LNCS, vol. 8504, pp. 326–337. Springer, Cham (2014). doi:10.1007/978-3-319-07959-2_28
12. Gusfield, D.: Algorithms on Strings, Trees and Sequences: Computer Science and Computational Biology. Cambridge University Press, Cambridge (1997)
13. Kärkkäinen, J., Ukkonen, E.: Lempel-Ziv parsing and sublinear-size index structures for string matching. In: Proceedings of the WSP, pp. 141–155 (1996)
14. Kreft, S.: Self-index based on LZ77. Master's thesis, Department of Computer Science, University of Chile (2010)
15. Kreft, S., Navarro, G.: On compressing and indexing repetitive sequences. TCS **483**, 115–133 (2013)
16. Mäkinen, V., Navarro, G.: Succinct suffix arrays based on run-length encoding. In: Apostolico, A., Crochemore, M., Park, K. (eds.) CPM 2005. LNCS, vol. 3537, pp. 45–56. Springer, Heidelberg (2005). doi:10.1007/11496656_5
17. Mäkinen, V., et al.: Storage and retrieval of highly repetitive sequence collections. JCB **17**, 281–308 (2010)
18. Morrison, D.R.: PATRICIA – practical algorithm to retrieve information coded in alphanumeric. JACM **15**, 514–534 (1968)
19. Munro, J.I., Raman, V.: Succinct representation of balanced parentheses and static trees. SIAM J. Comput. **31**, 762–776 (2002)
20. Raffinot, M.: On maximal repeats in strings. IPL **80**, 165–169 (2001)

21. Sirén, J., Välimäki, N., Mäkinen, V., Navarro, G.: Run-length compressed indexes are superior for highly repetitive sequence collections. In: Amir, A., Turpin, A., Moffat, A. (eds.) SPIRE 2008. LNCS, vol. 5280, pp. 164–175. Springer, Heidelberg (2008). doi:10.1007/978-3-540-89097-3_17

22. Valenzuela, D.: CHICO: a compressed hybrid index for repetitive collections. In: Goldberg, A.V., Kulikov, A.S. (eds.) SEA 2016. LNCS, vol. 9685, pp. 326–338. Springer, Cham (2016). doi:10.1007/978-3-319-38851-9_22

23. Willard, D.E.: Log-logarithmic worst-case range queries are possible in space $\theta(n)$. IPL **17**, 81–84 (1983)

24. Ziv, J., Lempel, A.: A universal algorithm for sequential data compression. IEEE TIT **23**, 337–343 (1977)

Admissibles in Gaps

Merlin Carl[1,2]([✉]), Bruno Durand[3]([✉]), GrégoryLafitte[3]([✉]),
and Sabrina Ouazzani[4]

[1] Fachbereich Mathematik und Statistik, Universität Konstanz,
78457 Konstanz, Germany
merlin.carl@uni-konstanz.de
[2] Lehrstuhl für Theoretische Informatik, Universität Passau,
94032 Passau, Germany
[3] LIRMM, CNRS, Université de Montpellier,
161 Rue Ada, 34090 Montpellier, France
{bruno.durand,gregory.lafitte}@lirmm.fr
[4] LACL, Université Paris-Est,
61 Avenue du Général de Gaulle, 94010 Créteil, France
sabrina.ouazzani@lacl.fr

Abstract. We consider clockable ordinals for Infinite Time Turing
Machines (ITTMs), *i.e.*, halting times of ITTMs on the empty input.
It is well-known that, in contrast to the writable ordinals, the set of
clockable ordinals has 'gaps'. In this paper, we show several results on
gaps, mainly related to the admissible ordinals they may properly contain. We prove that any writable ordinal can occur as the order type of
the sequence of admissible ordinals in such a gap. We give precise information on their ending points. We also investigate higher rank ordinals
(recursively inaccessible, etc.). Moreover, we show that those gaps can
have any reasonably effective length (in the sense of ITTMs) compared
to their starting point.

1 Introduction

Infinite Time Turing Machines (ITTMs), invented by Hamkins and Kidder and
first introduced in [3], are the historically first of a number of machine models of
infinitary computability. Among these various models, ITTMs have been most
extensively studied. A topic that has received particular attention is the issue of
clockability.

An ordinal α is called 'clockable' if and only if there is an ITTM-program P
such that P halts on the empty input after exactly α many steps. As there are
only countable many programs, there are only countable many halting times.
Moreover, a cofinality argument shows that all halting times of ITTMs must be
countable ordinals. There is thus a countable supremum of the ITTM-halting
times, called γ_∞.

The authors would like to express their thanks to the anonymous referees, who made
numerous suggestions and interesting remarks.

J. Kari et al. (Eds.): CiE 2017, LNCS 10307, pp. 175–186, 2017.
DOI: 10.1007/978-3-319-58741-7_18

However, not every countable ordinal below γ_∞ is clockable: It was first observed and demonstrated in [3] that there are 'gaps' in the clockable ordinals, *i.e.*, ordinals $\alpha < \beta < \delta < \gamma_\infty$ such that α and δ are clockable, but β is not. Further information on those gaps and their distribution was obtained in [3,5–7]: The length of a gap is always a limit ordinal, the first gap is of length ω, gaps are always started by admissible ordinals and admissible ordinals are never clockable.

If one wants to obtain a characterization of those ordinals that start gaps, the role of admissibility thus seems to be a good starting point. These aforementioned results hence motivate a further study of the relation of admissible ordinals to gaps. In particular, one may ask the questions whether admissible ordinals always start gaps or whether they can also sit properly inside a gap, which order types can appear as the order type of the sequence of admissible ordinals in a gap and how long gaps can become relative to their starting point. In this paper, we deal with these questions. We also investigate gaps related with admissible of high rank and with rank-admissible (fixed points of the rank function).

In particular, we show that every writable ordinal is the order type of the sequence of admissible ordinals in some gap and that for any function f sending countable ordinals to countable ordinals that is ITTM-computable in an appropriate sense, there is an ordinal α starting a gap of length $\geq f(\alpha)$.

2   Preliminaries

Fix a natural enumeration $(P_i : i \in \omega)$ of the ITTM-programs. An ordinal α is *clockable* if and only if there is an ITTM-program P such that P halts on the empty input after exactly α many steps. We denote by γ_∞ the supremum of the clockable ordinals.

An ordinal α is *writable* if and only if there is an ITTM-program P such that P halts on the empty input with a real number x on the output tape that codes a well-ordering of length α. We denote by λ_∞ the supremum of the writable ordinals.

A *gap* is a non-empty interval $[\alpha, \beta)$ of ordinals that has no clockable element. The length of the gap $[\alpha, \beta)$ is the unique ordinal δ such that $\alpha + \delta = \beta$. The length of a gap is always a limit ordinal (*cf.* [3, Section 3]).

An ordinal α is *eventually writable* if some ITTM working on empty input has a code x for α on its output tape at a certain point and never changes it again; it is *accidentally writable* if some ITTM working on empty input has x at some point on its output tape, possibly changing it again later on.

Theorem 1 (Welch, $\lambda - \zeta - \Sigma$ Theorem [7]). *There are countable ordinals* $\lambda_\infty, \zeta_\infty, \Sigma_\infty$ *such that* $x \subseteq \omega$ *is writable if and only if* $x \in L_{\lambda_\infty}$, *x is eventually writable if and only if* $x \in L_{\zeta_\infty}$ *and x is accidentally writable if and only if* $x \in L_{\Sigma_\infty}$. *Here, λ_∞ is equal both to the supremum of the clockable and the supremum of the writable ordinals. Moreover, $(\lambda_\infty, \zeta_\infty, \Sigma_\infty)$ is characterized as the lexically minimal triple (α, β, δ) such that $L_\alpha \prec_{\Sigma_1} L_\beta \prec_{\Sigma_2} L_\delta$.*

Theorem 2. *(a)* λ_∞ *is an admissible limit of admissible ordinals and a limit of such; in fact, we have* $\omega_{\lambda_\infty}^{CK} = \lambda_\infty$

(b) ζ_∞ *is* Σ_2*-admissible; in particular, the claims about* λ_∞ *from (a) also hold for* ζ_∞.

(c) $\lambda_\infty = \gamma_\infty$

Proof. (a) is the Indescribability Theorem of Hamkins and Lewis [3], (b) is Lemma 23 from [7], we can see (c) as a corollary to Theorem 1.

Theorem 3. *(a) No admissible ordinal is clockable.*
(b) If $\alpha < \lambda_\infty$ *starts a gap, then* α *is admissible.*
Both statements relativise to arbitrary oracles.

Proof. For (a), see [3], for (b), see [5].

Corollary 1. *For every* $x \subseteq \omega$, *the smallest ordinal not clockable in the oracle* x *is* $\omega_1^{CK,x}$. *Moreover, if* x *codes an ordinal* α *and* β *is the next admissible ordinal* $> \alpha$ *such that* $x \in L_\beta$, *then* β *is the first ordinal not clockable in the oracle* x. *If* α *is ITTM-writable as* x *then* $\omega_1^{CK,x}$ *is the first admissible ordinal greater than* α.

Proof. The first two claims are immediate from Theorem 3. For the last claim, note that, by Σ_1-recursion in L_β, any ITTM-computation of length $< \beta$ is contained in L_β, hence so is its output.

Theorem 4 (Theorem 50 from [7]**).** *Every ordinal* $< \lambda_\infty$ *is writable. Thus every clockable ordinal is writable. In fact, if* α *is clockable, then there is an ITTM-program* P *that halts after at most* $\alpha + \omega$ *many steps with a code for* α *on the output tape.*

Proposition 1. *For every* $\alpha < \lambda_\infty$, *there exists* $x \subseteq \omega$ *that codes* α *such that* $\omega_1^{CK,x}$ *is the first admissible after the last gap such that its starting point is* $\leq \alpha$.

Note that if α is not in a gap (hence clockable), then the above proposition tells us that $\omega_1^{CK,x}$ is the next admissible after α and thus starts the next gap after α.

Proof. We use Corollary 1 and the fact that α is writable in time less than the end of the gap that contains it, and Theorem 4 applied to α if clockable, or else to the end of its gap.

Theorem 5. *There is an ITTM-program* $\mathcal{U}$ *such that, for every* $x \subseteq \omega$, $\mathcal{U}^x$ *simulates the computations of all ITTM-programs on* ω *many disjoint portions of the working tape (and we may assume that the ith portion consists of the cells with indices* $\{p(i+1,k) : k \in \omega\}$, *where* p *is Cantor's pairing function and the '+1' creates space for scratch work necessary for the simulation). Moreover,* $\mathcal{U}$ *can be chosen to almost work in 'ω-real-time', i.e., such that* $\mathcal{U}$ *simulates* ω *many steps of each machine in* ω *many steps of simulation time.*

It will occasionally be important to use codes for ordinals as 'stopwatches'. For a better use, such a program should both verify that the input is a code for a well-founded order, and halt in proper time.

Proposition 2 (Count-through program from [2]). *There is an ITTM-program $P_{stopwatch}$ such that, for every $x \subset \omega$ given as input, always halt, accepts x if and only if x codes a well-ordering. If the ordinal type of this well-ordering is α, the halting time is:*

- *exactly α if α is a limit ordinal and $\alpha \geq \omega \cdot 2$,*
- *$\alpha + n$ for some $n < \omega$ if α is a successor ordinal and $\alpha \geq \omega \cdot 2$,*
- *$\omega \cdot 2$ if $\alpha < \omega \cdot 2$.*

In a few words, the idea of the program is to check that x codes an order (requires ω time steps). In parallel it detects the smallest elements of the order by runs of ω ordinals and 'suppress' them in the order in time ω (thus real-time). The difficult point is to discover that all the order has been erased, which requires ω more steps in the literature's algorithms. This last procedure is not needed thanks to an *ad hoc* pre-treatment that uses $\omega \cdot 2$ steps, and a proper use of flashing cells.

Proposition 3. *There is an ITTM-program P_{gap} that never halts but such that, whenever α starts a gap, P produces a code for α on its output tape at time exactly α and keeps it until the end of the gap. This also holds relative to arbitrary oracles.*

Remark that after λ_∞, this program has a code for λ_∞ written on its output tape and keeps it forever, but never halts. The existence of this program proves that λ_∞ is eventually writable.

Proof. Let α start a gap. By Theorem 3, α is both admissible and a limit of clockables. We remark that the ordinal type of clockables below α is exactly α. Thus our program will construct a well-order that corresponds to those clockables. For this, the simplest way is to denote a clockable δ by the index of one of the machines that halts in time δ. Then if we have two integers i, j, in our order, $i < j$ if P_i halts before P_j. Of course, there are ω machines that halt at a given time. Thus we consider only the first program with a given halting time that is discovered while running the chosen universal ω-real-time machine (called $\mathcal{U}$ above).

Thus our program runs as follows: it simulates $\mathcal{U}$ and little by little constructs an order such that $i < j$ if i and j are considered and if P_i halts before P_j. The representation of the order must be adjusted carefully: when we add an element in this order, then ω values are changed, but we can propagate those changes in parallel by runs of ω. Thus if α is a limit ordinal, we have on the tape a representation of the order type of clockables $< \alpha$ and the proposition is proved.

Lemma 1. *For the specific program P_{gap} defined in the proof of Proposition 3, if we start a run at any clockable δ, whenever $\alpha > \delta$ starts a gap, then P produces a code for α on its output tape at time exactly α and keeps it until the end of the gap. This also holds relative to arbitrary oracles.*

Proof. Admissibles are well closed ordinal: the smallest ι such that $\delta + \iota = \alpha$ is exactly α. This implies that the order type of clockables between δ and α is α.

Most of our notations are standard. For $\iota \in \mathrm{On}$, $x \subseteq \omega$, $\omega_\iota^{\mathrm{CK},x}$ denotes the ιth admissible ordinal relative to x. We write $P^x(0) \downarrow = y$ to indicate that the ITTM-program P halts on input 0 in the oracle x with y on its output tape. For $\alpha < \lambda_\infty$, writing(α) denotes the minimal writing time of α, *i.e.*, the minimal length of an ITTM-computation that halts with a code for α on its output tape. For $\alpha \in \mathrm{On}$, α^+ denotes the smallest admissible ordinal $> \alpha$; if $\iota \in \mathrm{On}$, then $\alpha^{+\iota}$ denotes the ιth admissible ordinal $> \alpha$. We will frequently and tacitly make use of the fact that admissible ordinals are closed under ordinal addition, multiplication and exponentiation. We also make use of the equivalence of ITTMs with any finite number of scratch tapes and ITTMs with exactly one scratch tape without explicit mentioning.

3 Admissible Ordinals in Gaps

Theorem 6. *Within the clockable ordinals, there exists a gap properly containing an admissible.*

Proof. This theorem can be seen as a quite direct application of Theorem 1.

There are no clockable ordinals $> \lambda_\infty = \gamma_\infty$. However, as ζ_∞ is recursively inaccessible, there are many admissible ordinals between λ_∞ and $\zeta_\infty < \Sigma$. Thus, $L_\Sigma \models$ "There are ordinals α, β, δ such that $\alpha < \beta < \delta$, β is admissible and L_δ contains no ITTM-computation on input 0 whose length is contained in the interval $[\alpha, \beta)$" (*i.e.*, $[\alpha, \beta)$ contains no clockable).

This statement is clearly Σ_1; as $L_{\lambda_\infty} \prec_{\Sigma_1} L_\Sigma$ by Theorem 1, it holds in L_{λ_∞}. Let $\alpha, \beta, \delta < \lambda_\infty$ witness this. So L_δ believes that β is an admissible ordinal properly (because β is not clockable) inside a gap. However, by Σ_1-recursion in KP and admissibility of β, any ITTM-computation on the empty input of length $< \beta$ is already contained in L_β; thus $[\alpha, \beta)$ is indeed inside a gap which properly contains the admissible ordinal β.

To better understand Theorem 6, we give the sketch of an alternative (more algorithmic) proof. We first design an algorithm that checks if a real x given as input is a code for an ordinal which is the beginning of a gap containing $\omega_1^{\mathrm{CK},x}$. The property "there exists an x such that x is accepted by the algorithm above" is clearly Σ_1. Now remark that this algorithm accepts any code for λ_∞. Thus by Theorem 1, we also have a witness of this property in L_{λ_∞}, which has to be the code of an admissible ordinal less than λ_∞, and that begins a gap with an admissible inside.

More care is required if one further wants to control the number of admissible ordinals inside a gap.

Theorem 7. *The first gap with an admissible inside ends ω steps after this admissible, thus there is exactly one admissible inside.*

Proof. We consider the following algorithm on input 0, which is a variation of the algorithm explained in the second proof of Theorem 6, and we analyse its halting time.

Run the algorithm P_{gap} from Proposition 3. Whenever P_{gap} writes a new code c for an ordinal α starting a gap to its output tape, we continue to run P_{gap}, while in parallel running[1] the following routine R on some reserved portion of the scratch tape:

Use Theorem 5 to simulate all ITTM-programs in the oracle c and wait for a gap. By Corollary 1, the first such gap will start at time $w_1^{\text{CK},x} = \alpha^+$ and it will be detected at time $\alpha^+ + \omega$. If one of the simulated programs in the execution of P_{gap} halts, the execution of R is stopped, the reserved portion of the scratch tape is erased and the execution of P_{gap} just continues. On the other hand, if R detects a gap before any of the simulated programs in the execution of P_{gap} halts, the whole program stops.

Let α be the minimal admissible ordinal such that $[\alpha, \alpha^+)$ is a prefix of a gap (α starts a gap and no clockable can be found in the interval). Then the algorithm just described will first run for α many steps, at which point R will be started with a code c for α in the oracle. By assumption on α, no simulated program in the execution of P_{gap} will halt between times α and α^+, so R will run until the first gap in the oracle c is detected at time $\alpha^+ + \omega$, at which time the whole program halts.

It follows that α^+ is properly contained in the gap started by α. Moreover, as the algorithm just described halts at time $\alpha^+ + \omega$, α^+ is the only admissible ordinal properly contained in this gap. Finally, as gaps always have limit length, no ITTM-program can halt between times α^+ and $\alpha^+ + \omega$, hence the gap starting with α indeed ends exactly at $\alpha^+ + \omega$.

Remark: The last proof also yields a direct proof of Theorem 7 that does not use Theorem 1: indeed if no gap properly containing an admissible exists before λ_∞ then the algorithm just described halts at time $\lambda_\infty^+ + \omega$ which is impossible since it works on input 0 and λ_∞ is the supremum of the clockable ordinals.

It is rather easy to imagine how to modify the program in order to get 2, 3,... admissible ordinals in a gap. Our goal now is to extend this to arbitrary writable order types, but there are some difficulties to deal with recursively inaccessibles that may appear in limit cases.

Theorem 8. *Let α be a writable ordinal. Let $\beta > writing(\alpha)$ be minimal such that β starts a gap such that the set S of admissible ordinals properly contained in this gap is of order type $\beta \geq \alpha$. Then $\beta = \alpha$.*

In other terms, after the writing time for α, the first gap with at least α admissibles inside has exactly α admissibles inside (thus occurs before λ_∞). We first give a proof when α is a limit ordinal, and then a more general and more complex proof when α is a successor ordinal.

---

[1] Here, by 'parallel', we mean that we alternately perform one step of the first and one of the second algorithm. After ω many steps of the parallel execution, we will thus have performed ω many steps of both algorithms.

Proof (Proof for limit cases). We assume here that α is a limit ordinal. First, write a code d for α on some scratch tape. For $\iota < \alpha$, we denote by d_ι the natural number coding ι in d. We consider two further scratch tapes, the first of which we organize into ω many disjoint portions. We also reserve a fourth scratch tape for 'bookkeeping'. Initially, this tape will be empty.

Now consider the following algorithm:

On a fifth scratch tape, run the program P_{gap} from Proposition 3. When a gap is detected starting at an ordinal α_0 coded by c_0, write c_0 to the $d(0)$th portion of the second tape and mark the $d(0)$th cell of the fourth tape.

We now use the third tape to carry out $P_{\text{gap}}^{c_0}$ in parallel with P_{gap}. If any program simulated in the execution of P_{gap} halts before $P_{\text{gap}}^{c_0}$ detects a gap, we erase the first, second, third and fourth tape and continue running P_{gap}. On the other hand, if none of these programs halts before $P_{\text{gap}}^{c_0}$ detects a gap, $P_{\text{gap}}^{c_0}$ will write a code c_1 for α_0^+. We then write c_1 to the $d(1)$th portion of the second tape and mark the $d(1)$th cell of the fourth tape.

We now proceed in ι phases for $\iota < \alpha$:

If $\iota = \bar{\iota} + 1$ is a successor ordinal, and if we have a code for $c_{\bar{\iota}}$ written on the $d(\bar{\iota})$th portion of the second tape, we can use the third tape to run $P_{\text{gap}}^{c_{\bar{\iota}}}$. If any program simulated in the execution of P_{gap} halts before $P_{\text{gap}}^{c_{\bar{\iota}}}$ detects an ordinal α_ι starting a gap, we erase the first four scratch tapes and continue with P_{gap}. If not, we take the output c_ι of P_{gap} (which will code $\alpha_\iota = \alpha_{\bar{\iota}}^+$), write it to the $d(\iota)$th portion of the second tape and mark the $d(\iota)$th cell of the third tape.

On the other hand, if ι is a limit ordinal, then codes have been written to all portions of the second tape with index $< \iota$ and no program simulated by P_{gap} has halted so far. We compute in ω-real time the sum of all of these ordinals to the δth portion.

Thus we get a representation of the supremum of all ι's admissibles in the gap.

Now let us define the halting case for this loop on ι, i.e., $\iota = \alpha$. As soon as the α's storage is filled, we halt. This is implemented by using an improved version of the count-through algorithm $P_{\text{stopwatch}}$ which does not use ω extra time to check that α's storage has been filled: we halt immediately after having filled the α's list. But in our program we could miss some admissibles: exactly those that are limits of admissibles (supremum of ι's admissibles in the gap for all $\iota < \alpha$ that are limit ordinals). We can ignore them because those recursively inaccessible ordinals are rare enough: when α is a limit ordinal, if you consider any sequence of consecutive α ordinals, the subsequence of those that are not recursively inaccessible has the same ordinal type α.

Thus the halting time for the present algorithm is clockable and so is the supremum of the α's admissibles which is not admissible. Hence we have exactly α admissibles in the gap.

To prove the case where α is a successor ordinal, we must be more precise: if we denote by $\bar{\alpha}$ the largest limit ordinal $< \alpha$, then we have to check whether the $\bar{\alpha}$'s ordinal of the gap is recursively inaccessible or not. Remark that for other admissibles of the list we do not need to check for recursive enumerability (as explained above), but our program will take them into account anyway.

Thus, we need an algorithmic characterisation for recursively inaccessible ordinals :

Lemma 2. *Let $\beta < \lambda_\infty$ be a limit of admissibles α_i. Let x_i be a code for α_i in L_β. The ordinal β is admissible (hence recursively inaccessible) if and only if β is in a gap for all computations with oracle x_i.*

Proof. Of course, if β is admissible, then it is not clockable for any x_i-oracle computation with $x_i \in L_\beta$. Conversely, if β is not admissible, then there will be a Σ_1-definable surjection from ω to β, and this surjection can be computed by an algorithm using as oracle some $\delta < \beta$. Thus we can consider $\alpha_i > \delta$, compute this surjection from ω and halt.

Proof (Proof for successor cases). Now α is a successor ordinal. We explain below how we improve the program defined in the limit case above. The aim of our improvement is to check when ι is a limit ordinal, whether the sup of the first ι admissibles is admissible or not. As soon as we introduce a new admissible in the α list, then we start the machine $\mathcal{U}$ on this input, in order to check for gaps with this admissible as oracle. Of course, what is subtle is to arrange all these computations, both in space and in time so that it is realized in ω-real time. The argument for space is that we can arrange an extra working tape as a ω^2 data structure, and for time, the argument is that making n more steps of the n first machines requires only a finite amount of time.

Now, we can use Lemma 2 directly and when ι is a limit ordinal determine whether the sup of the first ι admissibles is admissible after only ω steps by flashing a common cell when a halting program is discovered by one of the oracle computations. When we thus find a recursively inaccessible ordinal, we introduce it in the α-storage.

The program halts as previously.

Corollary 2. *If α is a writable ordinal, then there is a gap such that the set of admissible ordinals properly contained in this gap has order type α.*

Proof. By Theorem 8, this follows if there is a gap such that the set of ordinals properly contained in it has order type $\geq \alpha$.

However, if this is not the case, then the algorithm described in the proof of Theorem 8 will run up to time λ_∞, after which no further clockable ordinals can appear; therefore, the algorithm will then continue by running until the working time has passed through a set of admissible ordinals of order type α and then stop.

Thus, we have found an ITTM-program that halts after more than λ_∞ many steps on input 0, a contradiction.

Now we provide a lower bound on the ending points of those gaps with exactly α admissibles.

Corollary 3. *Let α be a writable ordinal. After the writing time for α, the first gap with α admissibles inside ends*

- *if α is a limit ordinal, then exactly at the sup of those α admissibles in the gap,*
- *if α is a successor ordinal, then exactly ω steps after the last admissible of the gap.*

Proof. The proof consists in checking carefully the halting time of our program given in the proofs of Theorem 8. If α is a limit ordinal, then the situation is clear: our algorithm halts immediately thanks to our use of an improved version of the count through program and this result has been used to prove that the supremum of the α admissibles is not admissible.

For the successor case, then there are two different possible situations. Let us denote by $\overline{\alpha}$ the greatest limit ordinal $< \alpha$. The easy situation is when the supremum of the first $\overline{\alpha}$'s admissibles of the gap is not admissible. Then before halting our algorithm finds the last admissibles by the procedure of computing the next admissible through the computation of ω_1^{CK} relativised to the last admissible found and the procedures halts ω steps after for the same reason that we can find in the proof that the gap starting at ω_1^{CK} is of size ω. If the supremum of the first $\overline{\alpha}$'s admissibles of the gap is admissible and $\alpha > \overline{\alpha} + 1$, the last admissible is found by the same procedure and the program halts ω steps after it. But in this case, if $\alpha = \overline{\alpha} + 1$, then we should find the exact time when we discover that the supremum is admissible. Our programs checks the hypothesis of Lemma 2, all the parallel oracle computations look for a clockable but do not find any, thus we get this result just ω steps after the supremum, which is the last admissible among the α's.

We note that when we halt ω steps after an admissible, then we get directly an optimal bound since ω is the minimal time between an admissible and the end of its gap.

Remark that the previous theorems start with "After the writing time for α..." We can modify this as follows: "Let δ be any clockable such that $\delta \geq \text{writing}(\alpha)$. After the writing time for δ...". The proofs are slightly modified: we run the programs as previously but start the gap detection only after having clocked δ.

3.1 Admissibles of Higher Rank

We define the *rank* $\text{rk}(x)$ of an admissible ordinal α as follows:

- If $\alpha = \beta^+$ for some β, then $\text{rk}(\alpha) = 0$.
- If β is maximal such that α is a limit of admissible ordinals of rank δ for every $\delta < \beta$, then $\text{rk}(\alpha) = \beta$.

We call *rank-admissible* an admissible α of rank α, *i.e.*, and admissible α with $\text{rk}(\alpha) = \alpha$. Note that λ_∞ is a rank-admissible, but it is not the smallest one, as follows quite easily from Theorem 2.

Lemma 3. *For any $\delta < \lambda_\infty$, there is an admissible α with $rk(\alpha) = \delta$ that starts a gap. Moreover, there is an α with $rk(\alpha) = \alpha$ that starts a gap (and more).*

Proof. This follows as in the first proof of Theorem 6: As λ_∞ has all of the required rank properties and there are cofinally many clockable ordinals below λ_∞, these properties are Σ_1-expressible and there are no clockable ordinals $>$ λ_∞, L_{ζ_∞} believes in the existence of ordinals with the required rank properties with cofinally many clockable ordinals below. This is then reflected to L_{λ_∞}, thus L_{λ_∞} also contains such ordinals, but since λ_∞ is a limit of clockable ordinals, these must start gaps.

Once again, we can have some more information on the size of the gap through a more algorithmic version of this proof.

Theorem 9. *Consider $\alpha < \lambda_\infty$. Let δ be any clockable such that $\delta \geq writing(\alpha)$. Then after δ, the first gap starting with an admissible of rank α is of size ω.*

Proof. (Sketch) We start gap detection as usual, using P_{gap}. In parallel, we write α and clock δ. Let x be a code for α, written to some scratch tape at time $writing(\alpha)$. For $\iota < \alpha$, let $d(\iota)$ denote the natural number representing ι in the sense of x.

We consider ω cells, each of them will correspond to a level in α's representation, we call them "cells in α". (This construction requires ω steps but it does not matter since there is no admissible nearby). For each clockable discovered, we flash a special "clockable" cell thus at each limit ordinal, we know whether it is a limit of clockable ordinals or not. For each limit ordinal that is a limit of clockable ordinals, we test in parallel:

– whether it starts a gap (requires ω steps),
– what is the smallest (in the order of α) cell in α with value 0 (requires ω steps).

If the limit does not start a gap, then it is clockable and we erase all the cells in α. If it starts a gap, then we flash the smallest cell in α order with value 0. If there is no such cell, we halt.
Note: the algorithm halts exactly ω steps after the first admissible of rank α found after δ. We get some extra information: for each limit ordinal that is a limit of clockable ordinals, its potential rank (if it is admissible) is exactly the smallest cell at 0 in α.

Using the extra information provided by the program described in the previous proof we easily obtain the following result.

Proposition 4. *There is an ITTM-program P such that, whenever x and y are codes for ordinals α and β, then $P(x, y)$ halts with output 1 if β is an admissible of rank α and otherwise halts with output 0.*

Proof. This is basically just a recursive algorithm, calling itself for parts of x that code ordinals $< \alpha$, combined with the admissibility checking with oracles used in the proof of Theorem 8.

As a special case, when we run $P(x, x)$ we get the following result.

Corollary 4. *There is an ITTM-program Q such that $Q(x)$ halts with output 1 if and only if x codes a rank-admissible.*

We then get an improved version of the existence of rank-admissibles by combining this result with Theorem 9.

Corollary 5. *The smallest admissible ordinal α with $rk(\alpha) = \alpha$ starts a gap of length ω.*

Theorem 10. *Let us call β_0 the first admissible that starts a gap of size equal to its starting point. The ordinal β_0 is a rank-admissible but not the first rank-admissible.*

Proof. Assume that β_0 is of rank $\alpha < \beta_0$, then writing$(\alpha) < \beta_0$. Remark that β_0 is a limit of clockables as a starting point of a gap. Thus there exists a clockable $\delta < \beta_0$ such that all admissibles in the interval $[\delta, \beta_0]$ have rank $< \alpha$ (by the definition of rank). Thus β_0 is the first ordinal of rank α after this clockable δ which contradicts the size of its gap. Hence it is rank-admissible, but not the first one since its gap is too large by Corollary 5.

Please note that the same proof can be used for γ_0 instead of β_0, where γ_0 is the first admissible that starts a gap with as many admissible inside as its starting point. This γ_0 is a rank-admissible.

3.2   Long Gaps

We finish with a rather general theorem on the size of gaps relative to their starting point. Let $f : \omega_1 \to \omega_1$. Then f is called *ITTM-computable* if and only if there is an ITTM-program P such that, whenever x is a code for $\alpha < \omega_1$, then $P^x(0) \downarrow = y$, where y is a code for $f(\alpha)$.

If $\theta : \omega_1 \to \omega_1$ and $f : \omega_1 \to \omega_1$ is an ITTM-computable function that is computed by the program P, we say that θ is the *running time* of P if and only if, for every $\alpha < \omega_1$ and every code x for α, $P^x(0)$ halts in exactly $\theta(\alpha)$ many steps.

In general, an ITTM-program computing a function $f : \omega_1 \to \omega_1$ need not have a running time, as the computation time on α may turn out to depend on the code for α. However, many functions on ordinals —such as the successor function— can be computed in a way that does not depend on the code.

Theorem 11. *Let $f : \omega_1 \to \omega_1$ be ITTM-computable. Then there is an ordinal α starting a gap of size $\geq f(\alpha)$.*

Proof. Run P_{gap} as usual. When a code c for an ordinal α starting a gap is written, use it to compute a code c' for $f(\alpha)$, while continuing to run P_{gap} in parallel. If any simulated program halts in the meantime, stop the computation of $f(\alpha)$ and continue with P_{gap}. Otherwise, use $P^{c'}_{\text{stopwatch}}$ to run for another

$f(\alpha)$ many steps in parallel with P_{gap}, again stopping if any simulated program stops along the way. On the other hand, if $P^{c'}_{\text{stopwatch}}$ stops, we halt.

Clearly, this program halts when P_{gap} writes a code for an ordinal α such that no clockable ordinal exists in the interval $[\alpha, f(\alpha))$. As such an α cannot exist below γ_∞, P_{gap} would halt after at least $\gamma_\infty + f(\gamma_\infty) > \gamma_\infty$ many steps, which is a contradiction. Hence such a gap must exist.

With a bit more work, one can also control the exact sizes of such gaps:

Theorem 12. *Let $f : \omega_1 \to \omega_1$ be an ITTM-computable function, that maps limit ordinals to limit ordinals, and denote by $\theta(\alpha)$ the computation time of f on α. Then there is some ordinal α starting a gap of size exactly $\theta(\alpha) + f(\alpha) + \omega$.*

Proof. Using the fact that clockable ordinals are writable, it is not hard to see that θ is computable. By Theorem 11, there is therefore an α starting a gap of length at least $\theta(\alpha) + f(\alpha)$.

To see that there is gap of exactly the size in question, we consider the following algorithm:

First, run P_{gap}; whenever P_{gap} outputs a real number x coding an ordinal α that starts a gap, we run the following parallel to the further execution of P_{gap}: First compute $f(\alpha)$, which takes $\theta(\alpha)$ many steps. After this, a code c for $f(\alpha)$ will be written. Then run $P^c_{\text{stopwatch}}$. If any of the simulated programs in P_{gap} halt in the course of this, we erase the tapes used for the computation of $f(\alpha)$ and the run of $P^c_{\text{stopwatch}}$ and continue with P_{gap}. On the other hand, if $P^c_{\text{stopwatch}}$ halts, the whole computation halts.

When this algorithm first checks some ordinal β starting a gap, it has taken β many steps. After encountering the first α starting a gap of size $\geq \theta(\alpha) + f(\alpha)$, the algorithm runs for another $f(\alpha) + \omega$ many steps and then halts. Also, by the definition of the algorithm, there are no halting times between α and $\theta(\alpha) + f(\alpha) + \omega$. Thus α is as desired.

References

1. Barwise, J.: Admissible Sets and Structures: An Approach to Definability Theory. Perspectives in Mathematical Logic, vol. 7. Springer, Heidelberg (1975)
2. Durand, B., Lafitte, G.: A constructive swiss knife for infinite time turing machines (2016)
3. Hamkins, J.D., Lewis, A.: Infinite time turing machines. J. Symbolic Log. **65**(2), 567–604 (2000)
4. Welch, P.D.: Eventually infinite time turing degrees: Infinite time decidable reals. J. Symbolic Log. **65**(3), 1193–1203 (2000)
5. Welch, P.D.: The length of infinite time turing machine computations. Bull. London Math. Soc. **32**(2), 129–136 (2000)
6. Welch, P.D.: The transfinite action of 1 tape turing machines. In: Cooper, S.B., Löwe, B., Torenvliet, L. (eds.) CiE 2005. LNCS, vol. 3526, pp. 532–539. Springer, Heidelberg (2005). doi:10.1007/11494645_65
7. Welch, P.D.: Characteristics of discrete transfinite time turing machine models: Halting times, stabilization times, and normal form theorems. Theoret. Comput. Sci. **410**, 426–442 (2009)

Koepke Machines and Satisfiability for Infinitary Propositional Languages

Merlin Carl[1,2], Benedikt Löwe[3,4,5(✉)], and Benjamin G. Rin[6]

[1] Fachbereich Mathematik und Statistik,
Universität Konstanz, 78457 Konstanz, Germany
merlin.carl@uni-konstanz.de
[2] Fakultät für Informatik und Mathematik,
Universität Passau, Innstraße 33, 94032 Passau, Germany
[3] Institute for Logic, Language and Computation, Universiteit van Amsterdam,
Postbus 94242, 1090GE Amsterdam, The Netherlands
b.loewe@uva.nl
[4] Fachbereich Mathematik, Universität Hamburg,
Bundesstrasse 55, 20146 Hamburg, Germany
[5] Christ's College, Churchill College, and Faculty of Mathematics,
University of Cambridge, Wilberforce Road, Cambridge CB3 0WA, England
[6] Departement Filosofie En Religiewetenschap, Universiteit Utrecht,
Janskerkhof 13, 3512BL Utrecht, The Netherlands
b.g.rin@uu.nl

Abstract. We consider complexity theory for Koepke machines, also known as Ordinal Turing Machines (OTMs), and define infinitary complexity classes ∞-**P** and ∞-**NP** and the OTM analogue of the *satisfiability problem*, denoted by ∞-SAT. We show that ∞-SAT is in ∞-**NP** and ∞-**NP**-hard (i.e., the problem is ∞-**NP**-complete), but not OTM decidable.

1 Infinitary Computation and Its Running Times

1.1 Introduction

Various versions of Turing machines for infinitary computation have been proposed. They all have in common that they have ordinal-indexed tapes on which they can read and write symbols from a finite alphabet Σ, they run in ordinal-indexed steps of time, and follow the usual instructions for Turing machines for the successor ordinal steps. The first such type of machines were the *Hamkins-Kidder machines* or *Infinite Time Turing Machines* (ITTMs) defined in [5]. These machines have a regular tape of order type ω, but do not have to halt in finite time: instead, they can run through transfinite ordinal time steps. This results in an asymmetric situation between time and space as an ITTM can run through the class of all ordinals, but only has ω many cells to write on. In [10,11], Koepke symmetrised ITTMs and defined what are now known as *Koepke machines* or *Ordinal Turing Machines* (OTMs): OTMs have a class-sized tape indexed by ordinals and run through ordinal time. Other machine

© Springer International Publishing AG 2017
J. Kari et al. (Eds.): CiE 2017, LNCS 10307, pp. 187–197, 2017.
DOI: 10.1007/978-3-319-58741-7_19

concepts include machines that restrict the space to a given ordinal α but run through arbitrary ordinal time and machines where both time and space are symmetrically restricted to an ordinal α (cf. [3,12,14]).

The symmetry between space and time for Koepke machines reflects that of finitary Turing machines. The first author has argued in [1] that Koepke machines are the natural infinitary analogue for finitary computability theory. In this paper, we shall study deterministic and nondeterministic polynomial time computation for Koepke machines. In Sect. 1.2, we give the basic definitions of our model of computation and its running time analysis. Complexity theory for infinitary computation was introduced by Schindler in [15] in the context of Hamkins-Kidder machines; for Koepke machines, the definitions were discussed by the second author and Winter [13,16,17]. We give precise definitions in this tradition (introducing the complexity class ∞-**NP**) and discuss fundamental differences between finitary and infinitary computation in Sect. 2. In Sect. 3, we introduce the OTM analogue of the satisfiability problem ∞-SAT, show that it is in ∞-**NP** and that every problem in ∞-**NP** polynomially reduces to it (i.e., ∞-SAT is ∞-**NP**-hard). However, due to the phenomena discussed in Sect. 2.2, being ∞-**NP**-complete does not necessarily imply that ∞-SAT is OTM decidable: in Sect. 4, we show that it is not (and discuss a notable difference between the general decision problem ∞-SAT and its countable fragment).

1.2 Basic Definitions

In the following, we shall be working with Koepke machines, or OTMs, sometimes allowing ordinal parameters in our computations. For detailed definitions, we refer the reader to [10].

An *OTM input* is a function $X : \alpha \to \Sigma$ where $\mathrm{lh}(X) := \alpha$ is an ordinal called the *length of* X. We assume that the tape alphabet Σ contains a blank symbol $\lrcorner$ that allows us to express shorter OTM inputs as longer ones: if $\mathrm{lh}(X) = \alpha < \beta$, we consider $X^* : \beta \to \Sigma$ with $X^*(\gamma) := X(\gamma)$ for $\gamma < \alpha$ and $X^*(\gamma) = \lrcorner$ for $\gamma \geq \alpha$ and identify X and X^*. A class of OTM inputs is called an *OTM decision problem*. A Koepke machine M *decides an OTM decision problem C in parameter η* if it halts with parameter η for every OTM input X and outputs 1 if and only if $X \in C$; an OTM decision problem is called *OTM decidable in parameter η* if there is a Koepke machine deciding it in parameter η. Parameter-free computation is the special case where η is a recursive ordinal (e.g., 0). An OTM decision problem is called *bounded with bound λ* if for every OTM input $X \in C$, we have that $\mathrm{lh}(X) < \lambda$.

We emphasize that (as in the finitary case) most natural decision problems do not occur as classes of OTM inputs, but have to be formally encoded as OTM decision problems. Typically, they come in the form of some class D of objects of some particular kind (e.g., formulas, trees, graphs, etc.) together with a coding (class) function code such that for every (relevant) set Z, code(Z) is an OTM input, and $C := \{\mathsf{code}(Z) \, ; \, Z \in D\}$ is an OTM decision problem. We shall see below that the choice of coding is crucial in the infinitary case.

Let $f : \text{Ord} \to \text{Ord}$ be an increasing class function (in the following, we shall refer to these as *complexity functions*). A class function f is a *polynomial function* if there are ordinals $\alpha_n \geq \ldots \geq \alpha_0$ such that for all γ, we have

$$f(\gamma) = \gamma^{\alpha_n} + \gamma^{\alpha_n - 1} + \ldots + \gamma^{\alpha_0}.$$

We say that a Koepke machine M is a *time f machine* if the machine halts for every OTM input X in less than $f(\text{lh}(X))$ steps.

If C is an OTM decision problem, we write $C \in \textbf{Time}(f)$ if there is a time f machine deciding C and $C \in \infty\text{-}\textbf{P}$ if there is a polynomial function f such that $C \in \textbf{Time}(f)$. In the latter case, we say that it is *OTM polynomial time decidable*.

Similarly, if C and D are OTM decision problems, we say that C *is reducible to D in time f* if there is a time f machine that takes an OTM input X and produces an output Y such that $X \in C$ if and only if $Y \in D$. We say that C *is reducible to D in polynomial time* if there is a polynomial function f such that C is reducible to D in time f.

Proposition 1. *If C is a bounded OTM decision problem with bound $\lambda \geq \omega$, then exactly one of the following holds:*

(i) The problem C is not OTM decidable in parameter λ; or
(ii) the problem C is decided by a time c machine in parameter λ, where c is the constant function $\alpha \mapsto |\lambda|^+$.

Proof. Suppose that C is decidable by a Koepke machine M in parameter λ. In particular, for every input X, the machine M halts. If $\text{lh}(X) < \lambda$, a standard Löwenheim-Skolem argument using the absoluteness of computations shows that there is some $\alpha < |\lambda|^+$ such that the model $\mathbf{L}_\alpha[X]$ is a model of "the computation of M with input X and parameter λ halts". But then the computation must halt before $\alpha < |\lambda|^+$. Checking whether $\text{lh}(X) < \lambda$ can be done in time λ using the parameter λ. □

The proof of Proposition 1 has two immediate consequences for running times of infinitary computations:

First, while Koepke machines can in principle use the entire length of the class of ordinals for their computation time, halting Koepke machines (and these are the only ones that matter for decision problems) do not: they can never substantially outrun the size of the input (in the sense that, if the input has cardinality κ, then the computation will take less than $(\kappa^+)^{\mathbf{L}}$ many steps). This is a marked difference to the finitary case.

This immediately implies that the relevant operations for running time analysis of Koepke machines must necessarily be *ordinal operations* rather than *cardinal operations* since any non-trivial cardinal operations on infinite ordinals will move beyond the bounds of Proposition 1.

This in turn yields a second important consequence: as in the finitary case, infinitary complexity theory is sensitive to the encoding of the input; the more efficient the input encoding is, the harder it is to prove complexity bounds for a decision problem. But every ordinal $\kappa \leq \alpha \leq \kappa^+$ can be encoded by a set of order type κ, so there is a maximally efficient encoding in terms of input length.

We illustrate this phenomenon by showing that any OTM decision problem that is decidable in time f by a machine M has a re-coded version such that the re-coded version of M is not a time f machine. Let κ be a cardinal and C be an OTM decision problem that was obtained from some class D by means of a coding function code such that $C = \{\text{code}(Z)\,;\, Z \in D\}$. Let $f : \text{Ord} \to \text{Ord}$ be a complexity function and M a time f machine that decides C and for the sake of non-triviality, assume that $f(\kappa) < \kappa^+$. Now, if there is some OTM input $X = \text{code}(Z)$ with $\text{lh}(X) < \kappa^+$ such that M takes more than $f(\kappa)$ many steps before halting. Then let $\lambda := \max\{f(\kappa), \text{lh}(X)\}$, let π be a bijection between κ and λ, and let code* be the coding function corresponding to the combination of code and π. In particular, code$^*(Z)$ has length κ. If we write $C^* := \{\text{code}^*(Z)\,;\, Z \in D\}$, then the appropriately re-coded version of M does not decide C^* in time f since it will run for more than $f(\text{lh}(\text{code}^*(Z))) = f(\kappa)$ many steps.

Consequently, in infinitary computation, a sufficiently efficient coding for the input can potentially destroy the complexity properties of any machine. Hence, we need to assume that the encoding of the decision problem respects the *natural length* of the objects being coded.

2   Complexity Theory for Koepke Machines

2.1   Definitions

Nondeterministic complexity classes for infinitary computation were first introduced by Schindler in [15] for Hamkins-Kidder machines; Schindler's definition did not use nondeterministic Hamkins-Kidder machines, but defined the class **NP** in terms of checking a witness. This was linked in [13, Proposition 11] to nondeterministic Hamkins-Kidder machines. Schindler exploited the asymmetry between time and space for Hamkins-Kidder machines and showed that for his definitions of **P** and **NP** for Hamkins-Kidder machines, we get $\mathbf{P} \subsetneq \mathbf{NP}$. His results were later improved to $\mathbf{P} \subsetneq \mathbf{NP} \cap \mathbf{co\text{-}NP}$ in [4]; cf. also [6,13,16,17].

In the following, we give the corresponding definitions for Koepke machines that were essentially first developed by Winter in [16]. We call a class D of pairs of OTM inputs a *witnessed OTM decision problem*. If C is an OTM decision problem and D is a witnessed OTM decision problem, we say that C *is the projection of D* if

$$X \in C \iff \exists W((W, X) \in D);$$

i.e., if $(W, X) \in D$, we interpret W as a witness for the membership of X in C.

As usual, if X is an OTM input of limit ordinal length λ, we can consider X as a pair (X_0, X_1) of OTM inputs of length λ via $X_0(\mu + n) := X(\mu + 2n)$ and $X_1(\mu + n) := X(\mu + 2n + 1)$ (for limit ordinals $\mu < \lambda$). If $f : \text{Ord} \to \text{Ord}$ is a complexity function, we say that a Koepke machine M is a *-time f machine if the machine halts for every OTM input X in less than $f(\text{lh}(X_1))$ steps.

If D is a witnessed OTM decision problem and M is a Koepke machine, we say that M decides D if it halts on all OTM inputs and it outputs 1 on input X if and only if $(X_0, X_1) \in D$. If C is an OTM decision problem, we write $C \in \mathbf{NTime}(f)$ if there is a witnessed OTM decision problem D such that C is the projection of D and there is a *-time f machine deciding D. We write $C \in \infty\text{-}\mathbf{NP}$ if there is a polynomial function f such that $C \in \mathbf{NTime}(f)$.

2.2 The Fundamental Difference Between Finitary and Infinitary Computation

In the case of ordinary Turing machines and complexity functions $f : \mathbb{N} \to \mathbb{N}$, we can recapture nondeterministic computation by deterministic computation: suppose that you have some complexity function $f : \mathbb{N} \to \mathbb{N}$ and some decision problem C that is decided by a *-time f machine M. This means that there is a witnessed decision problem D such that C is the projection of D. On input (W, X) with $\text{lh}(X) = n$, the machine halts in less than $f(n)$ steps. In particular, it reads at most the first $f(n)$ many digits of W, so we can ignore the rest of the information in W. This allows us to run an *exhaustive brute force algorithm* that checks all possible witnesses: there are $\Sigma^{f(n)}$ many input sequences of length $f(n)$, so we can just run M on all of these in combination with X; if $\widehat{f}(n) := f(n) \cdot \Sigma^{f(n)} \cdot k$ for a sufficiently large $k \in \mathbb{N}$, then the brute force algorithm checks in time $\widehat{f}$ whether $X \in C$. This argument breaks down for infinitary computation, as will be shown in Proposition 2.

Since the state of a Koepke machine is absolute between transitive models of set theory, the content of the tape has to be constructible. Thus, for the discussion of brute force algorithms, it is sensible to work under the assumption of $\mathbf{V}{=}\mathbf{L}$.

In general, if M is any Koepke machine and f is any complexity function, we say that a machine $\widehat{M}$ is an *exhaustive brute force machine associated to M relative to f* if at input Y with $\text{lh}(Y) = \alpha$, the machine $\widehat{M}$ successively writes all OTM inputs Z of length $f(\alpha)$ on the scratch tape and then runs the machine M on input X with $X_0 = Z$ and $X_1 = Y$. It gives the output 0 if all of the runs of M produced output 0 and 1 if one of the runs of M produced 1. The above argument shows for finitary computation that $\widehat{M}$ is a time $\widehat{f}$-machine if M was a *-time f machine.

Proposition 2. *Assume* $\mathbf{V}{=}\mathbf{L}$. *Suppose that f is a complexity function such that there is some $\alpha \geq \omega$ with $f(\alpha) \geq |\alpha|$ and that M is a *-time f machine and $\widehat{M}$ is an exhaustive brute force machine associated to M relative to f. Then $\widehat{M}$ does not halt for any OTM input of length α.*

Proof. Let X be an OTM input of length α. The machine $\widehat{M}$ runs for at least $2^{|\alpha|} \geq |\alpha|^+$ many steps since the exhaustive brute force machine has to produce all OTM inputs of length $f(\alpha) \geq |\alpha|$. But Proposition 1 tells us that no machine can run for $|\alpha|^+$ many steps on input X and after that still halt.   □

The fact that for infinitary computation, decision problems that are nondeterministically decidable can be deterministically undecidable has been observed and used by Winter [16, p. 74]. We shall provide a concrete example for this in Sect. 4.

3   An Infinitary Analogue of SAT

In finitary computation, the decision problem SAT is the set of satisfiable propositional formulas.

We define the natural analogue of SAT in the context of Koepke machines in the style of infinitary languages (cf. [9]). Let Var $\subseteq \mathbf{L}$ be a class of propositional variables. We form *formulas* of the infinitary propositional language $\mathcal{L}_{\infty,0}$ with the unary operator $\neg$ corresponding to negation and the operator $\bigwedge$ that takes a set of formulas and produces its conjunction.

1. Every element of Var is in $\mathcal{L}_{\infty,0}$.
2. If $\varphi \in \mathcal{L}_{\infty,0}$, then so is $\neg\varphi$.
3. If Φ is a set of members of $\mathcal{L}_{\infty,0}$, then $\bigwedge \Phi$ is an element of $\mathcal{L}_{\infty,0}$.

As usual, we abbreviate $\neg \bigwedge \{\neg\varphi \, ; \, \varphi \in \Phi\}$ by $\bigvee \Phi$. Formulas of the language $\mathcal{L}_{\infty,0}$ naturally correspond to labelled well-founded trees (T, ℓ) where each node in T has a set of successors and ℓ is a function from the set of leaves of T to Var. We write $\mathrm{Var}_\varphi := \mathrm{ran}(\ell)$ for the set of variables occurring in φ. By a simple Mostowski collapse argument, we may assume that $\mathrm{Var}_\varphi \subseteq \mathbf{L}_\beta$ for some $\beta < |T|^+$. The width of the tree T, denoted by width(T) is the supremum of the cardinalities of the sets of successors of branching nodes; the height of the tree T, denoted by height(T) is defined by the usual recursion on the well-founded tree structure. By interpreting the leaves t of T as propositional variables $\ell(t)$, its branching nodes as infinite conjunctions, and its non-branching nodes as negations, we can identify a formula with a labelled well-founded tree. A function $v : \mathrm{dom}(v) \to \{0,1\}$ with $\mathrm{dom}(v) \subseteq \mathrm{Var}$ is called a *valuation*. As usual, if $\varphi = (T, \ell)$ is a labelled well-founded tree, any valuation v with $\mathrm{dom}(v) \supseteq \mathrm{Var}_\varphi$ uniquely extends to a map $\widehat{v} : T \to \{0,1\}$ via the following recursive definition:

1. if $t \in T$ is a leaf, then $\widehat{v}(t) := v(\ell(t))$;
2. if $t \in T$ is a non-branching node and t' is its unique successor in T, then $\widehat{v}(t) := 1 - \widehat{v}(t')$;
3. if $t \in T$ is a branching node and X is the set of its successors in T, then $\widehat{v}(t) := \min\{\widehat{v}(x) \, ; \, x \in X\}$.

We define $\widehat{v}(T, \ell)$ to be the value of $\widehat{v}$ at the root of the tree T.

Definition 3. *The problem* ∞-SAT *is to decide on input* $(T, \ell) \in \mathcal{L}_{\infty,0}$ *whether there is a valuation* v *such that* $\widehat{v}(T, \ell) = 1$.

Definition 3 does not define an OTM decision problem: in order to do so, we still need to specify in which way the formula (T, ℓ) is encoded as an OTM input.

As emphasized before, it is crucial in the realm of infinitary computation that this encoding respects the natural length of the input. We shall not specify a concrete encoding here (since it does not matter for anything that follows), but insist that the encoding function has the property that

$$\text{lh}(\text{code}(T, \ell)) = \max(\text{width}(T), \text{height}(T)).$$

With this requirement, the following results are straightforward adaptations of the classical arguments showing that SAT is **NP**-complete:

Theorem 4. *The OTM decision problem ∞-SAT is in ∞-**NP**.*

Theorem 5. *Every problem in ∞-**NP** reduces in polynomial time to ∞-SAT.*

Proof. The proof largely follows the same general structure as standard textbook proofs of the finitary Cook-Levin theorem, but with an additional component to accommodate the limit stages of machine computation.   □

4   The Undecidability of ∞-SAT

In Sect. 3, we proved that ∞-SAT is ∞-**NP**-complete. However, as pointed out in Sect. 2.2, in infinitary computation, being nondeterministically decidable does not imply deterministic decidability. In this section, we shall show that ∞-SAT is not OTM decidable. In fact, the decidability behaviour of ∞-SAT restricted to constructibly countable formulas is different from the general behaviour; this follows from a theorem by Jensen and Karp:

Theorem 6 (Jensen & Karp). *If x is a real and α is a limit of x-admissible ordinals, then $\Sigma_1(x)$-sentences are absolute between $\mathbf{V}_\alpha$ and $\mathbf{L}_\alpha[x]$.*

Proof. This is the relativised version of a theorem proved in [7, p. 162]. The relativisation is discussed in [2, Appendix].   □

Theorem 7. *Let $\varphi = (T, \ell) \in \mathbf{L}$ be a constructibly countable formula, i.e., $\mathbf{L} \models$ "T is a countable tree". Then there is an $\alpha < \omega_1^{\mathbf{L}}$ such that exactly one of the following holds:*

1. φ is not satisfiable, or
2. there is a $v \in \mathbf{L}_\alpha$ such that $\hat{v}(\varphi) = 1$.

Proof. Since φ is countable in $\mathbf{L}$, find a real c and some $\beta < \omega_1^{\mathbf{L}}$ such that $c \in \mathbf{L}_\beta$ and c encodes φ. Let α be a limit of c-admissibles above β. The sentence "there is a valuation v such that $\hat{v}(\varphi) = 1$" is $\Sigma_1(c)$ and hence by Theorem 6 absolute between $\mathbf{V}_\alpha$ and $\mathbf{L}_\alpha[c] = \mathbf{L}_\alpha$. So, if φ is satisfiable, then a witness of this lies in $\mathbf{L}_\alpha$.   □

In contrast, the conclusion of Theorem 7 is consistently false if we allow for formulas that are not countable in $\mathbf{L}$:

Theorem 8. *There is a constructible formula $\varphi \in \mathcal{L}_{\infty,0}$ such that*

1. *for all constructible valuations $v \in \mathbf{L}$, we have that $\widehat{v}(\varphi) = 0$, and*
2. *if $\omega_1^{\mathbf{L}} < \omega_1$, then there is a valuation v such that $\widehat{v}(\varphi) = 1$.*

Proof. For every $i \in \mathbb{N}$ and $\alpha < \omega_1^{\mathbf{L}}$, let $P_{i,\alpha}$ be a propositional letter. We define

$$\Phi := \bigwedge_{i \in \mathbb{N}} \bigwedge_{\alpha < \omega_1^{\mathbf{L}}} \bigwedge_{\substack{\beta < \omega_1^{\mathbf{L}} \\ \beta \neq \alpha}} \neg(P_{i,\alpha} \wedge P_{i,\beta}) \wedge \bigwedge_{\alpha \in \omega_1^{\mathbf{L}}} \bigvee_{i \in \mathbb{N}} P_{i,\alpha} \wedge \bigwedge_{i \in \mathbb{N}} \bigvee_{\alpha \in \omega_1^{\mathbf{L}}} P_{i,\alpha}.$$

If Φ is satisfiable, then there is a surjection from $\mathbb{N}$ onto $\omega_1^{\mathbf{L}}$, so clearly, Φ is satisfiable if and only if $\omega_1^{\mathbf{L}} < \omega_1$.   □

The formula φ of Theorem 8 has size $\aleph_1^{\mathbf{L}}$; by Theorem 7, it is impossible to have a smaller example. Theorem 8 allows us to refine the argument of Proposition 2: in Proposition 2, it was the exhaustivity of the machine $\widehat{M}$ that did not allow it to stop (since it would run for too long); we can now see that sometimes, even writing the witness itself can be too much to ask. If C is an OTM decision problem which is the projection of a witnessed OTM decision problem D, then we say that a Koepke machine M *decides C by producing a witness in D* if it halts on every input X and outputs either 0 or some sequence Z such that $(Z, X) \in D$. The following statement is a very weak version of our later main result, Theorem 10:

Corollary 9. *If $\omega_1^{\mathbf{L}} < \omega_1$, then ∞-SAT cannot be decided by producing a witness.*

Proof. By Theorem 8 and the assumption, we have a constructible satisfiable formula φ with no constructible witness. So, any machine that decides ∞-SAT by producing a witness will write a non-constructible valuation on the tape. But no Koepke machine can produce a non-constructible output on constructible input. Contradiction!   □

Theorem 10. *The OTM decision problem ∞-SAT is OTM undecidable.*

Proof. We shall describe a Koepke machine M that produces with parameter $\omega_1^{\mathbf{L}}$ on input $i \in \omega$ a formula $\Psi_i \in \mathcal{L}_{\infty,0}$ that is satisfiable if and only if the ith Koepke machine halts with parameter $\omega_1^{\mathbf{L}}$.

The proof of the theorem will then proceed by contradiction: Assume that there is a Koepke machine M' that decides ∞-SAT, then we can combine M and M' to get a Koepke machine with parameter $\omega_1^{\mathbf{L}}$ that decides the halting problem for Koepke machines with parameter $\omega_1^{\mathbf{L}}$. But such a machine cannot exist.

The main task in the proof is the construction of the formula Ψ_i; this requires coding of $\mathbf{L}$-structures. We recall that there is a sentence $\sigma \in \mathcal{L}_\in$ such that for any transitive N, we have $(N, \in) \models \sigma$ if and only if $N = \mathbf{L}_\gamma$ for some limit ordinal γ (cf., e.g., [8, Theorem 3.3]).

For every $\beta, \gamma < \omega_1^{\mathbf{L}}$, we fix a propositional variable $P_{\beta,\gamma}$. If v is a valuation, we can define a binary relation E_v on $\omega_1^{\mathbf{L}}$ by

$$E_v := \{(\beta,\gamma)\,;\, v(P_{\beta,\gamma}) = 1\}$$

and prove the following translation from first-order logic into infinitary logic:

Lemma 11. *If $\varphi \in \mathcal{L}_\in$, then there is $\Phi_\varphi \in \mathcal{L}_{\infty,0}$ such that for every valuation v, we have $\hat{v}(\Phi_\varphi) = 1$ if and only if $(\omega_1^{\mathbf{L}}, E_v) \models \varphi$.*

Proof. This is an easy induction on the formula complexity of φ. The induction steps are trivial for propositional connectives. Concerning the quantifiers, we use infinite conjunctions to express universal quantifiers and infinite disjunctions for existential quantifiers in the obvious way.   □

We now add additional propositional variables $B_{\beta,\gamma}$ for every $\beta, \gamma < \omega_1^{\mathbf{L}}$ to encode an embedding from $(\omega_1^{\mathbf{L}} + 1, \in)$ into $(\omega_1^{\mathbf{L}}, E_v)$. If v is any valuation, we define a second binary relation

$$\pi_v := \{(\beta,\gamma)\,;\, v(B_{\beta,\gamma}) = 1\}$$

similar to E_v, but based on the values of $B_{\beta,\gamma}$ instead of the values of $P_{\beta,\gamma}$.

Lemma 12. *There is a formula Ξ such that for all valuations v, we have have $\hat{v}(\Xi) = 1$ if and only if π_v is a structure-preserving embedding from $(\omega_1^{\mathbf{L}} + 1, \in)$ into $(\omega_1^{\mathbf{L}}, E_v)$.*

Proof. Similar to the formula in the proof of Theorem 8.   □

We write ψ_i for the formula expressing "the ith Koepke machine with parameter $\omega_1^{\mathbf{L}}$ halts" and notice that this is first-order expressible in all structures $\mathbf{L}_\eta$ for $\eta > \omega_1^{\mathbf{L}}$ (by condensation, in any such $\mathbf{L}_\eta$, the ordinal $\omega_1^{\mathbf{L}}$ is definable as the smallest ordinal that does not have a real coding it).

Let $S := \{s \in \mathbf{L}\,;\, s : \omega \to \omega_1^{\mathbf{L}}\}$. An easy condensation argument shows that $S \subseteq \mathbf{L}_{\omega_1^{\mathbf{L}}}$. We now write

$$\Psi_i := \Xi \wedge \Phi_{\sigma \wedge \psi_i} \wedge \bigwedge_{s \in S} \bigvee_{i \in \omega} \neg P_{s(i+1),s(i)}.$$

Lemma 13. *The formula Ψ_i is satisfiable if and only if the ith Koepke machine with parameter $\omega_1^{\mathbf{L}}$ halts.*

Proof. "$\Leftarrow$". If the ith Koepke machine with parameter $\omega_1^{\mathbf{L}}$ halts, then by the proof of Proposition 1, it has halted at some time $\omega_1^{\mathbf{L}} < \eta < \omega_2^{\mathbf{L}}$. Find a bijection $j : \omega_1^{\mathbf{L}} \to \mathbf{L}_\eta$. Since $\eta > \omega_1^{\mathbf{L}}$, we find $\gamma_\beta \in \omega_1^{\mathbf{L}}$ such that $j(\gamma_\beta) = \beta$ for every $\beta < \omega_1^{\mathbf{L}} + 1$. We define a valuation as follows: $v(P_{\beta,\gamma}) = 1$ if and only if $j(\beta) \in j(\gamma)$ and $v(B_{\beta,\gamma}) = 1$ if and only if $\gamma = \gamma_\beta$. It is easy to check that $\hat{v}(\Psi_i) = 1$.

"⇒". If $\widehat{v}(\Psi_i) = 1$, then define E_v and π_v as above. The structure $(\omega_1^{\mathbf{L}}, E_v)$ satisfies $\sigma \wedge \psi_i$ by Lemma 11 and is well-founded by $\bigwedge_{s \in S} \bigvee_{i \in \omega} \neg P_{s(i+1),s(i)}$ (there are no descending E_v-sequences). So by Mostowski's Collapsing Lemma, it is isomorphic to a transitive structure $(N, \in) \models \sigma \wedge \psi_i$. This means that $N = \mathbf{L}_\eta$ for some limit ordinal η. But Lemma 12 shows that $\omega_1^{\mathbf{L}} + 1$ embeds into $\mathbf{L}_\eta$, and hence, $\eta > \omega_1^{\mathbf{L}}$. Therefore, $\mathbf{L}_\eta$ sees that the ith Koepke machine with parameter $\omega_1^{\mathbf{L}}$ halts. Now the claim follows from absoluteness. □

Clearly, there is a Koepke machine that produces, with parameter $\omega_1^{\mathbf{L}}$, upon input $i \in \omega$, the formula Ψ_i. As mentioned before, this finishes the proof of Theorem 10 by contradiction. □

We note that the proof of Theorem 10 can be generalised to show that there is no ordinal α such that ∞-SAT is OTM decidable in the ordinal parameter α.

We also mention that it is possible to define OTM analogues of other classical **NP**-complete problems such as 3SAT and Subset Sum and prove their ∞-**NP**-completeness as well as an OTM analogue of Ladner's theorem ("there are OTM decision problems that are in ∞-**NP**, but neither in ∞-**P** nor ∞-**NP**-complete"). We shall describe these results in future work.

References

1. Carl, M.: Towards a Church-Turing-Thesis for infinitary computation (2013) preprint. arXiv:1307.6599
2. Carl, M.: Infinite time recognizability from random oracles and the recognizable jump operator. Computability (to appear)
3. Dawson, B.: Ordinal time Turing Computation. Ph.D. thesis, University of Bristol (2009)
4. Deolalikar, V., Hamkins, J.D., Schindler, R.: $\mathbf{P} \neq \mathbf{NP} \cap \mathbf{co\text{-}NP}$ for infinite time Turing machines. J. Log. Comput. **15**(5), 577–592 (2005)
5. Hamkins, J.D., Lewis, A.: Infinite time turing machines. J. Symb. Log. **65**(2), 567–604 (2000)
6. Hamkins, J.D., Welch, P.D.: $\mathbf{P}^f \neq \mathbf{NP}^f$ for almost all f. Math. Log. Q. **49**(5), 536–540 (2003)
7. Jensen, R.B., Karp, C.: Primitive recursive set functions. In: Axiomatic Set Theory. Proceedings of the Symposium in Pure Mathematics of the American Mathematical Society held at the University of California, Los Angeles, California, 10 July–5 August, vol. XIII/I of Proceedings of Symposia in Pure Mathematics, pp. 143–176. American Mathematical Society (1971)
8. Kanamori, A.: The Higher Infinite. Large Cardinals in Set Theory from Their Beginnings. Springer Monographs in Mathematics, 2nd edn. Springer, Heidelberg (2003)
9. Karp, C.: Languages with Expressions of Infinite Length. North-Holland, Amsterdam (1964)
10. Koepke, P.: Turing computations on ordinals. Bull. Symb. Log. **11**(3), 377–397 (2005)
11. Koepke, P.: Ordinal computability. In: Ambos-Spies, K., Löwe, B., Merkle, W. (eds.) CiE 2009. LNCS, vol. 5635, pp. 280–289. Springer, Heidelberg (2009). doi:10.1007/978-3-642-03073-4_29

12. Koepke, P., Seyfferth, B.: Ordinal machines and admissible recursion theory. Ann. Pure Appl. Log. **160**, 310–318 (2009)
13. Löwe, B.: Space bounds for infinitary computation. In: Beckmann, A., Berger, U., Löwe, B., Tucker, J.V. (eds.) CiE 2006. LNCS, vol. 3988, pp. 319–329. Springer, Heidelberg (2006). doi:10.1007/11780342_34
14. Rin, B.: The computational strengths of α-tape infinite time turing machines. Ann. Pure Appl. Log. **165**(9), 1501–1511 (2014)
15. Schindler, R.: **P** $\neq$ **NP** infinite time turing machines. Monatsh. Math. **139**, 335–340 (2003)
16. Winter, J.: Space complexity in infinite time Turing machines. Master's thesis, Universiteit van Amsterdam. ILLC Publications MoL-2007-14 (2007)
17. Winter, J.: Is **P** = **PSPACE** for Infinite time turing machines? In: Archibald, M., Brattka, V., Goranko, V., Löwe, B. (eds.) ILC 2007. LNCS, vol. 5489, pp. 126–137. Springer, Heidelberg (2009). doi:10.1007/978-3-642-03092-5_10

The Recognizability Strength of Infinite Time Turing Machines with Ordinal Parameters

Merlin Carl[1,2,3]($\boxtimes$) and Philipp Schlicht[1,2,3]

[1] Fachbereich Mathematik und Statistik, Universität Konstanz, Konstanz, Germany
merlin.carl@uni-konstanz.de
[2] Lehrstuhl für Theoretische Informatik, Universität Passau, Passau, Germany
[3] Mathematisches Institut, Universität Bonn, Bonn, Germany

Abstract. We study infinite time Turing machines that attain a special state at a given class of ordinals during the computation. We prove results about sets that can be recognized by these machines. For instance, the recognizable sets of natural numbers with respect to the cardinal-detecting infinite time Turing machines introduced in [Hab13] are contained in a countable level of the constructible hierarchy, and the recognizable sets of natural numbers with respect to finitely many ordinal parameters are constructible.

1 Introduction

Since the introduction of Infinite Time Turing Machines in [HL00], a variety of machine models of infinitary computability has been introduced: Turing machines that work with time and space bounded by an ordinal α, or with tape of length α but no time limit, or with tape and time both of length On; register machines that work in transfinite time and can store natural numbers or arbitrary ordinals in their registers; and so on.

A common feature to all these machine models of infinitary computations is that they are strongly linked to Gödel's constructible universe L. Since their operations are absolute between V and L, all objects that are writable by such machines are constructible. However, as was shown in [CSW], these machines can in a sense deal with objects far beyond L when one considers recognizability instead of computability, i.e. the ability of the machine to identify some real number x given in the oracle instead of producing x on the empty input. The model considered in [CSW] were Koepke's ordinal Turing machines (OTMs) with ordinal parameters.

A natural next step is then to determine how strong a machine type has to be to allow the recognizability of non-constructible real numbers. This motivates the question whether other of these models also have such strong properties.

It is relatively easy to deduce from Shoenfield's absoluteness theorem that, without ordinal parameters, recognizability for all machine types is restricted to a certain countable level L_σ of Gödel's constructible hierarchy L. The question must hence be what happens when we equip the other models with ordinal

© Springer International Publishing AG 2017
J. Kari et al. (Eds.): CiE 2017, LNCS 10307, pp. 198–209, 2017.
DOI: 10.1007/978-3-319-58741-7_20

parameters. For $OTMs$, an ordinal parameter α is given to the machine by simply marking the αth cell of the working tape.

For $ITTMs$ with the tape length ω, ordinal parameters cannot be introduced in this way. However, there is a rather natural way to make $ITTMs$ work with ordinal parameters, first introduced in [Hab13]. Namely, we introduce a new inner machine state that is assumed whenever the current time is an element of a given class X of ordinals, and one hence marks one or several points of time instead of tape cells. In this way, an $ITTM$ can be made to work relative to an arbitrary class of ordinals, where singletons correspond to single ordinal parameters. In this case, we will speak of X-$ITTMs$ or *ordinal-detecting ITTMs*. For the class of cardinals X, these are the *cardinal-detecting ITTMs* studied in [Hab13].

In this paper, we study the recognizability strength of cardinal detecting $ITTMs$ and more generally of ordinal-detecting $ITTMs$. We show that every recognizable set of natural numbers with respect to cardinal-detecting $ITTMs$ is an element of L_σ, the first level of the constructible hierarchy where every Σ_1-statement that is true in L is already true (Theorem 9). However, these machines can recognize more real numbers than mere $ITTMs$ (Lemma 12). Moreover, we show that every recognizable set of natural numbers with respect to $ITTMs$ with finitely many ordinal parameters is constructible (Theorem 21). However, these machines recognize some sets of natural numbers outside of L_σ for certain ordinals α (Lemma 19). We conclude that even with ordinal parameters, $ITTM$-recognizability does not lead out of L.

2   Basic Notions and Results

Infinite Time Turing Machines, introduced by Hamkins and Kidder (see [HL00]), generalize Turing computability to transfinite working time. Their computations work like ordinary Turing computations at successor times, while the tape content at limit times is obtained as a cell-wise inferior limit of the sequence of earlier contents and the inner state at limit times is a special limit state. For details, we refer to [HL00].

There are various notions of computability associated with $ITTMs$.

Definition 1. Suppose that x and y are subsets of ω.

1. x is *writable* in the oracle y if and only if there is an $ITTM$-program P such that $P^y \downarrow = x$, i.e. P, run in the oracle y, halts with x on the output tape.
2. x is *eventually writable* in the oracle y if and only there is an $ITTM$-program P such that P, when run in the oracle y on the empty input, eventually has x on its output tape and never changes it again.
3. x is *accidentally writable* in the oracle y if and only there is an $ITTM$-program P such that P, when run in the oracle y on the empty input, has x on its output tape at some point, but may overwrite it later on.

The *ITTM*-recognizable sets are defined as follows.

Definition 2. Suppose that x and y are subsets of ω.

1. x is *ITTM-recognizable* or simply *recognizable* relative to y if and only if there is an *ITTM*-program P such that, for all subsets z of ω, $P^{z \oplus y} \downarrow = \delta_{z,x}$, where δ is the Kronecker symbol.
2. x is *non-deterministically ITTM-recognizable* if and only if there is a subset y of ω such that $x \oplus y$ is *ITTM*-recognizable.
3. The *recognizable closure*, denoted by $\mathcal{R}$, is the closure of the empty set under relativized recognizability.

We will call sets of natural numbers *reals*. The following alternative characterization of the recognizable closure works rather generally for models of infinite computation.

Lemma 3. The non-deterministically *ITTM*-recognizable reals are exactly those in $\mathcal{R}$.

Proof. If $x \oplus y$ is *ITTM*-recognizable, then clearly $x \in \mathcal{R}$. Suppose on the other hand that $x \in \mathcal{R}$. Then there is a sequence $\langle x_0, \ldots, x_n \rangle$ with $x = x_0$ such that x_n is *ITTM*-recognizable and x_i is *ITTM*-recognizable relative to x_{i+1} for all $i < n$. It is easy to see that the join $\bigoplus_{i \leq n} x_i$ is *ITTM*-recognizable by first identifying the last component and then successively the previous components.

It was observed in [CSW, Lemma 3.2] that $\mathcal{R} = L_\sigma \cap \mathcal{P}(\omega)$, where σ is least with the property that $L_\sigma \prec_{\Sigma_1} L$ or equivalently least with the property that every Σ_1-statement that it true in L is already true in L_σ.

We will need the following results.

Theorem 4. Suppose that y is a subset of ω.

1. [Wel09, Fact 2.4 & Fact 2.5 & Fact 2.6] There are countable ordinals $\lambda^y, \zeta^y, \Sigma^y$ with the following properties for all subsets x of ω.
   (a) x is writable relative to y if and only if $x \in L_{\lambda^y}[y]$.
   (b) x is eventually writable relative to y if and only if $x \in L_{\zeta^y}[y]$.
   (c) x is accidentally writable relative to y if and only if $x \in L_{\Sigma^y}[y]$.
2. [Wel14, p.11-12] An ITTM-program in the oracle y will either halt in strictly less than λ^y many steps, or it will run into an ever-repeating loop, repeating the sequence of configurations between ζ^y and Σ^y, which is of order-type Σ^y, from Σ^y on.
3. [Wel09, Theorem 1 & Corollary 2] [Wel14, Theorem 3] The triple $\langle \lambda^y, \zeta^y, \Sigma^y \rangle$ is the lexically least triple $\langle \alpha, \beta, \gamma \rangle$ of distinct ordinals with $L_\alpha[y] \prec_{\Sigma_1} L_\beta[y] \prec_{\Sigma_2} L_\gamma[y]$.

Remark 5. An important result for *ITTMs*, and many other models of infinite computation, is the existence of *lost melodies*, i.e. real numbers that are *ITTM*-recognizable, but not writable. The existence of lost melodies for *ITTMs* was proved in [HL00]. For more on lost melodies for other machine types, see [Car14a, Car14b, Car15, CSW].

We now define how an infinite time Turing machine works relative to a class of ordinals.

Definition 6. For classes X of ordinals, an X-$ITTM$ works like an $ITTM$ with the modification that whenever the running time is an element of X, the machine state is set to a special reserved state. If an $ITTM$-program P is run relative to a class X, we will write X-P instead of P. When X consists of the ordinals in $\alpha = \langle \alpha_0, \ldots, \alpha_n \rangle$, we will write α-$ITTM$ for the machine with special states at times in α, and for $\alpha = \langle \alpha \rangle$ simply α-$ITTM$.

We thus obtain the *cardinal-detecting* $ITTMs$ of [Hab13], called *cardinal-recognizing* $ITTMs$ there, for the class of all cardinals. Note that there are several variants of $ITTMs$, for instance the original definition in [HL00] with three tapes of input, output and scratch, that we use here, its variant with only one tape, and the variant where at limit times, the head is set to the inferior limit of the previous head positions, instead of moving to the first tape cell. All proofs in this paper can be easily modified to work for each of these variants.

We take the opportunity to answer in the negative [Hab13, Question 10], which asked whether every real number accidentally writable by a cardinal-recognizing $ITTM$ is also accidentally writable by a plain $ITTM$.

Theorem 7. There is a cardinal-recognizing $ITTM$-program that writes a code for Σ.

Proof. Let U denote an $ITTM$-program that simulates all $ITTM$-programs simultaneously. The configuration c of U at time ω_1, when the special state is assumed for the first time, is the same as at time Σ and at time ζ [Wel14, p.11-12]. If c would occur prior to ζ, then U would start looping before time ζ, contradicting the assumption that U simulates all $ITTM$-programs. Hence c is accidentally writable, but not eventually writable. The L-least code for Σ is $ITTM$-writable from every real that is accidentally writable, but not eventually writable by the proof of [CH11, Proposition 4.6]. Hence there is an $ITTM$-program P that computes a code for Σ from c. We now run P on the tape content of U when the special state is assumed for the first time. This program will halt with a code for Σ on the output tape, as required.

3 Recognizable Reals Relative to Cardinals

In this section, we will determine the recognizable closure for cardinal-detecting $ITTMs$. It is easy to see that the parameter ω does not add recognizability strength. We begin by considering $ITTMs$ with uncountable parameters.

Theorem 8. Every subset x of ω that is $\omega_1 \alpha$-recognized by P for some ordinal α is an element of L_σ.

Proof. Suppose that P is a program that recognizes x and $\omega_1\alpha$-P^x halts with the final state s. For any subset y of ω, after time Σ^y, the computation repeats a loop of length Σ^y by Theorem 4, and thus the state of P^y at time Σ_y is the same as at time $\omega_1\alpha$. Consequently, the computation will continue exactly the same whether the new inner state s is assumed at time Σ^y or at time $\omega_1\alpha$. Since P recognizes x, Σ^y-P^y and $\omega_1\alpha$-P^y both halt with the same output and the same final state s.

Let $c_{y,\beta}$ denote the $L[y]$-least code for β, if β is countable in $L[y]$. We argue that the halting time of Σ^y-P^y is strictly less than $\lambda^{y\oplus c}$ for $c = c_{y,\Sigma^y}$. The tape content z of P^y at time Σ^y is accidentally writable in y and hence z is an element of $L_{\Sigma^y}[y]$. Therefore z is writable from $y \oplus c$ and $\lambda^z \leq \lambda^{y\oplus c}$. Then the halting time of Σ^y-P^y is strictly less than $\lambda^{y\oplus c}$.

We can thus characterize x as the unique real y with $\phi(y)$, where $\phi(y)$ is the statement that Σ^y-$P^y \downarrow= 1$ holds in $L_{\lambda^{y\oplus c_{y,\Sigma^y}}}[y]$. To see that $\phi(y)$ is a Σ_1-statement, we call a triple $\alpha = \langle \alpha_0, \alpha_1, \alpha_2 \rangle$ a *y-triple* if $\alpha_0 < \alpha_1 < \alpha_2$ and $L_{\alpha_0}[y] \prec_{\Sigma_1} L_{\alpha_1}[y] \prec_{\Sigma_2} L_{\alpha_2}[y]$. Then $\phi(y)$ is equivalent to the Σ_1-statement that there is some γ such that in L_γ, the lexically least y-triple $\alpha = \langle \alpha_0, \alpha_1, \alpha_2 \rangle$ and the lexically least $y \oplus c_{y,\alpha_2}$-triple $\beta = \langle \beta_0, \beta_1, \beta_2 \rangle$ exist and α_2-$P^y \downarrow= 1$ holds in $L_{\beta_0}[y]$.

Since $\phi(x)$ holds, there is some $y \in L$ such that $\phi(y)$ holds in L by Shoenfield absoluteness. Then there is some z in L_σ such that $\phi(z)$ holds. This implies that $\omega_1\alpha - P^z \downarrow= 1$, so $z = x$ and $x \in L_\sigma$.

Theorem 9. If x is a subset of ω that is recognized by an X-$ITTM$, where X is a closed class of ordinals of the form $\omega_1\alpha$, then $x \in L_\sigma$. In particular, this holds for subsets of ω recognized by a cardinal-detecting $ITTM$.

Proof. The proof is a variation of the proof of Theorem 8. We first assume that X is a proper class.

Suppose that X-P recognizes x. A computation by P in the oracle x will assume its special state at the times $\langle \alpha_i \mid i < \gamma \rangle$ for some ordinal γ. Let x_i denote the tape contents at time α_i. Between the times α_i and α_{i+1}, we have an ordinary $ITTM$-computation in the oracle x with input x_i on the tape. Such a computation will either halt or cycle from the time $\Sigma^{x\oplus x_i}$ with a loop of length $\Sigma^{x\oplus x_i}$. Now any α_i for $i \geq 1$ is a multiple of Σ^y for all reals y. In particular, the tape contents at time $\alpha_i + \Sigma^{x\oplus x_i}$ will be the same as at time α_{i+1}.

We can hence characterize x as the unique real y with $\phi(y)$, where $\phi(y)$ is the following statement. There is an ordinal δ, a continuous sequence $\langle \alpha_i \mid i < \delta \rangle$ of ordinals and a sequence $\langle y_i \mid i < \delta \rangle$ of real numbers such that for all i with $i + 1 < \delta$, y_{i+1} is the tape contents of the computation of P with oracle y and input y_i at time $\Sigma^{y\oplus y_i}$, $\alpha_{i+1} = \alpha_i + \Sigma^{y\oplus y_i}$ and the computation P^y with special state at the elements of the sequence $\langle \alpha_i \mid i < \delta \rangle$ halts with output 1.

As in the proof of Theorem 8, $\phi(y)$ is a Σ_1-statement. Since $\phi(x)$ is valid, $\phi(y)$ holds for some real y in L_σ. Then $y = x$ and $x \in L_\sigma$.

We now assume that X is a set. Let γ be the supremum of halting times of $P(x)$ for all subsets x of ω and let Y be a closed proper class of ordinals of the form $\omega_1\alpha$ with $Y \cap \gamma = X$. Then the claim follows from the previous argument applied to Y.

The assumption that X is closed is necessary in the previous theorem, since every subset of ω is recognizable by an X-$ITTM$ for some set X of ordinals of the form $\omega_1\alpha$, as one can easily show.

Theorem 10. The recognizable closure both for $\omega_1\alpha$-$ITTMs$ for any ordinal α and for cardinal-detecting $ITTMs$ is $L_\sigma \cap \mathcal{P}(\omega)$.

Proof. We first argue that the recognizable closure is contained in L_σ. Suppose that y is an element of L_σ and x is $ITTM$-recognizable from y in either machine type. By a relativization of the proofs of Theorems 8 and 9, x has a Σ_1-characterization in the parameter y. Since y is an element of L_σ, it is Σ_1-definable in L without parameters and hence can be eliminated from the definition of x. It follows that there is a Σ_1-formula ϕ such that x is the only witness for ϕ in L. Hence $x \in L_\sigma$.

Moreover, $L_\sigma \cap \mathcal{P}(\omega)$ is contained in the recognizable closure for plain $ITTMs$ by [CSW, Lemma 3.2] and hence recognizable closures for both machine types are equal to $L_\sigma \cap \mathcal{P}(\omega)$.

The recognizability strength of cardinal-detecting $ITTMs$ is strictly higher than that of $ITTMs$ by the next results. The next lemma shows that not every real in the recognizable closure for $ITTMs$ is itself recognizable.

Lemma 11. If $x \in L_\Sigma \setminus L_\lambda$, then x is not $ITTM$-recognizable.

Proof. Suppose that $x \in L_\Sigma$ and x is $ITTM$-recognizable. We consider an $ITTM$-program P which writes every accidentally writable real at some time. If x is $ITTM$-recognizable by a program Q, we can write x by letting P run and checking in each step with Q whether the contents of the output tape is equal to x, and in this case stop. Then x is writable and hence $x \in L_\lambda$.

Lemma 12. There is a real number that is $ITTM$-recognizable by a cardinal-detecting $ITTM$ and by an α-$ITTM$ for every $\alpha \geq \lambda$, but not $ITTM$-recognizable.

Proof. Let $0^\nabla = \{\varphi \mid \varphi(0) \downarrow\}$ denote the *halting problem* or *jump* for $ITTMs$. Since 0^∇ is Σ_1-definable over L_λ, but certainly not $ITTM$-writable, we have $0^\nabla \in L_\Sigma \setminus L_\lambda$ and hence 0^∇ is not $ITTM$-recognizable. We now argue that 0^∇ is writable by a cardinal-detecting $ITTM$ and α-$ITTM$-writable, hence it is recognizable with respect to these machines. This was already observed in [Hab13] for cardinal-detecting machines. We can simulate all $ITTM$-programs simultaneously and write 1 in the n-th place of the output tape when the n-th program has stopped. As all halting times are countable, the output tape will contain 0^∇ at time α, and the special state at time α allows us to stop.

4   Recognizable Reals Relative to Finitely Many Ordinals

In this section, we consider what happens when we allow the machine to enter a special state at an ordinal time α. We first determine the writability strength of such machines. The next result follows from [CH11, Proposition 4.6]. We give a short proof from the λ-ζ-Σ theorem for the reader.

Lemma 13. Let λ^α denote the supremum of the halting times of α-$ITTMs$.

1. $\lambda^x > \Sigma$ for every real x with $\lambda^x \geq \zeta$.
2. $\lambda^\alpha > \Sigma$ for every $\alpha \geq \zeta$.

Proof. To prove the first claim, we first suppose that $\Sigma = \Sigma^x$. Since $\zeta \leq \lambda^x < \zeta^x$, this implies $L_\zeta \prec_{\Sigma_2} L_{\zeta^x} \prec_{\Sigma_2} L_\Sigma$ and this contradicts the minimality of Σ. Second, suppose that $\Sigma < \Sigma^x$. Then there is a triple $\langle \alpha, \beta, \gamma \rangle$ in L_{Σ^x} with $L_\alpha \prec_{\Sigma_1} L_\beta \prec_{\Sigma_2} L_\gamma$, namely $\langle \lambda, \zeta, \Sigma \rangle$. Since $L_{\lambda^x}[x] \prec_{\Sigma_1} L_{\Sigma^x}[x]$, we have $L_{\lambda^x} \prec_{\Sigma_1} L_{\Sigma^x}$ and therefore, there is such a triple in L_{λ^x}. Since Σ is the least value for γ for such triples $\langle \alpha, \beta, \gamma \rangle$, we have $\Sigma \leq \gamma < \lambda^x$.

The second claim is clear if $\alpha \geq \Sigma$. If $\alpha < \Sigma$, we can write an accidentally writable, but not eventually writable real x with an α-$ITTM$. For instance, we can write a code for an ordinal $\beta \geq \zeta$ by adding all ordinals written by a universal machine at time α. Since we can search for an L-level containing x and halt, it follows that $\lambda^x \geq \zeta$. Hence $\lambda^\alpha \geq \lambda^x > \Sigma$ by the first claim.

We remark that the assumption of Lemma 13 cannot be weakened to $\lambda^x > \lambda$ by the following counterexample. Let x be the L-minimal code for λ. Then clearly $\lambda^x > \lambda$. On the other hand, x is eventually writable, the L-minimal code for λ^x is eventually writable relative to x and the eventually writable reals are closed under eventual writability. Hence the L-minimal code for λ^x is eventually writable and therefore $\lambda^x < \zeta < \Sigma$.

Lemma 14. The following statements are equivalent for a real x.

1. x is α-$ITTM$-writable for some ordinal α.
2. x is $ITTM$-writable from some accidentally writable real number.
3. x is $ITTM$-writable from every accidentally writable real number that is not eventually writable.
4. x is an element of L_{λ^z}, where z is the L-least code for ζ.

Proof. Suppose that x is α-$ITTM$-writable for some ordinal α by a computation of a program P. Up to time α, this computation is just an ordinary $ITTM$-computation and hence at time α, the tape will contain some accidentally writable real number y. The rest of the computation will again be an ordinary $ITTM$-computation with the input y and thus the output will be $ITTM$-writable from y.

Every accidentally writable x that is not eventually writable has $\lambda^x > \zeta$ and hence $\lambda^x > \Sigma$ by Lemma 13, so every accidentally writable is writable from x.

Suppose that x is $ITTM$-writable by a program P from some accidentally writable real number y. Suppose that Q is a program that has y on its tape at time α. If we run Q up to time α and then run P, this will write x.

Since the L-least code z for ζ is accidentally writable, the remaining implications follow from Lemma 13.

We obtain the following generalization of Lemma 14.

Lemma 15. The following statements are equivalent for a real x.

1. x is α-$ITTM$-writable for some sequence α of length n.
2. x is $ITTM$-writable from x_{n-1} for some sequence $\boldsymbol{x} = \langle x_0, \ldots, x_{n-1} \rangle$, where x_j is accidentally writable from $\bigoplus_{i<j} x_i$ for all $j < n$.
3. x is an element of $L_{\lambda^{z_{n-1}}}$, where $z_0 = 0$ and z_{i+1} is the L-least code for ζ^{z_i} for all $i < n - 1$.

Proof. The implications follow by iterated application of Lemma 14.

To show that the recognizability strength of $ITTMs$ with arbitrary ordinal parameters is beyond L_σ, we need the next definition and two well-known results. For technical convenience, we work with Jensen's J-hierarchy instead of Gödel's L-hierarchy. Note that $J_\alpha = L_\alpha$ if α takes one of the values λ, ζ, Σ or σ.

Definition 16. An ordinal α is an *index* if there is a real in $J_{\alpha+1} \setminus J_\alpha$.

Lemma 17. (Jensen) If α is an index, then there is a surjection from ω onto J_α that is definable over J_α and hence there is a code for J_α in $J_{\alpha+1}$.

Proof. This follows from the fact that $\langle J_\alpha \mid \alpha \in \mathrm{Ord} \rangle$ is acceptable by [Zem02, Lemma 1.10.1].

Lemma 18 (folklore). There are unboundedly many admissible indices α below ω_1^L.

Proof. There are unboundedly many indices below ω_1^L, since there are ω_1^L many reals in L. Suppose that α is an index. Let c denote the L-least code for α. Since α is an index, $c \in J_{\alpha+1}$ by Lemma 17. Suppose that $\beta = \omega_1^c$ is the least c-admissible ordinal. Then β is admissible and it remains to show that β is an index. Since $c \in J_\beta$ and J_β is the Skolem hull of c in J_β, there is a surjection from ω onto J_β that is definable over J_β. There is a real x in $J_{\beta+1}$ that codes this surjection. Since J_β is admissible, x cannot be an element of J_β and hence β is an index.

The next result shows that there are $ITTM$-recognizable reals with respect to ordinal parameters beyond L_σ.

Lemma 19. Suppose that $\alpha = \omega\beta$ is an index and c is the L-least code for J_α in $J_{\alpha+1} \setminus J_\alpha$. Then c is α-$ITTM$-recognizable.

Proof. The claim is easy to see for $\alpha = \omega$, so we assume that $\alpha > \omega$. Suppose that the input is x. We first check whether x codes a set with an extensional relation and otherwise reject x. We then count through the ordinals of the set coded by x, i.e. in each step we search for the least next ordinal, while simultaneously searching for infinite strictly decreasing sequences of ordinals above. This is possible by keeping markers at all previous ordinals in every step. If we have exhausted the ordinals or if we find an infinite strictly decreasing sequence of ordinals in the structure coded by x before time α, then we reject x. This algorithm is carried out up to time α. After time α, we check if the structure coded by x is well-founded, and reject x if this is not the case. If the structure is well-founded, we check whether it is isomorphic to some J_β. In this case, we write a code for $J_{\beta+1}$ and check whether it contains a code for J_β. If it does, we determine the least such code in $J_{\beta+1} \setminus J_\beta$ and check if it is equal to c. If it is equal to c, then we accept x, and otherwise reject x. There is a code for J_α in $J_{\alpha+1} \setminus J_\alpha$ by Lemma 17. Therefore this algorithm accepts a real x if and only if $x = c$.

To show that every real that is *ITTM*-recognizable relative to finitely many ordinal parameters is in L, we need the following result.

Lemma 20. Suppose that x is a subset of ω that is *ITTM*-recognizable from n ordinal parameters. Then x is *ITTM*-recognizable from finitely many ordinal parameters strictly below $\omega_1 \cdot (n+1)$.

Proof. Suppose that x is *ITTM*-recognizable from $\boldsymbol{\alpha} = \langle \alpha_0, \ldots, \alpha_{n-1} \rangle$ and $\boldsymbol{\alpha}$ is strictly increasing. We can assume that α_{n-1} is uncountable. Suppose that α_i^* is the remainder of the division of α_i by ω_1 for $i < n$. Suppose that $\boldsymbol{k} = \langle k_0, \ldots, k_l \rangle$ is the unique sequence such that $k_0 < n$ is least such that α_{k_0} is uncountable and for all $i < l$, $k_{i+1} < n$ is least such that the unique ordinal α with $\alpha_{k_i} + \alpha = \alpha_{k_{i+1}}$ is uncountable. Let $\beta_i = \alpha_i$ for $i < k_0$, $\beta_i = \omega_1 j + \alpha_i^*$ for $k_j \leq i < k_{j+1}$ and $j < l$, and let $\beta_i = \omega_1 l + \alpha_i^*$ for $k_l \leq i < n$.

Suppose that a program P recognizes x from the parameters $\alpha_0 < \cdots < \alpha_n$. For every input y, P^y will cycle from time Σ^y between multiples of Σ^y and hence of ω_1 by [Wel09]. This implies that for every input y, P^y will halt with the same state for the parameters $\alpha_0, \ldots, \alpha_n$ and the parameters $\beta_0, \ldots, \beta_n$. Hence P recognizes x from the parameters $\beta_0, \ldots, \beta_n$.

Theorem 21. Every subset of ω that is *ITTM*-recognizable from finitely many ordinal parameters is an element of L.

Proof. Suppose that x is *ITTM*-recognizable from finitely many ordinal parameters. Then x is recognized by a program P from finitely many ordinal parameters strictly below $\omega_1 \cdot (n+1)$ by Lemma 20.

We can assume that x is recognized by a single ordinal parameter δ with $\omega_1 \leq \delta < \omega_1 2$. The proof of the general case is analogous. Suppose that P^x with the special state at time δ halts at time η. Let $\delta = \omega_1 + \delta^*$ and $\eta = \omega_1 + \eta^*$. We consider the Σ_1-statement $\psi(\bar{x})$ stating that $P^{\bar{x}}$ with the special state at time $\Sigma^{\bar{x}} + \delta^*$ halts at time $\Sigma^{\bar{x}} + \eta^*$ and accepts $\bar{x}$. Since the program will cycle from time $\Sigma^{\bar{x}}$ in intervals of length $\Sigma^{\bar{x}}$ by Theorem 4, $\psi(\bar{x})$ is equivalent to the statement that $P^{\bar{x}}$ with the special state at time $\Sigma^{\bar{x}} \cdot \alpha + \delta^*$ halts at time $\Sigma^{\bar{x}} \cdot \alpha + \eta^*$ for some $\alpha \geq 1$, or equivalently for all $\alpha \geq 1$.

The statement $\psi(x)$ holds in V and in every generic extension of V. In particular, $\psi(x)$ holds in every $\mathrm{Col}(\omega, \zeta)$-generic extension $V[G]$ of V, where ζ is a countable ordinal with $\delta^*, \eta^* \leq \zeta$ and $\mathrm{Col}(\omega, \zeta)$ is the standard collapse forcing to make ζ countable. Moreover $\exists \bar{x} \psi(\bar{x})$ holds in $L[G]$ by Shoenfield absoluteness, since $\exists \bar{x} \psi(\bar{x})$ is a Σ_1-statement and the parameters δ^* and η^* are countable in $L[G]$.

Suppose that θ is an L-cardinal such that L_θ is sufficiently elementary in L. Suppose that $M \prec L_\theta$ is countable with $\zeta + 1 \subseteq M$ and $\bar{M}$ is the transitive collapse of M. Suppose that g, h are mutually $\mathrm{Col}(\omega, \zeta)$-generic over $\bar{M}$ in V. The statement $\exists \bar{x} \psi(\bar{x})$ is forced over L_θ and $\bar{M}$ and therefore holds in $\bar{M}[g]$ and in $\bar{M}[h]$, witnessed by some reals x_g and x_h. Since P recognizes x with the special state at time δ, the uniqueness of x implies that $x_g = x_h = x$. Since g and h are mutually generic over $\bar{M}$, we have $\bar{M}[g] \cap \bar{M}[h] = \bar{M}$ and hence $x \in \bar{M}$. Since $\bar{M}$ is a subset of L, this implies $x \in L$.

This allows us to determine the recognizable closure with respect to ordinal parameters.

Theorem 22. The recognizable closure for *ITTMs* with single ordinal parameters and for *ITTMs* with finitely many ordinal parameters is $P(\omega)^L$.

Proof. This follows from Lemmas 18 and 19 and Theorem 21.

Note that Theorem 21 cannot be extended to countable sets of ordinal parameters, since it is easy to see that every real is writable from a countable set of ordinal parameters. The previous results suggest the question whether the number of ordinal parameters is relevant for the recognizability strength. The next result shows that this is the case.

Theorem 23. For every n, there is a subset x of ω that is *ITTM*-recognizable from $n + 1$ ordinals, but not from n ordinals.

Proof. We define $\boldsymbol{x}_n = \langle x_0, \ldots, x_n \rangle$, $\boldsymbol{\lambda}_n = \langle \lambda_0, \ldots \lambda_n \rangle$, $\boldsymbol{\zeta}_n = \langle \zeta_0, \ldots \zeta_n \rangle$, and $\boldsymbol{\Sigma}_n = \langle \Sigma_0, \ldots \Sigma_n \rangle$ as follows for all n. Let $\zeta_0 = \zeta$ and $\zeta_{i+1} = \zeta^{x_i}$, where x_0 is the L-least code for ζ and x_{i+1} is the $L[x_i]$-least code for ζ_{i+1}. Moreover, let $\lambda_0 = \lambda$, $\lambda_{i+1} = \lambda^{x_i}$, $\Sigma_0 = \Sigma$ and $\Sigma_{i+1} = \Sigma^{x_i}$. Then $\lambda^{x_i} > \Sigma_{i+1}$ for all i by the relativized version of Lemma 13. Moreover, let $\boldsymbol{\lambda}_n^y$, $\boldsymbol{\zeta}_n^y$ and $\boldsymbol{\Sigma}_n^y$ denote the relativized versions of $\boldsymbol{\lambda}_n$, $\boldsymbol{\zeta}_n$ and $\boldsymbol{\Sigma}_n$ for any real y.

Claim 24. A Cohen real x over $L_{\Sigma_n + 1}$ is not *ITTM*-recognizable from n ordinals.

Proof. Suppose that x is recognized by a program P in the parameter $\gamma = \langle \gamma_0, \ldots, \gamma_{n-1} \rangle$, where γ is strictly increasing. We define $\gamma^* = \langle \gamma_0^*, \ldots, \gamma_{n-1}^* \rangle$

as follows. Let $\gamma_0^* = \gamma_0$ if $\gamma_0 < \Sigma$ and $\gamma_0^* = \zeta_0 + \delta_0$ if $\gamma_0 \geq \Sigma$, where δ_0 is the remainder of the division of γ_0 by Σ. For all i with $i + 1 < n$, let $\gamma_{i+1}^* = \gamma_{i+1}$ if $\gamma_{i+1} < \Sigma_{i+1}$ and $\gamma_{i+1}^* = \zeta_{i+1} + \delta_{i+1}$ if $\gamma_{i+1} \geq \Sigma_{i+1}$, where δ_{i+1} is the remainder of the division of γ_{i+1} by Σ_{i+1}.

A computation with input y cycles from ζ^y in intervals of length Σ^y by Theorem 4. For every Cohen real y over L_{Σ_n+1}, $\lambda_n^y = \lambda_n$, $\zeta_n^y = \zeta_n$ and $\Sigma_n^y = \Sigma_n$ by the variant of [CS, Lemma 3.12] for Cohen forcing (see also [Wel99, p.11]). Since P^x with special states at γ accepts x, this implies that P^x with special states at γ^* accepts x as well.

Since $L_{\Sigma_n}[y]$ is a union of admissible sets by the variant of [CS, Lemma 2.11] for Cohen forcing, the run of P^y is an element of $L_{\Sigma_n}[y]$. Let σ be a name in L_{Σ_n} for the run of P^x with special states at γ^*. The statement that P^x with special states at γ^* accepts x is forced for σ by a condition p in Cohen forcing over L_{Σ_n} by the variant of [CS, Lemma 2.7] for Cohen forcing. Suppose that y is a Cohen generic over L_{Σ_n+1} with $x \neq y$ that extends the condition p. Then P^y accepts y by the truth lemma in the variant of [CS, Lemma 2.8] for Cohen forcing. This contradicts the uniqueness of x.

Claim 25. The L-least Cohen real over L_{Σ_n+1} is writable from ζ_n.

Proof. By the relativized version of Lemma 13 applied to $x_0, \ldots, x_n$, we can successively compute $x_0, \ldots, x_n$ from ζ_n. We can then compute codes for Σ_n, L_{Σ_n}, L_{Σ_n+1} and hence the L-least Cohen real over L_{Σ_n} in L_{Σ_n+1} from x_n.

Remark 26. For finitely many parameters, the writability and recognizability strengths do not change if we allow more than one special state, since such a program can be simulated with a single special state by coding the special states into tape cells.

5   Conclusion and Open Questions

We have seen that equipping *ITTMs* with the power to recognize one particular or all uncountable cardinals increases the set of *ITTM*-recognizable real numbers, but not the recognizable closure, which remains L_σ. Moreover, certain ordinals parameters enable an *ITTM* to recognize real numbers outside of L_σ, but *ITTM*-recognizability with finitely many ordinal parameters does not lead out of the constructible universe.

We conclude with the following open questions. It is open whether the ordinals in Lemma 20 can be chosen to be countable.

Question 27. Is every real x that is *ITTM*-recognizable from an ordinal already *ITTM*-recognizable from a countable ordinal?

Let $\mathcal{R}_\alpha$ denote the recognizable closure with respect to *ITTMs* with the parameter α and let $\sigma(\alpha)$ denote the least ordinal $\gamma > \alpha$ with $L_\gamma \prec_{\Sigma_1} L$. It is open what is $\mathcal{R}_\alpha$ and whether there is a relationship between $\mathcal{R}_\alpha$ and $L_{\sigma(\alpha)}$.

Question 28. What is $\mathcal{R}_\alpha$ for arbitrary ordinals α?

The notion of *semi-recognizable reals* is defined by asking that the program halts for some input x and diverges for all other inputs. The notion of *anti-recognizable reals* is defined by asking the program diverges for some input x and halts for all other inputs. The following question seems fundamental.

Question 29. Are there semi-recognizable reals and anti-recognizable reals that are not recognizable?

References

[Car14a] Carl, M.: The distribution of $ITRM$-recognizable reals. Ann. Pure Appl. Logic **165**(9), 1403–1417 (2014)

[Car14b] Carl, M.: The lost melody phenomenon. In: Infinity, computability, and metamathematics, vol. 23 of Tributes, pp. 49–70. College Publications, London (2014)

[Car15] Carl, M.: Optimal results on recognizability for infinite time register machines. J. Symb. Log. **80**(4), 1116–1130 (2015)

[CH11] Coskey, S., Hamkins, J.D.: Infinite time decidable equivalence relation theory. Notre Dame J. Formal Log. **52**(2), 203–228 (2011)

[CS] Carl, M., Schlicht, P.: Randomness via infinite time machines and effective descriptive set theory (2016). Submitted

[CSW] Carl, M., Schlicht, P., Welch, P.: Recognizable sets and Woodin cardinals (2016). Submitted

[Hab13] Habič, M.E.: Cardinal-recognizing infinite time turing machines. In: Bonizzoni, P., Brattka, V., Löwe, B. (eds.) CiE 2013. LNCS, vol. 7921, pp. 231–240. Springer, Heidelberg (2013). doi:10.1007/978-3-642-39053-1_27

[HL00] Hamkins, J.D., Lewis, A.: Infinite time turing machines. J. Symbolic Log. **65**(2), 567–604 (2000)

[Wel99] Welch, P.D.: Minimality arguments for infinite time turing degrees. In: Sets and proofs (Leeds 1997). London Mathematical Society Lecture Note Series, vol. 258, pp. 425–436. Cambridge University Press, Cambridge (1999)

[Wel09] Welch, P.D.: Characteristics of discrete transfinite time Turing machine models: halting times, stabilization times, and normal form theorems. Theoret. Comput. Sci. **410**(4–5), 426–442 (2009)

[Wel14] Welch, P.D.: Transfinite machine models. In: Turing's Legacy: Developments from Turing's Ideas in Logic. Lecture Notes in Logic, vol. 42, pp. 493–529. Association Symbolic Logic, La Jolla, CA (2014)

[Zem02] Zeman, M.: Inner Models and Large Cardinals. De Gruyter Series in Logic and its Applications, vol. 5. Walter de Gruyter & Co., Berlin (2012)

New Bounds on the Strength of Some Restrictions of Hindman's Theorem

Lorenzo Carlucci[1]([✉]), Leszek Aleksander Kołodziejczyk[2], Francesco Lepore[1], and Konrad Zdanowski[3]

[1] Department of Computer Science, University of Rome I, Rome, Italy
carlucci@di.uniroma1.it, leporefc@gmail.com
[2] Institute of Mathematics, University of Warsaw, Warsaw, Poland
lak@mimuw.edu.pl
[3] Faculty of Mathematics and Natural Sciences, Cardinal Stefan Wyszyński
University in Warsaw, Warsaw, Poland
k.zdanowski@uksw.edu.pl

Abstract. We prove upper and lower bounds on the effective content and logical strength for a variety of natural restrictions of Hindman's Finite Sums Theorem. For example, we show that Hindman's Theorem for sums of length at most 2 and 4 colors implies ACA_0. An emerging *leitmotiv* is that the known lower bounds for Hindman's Theorem and for its restriction to sums of at most 2 elements are already valid for a number of restricted versions which have simple proofs and better computability- and proof-theoretic upper bounds than the known upper bound for the full version of the theorem. We highlight the role of a sparsity-like condition on the solution set, which we call apartness.

1 Introduction and Motivation

The Finite Sums Theorem by Neil Hindman [15] (henceforth denoted HT) is a celebrated result in Ramsey Theory stating that for every finite coloring of the positive integers there exists an infinite set such that all the finite non-empty sums of distinct elements from it have the same color. Thirty years ago Blass, Hirst and Simpson proved in [2] that *all* computable instances of HT have *some* solutions computable in $\emptyset^{(\omega+1)}$ and that for *some* computable instances of HT *all* solutions compute $\emptyset'$. In terms of Reverse Mathematics, they showed that $\mathsf{ACA}_0^+ \vdash \mathsf{HT}$ and that $\mathsf{RCA}_0 \vdash \mathsf{HT} \to \mathsf{ACA}_0$ (see [17,20] for the definition of these systems). Both bounds hold for the particular case of colorings in two colors. Closing the gap between the upper and lower bound is one of the major open problems in Computable and Reverse Mathematics (see, e.g., [19]).

Part of this work was done while the first author was visiting the Institute for Mathematical Sciences, National University of Singapore in 2016. The visit was supported by the Institute. The second author was partially supported by Polish National Science Centre grant no. 2013/09/B/ST1/04390. The fourth author was partially supported by University Cardinal Stefan Wyszyński in Warsaw grant UmoPBM-26/16.

© Springer International Publishing AG 2017
J. Kari et al. (Eds.): CiE 2017, LNCS 10307, pp. 210–220, 2017.
DOI: 10.1007/978-3-319-58741-7_21

Blass advocated the study of restrictions of Hindman's Theorem in which a bound is put on the length (i.e., number of distinct terms) of sums for which monochromaticity is guaranteed [1], conjecturing that the complexity of Hindman's Theorem grows as a function of the length of sums. Recently Dzhafarov, Jockusch, Solomon and Westrick showed (see Corollary 3.4 in [12]) that the known $\emptyset'$ (ACA$_0$) lower bound on Hindman's Theorem holds for the restriction to sums of at most 3 terms (with no repetitions, as is the case throughout the paper), and 3 colors (henceforth denoted by HT$_3^{\leq 3}$). They also established that the restriction to sums of at most 2 terms, and 2 colors (denoted HT$_2^{\leq 2}$), is unprovable in RCA$_0$ (Corollary 2.3 in [12]) and implies SRT$_2^2$ (the Stable Ramsey's Theorem for pairs and 2 colors) over RCA$_0 + B\Sigma_2^0$ (Corollary 2.4 in [12]). This prompted the first author to look into direct combinatorial reductions yielding, e.g., a direct implication from HT$_5^{\leq 2}$ to the Increasing Polarized Ramsey's Theorem for pairs of Dzhafarov and Hirst [11], which is strictly stronger than SRT$_2^2$ (see Sect. 4 for details).

It should be stressed that no upper bound other than the $\emptyset^{(\omega+1)}$ (ACA$_0^+$) upper bound on the full Finite Sums Theorem is known to hold for the restrictions of the theorem to sums of length (i.e., number of terms) ≤ 2 or ≤ 3. It is indeed a long-standing open question in Combinatorics whether the latter restrictions admit a proof that does not establish the full Finite Sums Theorem (see, e.g., [16], Question 12). On the other hand, Hirst investigated in [18] an apparently slight variant of the Finite Sums Theorem and proved it *equivalent* to $B\Sigma_2$. This prompted the first author to investigate versions of HT for which an upper bound better than $\emptyset^{(\omega+1)}$ (ACA$_0^+$) could be established, while retaining as strong a lower bound as possible. In [4] (resp. [3]) such restrictions were isolated and proved to attain the known lower bounds for HT (resp. HT$_5^{\leq 2}$), while being provable from ACA$_0$ (resp. RT$_2^2$).

We present new results along these lines of research. In Sect. 3 we prove an ACA$_0$ lower bound for HT$_4^{\leq 2}$, and an equivalence with ACA$_0$ for some principles from [4]. In Sect. 4 we establish combinatorial implications from other restrictions of Hindman's Theorem to the Increasing Polarized Ramsey's Theorem for Pairs. These reductions imply unprovability-in-WKL$_0$ results and also yield strong computable reducibility of IPT$_2^2$ to some Hindman-type theorem. We highlight the role of a sparsity-like condition on the solution set which we call the apartness condition, which is crucial in earlier work ([3,4,12,15]).

2   Restricted Hindman and the Apartness Condition

Let us fix some notation. For technical convenience and to avoid trivial cases we will deal with colorings of the positive integers. We use $\mathbf{N}$ to denote the positive integers. If $a \in \mathbf{N}$ and B is a set we denote by $FS^{\leq a}(B)$ (resp. $FS^{=a}(B)$) the set of non-empty sums of at most (resp. exactly) a-many distinct elements from B. More generally, if A and B are sets we denote by $FS^A(B)$ the set of all sums of j-many distinct terms from B, for all $j \in A$. By $FS(B)$ we denote $FS^{\mathbf{N}}(B)$.

We use the notation $X = \{x_1, x_2, \ldots\}_<$ to indicate that $x_1 < x_2 < \ldots$. Let us recall the statement of Hindman's Finite Sums Theorem [15].

Definition 1 (Hindman's Finite Sums Theorem). HT *is the following assertion: For every* $k \in \mathbf{N}$, *for every coloring* $f : \mathbf{N} \to k$ *there exists an infinite set* $H \subseteq \mathbf{N}$ *such that* $FS(H)$ *is monochromatic for* f.

We define below two restrictions of Hindman's Theorem that will feature prominently in the present paper. We then discuss a sparsity-like condition that will be central to our results.

2.1 Hindman's Theorem with Bounded-Length Sums

The following principles were discussed in [1] (albeit phrased in terms of finite unions instead of sums) and first studied from the perspective of Computable and Reverse Mathematics in [12].

Definition 2 (Hindman's Theorem with bounded-length sums). *Fix* $n, k \geq 1$.

1. $\mathsf{HT}_k^{\leq n}$ *is the following principle: For every coloring* $f : \mathbf{N} \to k$ *there exists an infinite set* $H \subseteq \mathbf{N}$ *such that* $FS^{\leq n}(H)$ *is monochromatic for* f.
2. $\mathsf{HT}_k^{=n}$ *is the following principle: For every coloring* $f : \mathbf{N} \to k$ *there exists an infinite set* $H \subseteq \mathbf{N}$ *such that* $FS^{=n}(H)$ *is monochromatic for* f.

The principle $\mathsf{HT}_2^{\leq 2}$ is the topic of a long-standing open question in Combinatorics: Question 12 of [16] asks whether there exists a proof of $\mathsf{HT}_2^{\leq 2}$ that does not also prove the full Finite Sums Theorem. On the other hand, the principle $\mathsf{HT}_2^{=2}$ easily follows from Ramsey's Theorem for pairs: given an instance $f : \mathbf{N} \to 2$ of $\mathsf{HT}_2^{=2}$, define $g : [\mathbf{N}]^2 \to 2$ by setting $g(x, y) := f(x+y)$. A solution for Ramsey's Theorem for pairs for g is a solution for $\mathsf{HT}_2^{=2}$ for f.

Dzhafarov, Jockusch, Solomon and Westrick recently proved in [12] that $\mathsf{HT}_3^{\leq 3}$ implies ACA_0 over RCA_0 (Corollary 3.4 of [12]) and that $\mathsf{HT}_2^{\leq 2}$ implies SRT_2^2 (the Stable Ramsey's Theorem for pairs) over $\mathsf{RCA}_0 + B\Sigma_2^0$ (Corollary 2.4 of [12]).[1]

The first author proved that $\mathsf{HT}_5^{\leq 2}$ implies IPT_2^2 (the Increasing Polarized Ramsey's Theorem for pairs) over RCA_0 (see [5]).

2.2 The Apartness Condition

We discuss a property of the solution set – which we call the apartness condition – that is crucial in Hindman's original proof and in the proofs of the $\emptyset'$ (ACA_0) lower bounds in [2,4,12]. We use the following notation: Fix a base $t \geq 2$. For $n \in \mathbf{N}$ we denote by $\lambda_t(n)$ the least exponent of n written in base t, by $\mu_t(n)$ the

---

[1] The principle $B\Sigma_2^0$ is used in the proof of the implication Corollary 2.4 in [12], as indicated in the final version of Dzhafarov et al. paper – our reference [12].

largest exponent of n written in base t, and by $i_t(n)$ the coefficient of the least term of n written in base t. Our results are in terms of 2-apartness except in one case (Lemma 1 below) where we have to use 3-apartness for technical reasons. We will drop the subscript when clear from context.

Definition 3 (Apartness Condition). *Fix $t \geq 2$. We say that a set $X \subseteq \mathbf{N}$ satisfies the t-apartness condition (or is t-apart) if for all $x, x' \in X$, if $x < x'$ then $\mu_t(x) < \lambda_t(x')$.*

Note that the apartness condition is inherited by subsets. In Hindman's original proof 2-apartness can be ensured (Lemma 2.2 in [15]) by a simple counting argument (Lemma 2.2 in [14]), under the assumption that we have a solution to the Finite Sums Theorem, i.e. an infinite H such that $FS(H)$ is monochromatic. For a Hindman-type principle P, let "P *with t-apartness*" denote the corresponding version in which the solution set is required to satisfy the t-apartness condition.

As will be observed below, it is significantly easier to prove lower bounds on P with t-apartness than on P in all the cases we consider. Moreover, for *all* restrictions of Hindman's Theorem for which a proof is available that does not also establish the full theorem, the t-apartness condition (for $t > 1$) can be guaranteed by construction (see, e.g., [3,4]). This is the case, e.g., for the principle $\mathsf{HT}_2^{=2}$: the proof from Ramsey's Theorem for pairs sketched above yields t-apartness for any $t > 1$ simply by applying Ramsey's Theorem relative to an infinite t-apart set. In *some* cases the apartness condition can be ensured at the cost of increasing the number of colors. This is the case of $\mathsf{HT}_k^{\leq n}$ as illustrated by the next lemma. The idea of the proof is from the first part of the proof of Theorem 3.1 in [12], with some needed adjustments.

Lemma 1 (RCA_0). *For all $n \geq 2$, for all $d \geq 1$, $\mathsf{HT}_{2d}^{\leq n}$ implies $\mathsf{HT}_d^{\leq n}$ with 3-apartness.*

Proof. We work in base 3. Let $f : \mathbf{N} \to d$ be given. Define $g : \mathbf{N} \to 2d$ as follows.

$$g(n) := \begin{cases} f(n) & \text{if } i(n) = 1, \\ d + f(n) & \text{if } i(n) = 2. \end{cases}$$

Let H be an infinite set such that $FS^{\leq n}(H)$ is homogeneous for g of color k. For $h, h' \in FS^{\leq n}(H)$ we have $i(h) = i(h')$. Then we claim that for each $m \geq 0$ there is at most one $h \in H$ such that $\lambda(h) = m$. Suppose otherwise, by way of contradiction, as witnessed by $h, h' \in H$. Then $i(h) = i(h')$ and $\lambda(h) = \lambda(h')$. Therefore $i(h + h') \neq i(h)$, but $h + h' \in FS^{\leq n}(H)$. Contradiction. Therefore we can computably obtain a 3-apart infinite subset of H. $\qquad\square$

3   Restricted Hindman and Arithmetical Comprehension

We prove a new ACA_0 lower bound and a new ACA_0 equivalence result for restrictions of Hindman's Theorem. The lower bound proof is in the spirit of the proof by Blass, Hirst and Simpson that Hindman's Theorem implies ACA_0 – on which the proof of Theorem 3.1 of [12] is also based – with extra care to work with sums of length at most two. The upper bound proof is in the spirit of [4].

3.1 $\mathsf{HT}_4^{\leq 2}$ implies ACA_0

We show that $\mathsf{HT}_4^{\leq 2}$ implies ACA_0 over RCA_0. This is to be compared with Corollary 2.3 and Corollary 3.4 of [12], showing, respectively, that $\mathsf{RCA}_0 \nvdash \mathsf{HT}_2^{\leq 2}$ and that $\mathsf{RCA}_0 \vdash \mathsf{HT}_3^{\leq 3} \to \mathsf{ACA}_0$. Blass, towards the end of [1], states without giving details that inspection of the proof of the $\emptyset'$ lower bound for HT in [2] shows that these bounds are true for the restriction of the Finite Unions Theorem to unions of at most two sets.[2] Note that the Finite Unions Theorem has a built-in apartness condition. Blass indicates in Remark 12 of [1] that things might be different for restrictions of the Finite Sums Theorem, as those considered in this paper. Also note that the proof of Theorem 3.1 in [12], which stays relatively close to the argument in [2], requires sums of length 3.

Proposition 1 (RCA_0). *For any fixed* $t \geq 2$, $\mathsf{HT}_2^{\leq 2}$ *with* t-*apartness implies* ACA_0.

Proof. We write the proof for $t = 2$. Assume $\mathsf{HT}_2^{\leq 2}$ with 2-apartness and consider $f \colon \mathbf{N} \to \mathbf{N}$. We have to prove that the range of f exists.

For a number n, written as $2^{n_0} + \cdots + 2^{n_r}$ in base 2, with $n_0 < \cdots < n_r$, we call $j \in \{0, \ldots, r\}$ *important in* n if some value of $f \restriction [n_{j-1}, n_j)$ is below n_0. Here $n_{-1} = 0$. The coloring $c \colon \mathbf{N} \to 2$ is defined by

$$c(n) := \mathrm{card}\{j : j \text{ is important in } n\} \bmod 2.$$

By $\mathsf{HT}_2^{\leq 2}$ with 2-apartness, there exists an infinite set $H \subseteq \mathbf{N}$ such that H is 2-apart and $\mathrm{FS}^{\leq 2}(H)$ is monochromatic w.r.t. c. We claim that for each $n \in H$ and each $x < \lambda(n)$, $x \in \mathrm{rg}(f)$ if and only if $x \in \mathrm{rg}(f \restriction \mu(n))$. This will give us a Δ_1^0 definition of $\mathrm{rg}(f)$: given x, find the smallest $n \in H$ such that $x < \lambda(n)$ and check whether x is in $\mathrm{rg}(f \restriction \mu(n))$.

It remains to prove the claim. In order to do this, consider $n \in H$ and assume that there is some element below $n_0 = \lambda(n)$ in $\mathrm{rg}(f) \setminus \mathrm{rg}(f \restriction \mu(n))$. By the consequence of Σ_1^0-induction known as *strong* Σ_1^0-*collection* (see Exercise II.3.14 in [20], Theorem I.2.23 and Definition I.2.20 in [13]), there is a number ℓ such that for any $x < \lambda(n)$, $x \in \mathrm{rg}(f)$ if and only if $x \in \mathrm{rg}(f \restriction \ell)$. By 2-apartness, there is $m \in H$ with $\lambda(m) \geq \ell > \mu(n)$. Write $n + m$ in base 2 notation,

$$n + m = 2^{n_0} + \cdots + 2^{n_r} + 2^{n_{r+1}} + \cdots + 2^{n_s},$$

where $n_0 = \lambda(n) = \lambda(n + m)$, $n_r = \mu(n)$, and $n_{r+1} = \lambda(m)$. Clearly, $j \leq s$ is important in $n + m$ if and only if either $j \leq r$ and j is important in n or $j = r + 1$; hence, $c(n) \neq c(n + m)$. This contradicts the assumption that $\mathrm{FS}^{\leq 2}(H)$ is monochromatic, thus proving the claim. □

---

[2] The Finite Unions Theorem states that every coloring of the finite non-empty sets of $\mathbf{N}$ admits an infinite and pairwise unmeshed family H of finite non-empty sets (sometimes called a block sequence) such that every finite non-empty union of elements of H is of the same color. Two finite non-empty subsets x, y of $\mathbf{N}$ are unmeshed if either $\max x < \min y$ or $\max y < \min x$. Note that Hindman's Theorem is equivalent to the Finite Unions Theorem only if the pairwise unmeshed condition is present.

Theorem 1 (RCA$_0$). HT$_4^{\leq 2}$ *implies* ACA$_0$.

Proof. By Proposition 1 and Lemma 1.   □

3.2 Equivalents of ACA$_0$

In [4], a family of natural restrictions of Hindman's Theorem was isolated such that each of its members admits a simple combinatorial proof, yet each member of a non-trivial sub-family implies ACA$_0$. The weakest principle of the latter kind considered in [4] is the following, called the Hindman-Brauer Theorem: Whenever **N** is 2-colored there is an infinite set $H \subseteq \mathbf{N}$ and *there exist* positive integers a, b such that $FS^{\{a,b,a+b,a+2b\}}(H)$ is monochromatic. It was proved in [4] that the Hindman-Brauer Theorem with 2-apartness is equivalent to ACA$_0$. We show that the same holds for the following apparently weaker principle.

Definition 4. HT$_2^{\exists\{a,b\}}$ *is the following principle: For every coloring* $f : \mathbf{N} \to 2$ *there is an infinite set* $H \subseteq \mathbf{N}$ *and positive integers* $a < b$ *such that* $FS^{\{a,b\}}(H)$ *is monochromatic.*

Theorem 2. HT$_2^{\exists\{a,b\}}$ *with 2-apartness is equivalent to* ACA$_0$ *over* RCA$_0$.

Proof. We first prove the upper bound. Given $c : \mathbf{N} \to 2$ let $g : [\mathbf{N}]^3 \to 8$ be defined as follows:

$$g(x_1, x_2, x_3) := \langle c(x_1), c(x_1 + x_2), c(x_1 + x_2 + x_3) \rangle.$$

Fix an infinite and 2-apart set $H_0 \subseteq \mathbf{N}$. By RT$_8^3$ relativized to H_0 we get an infinite (and 2-apart) set $H \subseteq H_0$ monochromatic for g. Let the color be $\sigma = (c_1, c_2, c_3)$, a binary sequence of length 3. Then, for each $i \in \{1, 2, 3\}$, g restricted to $FS^{=i}(H)$ is monochromatic of color c_i. Obviously for some positive integers a, b such that $a < b \leq 3$ it must be that $c_a = c_b$. Then $FS^{\{a,b\}}(H)$ is monochromatic of color c_a.

The lower bound is proved by a minor adaptation of the proof of Proposition 1. As the n in that proof take an a-term sum. Then take a $(b - a)$-term sum as the m.   □

The same proof yields that the following Hindman-Schur Theorem with 2-apartness from [4] implies ACA$_0$: Whenever **N** is 2-colored there is an infinite 2-apart set H and *there exist* positive integers a, b such that $FS^{\{a,b,a+b\}}(H)$ is monochromatic. It was shown in [4] to be provable in ACA$_0$.

4 Restricted Hindman and Polarized Ramsey

In this section we establish new lower bounds for restricted versions of Hindman's Theorem, most of which do not imply ACA$_0$ and are therefore provably weaker than HT. Lower bounds are established by reduction to the Increasing Polarized Ramsey's Theorem for pairs [11]. In particular we obtain unprovability in WKL$_0$.

All proofs in the present section yield strongly computable reductions in the sense of [10], not just implications. P is *strongly computably reducible* to Q, written $P \leq_{sc} Q$, if every instance X of P computes an instance X^* of Q, such that if Y^* is any solution to X^* then there is a solution Y to X computable from Y^*.

Definition 5 (Increasing Polarized Ramsey's Theorem). *Fix $n, k \geq 1$.* IPT_k^n *is the following principle: For every $f : [\mathbf{N}]^n \to k$ there exists a sequence $(H_1, \ldots, H_n)$ of infinite sets such that all increasing tuples $(x_1, \ldots, x_n)$ in $H_1 \times \cdots \times H_n$ have the same color under f. The sequence $(H_1, \ldots, H_n)$ is called increasing polarized homogeneous (or increasing p-homogeneous) for f.*

Note that IPT_2^2 is strictly stronger than SRT_2^2. On the one hand, $\mathsf{RCA}_0 \vdash \mathsf{IPT}_2^2 \to \mathsf{D}_2^2$ by Proposition 3.5 of [11], and $\mathsf{RCA}_0 \vdash \mathsf{D}_2^2 \to \mathsf{SRT}_2^2$ by Theorem 1.4 of [8].[3] However, $\mathsf{RCA}_0 + \mathsf{SRT}_2^2 \nvdash \mathsf{IPT}_2^2$: Theorem 2.2 in [9] showed that there is a non-standard model of $\mathsf{SRT}_2^2 + B\Sigma_2^0$ having only low sets in the sense of the model. Lemma 2.5 in [11] can be formalized in RCA_0 and shows that no model of IPT_2^2 can contain only Δ_2^0 sets.[4]

4.1 $\mathsf{HT}_2^{=2}$ with 2-apartness implies IPT_2^2

We show that $\mathsf{HT}_2^{=2}$ with 2-apartness implies IPT_2^2 by a combinatorial reduction. This should be contrasted with the fact that no lower bounds on $\mathsf{HT}_2^{=2}$ without apartness are known.

Theorem 3 (RCA_0). *$\mathsf{HT}_2^{=2}$ with 2-apartness implies IPT_2^2.*

Proof Let $f : [\mathbf{N}]^2 \to 2$ be given. Define $g : \mathbf{N} \to 2$ as follows.

$$g(n) := \begin{cases} 0 & \text{if } n = 2^m, \\ f(\lambda(n), \mu(n)) & \text{if } n \neq 2^m. \end{cases}$$

Note that g is well-defined since $\lambda(n) < \mu(n)$ if n is not a power of 2. Let $H = \{h_1, h_2, \ldots\}_<$ witness $\mathsf{HT}_2^{=2}$ with 2-apartness for g. Let the color be $k < 2$. Let

$$H_1 := \{\lambda(h_{2i-1}) : i \in \mathbf{N}\}, \; H_2 := \{\mu(h_{2i}) : i \in \mathbf{N}\}.$$

We claim that (H_1, H_2) is increasing p-homogeneous for f.

First observe that we have

$$\lambda(h_1) < \lambda(h_3) < \lambda(h_5) < \ldots,$$

and

$$\mu(h_2) < \mu(h_4) < \mu(h_6) < \ldots.$$

---

[3] Note that the latter result is not present in the diagram in [11]. D_2^2, defined in [7], is the following assertion: For every $0, 1$-valued function $f(x, s)$ for which a $\lim_{s \to \infty} f(x, s)$ exists for each x, there is an infinite set H and a $k < 2$ such that for all $h \in H$ we have $\lim_{s \to \infty} f(h, s) = k$.

[4] We thank Ludovic Patey for pointing out to us the results implying strictness.

This is so because $\lambda(h_1) \leq \mu(h_1) < \lambda(h_2) \leq \mu(h_2) < \ldots$ by the 2-apartness condition. Then we claim that $f(x_1, x_2) = k$ for every increasing pair $(x_1, x_2) \in H_1 \times H_2$. Note that $(x_1, x_2) = (\lambda(h_i), \mu(h_j))$ for some $i < j$ (the case $i = j$ is impossible by construction of H_1 and H_2). Then we have

$$k = g(h_i + h_j) = f(\lambda(h_i + h_j), \mu(h_i + h_j)) = f(\lambda(h_i), \mu(h_j)) = f(x_1, x_2),$$

since $FS^{=2}(H)$ is monochromatic for g with color k. This shows that (H_1, H_2) is increasing p-homogeneous of color k for f. □

The proof of Theorem 3 yields that $\mathsf{IPT}_2^2 \leq_{sc} \mathsf{HT}_2^{=2}$ with 2-apartness, and, with minor adjustments, that $\mathsf{IPT}_2^2 \leq_{sc} \mathsf{HT}_4^{\leq 2}$ (a self-contained proof appeared in [5]).

4.2 IPT_2^2 and the Increasing Polarized Hindman's Theorem

We define an (increasing) polarized version of Hindman's Theorem. We prove that its version for pairs and 2 colors with an appropriately defined notion of 2-apartness is equivalent to IPT_2^2.

Definition 6 ((Increasing) Polarized Hindman's Theorem). *Fix $n \geq 1$. PHT_2^n (resp. IPHT_2^n) is the following principle: For every $f : \mathbf{N} \to 2$ there exists a sequence $(H_1, \ldots, H_n)$ of infinite sets such that for some color $k < 2$, for all (resp. increasing) $(x_1, \ldots, x_n) \in H_1 \times \cdots \times H_n$, $f(x_1 + \cdots + x_n) = k$.*

We impose a t-apartness condition on a solution $(H_1, \ldots, H_n)$ of IPHT_2^n by requiring that the union $H_1 \cup \cdots \cup H_n$ is t-apart. We denote by "IPHT_2^n with t-apartness" the principle IPHT_2^n with this t-apartness condition on the solution set.

Theorem 4. IPT_2^2 *and* IPHT_2^2 *with 2-apartness are equivalent over* RCA_0.

Proof. We first prove that IPT_2^2 implies IPHT_2^2 with 2-apartness. Given $c : \mathbf{N} \to 2$ define $f : [\mathbf{N}]^2 \to 2$ in the obvious way setting $f(x, y) := c(x + y)$. Fix two infinite disjoint sets S_1, S_2 such that $S_1 \cup S_2$ is 2-apart. By Lemma 4.3 of [11], IPT_2^2 implies over RCA_0 its own relativization: there exists an increasing p-homogeneous sequence (H_1, H_2) for f such that $H_i \subseteq S_i$. Therefore $H_1 \cup H_2$ is 2-apart by construction. Let the color be $k < 2$. Obviously we have that for any increasing pair $(x_1, x_2) \in H_1 \times H_2$, $c(x_1 + x_2) = f(x_1, x_2) = k$. Therefore (H_1, H_2) is an increasing p-homogeneous pair for c.

Next we prove that IPHT_2^2 with 2-apartness implies IPT_2^2. Let $f : [\mathbf{N}]^2 \to 2$ be given. Define as usual $c : \mathbf{N} \to 2$ by setting $c(n) := f(\lambda(n), \mu(n))$ if n is not a power of 2 and $c(n) := 0$ otherwise. Let (H_1, H_2) be a 2-apart solution to IPHT_2^2 for c, of color $k < 2$. By, possibly, recursively thinning out H_1 and H_2 we can assume without loss of generality that $H_1 \cap H_2 = \varnothing$. Let $H_1 = \{h_1, h_2, \ldots\}_<$ and $H_2 = \{h'_1, h'_2, \ldots\}_<$. Then set $H_1^+ := \{\lambda(h) : h \in H_1\}$ and $H_2^+ := \{\mu(h) : h \in H_2\}$. We claim that (H_1^+, H_2^+) is a solution to IPT_2^2 for f.

Let $(x_1, x_2) \in H_1^+ \times H_2^+$ be an increasing pair. Then for some $h \in H_1$ and $h' \in H_2$, $\lambda(h) = x_1$ and $\mu(h') = x_2$. Also, since $H_1 \cup H_2$ is apart and $H_1 \cap H_2 = \emptyset$, it must be the case that $h < h'$. Therefore (h, h') is an increasing pair in $H_1 \times H_2$ and the following holds:

$$k = c(h + h') = f(\lambda(h + h'), \mu(h + h')) = f(\lambda(h), \mu(h')) = f(x_1, x_2). \qquad \square$$

4.3   Hindman's Theorem for Exactly Large Sums

We present here some preliminary results on a restriction of Hindman's Theorem to exactly large sums. A finite set $S \subseteq \mathbf{N}$ is *exactly large*, or *!ω-large*, if $|S| = \min(S) + 1$. We denote by $[X]^{!\omega}$ the set of exactly large subsets of X and by $FS^{!\omega}(X)$ the set of positive integers that can be obtained as sums of terms of an exactly large subset of X. We call sums of this type *exactly large sums* (from X). Ramsey's Theorem for exactly large sums ($\mathsf{RT}_2^{!\omega}$) asserts that every 2-coloring f of the exactly large subsets of an infinite set $X \subseteq \mathbf{N}$ admits an infinite set $H \subseteq X$ such that f is constant on $[H]^{!\omega}$. It was studied in [6] and there proved equivalent to ACA_0^+. We introduce an analogue for Hindman's Theorem.

Definition 7 (Hindman's Theorem for Large Sums). $\mathsf{HT}_2^{!\omega}$ *denotes the following principle: For every coloring* $c : \mathbf{N} \to 2$ *there exists an infinite set* $H \subseteq \mathbf{N}$ *such that* $FS^{!\omega}(H)$ *is monochromatic under* c.

$\mathsf{HT}_2^{!\omega}$ (with t-apartness, for any $t > 1$) is a consequence of HT, but also admits an easy proof from $\mathsf{RT}_2^{!\omega}$. Given $c : \mathbf{N} \to 2$ just set $f(S) := c(\sum S)$, for S an exactly large set (to get t-apartness, restrict f to an infinite t-apart set). By results from [6] this reduction yields an upper bound of $\emptyset^{(\omega)}$ on $\mathsf{HT}_2^{!\omega}$.

Proposition 2 (RCA_0). $\mathsf{HT}_2^{!\omega}$ *with 2-apartness implies* IPHT_2^2 *with 2-apartness.*

Proof. Let $f : \mathbf{N} \to 2$ be given, and let $H = \{h_0, h_1, h_2, \dots\}_<$ be an infinite 2-apart set such that $FS^{!\omega}(H)$ is monochromatic for f of color $k < 2$. Let $H_s = \{s_1, s_2, \dots\}_<$ be a 2-apart set whose elements are exactly large sums of consecutive elements from H. Let $H_t = \{t_1, t_2, \dots\}_<$ be the set of elements from H_s minus their largest term (when written as $!\omega$-sums). Note that distinct elements of H_s share no term, because H_s is 2-apart. Let $H_1 := H_t$ and let $H_2 := \{s_i - t_i : i \in \mathbf{N}\}$. Then (H_1, H_2) is a 2-apart solution for IPHT_2^2: $\qquad \square$

From Proposition 3 we get that $\mathsf{HT}_2^{!\omega}$ with 2-apartness implies IPT_2^2. In particular it is unprovable in WKL_0. Other results on $\mathsf{HT}_2^{!\omega}$ have been proved by the third author in his BSc. Thesis. E.g., over RCA_0, $\mathsf{HT}_2^{!\omega}$ with 2-apartness implies $\forall n \mathsf{HT}_2^{=2^n}$, and $\mathsf{HT}_2^{!\omega}$ implies $\forall n \mathsf{PHT}_2^n$ (see Definition 6).

5   Conclusion

We contributed to the study of restricted versions of Hindman's Theorem by proving implications from (and equivalence of) some such restrictions to ACA_0 and to the Increasing Polarized Ramsey's Theorem for Pairs. Our results improve and integrate the recent results by Dzhafarov, Jockusch, Solomon and Westrick [12]. In many cases they confirm that the known lower bounds on Hindman's Theorem hold for restricted versions of Hindman's Theorem for which — contrary to the restrictions studied in [12] — the upper bound lies strictly below $\emptyset^{(\omega+1)}$ (most being consequences of ACA_0 or even of RT_2^2). This also complements the results of [3,4] and might be an indication that the known lower bounds for Hindman's Theorem are sub-optimal. We highlighted the role of the apartness condition on the solution set (Table 1).

Table 1. Summary of results

Principles	Lower Bounds	Upper Bounds
$\mathsf{HT}_2^{\leq 2}$	$\mathsf{RCA}_0 \nvdash$ ([12])	$\emptyset^{(\omega+1)}$, ACA_0^+ ([2])
$\mathsf{HT}_2^{\leq 2} + B\Sigma_2^0$	SRT_2^2 ([12])	$\emptyset^{(\omega+1)}$, ACA_0^+ ([2])
$\mathsf{HT}_2^{\leq 2}$ with 2-apartness	ACA_0 (Proposition 1)	$\emptyset^{(\omega+1)}$, ACA_0^+ ([2])
$\mathsf{HT}_4^{\leq 2}$	ACA_0 (Theorem 1)	$\emptyset^{(\omega+1)}$, ACA_0^+ ([2])
$\mathsf{HT}_2^{\exists\{a,b\}}$	?	$\emptyset'$, ACA_0 ([4])
$\mathsf{HT}_2^{\exists\{a,b\}}$ with 2-apartness	ACA_0 (Theorem 2)	$\emptyset'$, ACA_0 (Theorem 2)
$\mathsf{HT}_2^{=2}$	?	RT_2^2 (folklore)
$\mathsf{HT}_2^{=2}$ with 2-apartness	IPT_2^2 (Theorem 3)	RT_2^2 (folklore)
IPHT_2^2 with 2-apartness	IPT_2^2 (Theorem 4)	IPT_2^2 (Theorem 4)
$\mathsf{HT}_2^{!\omega}$	?	$\emptyset^{(\omega)}$, ACA_0^+ ([6])
$\mathsf{HT}_2^{!\omega}$ with 2-apartness	IPT_2^2 (Proposition 2)	$\emptyset^{(\omega)}$, ACA_0^+ ([6])

Note: We have improved some of the above results and obtained some new results. E.g., both $\mathsf{HT}_2^{=3}$ with 2-apartness and $\mathsf{HT}_2^{!\omega}$ imply ACA_0 over RCA_0. These and further results will be presented in an extended version of this paper.

References

1. Blass, A.: Some questions arising from Hindman's theorem. Sci. Math. Japonicae **62**, 331–334 (2005)
2. Blass, A.R., Hirst, J.L., Simpson, S.G.: Logical analysis of some theorems of combinatorics and topological dynamics. In: Logic and Combinatorics (Arcata, California, 1985), Contemporary Mathematics, vol. 65, pp. 125–156. American Mathematical Society, Providence, RI (1987)
3. Carlucci, L.: A weak variant of Hindman's Theorem stronger than Hilbert's Theorem. Preprint (2016). https://arxiv.org/abs/1610.05445

4. Carlucci, L.: Weak yet strong restrictions of Hindman's finite sums theorem. In: Proceedings of the American Mathematical Society. Preprint (2016). https://arxiv.org/abs/1610.07500. Accepted with minor revision for publication
5. Carlucci, L.: Bounded Hindman's theorem and increasing polarized Ramsey's theorem. In: Nies, A. (ed.) Logic Blog, Part 4, Section 9 (2016). https://arxiv.org/abs/1703.01573
6. Carlucci, L., Zdanowski, K.: The strength of Ramsey's theorem for coloring relatively large sets. J. Symbolic Logic **79**(1), 89–102 (2014)
7. Cholak, P.A., Jockusch, C.G., Slaman, T.A.: On the strength of Ramsey's theorem for pairs. J. Symbolic Logic **66**(1), 1–55 (2001)
8. Chong, C.T., Lempp, S., Yang, Y.: On the role of the collection principle for Σ_2^0 formulas in second-order reverse mathematics. Proc. Am. Math. Soc. **138**, 1093–1100 (2010)
9. Chong, C.T., Slaman, T.A., Yang, Y.: The metamathematics of the stable Ramsey's theorem for pairs. J. Am. Math. Soc. **27**, 863–892 (2014)
10. Dzhafarov, D.D.: Cohesive avoidance and strong reductions. Proc. Am. Math. Soc. **143**, 869–876 (2015)
11. Dzhafarov, D.D., Hirst, J.L.: The polarized Ramsey's theorem. Arch. Math. Logic **48**(2), 141–157 (2011)
12. Dzhafarov, D.D., Jockusch, C.G., Solomon, R., Westrick, L.B.: Effectiveness of Hindman's theorem for bounded sums. In: Day, A., Fellows, M., Greenberg, N., Khoussainov, B., Melnikov, A., Rosamond, F. (eds.) Computability and Complexity. LNCS, vol. 10010, pp. 134–142. Springer, Cham (2017). doi:10.1007/978-3-319-50062-1_11
13. Hàjek, P., Pudlàk, P.: Metamathematics of First-Order Arithmetic. Perspectives in Mathematical Logic. Springer, Heidelberg (1993)
14. Hindman, N.: The existence of certain ultrafilters on N and a conjecture of Graham and Rothschild. Proc. Am. Math. Soc. **36**(2), 341–346 (1972)
15. Hindman, N.: Finite sums from sequences within cells of a partition of N. J. Comb. Theory Ser. A **17**, 1–11 (1974)
16. Hindman, N., Leader, I., Strauss, D.: Open problems in partition regularity. Comb. Probab. Comput. **12**, 571–583 (2003)
17. Hirschfeldt, D.R.: Slicing the Truth (On the Computable and Reverse Mathematics of Combinatorial Principles). Lecture Notes Series, vol. 28, Institute for Mathematical Sciences, National University of Singapore (2014)
18. Hirst, J.: Hilbert vs. Hindman. Arch. Math. Logic **51**(1–2), 123–125 (2012)
19. Montalbán, A.: Open questions in Reverse Mathematics. Bull. Symbolic Logic **17**(3), 431–454 (2011)
20. Simpson, S.: Subsystems of Second Order Arithmetic, 2nd edn. Cambridge University Press, New York (2009). Association for Symbolic Logic

Infinite Time Busy Beavers

Oscar Defrain, Bruno Durand[(⊠)], and Grégory Lafitte[(⊠)]

LIRMM, CNRS, Université de Montpellier, 161 Rue Ada,
34095 Montpellier Cedex 5, France
{oscar.defrain,bruno.durand,gregory.lafitte}@lirmm.fr

Abstract. In 1962, Hungarian mathematician Tibor Radó introduced in [8] the busy beaver competition for Turing machines: in a class of machines, find one which halts after the greatest number of steps when started on the empty input. In this paper, we generalise the busy beaver competition to the infinite time Turing machines (ITTMs) introduced in [6] by Hamkins and Lewis in 2000. We introduce two busy beaver functions on ITTMs and show both theoretical and experimental results on these functions. We give in particular a comprehensive study, with champions for the busy beaver competition, of the classes of ITTMs with one or two states (in addition to the *halt* and *limit* states). The computation power of ITTMs is humongous and thus makes the experimental study of this generalisation of Radó's competition and functions a daunting challenge. We end this paper with a characterisation of the power of those machines when the use of the tape is restricted in various ways.

1 Introduction

Infinite time Turing machines (ITTMs) were defined by Hamkins and Kidder in 1989 and introduced in 2000 in a seminal paper [6] by Hamkins and Lewis. They are a generalisation of classical Turing machines to infinite ordinal time — the only difference is the behaviour at limit ordinal stages: the head moves back to the origin, the machine enters a special *limit* state, and each cell takes as value the *supremum limit* (abbreviated to *limsup*) of its previous values.

The busy beaver competition for Turing machines was presented by Hungarian mathematician Tibor Radó in 1962 [8]: in a class of machines, find one which halts after the greatest number of steps when started on the empty input. Radó also considered other complexity criteria, *e.g.*, the number of non-zero cells.

In this paper, we generalise the busy beaver competition to ITTMs and ordinal time. We introduce several busy beaver functions on ITTMs and show both theoretical and experimental results on these functions. The research in this paper is thus an original mixture of theoretical and experimental approaches,

The research for this paper has been done thanks to the support of the *Agence nationale de la recherche* through the RaCAF ANR-15-CE40-0016-01 grant.

The authors would also like to express their thanks to the anonymous referees, who made numerous suggestions and interesting remarks.

© Springer International Publishing AG 2017
J. Kari et al. (Eds.): CiE 2017, LNCS 10307, pp. 221–233, 2017.
DOI: 10.1007/978-3-319-58741-7_22

feeding each other. There are several subtle theoretical results helping the brute-force experimentation and the experimental results point in the direction of theoretical results which then need to be proved.

It is important to mention that the brute-force experimentation is very different from the kind of experimentation one comes across in the classical busy beaver case. In the classical case, the busy beaver functions that we try to "compute" are non-recursive. In the infinite-time case, our functions are not only non-recursive, they are not even arithmetic. We are able to go beyond this difficulty by inventing different heuristics and by having the program ask the user his analysis of the values of the cells for certain limit ordinal stages. In the classical case, if a machine halts, one just needs to run it for a long enough time to end up witnessing that it halts. In the infinite-time case, it is not obvious to simulate the machine for an unbounded ordinal number of steps, one needs to simulate the different kinds of ordinal limits to go higher up in the ordinal ladder. On the other hand, in the classical case, the number of steps reached are very large numbers without a clear meaning, but in the infinite-time case, the ordinals reached speak a lot more for themselves.

As in the classical case, one of the busy beaver complexity criteria for our infinite-time case is the number of ordinal steps. ITTMs have the particular property that every diverging machine eventually reaches a periodic behaviour, a *final loop* from which it will never escape. One of the other busy beaver complexity criteria is thus the stage at which a diverging machine enters this final loop. Another one is the *period* of that final loop. Each of these complexity criteria give rise to a distinct busy beaver function.

In Sect. 3, we give a comprehensive study of these busy beaver functions for classes of ITTMs with one or two states (in addition to the *halt* and *limit* states). We analyse completely and in detail the behaviour of the machines of those two classes when started with an empty input and classify those machines in different categories, similar to the study made by Allen H. Brady for the first classes of the classical case [3] in the 1970's. The ITTM simulation itself is described at the beginning of Sect. 3. The experimental simulation can be carried out and the different busy beaver functions are identified thanks to the theoretical results from Sect. 2. By this experimental study, we obtain the champions for those two classes of these various busy beaver competitions.

The busy beaver functions are also studied from a degree-theoretic point of view in Sect. 2 and we end this paper with characterisations of the power of those machines when the use of the tape is limited in various ways, characterisations hinted by the experimental study.

2   ITTMs and Busy Beaver Functions

2.1   Infinite Time Turing Machines

An *infinite time Turing machine* (ITTM for short) is a three-tape (input, scratch, output) Turing machine on alphabet $\{0, 1\}$ with one *halt* state enhanced with a special *limit* state.

The machine halts when it has reached its *halt* state. The *limit* state makes it possible for a Turing machine to carry on for an infinite ordinal time. At a limit ordinal stage, an ITTM enters the special *limit* state, moves its head back to the first cell, and updates its tapes such that each cell is equal to the *limsup* of its previous values. After reaching this special *limit* state, it carries on its computation as a classic Turing machine before entering again the special *limit* state, and so on. We can thus let the machine run for an arbitrary large ordinal time.

Fig. 1. ITTM loops

The *limsup* rule for the cells implies that if the value of a cell at a limit stage is 0, then there must be some earlier stage at which the cell had the value 0 and never subsequently changes. If the cell has the value 1 at a limit stage, then there are two possibilities, either at some earlier stage the cell obtained the value 1 and never subsequently changed, or the value of the cell alternated unboundedly often between the values 0 and 1 before that limit stage. To summarize, at a limit stage, a cell has either *stabilized* (to 0 or 1) or has value 1 because it *flashed*, *i.e.*, the cell alternated unboundedly often between values 0 and 1.

The power of these machines can be grasped through various ordinals. An ordinal α is *clockable* if there is an ITTM halting at this stage α starting with an empty tape.[1] An ordinal α is *writable* if there is an ITTM halting after having written a real coding the characteristic sequence of a well-order on ω of ordinal type α. The supremum of the clockable ordinals, denoted by γ_∞, and the supremum of the writable ordinals, denoted by λ_∞, can be shown to be equal and quite a large countable ordinal, well beyond ω_1^{CK} as it is the $\lambda_\infty{}^{th}$ admissible ordinal, the $\lambda_\infty{}^{th}$ admissible limit of admissible ordinals (recursively inaccessible), the $\lambda_\infty{}^{th}$ admissible limit of recursively inaccessibles, and so on.[2]

Two other ordinals which play an important role for these machines are ζ_∞ and Σ_∞ defined respectively as the supremum of eventually[3] writable and accidentally(see Footnote 3) writable ordinals. Philip Welch gave in [10] the characterisation for these ordinals: the triple λ_∞, ζ_∞, Σ_∞ is the lexicographically least triple α, β, γ such that $L_\alpha \prec_1 L_\beta \prec_2 L_\gamma$.

For classical Turing machines, when a machine enters a loop, it can never escape it. It is important to note that the behaviour of infinite-time Turing machines is completely different in that respect: a machine can enter a loop and

---

[1] The last transition is not counted to allow for clockable limit ordinals [6].

[2] Concerning admissible sets and ordinals, the interested reader is referred to [1].

[3] A real x is *eventually writable* if there is a non-halting infinite time computation, on input 0, which eventually writes x on the output tape and never changes again. A real x is *accidentally writable* if it appears on one of the tapes during a computation, possibly changing it again later on. An ordinal α is *eventually* (resp. *accidentally*) writable if the real coding a well-order on ω of order-type α is eventually (resp. accidentally) writable.

finally escape it thanks to the *supremum limit* rule of the cell values. A loop from which a machine cannot escape will be called a *final loop*. An example of a machine escaping a first loop and then being caught in an other loop, which is final, is given with parts of its space-time diagram [4] in Fig. 1. The following theorem states that every diverging machine eventually enters a final loop. Moreover, every ITTM either halts or repeats itself in countably many steps ([6, Theorem 1.1]). It also gives some explanations on how an ITTM can diverge. We sketch the proof of the first part because it gives fundamental insights in the workings of ITTMs and is essential to understand the rest of this paper.

Theorem 1 (Final Loop [6,10]). *Any halting ITTM halts at a countable stage. Moreover, any diverging ITTM eventually enters, at a countable stage, a final loop from which it will never escape. If the diverging machine runs on an empty tape, its configurations at time ω_1, Σ_∞ and ζ_∞ are equal. If the diverging machine runs on input x, its configurations at time ω_1, Σ_∞^x and ζ_∞^x are equal, where the two latter ordinals represent the relativised versions.*

The proof of the first part of this theorem can be found in [6]. It considers the configuration[5] C reached at stage ω_1. This configuration already appears at a countable stage γ and will be repeated unboundedly often. It is important to note that none of the cells which have stabilized to the value 0 at stage γ, and thus also at stage ω_1, will ever again turn to 1. So, the configuration at a limit stage before which the configuration C was repeated unboundedly often will again be the configuration C. The computation will thus repeat endlessly this *final loop* which started at stage γ and never halt. We consider the supremum of the enter-stages of these loops and using arguments from [10], we can show that it is in fact equal to ζ_∞. The supremum of the loop periods (and end-stages) of final loops can also be shown to be equal to Σ_∞.

This bound is reached by a variant of a universal machine: a machine $\mathcal{U}$ that simulates the computations of all ITTMs on the empty input, on ω many disjoint portions of the scratch tape. Moreover, $\mathcal{U}$ can be chosen to almost work in 'ω-real-time', *i.e.*, such that $\mathcal{U}$ simulates ω many steps of each machine in ω many steps of simulation time.

In the sequel, we say that an ordinal α is *looping* if there exists an ITTM that enters its final loop at stage exactly α. In other terms, the ω_1-configuration of this machine appears for the first time at stage α. A final loop example is given in Fig. 1, with a machine entering its final loop at stage ω^2, thus making this ordinal looping.

Now we can formulate some remarks on final loops that will be very useful for our experimental study. Let us consider the configuration of an ITTM when

---

[4] $^\omega 00|10^\omega$ denotes a tape with its left part filled with 0's and its right part filled with a 1 (the origin) and then 0's. In this notation, the origin is the first cell on the right of the symbol |, here a 1.

[5] The *configuration* of a computation at a given stage is a complete description of the *state* of the machine at this stage. It comprises the state, the position of the head and the complete content of the tapes.

it is entering a final loop. Its 0's will remain at 0 forever while its 1's may change. Indeed, if a cell at 0 changes to 1 at least once during a final loop, consider this cell after ω loops. Its value is 1, which contradicts that the loop was chosen final. As a consequence, if two configurations at distinct limit stages α and β of an ITTM M are the same and if none of the 0's ever flash between those two stages, then M is diverging and the loop between α and β is final. This property will be used to decide divergence in our simulator.

2.2 Infinite Time Busy Beaver Functions

We define here an infinite time version of the well known *busy beaver* competition introduced by Tibor Radó in 1962 [8] for classical Turing machines.

The model considered in [6] has three tapes. For a more natural definition of busy beavers, we limit our model to one tape (with the *supremum limit* behaviour for cell values at limit stages). It is easy to see that restricting our model to one tape changes the clockable ordinal gaps, but that the supremums of clockable, eventually writable and accidentally writable ordinals remain the same.

Informally, the *infinite time busy beaver* competition consists, for a given n, in selecting the machine among the ITTMs with n states in addition to the *halt* and *limit* states (called the BB-*category* n of the competition) which *produces* the greatest ordinal when run from an empty input. Several busy beaver functions can be devised according to the different ways these ordinals can be produced.

We define three natural generalisations of the classical case, derived from the notions of clockable ordinals and final loops.[6]

We first define $\Gamma : \omega \to \gamma_\infty$, $\Lambda : \omega \to \zeta_\infty + 1$, and $\Pi : \omega \to \Sigma_\infty + 1$ as the functions that consider as input an ITTM and produce the ordinal stage at which it reaches respectively the *halt* state (if it halts), its final loop and the period of its final loop (if it diverges)[7].

We now define the BB-functions Γ_{BB}, Λ_{BB} and Π_{BB} as the functions which maximize the functions Γ, Λ, and Π on each BB-category. The first BB-function $\Gamma_{\text{BB}} : \omega \to \gamma_\infty$ maps a given BB-category n to the largest clockable ordinal *produced* by an ITTM of that category, *etc.*

In order to compare the *power* of these functions and of related ordinals, we use the ITTM-reduction[8], denoted by $\preccurlyeq_\infty$ and defined in [6], and assume that the ordinals considered (also as output of these functions) are coded in reals with an appropriate coding. For each ordinal α, there are continuum many possible choices for r_α, the real coding α. We would like to consider codes that do not contain extra information, in particular with regards to $\preccurlyeq_\infty$. r_α is thus chosen to be the L-least real coding a well-order on ω of order-type α. For the ordinals considered in this paper, this means that the code r_α considered for α will be such that $r_\alpha \in L_{\alpha+1}$.

---

[6] Loops can be considered as transient steps followed by an eventual final loop, and it is natural to measure the stage at which it appears and the length of that final loop.

[7] By Theorem 1, we know that the *sup* of looping ordinals, ζ_∞, and periods, Σ_∞, are reached by the universal machine. Λ and Π can thus reach these maximum values.

[8] A is ITTM computable from B, written $A \preccurlyeq_\infty B$, if the characteristic function of A is infinite time computable with oracle B. Reals are seen here as subsets of ω.

Remark that the functions Λ_{BB} and Π_{BB} (but not Γ_{BB}) will become eventually constant when the BB-category reaches a level containing a universal machine.

Theorem 2. $r_{\gamma_\infty} \equiv_\infty 0^\triangledown \equiv_\infty \Gamma_{\text{BB}} \equiv_\infty \Gamma$ *(using the notations[9] of [6])*
$$\prec_\infty \Lambda_{\text{BB}} \equiv_\infty r_{\zeta_\infty} \equiv_\infty r_{\Sigma_\infty} \equiv_\infty \Lambda \equiv_\infty \Pi_{\text{BB}} \equiv_\infty \Pi \prec_\infty 0^\blacktriangledown$$

Proof. Let's first prove that $r_{\gamma_\infty} \preccurlyeq_\infty 0^\triangledown$. The idea is to build an order such that $i < j$ if M_i halts before M_j. As ω machines halt at a given time, we consider only the first program with a given halting time that is discovered (while running the chosen universal ω-real-time machine $\mathcal{U}$). Our program runs as follows: it simulates $\mathcal{U}$ and little by little constructs an order such that $i < j$ if i and j are considered and if M_i halts before M_j. The representation of the order must be adjusted carefully : when we add an element in this order, then ω values are changed, but we can propagate those changes in parallel by runs of ω. In parallel we check whether all elements of $0^\triangledown$ have been observed halting in the $\mathcal{U}$-simulation. If yes, we halt (at time γ_∞). We thus get x that codes γ_∞, and by minimality we get $r_{\gamma_\infty} \preccurlyeq_\infty x \preccurlyeq_\infty 0^\triangledown$.

$0^\triangledown \preccurlyeq_\infty r_{\gamma_\infty}$: We consider e as input, simulate the computation of M_e on 0, and count-through r_{γ_∞}. Either we observe first that $M_e(x)$ halts and then $e \in 0^\triangledown$, or we observe that the count-through halts and then $e \notin 0^\triangledown$.

$0^\triangledown \preccurlyeq_\infty \Gamma$ is trivial since Γ contains $0^\triangledown$ and in addition, contains the halting time for all machines which halt.

$\Gamma \preccurlyeq_\infty 0^\triangledown$ corresponds to an important result of the literature [10]: any clockable ordinal is writable. In the proof of this theorem (see also [5] for an alternative construction) a single program transforms M_e that halts in α steps into a (minimal) code for α. We thus get the required inequality.

$\Gamma_{\text{BB}} \equiv_\infty \Gamma$: the proof is analogous to the Turing case. To ITTM-compute Γ_{BB} from Γ, one just needs to compute a maximum. In the other direction, one simulates all machines of the same class up to the time given by Γ_{BB} and as previously compute the *ad hoc* halting time if any.

We obviously have $r_{\gamma_\infty} \preccurlyeq_\infty r_{\zeta_\infty} \preccurlyeq_\infty r_{\Sigma_\infty}$. $r_{\gamma_\infty} \not\equiv_\infty r_{\zeta_\infty}$ since $\lambda^{r_{\lambda_\infty}}_\infty$ is eventually writable: we eventually compute r_{λ_∞} and in parallel, we simulate $\mathcal{U}$. When we observe that we are in a gap, we start computing $\mathcal{U}$ on the ordinal that starts the gap (obtained while eventually-computing r_{λ_∞}). With this latter simulation of $\mathcal{U}$, we build the order-type of the set of its clockables as usual. This process stabilizes when the input for $\mathcal{U}$ is r_{λ_∞} and eventually-computes $\lambda^{r_{\lambda_\infty}}_\infty$.

$r_{\Sigma_\infty} \preccurlyeq_\infty r_{\zeta_\infty}$ since once we have found the configuration at stage ζ_∞ of $\mathcal{U}$, we just need to wait for the second occurrence to grasp Σ_∞.[10] $\Lambda_{\text{BB}} \equiv_\infty r_{\zeta_\infty}$ is rather trivial since Λ_{BB} consists of a finite number of ordinals $< \zeta_\infty$ followed by r_{ζ_∞} itself. The first ones are just encoded as an integer in r_{ζ_∞} thus we add only a finite information that we can produce by a constant program. □

---

[9] We denote by M_e the eth ITTM in a standard recursive enumeration. The two jumps are defined as $0^\triangledown = \{e \mid M_e(0) \downarrow\}$ and $0^\blacktriangledown = \{\langle e, x \rangle \mid M_e(x) \downarrow\}$.

[10] A corollary is that Σ_∞ is not admissible, as is already proven in [9, Corollary 3.4].

3 Experimental Study of the Busy Beaver Functions

3.1 Experimental Simulation of an Infinite Time Turing Machine

The experimental simulation of an ITTM is based on three parts. The first part is the simulation of the machine for one step. It simulates the read-write-move mechanism of the ITTM as any classical Turing machine simulator would do. The second part consists in guessing when to stop the simulation on successor ordinals, in order to jump to the next limit stage and devise the limit tape. It is based on the number of steps of computation as well as the machine's head trajectory (a great number of steps can be needed to reach such a threshold). The next limit stage itself is found by the simulator depending on current stages. The last part of the simulation is the guessing of the cell values of the limit tape considering the cells that have been flashed at previous stages. It is based on several criteria. A major one is cell stability, *i.e.*, analysing for each cell the number of times its value changed (till the last limit stage) and the number of computation steps that has elapsed till the last value change. Other criteria are the head trajectory for limits of successor ordinals, previous cofinal flashes [11] for higher limit ordinals, areas of differences between consecutive limit stages, *etc.* All these elements together allow us to understand the behaviour of the machine and to devise the limit tape. Finally, human intervention is possible and sometimes necessary if the simulator cannot make a decision on how to move up in the limit ordinal ladder or what values the cells of the limit tape should have.

As our simulation has obviously only finite means, there are many limits to our simulation. Considering previous flashing cells on the tape to build the cells' values at a limit stage, the simulator could recognise a stable repetitive pattern when in reality it would have changed a few thousand stages of computation later. This is especially the case when dealing with machines that present a chaotic behaviour (already present in the second category, BB-2). A solution is to increase memory thresholds (number of steps, tape size, *etc.*), at the expense of simulation speed. Finally, a lot of user double-checking by hand allowed us to build quite a reliable simulator (at least for machines with few states).

Clockable ordinals are simply obtained when reading the halting transition. Diverging machines and looping ordinals are detected using the divergence remark at the end of Sect. 2.1.

3.2 Study of Whole Classes of Machines

The study of a whole class of machines mixes enumeration and simulation. We first explain our enumeration of machines, and give a summary of the various behaviours of these machines and largest clockable and looping ordinals found during the simulation of these classes.

---

[11] Finitely many flashes that occur at successor ordinals do not affect the next limit stage but later ones of higher order (*cf. cofinally flashing* behaviour in the sequel).

Enumerating a Class of Machines. We enumerate the machines by adding transitions on the fly, making sure that we only add a transition when needed (in the simulation) and avoiding the different symmetries that can arise in the transitions graph. When an ITTM halts or is found as diverging, we simply update the transition that has been last set following a certain order on transitions. This order allow us to simulate a class completely.

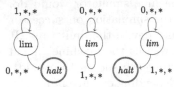

Fig. 2. ITTM specific filters

Moreover, filters are applied to get rid of non connected machines (in a sense of graph connectivity on the transitions graph), symmetric and trivial machines. This is very similar to what a classical Turing machine simulator would do. However, there is yet another filter specific to ITTMs which allows us to ignore patterns shown in Fig. 2 that lead either to a loop or to a halt at a stage that would not exceed [12] the limit ordinal $\omega \cdot 2$.

This enumeration on the fly allows us to find respectively 169 and 52189 distinct machines for the first two classes, BB-1 and BB-2.

Experimental Results. Experimental simulations gave us more material to appreciate how much ITTMs can be powerful even with only one or two states. Typical computational behaviours such as counters and fractal space-time diagrams already appear in the first and second categories, BB-1 and BB-2. Moreover, some behaviours of the classical busy beaver (such as *Xmas trees*, *cf.* [3]) appear with a smaller number of states in our ITTM competition, *e.g.*, Xmas trees with 1 state instead of 4.

We now give an overview of the first two classes with a selection of representative machines for each typical behaviour and their variants.

BB-1. This little class possesses 4 typical distinct behaviours, summarized in the following. The *Xmas tree* behaviour appears thanks to a previously written $^{\omega}0|1^{\omega}$ tape(see Footnote 4) after ω steps. The machine's head goes unboundedly often from left to right, drawing a Xmas tree (in its space-time diagram). The *cofinally flashing* behaviour is the ITTM's most natural behaviour: the machine flashes finitely many times a 0 cell and ends up stabilizing it to 0 (and thus without changing its value at the limit). Thus, if repeated on each limit, say ω, $\omega \cdot 2$, ..., these finitely many flashes directly affect the limit of limits, that is ω^2. The *noise expanding* behaviour consists in a machine generating noise at the origin, due to multiple limit jumps (and the head going directly to the origin).

---

[12] Indeed, suppose the machine did not halt for ω steps and reached the *limit* state. Either the machine will loop indefinitely on this *limit* state and is said to be diverging or will eventually halt excluding the middle pattern. Suppose it will eventually halt and go for more than ω steps taking the transition that loops on the *limit* state: either it will halt on the tape, reading a different symbol at a stage $\omega + c$ for an integer c, or reach the next *limit* stage at $\omega \cdot 2$ by reading a uniform tape. At this point, the machine halts if the looping transition modified the origin cell value, or loops indefinitely on the *limit* state.

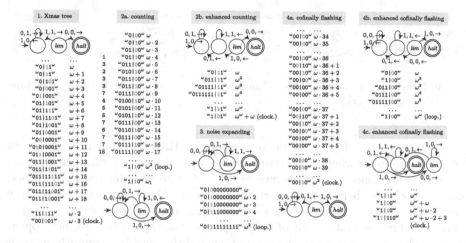

Fig. 3. Typical BB-1 behaviours (and their BB-2 extensions)

This noise grows and ends up taking all the tape at a higher limit. Finally, we denote by *counting* behaviour a machine that launches a counting procedure such as enumerating integers, in a certain order and written in binary, at limit tapes. These integers in binary can be viewed as patterns and their alternation affects limits of higher order.

The largest clockable ordinal for this class is produced by the *cofinally flashing* behaviour. The largest looping ordinals are produced by *cofinally flashing*, *noise expanding* and *counting* behaviours. For this first category BB-1, we get $\Gamma_{BB}(1) = \Lambda_{BB}(1) = \omega^2$ witnessed by the halting machine 4a and the diverging machines 2a ($\Pi(\cdot) = \omega^2$) and 3 ($\Pi(\cdot) = \omega$) in Fig. 3.

BB-2. Many enriched behaviours such as *moving* or *one-sided Xmas trees* are now possible, along with *fractal counting, etc.* Moreover, previous typical behaviours such as *cofinally flashing* and *counting* behaviours are now improved. Example machines for these enriched behaviours are given in Figs. 4 and 6. Improved aforementioned behaviours are given in Fig. 3, *e.g.*, the 'enhanced' counting and *cofinally flashing* behaviours.

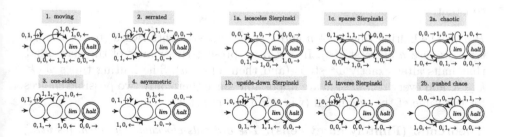

Fig. 4. Xmas BB-2 variants   **Fig. 5.** New BB-2 behaviours

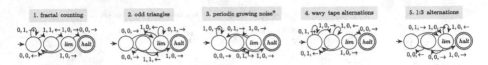

Fig. 6. Other BB-2 variants (*This halting machine ends up displaying a periodic tape $^\omega 0|(1110)^\omega$ at stage ω^2.)

Two new typical distinct behaviours appear in this class. The first one draws Sierpinski triangles when considering a history of all flashes performed to reach $\omega \cdot c$ stages. We denote this phenomenon by the *Sierpinski* behaviour. The second one is the *chaotic* behaviour that consists of a machine reducing or expanding concatenated groups of same values on each limit stage $\omega \cdot c$ for every integer c. This procedure generates a lot of flashes all over the tape. These flashes appear as being chaotic at a higher order limit stage and the machine promptly halts or diverges. Representative machines for these two behaviours are given in Fig. 5.

The largest clockable ordinal for this class is once again produced by the *cofinally flashing* behaviour with $\Gamma_{\mathrm{BB}}(2) = \omega^\omega + \omega \cdot 2 + 3$ witnessed by the machine 4c in Fig. 3. The largest looping ordinal is produced by *cofinally flashing* and *counting* behaviours with $\Lambda_{\mathrm{BB}}(2) = \omega^\omega$ witnessed by the machine 4b ($\Pi(\cdot) = \omega$) in Fig. 3.

4   Tape Use Restrictions

In this section, we investigate the consequences of several restrictions on the kind of tape that the machines are allowed to use.

4.1   Finite Tape Restriction

We say that an ITTM M *uses k cells on input x*, if M halts on the entry x, and if there exists a set of exactly k cells such that outside this set, the cells remain at value 0 during all the computation.

Proposition 1 (Only one cell). The possible halting times for machines using one cell on input 0 are exactly n, $\omega + n$, $\omega \cdot 2 + n$, or $\omega^2 + n$ where n is any finite constant.

Proof. The halting times $\omega + n$ and $\omega \cdot 2 + n$ are straightforward (to get the $+n$ term, we use n extra states). In Fig. 3 we present the winner of the class BB-1 that halts after exactly ω^2 steps (and then add $+n$). The halting times $\omega \cdot k$, $k \geq 3$, and every $\alpha > \omega^2 + \omega$ are impossible since there only two possible choices for the value of the origin cell (when in the *limit* state). □

Theorem 3 (Finitely many cells). *The ordinals clockable by machines using finitely many cells are exactly the ordinals $< \omega^\omega$.*

Proof. We assume that there exists an ITTM M that halts in ω^ω or more steps using k cells and obtain a contradiction. Consider the configuration at stage ω^ω and consider a window of finite size that contains all k cells touched by M.

There exists a stage γ after which all 0's in the window (identified as the *grey* cells) placed on the configuration at stage ω^ω have stabilized. Between γ and ω^ω, all limit configurations have the grey cells at 0. Among these limit configurations, let us consider pairs of equal configurations. We observe their number of 1's. It has a *supremum limit* denoted by S. If S is exactly the number of 1's of the configuration at stage ω^ω, then there exists a pair of configurations where all 0's from inside the window are grey cells. From the remarks at the end of Sect. 2.1, we conclude that M diverges. Thus S is less than the number of 1's of the configuration at stage ω^ω. Let us denote by β_1 and β_2 the stages corresponding to such a pair with S 1's. If all 0's of the window remain unchanged between stage β_1 and β_2, then again M diverges. Thus at least one non-grey 0 flashes at least once. Now let us consider ω times this (non-final) loop between β_1 and β_2. As β_2 is bounded by ω^a for some integer a, we get a limit configuration with at least $S+1$ 1's before ω^ω. After this configuration, we consider another pair β'_1 and β'_2 with S 1's. By the same argument we get another limit configuration with at least $S+1$ 1's before ω^ω. We iterate this process, and, having only a finite choice for these 1's in the window, we get two equal limit configurations with $S+1$ 1's which contradicts the definition of S.

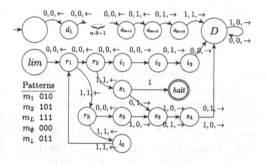

Fig. 7. Machine halting after ω^n steps

We now construct a machine that halts after time ω^n for any integer n using only a finite number of cells. This machine is based on the improved *counting* behaviour described in Sect. 3.2, and is intentionally left unoptimized for a better understanding. In the following, we denote by the term 'pattern' a set of 3 consecutive cells described in the tabular in Fig. 7. In the sequel, we explain how this machine uses them to iterate a counter up to ω^n.

First of all, a delimiter pattern $m_\downarrow$ is written $3n$ cells away at left from origin and the machine enters the D-state (*cf.* Fig. 7) that goes quietly to the ω limit without flashing any new cells. This is denoted by states $(d_1, \ldots, d_{3n+3})$ in Fig. 7. Then, a counting procedure is launched, based on pattern recognition and alternation. If the machine recognises an empty pattern $m_\emptyset$ (r_1, r_2, r_3 states), it remplaces it with a m_1 pattern (i_1, i_2, i_3) and jumps to the next limit. If one of the standard patterns (m_1 and m_2) are recognised, the pattern is switched from m_1 to m_2 or reciprocally (s_1, s_2, s_3) and the machine loops to the next limit. Although it is never written, a limit pattern m_L will be found after ω alternations of m_1 and m_2 and thus m_L appears for the first stage after ω^2 steps of computation. If m_L is read, then the machine ignores it and seeks the next pattern on the left (l_0). Using one state (D-state), we erase all m_L patterns after next patterns incrementation. When a limit pattern m_L is found for the first time in the kth pattern from the origin, the machine has computed ω^k steps. The whole counting procedure stops when the tape is full of m_L patterns between origin and $m_\downarrow$, which is discovered at stage $\omega^n + 3n + 2$ using $3n + 4$ cells. □

4.2 Ultimately Periodic and Recursive Tape Restrictions

We focus on those machines whose tapes are ultimately periodic at any computation time. We can define those machines as those for which their configurations at any limit stage is ultimately periodic.[13] One reason for considering these machines is that at any time, their tape can be finitely described.

Of course, the natural extension of these two tape-restriction notions is to ask for the tape at any stage to be recursive. This is also a constraint that only concerns limits stages: a point-to-point limit of those recursive tapes must be recursive.[14]

We conjecture the following generalisation of Theorem 3.

Conjecture. *Every ITTM which halts when started on the empty input and whose tapes are ultimately periodic (resp. recursive) halts after less than ε_0 (resp. ω_1^{CK}) steps.*

References

1. Barwise, J.: Admissible Sets and Structures: An Approach to Definability Theory. Perspectives in Mathematical Logic, vol. 7. Springer, Heidelberg (1975)
2. Brady, A.H.: The conjectured highest scoring machines for Radó's $\Sigma(k)$ for the value $k = 4$. IEEE Trans. Electron. Comput. EC **15**(5), 802–803 (1966)
3. Brady, A.H.: The determination of the value of Radó's noncomputable function $\Sigma(k)$ for four-state turing machines. Math. Comput. **40**(162), 647–665 (1983)
4. Brady, A.H.: The busy beaver game and the meaning of life. In: Herken, R. (ed.) The Universal Turing Machine: A Half-Century Survey, 2nd edn, pp. 237–254. Springer, New York (1995)
5. Durand, B., Lafitte, G.: A constructive swissknife for infinite time turing machines (2016)
6. Hamkins, J.D., Lewis, A.: Infinite time turing machines. J. Symbolic Log. **65**(2), 567–604 (2000)
7. Lafitte, G., Papazian, C.: The fabric of small turing machines. In: Proceedings of the Third Conference on Computability in Europe Computation and Logic in the Real World, CiE 2007, Siena, Italy, 18–23 June 2007, pp. 219–227 (2007)

---

[13] Note that knowing whether a machine has this property is not decidable. This was also the case with the finite tape restriction.

[14] The situation is not the same for stage ω than for compound limit ordinal stages, e.g., ω^2, since in the first case we take the limit of a recursive sequence of recursive tapes while in the latter, the sequence of tapes is not necessarily recursive.

8. Radó, T.: On non-computable functions. Bell Syst. Tech. J. **41**(3), 877–884 (1962)
9. Welch, P.D.: The length of infinite time turing machine computations. Bull. London Math. Soc. **32**, 129–136 (2000)
10. Welch, P.D.: Characteristics of discrete transfinite time turing machine models: Halting times, stabilization times, and normal form theorems. Theoret. Comput. Sci. **410**, 426–442 (2009)

Permutive One-Way Cellular Automata and the Finiteness Problem for Automaton Groups

Martin Delacourt[✉] and Nicolas Ollinger

Univ. Orléans, LIFO EA 4022, 45067 Orléans, France
{martin.delacourt,nicolas.ollinger}@univ-orleans.fr

Abstract. The decidability of the finiteness problem for automaton groups is a well-studied open question on Mealy automata. We connect this question of algebraic nature to the periodicity problem of one-way cellular automata, a dynamical question known to be undecidable in the general case. We provide a first undecidability result on the dynamics of one-way permutive cellular automata, arguing in favor of the undecidability of the finiteness problem for reset Mealy automata.

Keywords: Reset Mealy automata · One-sided cellular automata · Permutive cellular automata · Periodicity problem · Reversible computation

1 Introduction

Finite-state automata provide a convenient finite description for different kinds of behavior generated by their computations. As such, Mealy automata [10] provide a finite description for the family of automaton (semi)groups that has proven its usefulness to generate interesting counter examples in the field of group theory [3]. Several decision problems inspired by algebraic questions have been studied on automaton groups: the word problem is decidable whereas the conjugacy problem is undecidable [16]. The general case of the finiteness problem remains open [1] although special cases have been solved: Gillibert [9] proved that the problem is undecidable for semigroups and Klimann [15] that it is decidable for reversible Mealy automata with two states. The status of the finiteness problem remains open for the class of reset Mealy automata.

Cellular automata [14] provide a finite description for a family of discrete dynamical systems, the endomorphisms of the shift dynamical system [11]. Decision problems inspired by dynamical questions have been investigated on cellular automata since the work of Amoroso and Patt [2]. The computation nature of cellular automata lead to sophisticated construction techniques to establish the undecidability of various decision problems like the nilpotency problem [13] or more recently the periodicity problem [12]. The status of the periodicity problem remains open for one-way cellular automata.

Without much surprise, cellular automata can be a valuable tool to establish undecidability results on Mealy automata. Indeed, Gillibert's result is inspired by Kari's proof of the undecidability of the nilpotency problem.

© Springer International Publishing AG 2017
J. Kari et al. (Eds.): CiE 2017, LNCS 10307, pp. 234–245, 2017.
DOI: 10.1007/978-3-319-58741-7_23

In this paper, we study the computational power of reversible permutive one-way cellular automata [4,7]. Our first contribution is a precise formalization of the connection between both open problems: the finiteness problem for reset Mealy automata is decidable if and only if the periodicity problem for one-way cellular automata is decidable. Our second contribution is a technique to embed computation inside reversible one-way cellular automata using permutive automata. The technique is applied to prove a first undecidability result on these objects.

2   Definitions

For a detailed introduction on Mealy automata, the reader is referred to Bartholdi and Silva [3] and to Kari [14] for cellular automata.

2.1   Mealy Automata

A *Mealy automaton* is a deterministic complete 1-to-1 transducer $(A, \Sigma, \delta, \rho)$, where A is a finite set of states, Σ a finite alphabet, $\delta = (\delta_i : A \to A)_{i \in \Sigma}$ is the set of transition functions and $\rho = (\rho_x : \Sigma \to \Sigma)_{x \in A}$ the set of production functions. The transition $x \xrightarrow{i|\rho_x(i)} \delta_i(x)$ is depicted by

$$x \xrightarrow[\;i\;]{\rho_x(i)} \delta_i(x) \;.$$

The *production functions* naturally extend to functions on the set of finite words: $\rho = (\rho_x : \Sigma^* \to \Sigma^*)_{x \in A}$ with $\rho_x(au) = \rho_x(a)\rho_{\delta_a(x)}(u)$. The semigroup generated by the automaton is the set of all compositions of the production functions $H = \langle \rho_x : x \in A \rangle$. An *automaton semigroup* is a semigroup generated by a Mealy automaton.

A Mealy automaton is *invertible* if ρ_x is a permutation of Σ for every $x \in A$. Note that it implies that every ρ_x is also a permutation on Σ^k for every $k \in \mathbb{N}$. The group generated by a invertible Mealy automaton is $G = \langle \rho_x, \rho_x^{-1} : x \in A \rangle$. An *automaton group* is a group generated by a Mealy automaton. An invertible Mealy automaton generates a finite group if and only if it generates a finite semigroup [1].

Finiteness problem. *Given an invertible Mealy automaton, decide if the generated group is finite.*

A Mealy automaton is *reset* if, for each transition $x \xrightarrow{i|\rho_x(i)} \delta_i(x)$, the output state $\delta_i(x)$ depends only on the input letter i and not on the input state x, that is $\delta_i(x) = f(i)$ for some letter-to-state map $f : \Sigma \to A$. To simplify notations, such an automaton will be denoted as (A, Σ, f, ρ).

When studying the decidability of the finiteness problem restricted to reset automata, one can focus on the case $\Sigma = A$ and $f = Id$ as stated below.

Lemma 1. *The group generated by a reset Mealy automaton (A, Σ, f, ρ) is finite if and only if it is the case for the automaton $(\Sigma, \Sigma, 1, \rho')$ with $\rho'_x(i) = \rho_{f(x)}(i)$.*

Proof. Let G be the group generated by (A, Σ, f, ρ) and H be the group generated by $(\Sigma, \Sigma, 1, \rho')$. As every generator ρ'_x of H is a generator $\rho_{f(x)}$ of G then H is a subgroup of G. A generator ρ_y of G with $y \in A \setminus f(\Sigma)$ is not a generator of H, however as y can only be an initial state of a transition, it only impacts the size of G by a factor $n!$ where n is the size of Σ.   ∎

2.2   Cellular Automata

A *one-way cellular automaton* (OCA) $\mathcal{F}$ is a triple (X, r, δ) where X is the finite set of states, r is the radius and $\delta : X^{r+1} \to X$ is the local rule of the OCA. A *configuration* $c \in X^{\mathbb{Z}}$ is a biinfinite word on X. The *global function* $\mathcal{F} : X^{\mathbb{Z}} \to X^{\mathbb{Z}}$ synchronously applies the local rule: $\mathcal{F}(c)_i = \delta(c_i, \ldots, c_{i+r})$ for every $c \in X^{\mathbb{Z}}$ and $i \in \mathbb{Z}$. The *spacetime diagram* $\Delta : \mathbb{Z} \times \mathbb{N} \to X$ generated by an initial configuration c is obtained by iterating the global function: $\Delta(k, n) = \mathcal{F}^n(c)_k$ for every $k \in \mathbb{Z}$ and $n \in \mathbb{N}$. Following Hedlund [11] characterization of cellular automata as endomorphisms of the shift, we assimilate an OCA, with minimal radius, and its global function.

Notice that OCA are the restriction of classical cellular automata (CA) where a cell only depends on other cells on the right side. The *identity* function Id, the *left shift map* σ_l and the XOR rule are OCA, respectively encoded as $(X, 0, 1)$, $(X, 1, (x, y) \mapsto y)$ and $(\{0, 1\}, 1, \oplus)$, whereas the *right shift map* σ_r is not.

A state $x \in X$ is *quiescent* if $\delta(x, \ldots, x) = x$. A configuration is *finite* if it contains the same quiescent state x everywhere but on finitely many positions.

An OCA is (left) *permutive* if the map $x \mapsto \delta(x, x_1, \ldots, x_r)$ is a permutation of X for every $(x_1, \ldots, x_r) \in X^r$. An OCA is *periodic* of period $T > 0$ if $\mathcal{F}^T = \text{Id}$.

Periodicity problem. *Given a cellular automaton, decide if it is periodic.*

An OCA is *reversible* if its global function is bijective with an inverse that is also an OCA. A periodic OCA $\mathcal{F}$ of period T is reversible of inverse $\mathcal{F}^{T-1}$. Following Hedlund [11], the inverse of a bijective OCA is always a CA, however usually not one-sided. The following lemmas assert that this technical issue disappears by considering only permutive OCA.

Lemma 2. *Every reversible OCA is permutive.*

Proof. Let $\mathcal{F}$ be a reversible OCA. As both $\mathcal{F}$ and $\mathcal{F}^{-1}$ are OCA, $\mathcal{F}$ is bijective on $X^{\mathbb{N}}$ too. Let $(x, x', x_1, \ldots, x_r) \in X^{r+2}$ and $c = x_1 \cdots x_r x_r^{\omega}$. If $x \neq x'$, as $\mathcal{F}(xc) \neq \mathcal{F}(x'c)$, we have $\delta(x, x_1, \ldots, x_r) \neq \delta(x', x_1, \ldots, x_r)$.   ∎

Lemma 3. *Every bijective permutive OCA is reversible.*

Proof. Let $\mathcal{F}$ be a bijective permutive OCA. By permittivity $f : X \times X^{\mathbb{N}} \to X$ defined by $f(y, c) = x$ where $\mathcal{F}(xc) = y\mathcal{F}(c)$ is well defined and permutive in its first argument. Let $u, u' \in X^{-\mathbb{N}}$ and $v' \in X^{\mathbb{N}}$. Let $w \in X^{-\mathbb{N}}$ and $v \in X^{\mathbb{N}}$ be such that $\mathcal{F}^{-1}(u'v') = wv$. Let $w' \in X^{-\mathbb{N}}$ be defined recursively by

$w_i' = f(u_i, w_{i-1}' \cdots w_0' v)$. By construction $\mathcal{F}^{-1}(uv') = w'v$. As $\mathcal{F}^{-1}(uv')$ and $\mathcal{F}^{-1}(u'v')$ are equal on $\mathbb{N}$ for all u, u', v' the CA $\mathcal{F}^{-1}$ is an OCA. ∎

Note that the inverse of a bijective permutive OCA can have a larger radius. However, as bijectivity is decidable for cellular automata [2] and as bijectivity is preserved by grouping cells [8], when studying the decidability of the periodicity problem restricted to permutive OCA, one can focus on the case of reversible permutive OCA with radius 1 and inverse radius 1 syntactically characterized by the following lemma (already proven in [4]).

Lemma 4. *An OCA $(X, 1, \delta)$ is reversible with inverse radius 1 if and only if it is permutive and for all $x, y, x', y' \in X$, if $\delta(x, y) = \delta(x', y')$ then $\pi_x = \pi_{x'}$ where π_y maps x to $\delta(x, y)$.*

From now on, we only consider these OCA.

3   Linking Finiteness and Periodicity

Reset Mealy automata with $\Sigma = A$ and $f = Id$ and permutive OCA of radius 1 are essentially deterministic complete letter-to-letter transducers of a same kind, as depicted on Fig. 1. The following proposition formalizes this link.

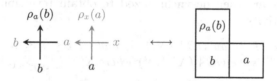

Fig. 1. Linking reset Mealy automata to permutive OCA

Proposition 1. *The group generated by a reset automaton $(\Sigma, \Sigma, 1, \rho)$ is finite if and only if the permutive OCA $(\Sigma, 1, \delta)$, where $\delta(x, y) = \rho_y(x)$, is periodic.*

Proof. First note that the following equations hold for all $u_0, u_1, \ldots, u_k \in \Sigma$:

$$\rho_{u_k}(u_{k-1}, u_{k-2}, \ldots, u_0) = \rho_{u_k}(u_{k-1})\rho_{u_{k-1}}(u_{k-2}) \cdots \rho_{u_1}(u_0)$$
$$\delta(u_0, u_1, \ldots, u_k) = \rho_{u_1}(u_0)\rho_{u_2}(u_1) \cdots \rho_{u_k}(u_{k-1})$$

By extension, for all $k > t > 0$ and for all words $u \in \Sigma^t$ and $v \in \Sigma^k$, the following equation holds: $\rho_u(v)_k = \delta^t(v_k, v_{k-1}, \ldots, v_{k-t})$.

Suppose now the group generated by the reset Mealy automaton is finite. Let $a \in \Sigma$ be any letter and let n be the order of ρ_a, then $\rho_{a^n} = (\rho_a)^n = Id$ thus $\delta^n = Id$, the OCA is periodic.

Conversely, let n be the period of the OCA. By previous remarks, for all $k > 0$ and words $u, v \in \Sigma^n$, $w \in \Sigma^k$, the image $\rho_u(vw)$ is $v'w$ for some $v' \in \Sigma^n$. The set of ρ_u generates a subgroup of permutations of Σ^n, which is finite, when u takes all possible values in Σ^n. The automaton group is finite. ∎

Corollary 1. *The finiteness problem restricted to reset Mealy automata is decidable if and only if the periodicity problem restricted to OCA is decidable.*

Notice that the situation described by this corollary is optimal: if the problem is decidable, Mealy automata is the right setting to prove this result and the decidability of the periodicity for OCA will be a consequence; if the problem is undecidable, cellular automata is the right setting to prove this result and the undecidability of the finiteness problem will be a consequence. The remainder of this paper is dedicated to prove that computational phenomena do appear inside the dynamics of permutive OCA, advocating for the undecidability of the problem.

Conjecture 1. The finiteness problem is undecidable.

4   Computing with Permutive OCA

As shown in Fig. 1, the time goes up in every representation of this paper.

Given a permutive OCA, we show how to build a reversible OCA that can simulate every spacetime diagram of the original. The idea is to slow down the computation by delaying each state using a fixed number of distinct copies per state and perform a transition only after going through every copy. Adjacent columns of states are then desynchronized to obtain reversibility as a consequence of permittivity.

Definition 1. *Let $\mathcal{F}$ be an OCA $(X, 1, \delta)$ and let $1 \leq k \leq n - 1$. The (n, k)-embedding $\mathcal{F}'$ of $\mathcal{F}$ is the OCA $(X', 1, \delta')$ where $X' = \bigcup_{1 \leq i \leq n} \{x^{(i)} : x \in X\}$ and such that:*

$$\forall x^{(\alpha)}, y^{(\beta)} \in X' \qquad \delta'\left(x^{(\alpha)}, y^{(\beta)}\right) = \begin{cases} \delta(x, y)^{(1)} & \textit{if } \alpha = n \textit{ and } \beta = k \\ x^{(1 + (\alpha \mod n))} & \textit{otherwise} \end{cases}$$

Figure 2 illustrates the embedding.

Lemma 5. *The (n, k)-embedding of a permutive OCA $\mathcal{F}$ is reversible.*

Proof. The local rule of the inverse OCA τ can be defined by

$- \tau(z^{(1)}, y^{(k+1)}) = x^{(n)}$ for all $x, y \in X$ such that $\delta(x, y) = z$;
$- \tau(x^{(i)}, *) = x^{((i-1 \mod n)+1)}$ otherwise.   ∎

The idea of the embedding is to desynchronize adjacent columns by shifting them vertically of some constant k between 1 and $n - 1$. When two consecutive columns are not correctly arranged, they do not interact.

Lemma 6. *There exists an injective transformation of spacetime diagrams of $\mathcal{F}$ (in $X^{\mathbb{Z} \times \mathbb{N}}$) into spacetime diagrams of $\mathcal{F}'$ (in $X'^{\mathbb{Z} \times \mathbb{Z}}$): for every $c \in X^{\mathbb{Z}}$, there exists a unique configuration $c' \in X'^{\mathbb{Z}}$ for $\mathcal{F}'$ with*

$$\forall m \in \mathbb{Z}, p \in \mathbb{N}, \forall 1 \leq i \leq n, \mathcal{F}'^{m(n-k)+pn+i-1}(c')_m = (\mathcal{F}^p(c)_m)^{(i)}$$

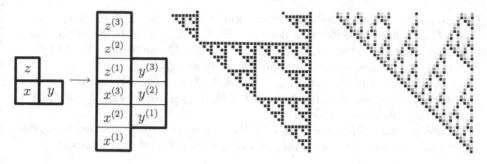

Fig. 2. The $(3,1)$-embedding is given by the transformation of the local rule (to the left). The space-time diagram of the XOR OCA with a unitary configuration (at the center) is transformed into a space-time diagram of a reversible OCA (to the right).

Remark 1. If the OCA has a particular state 0 such that $\delta(x, 0) = x$ for every $x \in X$, we can keep a unique version of state 0 by identifying the $0^{(i)}$ as a unique state 0 where

$$\delta'(0, y^{(k)}) = (\delta(0, y))^{(1)}$$
$$\delta'(0, *) = 0 \text{ where } * \text{ can be anything}$$
$$\delta'(x^{(i)}, 0) = x^{(1 + (i \mod n))}$$

This lemma allows to transfer results from $\mathcal{F}$ to $\mathcal{F}'$, in the sequel we prove the following result for permutive OCA and obtain it for reversible ones:

Reachability problem. *Given a reversible OCA with a quiescent state $0 \in X$ and two states $x, y \in X$, decide if y appears in the spacetime diagram generated by the initial configuration "$0.x0^\omega$".*

Theorem 1. *The reachability problem is undecidable for reversible OCA.*

5 Main Construction

To prove the theorem, we need to embed some Turing complete computation into permutive OCA. This goal is achieved by simulating multi-head walking automata. This section describes the simulation of these automata by permutive OCA.

5.1 Multi-head Walking Automata

A multi-head walking automaton consists of a finite number of heads on the discrete line, each one of them provided with a state out of a finite set. At each step, they can only interact (read the state) with the heads that share the same cell, update their state and move. Initially, all the heads are in position 0.

Definition 2. *A k-head walking automaton is defined by $(\Sigma, I, F, (f_i, g_i)_{1 \leq i \leq k})$ where Σ is a finite alphabet that does not contain $\bot$, $I \in \Sigma$ is the initial state, $F \subseteq \Sigma$ is a set of final states, $\forall i, f_i : (\Sigma \cup \{\bot\})^k \to \Sigma$ and $\forall i, g_i : (\Sigma \cup \{\bot\})^k \to \{-1, 0, 1\}$ are the update functions for the state and the position.*

A configuration of this automaton is a k-tuple $a = (a_i, s_i)_{1 \leq i \leq k} \in (\mathbb{Z} \times \Sigma)^k$ and its image configuration is $(a_i + g_i(b_i), f_i(b_i))_{1 \leq i \leq k}$ where $\forall 1 \leq i \leq k, b_i(j) = s_j$ if $a_i = a_j$ and $\bot$ otherwise.

Starting from configuration $(0, I)^k$, the automaton computes the successive images and stops only if the position of every head is 0 and their states are final.

A multi-head walking automaton can mimic a counter with 2 heads. Turing completeness is achieved by simulating 2-counter Minsky machines.

Proposition 2. *The halting problem of 4-head walking automata is undecidable.*

5.2   Finite Configurations

Every finite configuration can be written $c = {}^\omega 0.u_0 0^\omega$ for some $u_0 \in X^n, n \in \mathbb{N}$, hence its successive images can only be $\mathcal{F}^k(c) = {}^\omega 0 v_k.w_k 0^\omega$ for some $v_k \in X^k$, $w_k \in X^n$, that is: the finite non-quiescent word extends to the left, one cell at each step. We also have the following result.

Lemma 7 (Coven et al. [6]). *In the spacetime diagram associated to a finite configuration, every column is periodic.*

5.3   P-signals

Most cellular automata constructions use signals as elementary geometrical building blocks. Unfortunately, it is not possible to embed a classical signal in a permutive OCA nor is it possible to directly simulate multi-head automata. We provide P-signals as a technique to replace signals of speed k by an (n, k)-embedding of the front line of the spacetime diagram of the XOR with only one non-zero cell. This will effectively allow to send a signal through space in a reversible permutive OCA.

For our construction, we fix $n = 7$, hence we can use speeds from 1 to 6. Due to Remark 1, the states of these signals will be denoted $\mathbb{Z}_8 = \{0, 1, 2, 3, 4, 5, 6, 7\}$. Next lemma states that different speeds induce similar spacetime diagrams up to a vertical shift on each column.

Lemma 8. *Denote $\mathcal{F}_k$ and $\mathcal{F}_{k'}$ the OCA corresponding to signals of speeds k and k' and let c be the configuration ${}^\omega 0.10^\omega$ then we have*

$$\forall m \leq 0, \forall p \geq 0, \mathcal{F}_k^p(c)_m = \mathcal{F}_{k'}^{p-(k-k')m}(c)_m.$$

5.4   2-recognizability

Our construction relies on the ability to identify some specific sets of spacetime positions. We achieve this goal by considering products of independent P-signals. To avoid boring calculations, we rely on the following lemma to assert some regularity of P-signals and rely on the theory of p-recognizable sets of tuples of integers [5] to characterize these positions.

Lemma 9. *The spacetime sequence $\Delta : \mathbb{N}^2 \to \mathbb{Z}_{n+1}$ of the (n, k)-embedding $\mathcal{F}$ of the XOR, where $\Delta(x, y) = \mathcal{F}^y({}^\omega 0.10^\omega)_{-x}$, is 2-recognizable.*

Proof. The spacetime sequence Δ of the (n, k)-embedding $\mathcal{F}$ is generated by the 2-substitution $s : \mathbb{Z}_{n+1} \to \mathbb{Z}_{n+1}^2$ uniquely defined, as per Fig. 2, by

$$s(0) = \begin{pmatrix} 0\ 0 \\ 0\ 0 \end{pmatrix} \qquad s\begin{pmatrix} n \\ \vdots \\ 1 \end{pmatrix} = \begin{pmatrix} n \cdots (k+1)\ k \cdots 1\ n \cdots (k+1)\ \cdots\ 1 \\ 0 \cdots\ \ \ \ \ 0\ \ \ \ \ n \cdots \cdots \cdots \cdots\ \ 1\ \ \ \ 0 \cdots 0 \end{pmatrix}^{\mathsf{T}}$$

Indeed, the substitution rule is compatible with the local rule of the OCA.   ∎

5.5   Computation Windows

Let $\mathcal{F}$ denote the OCA we are constructing. We use 4 P-signals of speeds 2, 3, 5 and 6 as foundations of $\mathcal{F}$ — they allow us to build computation windows. Let c be the initial configuration which is null everywhere except for $c(0) = (1, 1, 1, 1)$. We use Lemma 9 twice for each of the following lemmas, that is, we have a characterization of the set of positions where some specific state pairs appear. Both lemmas are illustrated in Figs. 3, 4 and 5.

Lemma 10

$$\forall m \le 0, p \ge 0, \begin{cases} \mathcal{F}^p(c)_m = (1, 0, \_, \_) \\ \mathcal{F}^p(c)_{m+1} = (0, 3, \_, \_) \end{cases} \Leftrightarrow \begin{cases} \exists h \in \mathbb{N}^*, m = -8^h \\ p \equiv -3m - 1[-14m] \end{cases}$$

Lemma 11

$$\forall m \le 0, p \ge 0, \begin{cases} \mathcal{F}^p(c)_m = (\_, \_, 7, 6) \\ \mathcal{F}^p(c)_{m+1} = (\_, \_, 5, 0) \end{cases} \Leftrightarrow \begin{cases} \exists h \in \mathbb{N}^*, m = -8^h \\ p \equiv -12m - 1[-14m] \end{cases}$$

Hence we add a fifth layer using alphabet $\{0, 1\}$ with the rule:

$$\delta_5\left((1, 0, \_, \_, x), (0, 3, \_, \_, \_)\right) = 1 - x$$
$$\delta_5\left((\_, \_, 7, 6, x), (\_, \_, 5, 0, \_)\right) = 1 - x$$
$$\delta_5\left((\_, \_, \_, \_, x), (\_, \_, \_, \_, \_)\right) = x \quad \text{otherwise.}$$

Actually, the same arguments work for columns $-2 \cdot 8^h, h \in \mathbb{N}^*$ and $-4 \cdot 8^h, h \in \mathbb{N}^*$, hence we use them all. We therefore call computation windows the vertical segments in the spacetime diagram where the fifth layer contains 1:

$$(m, p) \text{ in a computation window} \Leftrightarrow \begin{array}{l} \exists h \in \mathbb{N}^*, m = -2^h, p' \equiv p[14 \cdot 2^h] \\ \text{and } 3 \cdot 2^h \le p' < 12 \cdot 2^h. \end{array}$$

 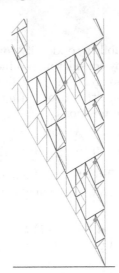 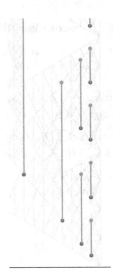

Fig. 3. Speeds 2 and 3 P-signals determine the positions of the lower points of the computation windows as stated in Lemma 10.

Fig. 4. Speeds 5 and 6 P-signals determine the positions of the upper points of the computation windows as stated in Lemma 11.

Fig. 5. In light gray, speeds 2, 3, 5 and 6 P-signals together allow to draw the whole computation windows.

5.6 Computing Heads

Every head of the walking automaton is simulated on a new layer. It is composed of a support that is mainly a speed 4 P-signal, and an internal state that belongs to the set of states Σ of the walking automaton. We need 4 heads, hence there are 4 additional layers (6 to 9).

The idea is that the heads move like speed 4 P-signal carrying internal states until they reach a computation window (as illustrated in Fig. 6). Then, depending on the context, they update their internal state and eventually shift upward or downward. Each shift corresponds to a move of the head of the walking automaton. Hence its position is encoded by the global shift applied to it, call it the *height* of the head.

More formally, each layer representing a head uses the alphabet $([1..7] \times \Sigma) \cup (0, \perp)$. Outside computation windows, the first component follows the rule of speed 4 signals, and the second component is maintained if possible or otherwise taken from the right neighbor.

Suppose now that we compute the new state of cell m at time p with (m, p) inside a computation window. Denote by x, y and z the states of cells (m, p), $(m + 1, p)$ and $(m, p + 1)$ in the spacetime diagram. The following rules apply to layer 6 (the same applies for other layers):

- if $y_6 = (0, \perp)$, then apply standard rules.

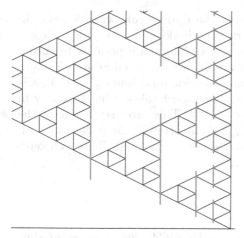 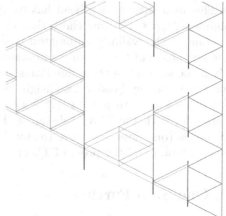

Fig. 6. The heads are supported by speed 4 P-signals that pass through the computation windows. The windows are large enough so that this property remains even after they are shifted up or down.

Fig. 7. When the head arrives in a computation window, it is here shifted upward, then upward again and downward in the third window. The state also changes then (red, later blue, orange and blue again). In light gray, another head whose position does not change. (Color figure online)

– if $y_6 \neq (0, \bot)$, look at every other layer where the support state is the same and apply f_1 and g_1, the update functions of the walking automaton, with these states. If the result of g_1 is -1 (resp. 0, 1), apply the rule of a speed 3 (resp. 4, 5) signal to the first component. If the support of z_6 is not 0, then the internal state is given by f_1.

5.7 Simulation

Finally, the OCA we build has 9 layers: 4 P-signals to determine the computation windows on the fifth layer, and 4 to simulate the 4 heads of the Turing universal walking automaton. The initial configuration is the finite configuration c with

$$c(0) = (1, 1, 1, 1, 0, (1, I), (1, I), (1, I), (1, I))$$
$$c(m) = (0, 0, 0, 0, 0, (0, \bot), (0, \bot), (0, \bot), (0, \bot)) \quad \text{elsewhere.}$$

First note that the height can vary of at most 1 each time the head crosses a column $-2^h, h \in \mathbb{N}$, hence:

Lemma 12. *Given any simulating head, while its height is between $-h$ and h in column $m = -2^h, h \in \mathbb{N}$, every non $(0, \bot)$ value of the layer in this column is inside a computation window.*

This means that each head has to apply the update rule when arriving in column $-2^h, h \in \mathbb{N}$. Now check that every time it does, one step of the computation of the walking automaton is simulated. The main point is to ensure that two heads of the walking automaton are on the same cell at step h if and only if the support of their simulating heads coincide on column $2^h - 1$. This is true since, for any head whose height is s, its support takes value 1 exactly in cells $(2^h - 1, p)$ with $p \equiv (4 \cdot (2^h - 1) + s) [7 \cdot 2^h]$. This property is due to the XOR rule, with the $(7, 4)$-embedding. Figure 7 illustrates this behaviour with two heads (one does not move to simplify the representation). This completes the simulation and the proof of Theorem 1.

6   One Step Further

Orbit periodicity problem. *Given a reversible OCA with a quiescent state $0 \in X$ and a state $x \in X$, decide if the spacetime diagram generated by the initial configuration "$0.x0^\omega$" is periodic.*

Theorem 2. *The orbit periodicity problem is undecidable for reversible OCA.*

If the computation halts, the windows and the 4 P-signals that help determine the computation heads keep progressing eternally. To reach periodicity, it is necessary (and enough thanks to Lemma 7) to kill them all. It is possible to do so by giving killing orders to the heads. At the halting step h_0, each head is given the responsibility to kill one of the P-signals while sacrificing itself. They have to slow down or speed up to meet the corresponding P-signal. Again, we use Lemma 9 to prove that the meeting can happen on column 2^{h_0+1} with a local context that does not happen elsewhere.

References

1. Akhavi, A., Klimann, I., Lombardy, S., Mairesse, J., Picantin, M.: On the finiteness problem for automaton (semi) groups. Int. J. Algebra Comput. **22**(06), 1250052 (2012)
2. Amoroso, S., Patt, Y.N.: Decision procedures for surjectivity and injectivity of parallel maps for tessellation structures. J. Comput. Syst. Sci. **6**(5), 448–464 (1972)
3. Bartholdi, L., Silva, P.V.: Groups defined by automata. In: AutoMathA Handbook (to appear). https://arxiv.org/abs/1012.1531
4. Boyle, M., Maass, A., et al.: Expansive invertible onesided cellular automata. J. Math. Soc. Jpn. **52**(4), 725–740 (2000)
5. Bruyere, V., Hansel, G., Michaux, C., Villemaire, R.: Logic and p-recognizable sets of integers. Bull. Belg. Math. Soc. Simon Stevin **1**(2), 191–238 (1994)
6. Coven, E., Pivato, M., Yassawi, R.: Prevalence of odometers in cellular automata. Proc. Am. Math. Soc. **135**(3), 815–821 (2007)
7. Dartnell, P., Maass, A., Schwartz, F.: Combinatorial constructions associated to the dynamics of onesided cellular automata. Theoret. Comput. Sci. **304**(1–3), 485–497 (2003)

8. Delorme, M., Mazoyer, J., Ollinger, N., Theyssier, G.: Bulking II: classifications of cellular automata. Theoret. Comput. Sci. **412**(30), 3881–3905 (2011)
9. Gillibert, P.: The finiteness problem for automaton semigroups is undecidable. Int. J. Algebra Comput. **24**(01), 1–9 (2014)
10. Glushkov, V.M.: The abstract theory of automata. Uspekhi Matematicheskikh Nauk **16**(5), 3–62 (1961)
11. Hedlund, G.A.: Endomorphisms and automorphisms of the shift dynamical systems. Math. Syst. Theory **3**(4), 320–375 (1969)
12. Kari, J., Ollinger, N.: Periodicity and immortality in reversible computing. In: Ochmański, E., Tyszkiewicz, J. (eds.) MFCS 2008. LNCS, vol. 5162, pp. 419–430. Springer, Heidelberg (2008). doi:10.1007/978-3-540-85238-4_34
13. Kari, J.: The nilpotency problem of one-dimensional cellular automata. SIAM J. Comput. **21**(3), 571–586 (1992)
14. Kari, J.: Theory of cellular automata: a survey. Theoret. Comput. Sci. **334**(1), 3–33 (2005)
15. Klimann, I.: Automaton semigroups: the two-state case. Theory Comput. Syst. **58**(4), 664–680 (2016)
16. Šunić, Z., Ventura, E.: The conjugacy problem in automaton groups is not solvable. J. Algebra **364**, 148–154 (2012)

Towards Computable Analysis
on the Generalised Real Line

Lorenzo Galeotti[1] and Hugo Nobrega[2(✉)]

[1] Fachbereich Mathematik, Universität Hamburg,
Bundesstraße 55, 20146 Hamburg, Germany
lorenzo.galeotti@gmail.com
[2] Institute for Logic, Language and Computation, Universiteit van Amsterdam,
Postbus 94242, 1090 Amsterdam, GE, The Netherlands
h.nobrega@uva.nl

Abstract. In this paper we use infinitary Turing machines with tapes of length κ and which run for time κ as presented, e.g., by Koepke & Seyfferth, to generalise the notion of type two computability to 2^κ, where κ is an uncountable cardinal with $\kappa^{<\kappa} = \kappa$. Then we start the study of the computational properties of $\mathbb{R}_\kappa$, a real closed field extension of $\mathbb{R}$ of cardinality 2^κ, defined by the first author using surreal numbers and proposed as the candidate for generalising real analysis. In particular we introduce representations of $\mathbb{R}_\kappa$ under which the field operations are computable. Finally we show that this framework is suitable for generalising the classical Weihrauch hierarchy. In particular we start the study of the computational strength of the generalised version of the Intermediate Value Theorem.

1 Introduction

The classical approach of computability theory is to define a notion of computability over ω and then extend that notion to any countable space via coding. A similar approach is taken in computable analysis, where one usually defines a notion of computability over Cantor space 2^ω or Baire space ω^ω by using the so-called type two Turing machines (T2TMs), and then extends that notion to spaces of cardinality at most the continuum via representations. Intuitively a T2TM is a Turing machine in which a successful computation is one that runs forever (i.e., for ω steps). Using these machines one can compute functions over 2^ω, by stipulating that a function $f : 2^\omega \to 2^\omega$ is computable if there is a T2TM which, when given $p \in \text{dom}(f)$ as input, writes $f(p)$ on the output tape in the long run. As an example, it is a classical result of computable analysis that, given the right representation of $\mathbb{R}$, the field operations are computable. For an introduction to computable analysis we refer the reader to [17].

Another classical application of T2TMs is the Weihrauch theory of reducibility (see, e.g., [2] for an introduction). The main aim of this theory is the study of the computational content of theorems of real analysis. Since many of these theorems are of the form $\forall x \in \mathcal{X} \exists y \in \mathcal{Y}\, \varphi(x, y)$, with $\varphi(x, y)$ a quantifier free

© Springer International Publishing AG 2017
J. Kari et al. (Eds.): CiE 2017, LNCS 10307, pp. 246–257, 2017.
DOI: 10.1007/978-3-319-58741-7_24

formula, they can be thought of as their own Skolem functions. Given representations of $\mathcal{X}$ and $\mathcal{Y}$, Weihrauch reducibility provides a tool for comparing the computational strength of such functions, and therefore of the theorems themselves. Using this framework, theorems from real analysis can be arranged in a complexity hierarchy analogous to the hierarchy of problems one has in classical computability theory.

Recently, the study of the descriptive set theory of the generalised Baire spaces κ^κ and Cantor spaces 2^κ for cardinals $\kappa > \omega$ has been of great interest to set theorists. In [11] the second author provided the foundational basis for the study of *generalised computable analysis*, namely the generalisation of computable analysis to generalised Baire and Cantor spaces. In particular, in [11] the second author introduced $\mathbb{R}_\kappa$, a generalised version of the real line, and proved a version of the intermediate value theorem (IVT) for that space.

This paper is a continuation of [10,11], strengthening their results and answering in the positive the open question from [11] of whether a natural notion of computability exists for 2^κ. We generalise the framework of type two computability to uncountable cardinals κ such that $\kappa^{<\kappa} = \kappa$. Then we use this framework to induce a notion of computability over the generalised real line $\mathbb{R}_\kappa$, showing that, as in the classical case, by using suitable representations, the field operations are computable. Finally we will generalise Weihrauch reducibility to spaces of cardinality 2^κ and extend a classical result by showing that the generalised version of the IVT introduced in [11] is Weihrauch equivalent to a generalised version of the boundedness principle B_I.

Throughout this paper κ will be a fixed uncountable cardinal, as usual assumed to satisfy $\kappa^{<\kappa} = \kappa$, which in particular implies that κ is a regular cardinal. The generalised Baire and Cantor spaces are equipped with their bounded topologies, i.e., the ones generated by the sets of the form $\{x \in \lambda^\kappa \; ; \; \sigma \subset x\}$ for $\sigma \in \lambda^{<\kappa}$ and $\lambda = 2$ or $\lambda = \kappa$, respectively.

2   The Surreal Numbers

The following definition as well as most of the results in this section are due to Conway [6] and have also been deeply studied by Gonshor in [12].

A *surreal number* is a function from an ordinal α to $\{+, -\}$, i.e., a sequence of pluses and minuses of ordinal length. We denote the class of surreal numbers by No, and the set of surreal numbers of length strictly less than α by $\mathrm{No}_{<\alpha}$. The *length* of a surreal number x, denoted $\ell(x)$, is its domain. For surreal numbers x and y, we define $x < y$ if there exists α such that $x(\beta) = y(\beta)$ for all $\beta < \alpha$, and (i) $x(\alpha) = -$ and either $\alpha = \ell(y)$ or $y(\alpha) = +$, or (ii) $\alpha = \ell(x)$ and $y(\alpha) = +$.

In Conway's original idea, every surreal number is generated by filling some gap between shorter numbers. The following theorem connects this intuition to the surreal numbers as we have defined them. First, given sets of surreal numbers X and Y, we write $X < Y$ if for all $x \in X$ and $y \in Y$ we have $x < y$.

Theorem 1 (Simplicity theorem). *If L and R are two sets of surreal numbers such that $L < R$, then there is a unique surreal x of minimal length such that*

$L < \{x\} < R$, denoted by $[L \mid R]$. Furthermore, for every $x \in$ No we have
$x = [L \mid R]$ for $L = \{y \in$ No $; \ x > y \wedge y \subset x\}$ and $R = \{y \in$ No $; \ x < y \wedge y \subset x\}$.
The pair $\langle L, R \rangle$ is called the canonical cut of x.

Using the simplicity theorem Conway defined the field operations $+_s$, $\cdot_s$, $-_s$,
and the multiplicative inverse over No and proved that these operations satisfy
the axioms of real closed fields. These operations satisfy the following, where
for any operation $*$, surreal z, and sets X, Y of surreals we use the notations
$z * X := \{z * x \ ; \ x \in X\}$ and $X * Y := \{x * y \ ; \ x \in X$ and $y \in Y\}$.

Theorem 2. Let $x = [L_x \mid R_x]$, $y = [L_y \mid R_y]$ be surreal numbers. We have

$$x +_s y = [L_x +_s y, x +_s L_y \mid R_x +_s y, x +_s R_y]$$
$$-_s x = [-_s R_x \mid -_s L_x] = [\{-_s x_R \ ; \ x_R \in R_x\} \mid \{-_s x_L \ ; \ x_L \in L_x\}]$$
$$x \cdot_s y = [L_x \cdot_s y +_s x \cdot_s L_y -_s L_x \cdot_s L_y, R_x \cdot_s y +_s x \cdot_s R_y -_s R_x \cdot_s R_y$$
$$\mid L_x \cdot_s y +_s x \cdot_s R_y -_s L_x \cdot_s R_y, R_x \cdot_s y +_s x \cdot_s L_y -_s R_x \cdot_s L_y]$$

Now let $z = [L_z \mid R_z]$ be a positive surreal number. Let $r_{\langle\rangle} := 0$ and recursively
for every $z_0, \ldots, z_n \in (L_z \cup R_z) \setminus \{0\}$ let $r_{\langle z_0, \ldots, z_n \rangle}$ be the solution for x of the
equation $(z -_s z_n) \cdot_s r_{\langle z_0, \ldots, z_{n-1} \rangle} +_s z_n \cdot_s x = 1$. Then we have $\frac{1}{z} = [L' \mid R']$,
where $L' = \{r_{\langle z_0, \ldots, z_n \rangle} \ ; \ n \in \mathbb{N}$ and $z_i \in L_z$ for even-many $i \leq n\}$ and $R' = \{r_{\langle z_0, \ldots, z_n \rangle} \ ; \ n \in \mathbb{N}$ and $z_i \in L_z$ for odd-many $i \leq n\}$.

On ordinals, the operations $+_s$ and $\cdot_s$ are the so-called *natural* or *Hessenberg*
operations. In particular, for any ordinal α and natural number n, we have
$\alpha +_s n = \alpha + n$.

3   The Generalised Real Line

A crucial property of the real line is its Dedekind completeness, forming the
cornerstone of many theorems in real analysis. However, it is a classical theorem
that there are no real closed proper field extensions of $\mathbb{R}$ which are Dedekind
complete (see, e.g., [5, Theorem 8.7.3]). We therefore need to replace Dedekind
completeness with a weaker property. This was done in [10, 11], and we repeat
the central definitions here.

Let X be an ordered set and κ be a cardinal. We say that X is an η_κ-*set* if
whenever $L, R \subseteq X$ are such that $L < R$ and $|L \cup R| < \kappa$, there is $x \in X$ such
that $L < \{x\} < R$. Let K be an ordered field. We call $\langle L, R \rangle$ a *cut* over K if
$L, R \subseteq K$ and $L < R$. Moreover we say that $\langle L, R \rangle$ is a *Veronese cut* if it is a cut
and L has no maximum, R has no minimum and for each $\varepsilon \in K^+$ there are $\ell \in L$
and $r \in R$ such that $r < \ell + \varepsilon$. We say that K is *Veronese complete* if for each
Veronese cut $\langle L, R \rangle$ there is $x \in K$ such that $L < \{x\} < R$. Note that Veronese
completeness is a reformulation of Cauchy completeness in terms of cuts (see,
e.g., [8,9]), so we can define the Cauchy completion of No$_{<\kappa}$ as follows.

Definition 3. $\mathbb{R}_\kappa = $ No$_{<\kappa} \cup \{[L \mid R] \ ; \ \langle L, R \rangle$ *is a Veronese cut over* No$_{<\kappa}\}$.

Theorem 4 (Galeotti [11]). *The field* $\mathbb{R}_\kappa$ *is the unique Cauchy-complete real closed field extension of* $\mathbb{R}$ *which is an* η_κ-*set of cardinality* 2^κ, *degree* κ, *and in which* $\mathrm{No}_{<\kappa}$ *can be densely embedded.*

In view of the previous theorem from now on we will call $\mathrm{No}_{<\kappa}$ the κ-*rational numbers* and we use the symbol $\mathbb{Q}_\kappa$ instead of $\mathrm{No}_{<\kappa}$.

The field $\mathbb{R}_\kappa$ is a suitable setting for generalising results from classical analysis. For example, a generalised version of the intermediate value theorem [11], a generalised version of the extreme value theorem [10], and recently a generalised version of the Bolzano-Weierstraß theorem (for κ weakly compact) [4] have been proved to hold for $\mathbb{R}_\kappa$. In this section we briefly recall some of the definitions from [11] which will be needed in the last part of this paper.

A κ-*topology* over a set X is a collection of subsets τ of X satisfying: $\emptyset, X \in \tau$; for any $\alpha < \kappa$, if $\{A_i\}_{i \in \alpha}$ is a collection of sets in τ then $\bigcup_{i<\alpha} A_i \in \tau$; and for all $A, B \in \tau$, we have $A \cap B \in \tau$. With κ-topologies one can define direct analogues of many topological notions. We refer to these with the prefix "κ-"; thus we have κ-open sets, κ-continuous functions, κ-topologies generated by families of subsets of a set, etc. Note that, unlike the classical case of the interval topology over $\mathbb{R}$, the interval κ-topologies over $\mathbb{R}_\kappa$ in which the intervals have endpoints in $\mathbb{R}_\kappa \cup \{-\infty, +\infty\}$ or in $\mathbb{Q}_\kappa \cup \{-\infty, +\infty\}$ are different in general. In what follows we will only consider the generalised real line $\mathbb{R}_\kappa$ equipped with the former.

Theorem 5 (IVT$_\kappa$ [11]). *Let* $a, b \in \mathbb{R}_\kappa$ *and* $f : [0, 1] \to \mathbb{R}_\kappa$ *be a* κ-*continuous function. Then for every* $r \in [f(0), f(1)]$ *there exists* $c \in [0, 1]$ *such that* $f(c) = r$.

4   Generalised Type Two Turing Machines

In this section we define a generalised version of type two Turing machines (T2TMs). We will only sketch the definition of κ-Turing machines, which were developed by several people (e.g., [7,14,16]); we are going to follow the definition of Koepke and Seyfferth [14, Sect. 2].

A κ-*Turing machine* has the following tapes of length κ: finitely many read-only tapes for the input, finitely many read and write scratch tapes and one write-only tape for the output. Each cell of each tape has either 0 or 1 written in it at any given time, with the default value being 0. These machines can run for infinite time of ordinal type κ; at successor stages of a computation a κ-Turing machine behaves exactly like a classical Turing Machine, while at limit stages the contents of each cell of each tape and the positions of the heads is computed using inferior limits.

As in the classical case $\kappa = \omega$, the difference between κ-Turing machines and type 2 κ-Turing machines is *not* on the machinery level, but rather on the notion of what it means for a machine to compute a function. A partial function $f : 2^{<\kappa} \to 2^{<\kappa}$ is *computed* by a κ-Turing machine M if whenever M is given $x \in \mathrm{dom}(f)$ as input, its computation halts after fewer than κ steps with $f(x)$ written on the output tape. A partial function $f : 2^\kappa \to 2^\kappa$ is *type two-computed* by a κ-Turing machine M, or *computed by the type 2 κ-Turing machine M*, or simply *computed* by M, if whenever M is given $x \in \text{dom}(f)$ as input, for every $\alpha < \kappa$ there exists a stage $\beta < \kappa$ of the computation at which $f(x) \restriction \alpha$ is written on the output tape. We abbreviate *type 2 κ-Turing machine* by $T2\kappa TM$. An *oracle $T2\kappa TM$* is a T2κTM with an additional read-only input tape of length κ, called its *oracle tape*. A partial function $f : 2^\kappa \to 2^\kappa$ is *computable with an oracle* if there exists an oracle T2κTM M and $x \in 2^\kappa$ such that M computes f when x is written on the oracle tape. Note that by minor modifications of classical proofs one can prove that T2κTMs are closed under recursion and composition, and that there is a universal T2κTM. In what follows, the term *computable* will mean *computable by a T2κTM*, unless specified otherwise.

Theorem 6. *A partial function $f : 2^\kappa \to 2^\kappa$ is continuous iff it is computable with some oracle.*

5 Represented Spaces

In this section we generalise the classical definitions of the theory of represented spaces to 2^κ (see, e.g., [15,17] for the classical case).

A *represented space* $\mathbf{X}$ is a pair (X, δ_X) where X is a set and $\delta_X : 2^\kappa \to X$ is a partial surjective function. As usual a multi-valued function between represented spaces is a multi-valued function between the underlying sets. Let $f : \mathbf{X} \rightrightarrows \mathbf{Y}$ be a partial multi-valued function between represented spaces. We call $F : 2^\kappa \to 2^\kappa$ a *realizer* of f, in symbols $F \vdash f$, if for every $x \in \text{dom}(\delta_X)$ we have that $\delta_Y(F(x)) \in f(\delta_X(x))$. Given a class Γ of functions between 2^κ and 2^κ, we say f is (δ_X, δ_Y)-Γ, or δ_X-Γ in case $\delta_X = \delta_Y$, if f has a realizer in Γ. For example, a function $f : X \to Y$ is (δ_X, δ_Y)-computable if it has a computable realizer.

Let f and g be two multi-valued functions between represented spaces. Then we say that f is *strongly topologically-Weihrauch reducible* to g, in symbols $f \leq^t_W g$, if there are two continuous functions $H, K : 2^\kappa \to 2^\kappa$ such that $H \circ G \circ K \vdash f$ whenever $G \vdash g$. If the functions H, K above can be taken to be computable, then we say f is *strongly Weihrauch reducible* to g, in symbols $f \leq_W g$.[1] As usual, if $f \leq^t_W g$ and $g \leq^t_W f$ then we say that f is *strongly topologically-Weihrauch equivalent* to g and write $f \equiv^t_W g$. The relation $\equiv_W$ is defined analogously.

Let $\delta : 2^\kappa \to X$ and $\delta' : 2^\kappa \to X$ be two representations of a space X. Then we say that δ *continuously reduces* to δ', in symbols $\delta \leq_t \delta'$, if there is a continuous function $h : 2^\kappa \to 2^\kappa$ such that for every $p \in \text{dom}(\delta)$ we have $\delta(p) = \delta'(h(p))$. Similarly we say δ *computably reduces* to δ', in symbols $\delta \leq \delta'$, if h above can be taken computable. If $\delta \leq_t \delta'$ and $\delta' \leq_t \delta$ we say that δ and δ' are continuously equivalent and write $\delta \equiv_t \delta'$, and similarly for the computable case. Note that as in classical computable analysis if $\delta \leq \delta'$ and f

---

[1] Carl has also introduced a notion of generalized (strong) Weihrauch reducibility in [3]. Because his goal is to investigate multi-valued (class) functions on V, the space of codes he uses is the class of ordinal numbers, considered with the ordinal Turing machines of Koepke [13]. Therefore his approach is significantly different from ours, and we do not know of any connections between the two.

is δ-computable then f is also δ'-computable. Finally, as in the classical case, given two represented spaces $\mathbf{X}$ and $\mathbf{Y}$, we can define canonical representations for the product space $\mathbf{X} \times \mathbf{Y}$, the union space $\mathbf{X} + \mathbf{Y}$ and the space of continuous functions $[\mathbf{X} \to \mathbf{Y}]$. In particular, as in classical computable analysis $[\mathbf{X} \to \mathbf{Y}]$ can be represented as follows: $\delta_{[X \to Y]}(p) = f$ iff $p = 0^n 1 p'$ with $p' \in 2^\kappa$ and $n \in \mathbb{N}$ is a code for an oracle T2κTM which (δ_X, δ_Y)-computes f when given the oracle p'.

Recall that the following relation is a well-ordering of the class of pairs of ordinal numbers: $\langle \alpha_0, \beta_0 \rangle \prec \langle \alpha_1, \beta_1 \rangle$ iff $\langle \max(\alpha_0, \beta_0), \alpha_0, \beta_0 \rangle$ is lexicographically-less than $\langle \max(\alpha_1, \beta_1), \alpha_1, \beta_1 \rangle$. The *Gödel pairing function* is given by $\mathfrak{g}(\alpha, \beta) = \gamma$ iff $\langle \alpha, \beta \rangle$ is the γth element in $\prec$. Given sequences $\langle w_\alpha \rangle_{\alpha < \kappa}$ and $\langle p_\alpha \rangle_{\alpha < \beta}$ of elements in $2^{<\kappa}$ and 2^κ, respectively, we define elements $q := [w_\alpha]_{\alpha < \kappa}$ and $p := (p_\alpha)_{\alpha < \kappa}$ in 2^κ by letting q be the concatenation of the w_α and $p(\mathfrak{g}(\alpha, \beta)) = p_\alpha(\beta)$.

We fix the following representations of κ and κ^κ: $\delta_\kappa(p) = \alpha$ iff $p = 0^\alpha 10$, where $\mathbf{0}$ is the constant 0 κ-sequence, $\delta_{\kappa^\kappa}(p) = x$ iff $p = [0^{\alpha_\beta + 1} 1]_{\beta < \kappa}$ and $x = \langle \alpha_\beta \rangle_{\beta < \kappa}$. It is straightforward to see that a function $f : \kappa \to \kappa$ is δ_κ-computable iff it is computable by a κ-machine as in [14, Definition 2].

Lemma 7. *The restriction of $\mathfrak{g}$ to $\kappa \times \kappa$ is a δ_κ-computable bijection between $\kappa \times \kappa$ and κ, and has a δ_κ-computable inverse.*

Proposition 8. *δ_{κ^κ} is $\leq$-maximal among the continuous representations of κ^κ.*

6   Representing $\mathbb{R}_\kappa$

In classical computable analysis one can show that many of the natural representations of $\mathbb{R}$ are well behaved with respect to type two computability. In this section we show that some of these results naturally extend to the uncountable case. First we introduce representations for generalised rational numbers, which will serve as a starting point to representing $\mathbb{R}_\kappa$. As we have seen in the introduction, surreal numbers can be expressed as binary sequences and, because of the simplicity theorem, as cuts. It is then natural to introduce two representations which reflect this fact. Let $p \in 2^\kappa$ and $q \in \mathbb{Q}_\kappa$. We define $\delta_{\mathbb{Q}_\kappa}(p) = q$ iff $p = [w_\alpha]_{\alpha < \kappa}$ where $w_\alpha := 00$ if $\alpha \in \operatorname{dom}(q)$ and $q(\alpha) = -$, $w_\alpha := 01$ if $\alpha \notin \operatorname{dom}(q)$ and finally $w_\alpha := 11$ if $\alpha \in \operatorname{dom}(q)$ and $q(\alpha) = +$. It is not hard to see that since every rational is a sequence of $+$ and $-$ of length less than κ the function $\delta_{\mathbb{Q}_\kappa}$ is indeed a representation of $\mathbb{Q}_\kappa$. Now we define a representation based on cuts by recursion on the simplicity structure of the surreal numbers. We define $\delta^0_{\mathrm{Cut}_{\mathbb{Q}_\kappa}}(p) = 0$ iff $p = (p_\alpha)_{\alpha < \kappa}$ and $p_\alpha = [10]_{\beta < \kappa}$ for every $\alpha < \kappa$. For $\alpha > 0$ we define $\delta^\alpha_{\mathrm{Cut}_{\mathbb{Q}_\kappa}}(p) = [\,L \mid R\,]$ where $p = (p_\alpha)_{\alpha < \kappa}$ and:

1. $p_\alpha \in \operatorname{dom}(\bigcup_{\gamma < \alpha} \delta^\gamma_{\mathbb{Q}_\kappa}) \cup \{[10]_{\beta < \kappa}\}$ for every $\alpha < \kappa$,
2. for all even[2] $\alpha < \kappa$, if $p_\alpha = [10]_{\beta < \kappa}$ then for all even $\beta > \alpha$ we have $p_\beta = [10]_{\beta < \kappa}$,

---

[2] We call an ordinal α *even* if $\alpha = \lambda + 2n$ for some limit λ and natural n, *odd* otherwise.

3. for all odd $\alpha < \kappa$, if $p_\alpha = [10]_{\beta < \kappa}$ then for all odd $\beta > \alpha$ we have $p_\beta = [10]_{\beta < \kappa}$,
4. finally: $L = \{\delta^\gamma_{\mathrm{Cut}_{\mathbb{Q}_\kappa}}(p_\beta) \; ; \; \gamma < \alpha, \beta < \kappa$ is even and $p_\beta \in \mathrm{dom}(\delta^\gamma_{\mathrm{Cut}_{\mathbb{Q}_\kappa}})\}$ and
   $R = \{\delta^\gamma_{\mathrm{Cut}_{\mathbb{Q}_\kappa}}(p_\beta) \; ; \; \gamma < \alpha, \beta < \kappa$ is odd and $p_\beta \in \mathrm{dom}(\delta^\gamma_{\mathrm{Cut}_{\mathbb{Q}_\kappa}})\}$.

Then we define $\delta_{\mathrm{Cut}_{\mathbb{Q}_\kappa}} := \bigcup_{\gamma < \kappa} \delta^\gamma_{\mathrm{Cut}_{\mathbb{Q}_\kappa}}$.

Note that $\delta_{\mathrm{Cut}_{\mathbb{Q}_\kappa}}$ is surjective, since for every $x \in \mathbb{Q}_\kappa$ there exists $p \in \mathrm{dom}(\delta_{\mathrm{Cut}_{\mathbb{Q}_\kappa}})$ such that $\delta_{\mathrm{Cut}_{\mathbb{Q}_\kappa}}(p)$ is the canonical cut for x. Therefore $\delta_{\mathrm{Cut}_{\mathbb{Q}_\kappa}}$ is indeed a representation of $\mathbb{Q}_\kappa$.

Lemma 9. $\delta_{\mathbb{Q}_\kappa} \equiv \delta_{\mathrm{Cut}_{\mathbb{Q}_\kappa}}$.

Proof. First we show that $\delta_{\mathbb{Q}_\kappa} \leq \delta_{\mathrm{Cut}_{\mathbb{Q}_\kappa}}$. Let $p \in \mathrm{dom}(\delta_{\mathbb{Q}_\kappa})$. The conversion can be done recursively. If p is a code for the empty sequence[3] we just return a representation for $[\emptyset \mid \emptyset]$. Otherwise we compute two subsets $L_s := \{p'01 \; ; \; p'11 \sqsubset p\}$ and $R_s := \{p'01 \; ; \; p'00 \sqsubset p\}$. Then we compute recursively the cuts for the elements of L_s and R_s and return them respectively as the left and right sets of the cut representation of p. It easy to see that the algorithm computes a code for the canonical cut of $\delta_{\mathbb{Q}_\kappa}(p)$.

Now we will show that $\delta_{\mathrm{Cut}_{\mathbb{Q}_\kappa}} \leq \delta_{\mathbb{Q}_\kappa}$. Let $p \in \mathrm{dom}(\delta_{\mathrm{Cut}_{\mathbb{Q}_\kappa}})$. If p is a code for the cut $[\emptyset \mid \emptyset]$ we return a representation of the empty sequence. If p is the code for the cut $[L \mid R] \neq [\emptyset \mid \emptyset]$. We first recursively compute the sequences for the element of L and R, call the sets of these sequences L_s and R_s. Now suppose $\alpha < \kappa$ is even and we want to compute the value at α and $\alpha + 1$ of the output sequence. We first compute M_L and m_R respectively the minimal and maximal in $\{00, 01, 11\}$ such that for every $p' \in L_s$ and $p'' \in R_s$ we have $p'(\alpha)p'(\alpha + 1) \leq M_L$ and $m_R \leq p''(\alpha)p''(\alpha + 1)$. Then by a case distinction on M_L and m_R we can decide the ith sign of the output. For example if the output is already smaller than R_s, $M_L = 00$ (i.e. $-$) and $m_R = 00$ (i.e. $-$) then we can output the sequence 01 (i.e. undefined). All the other combinations can be treated similarly.

Lemma 10. *The operations* $+_s$, $-_s$, $\cdot_s$, $\frac{1}{x}$ *and the order* $<$ *are* $\delta_{\mathrm{Cut}_{\mathbb{Q}_\kappa}}$-*computable.*

Proof. We will only prove the lemma for $+_s$. Given $q, q' \in \mathbb{Q}_\kappa$ we want to $\delta_{\mathrm{Cut}_{\mathbb{Q}_\kappa}}$-compute $q +_s q'$. The algorithm is given by recursion. If $q = 0$ (similarly for $q' = 0$)[4] copy the code of q' on the output tape. If neither q nor q' are 0 then by using Theorem 2 we compute a representation for $q +_s q'$ (note that this involves the computation of less than κ many rational sums of shorter length). Finally, since the resulting code would not in general be in $\mathrm{dom}(\delta_{\mathrm{Cut}_{\mathbb{Q}_\kappa}})$, we use the algorithms of the previous lemma to convert $q +_s q'$ to a sign sequence code and than we convert it back to an element in $\mathrm{dom}(\delta_{\mathrm{Cut}_{\mathbb{Q}_\kappa}})$. By using the second

---

[3] Note that this can be checked just by looking at the first two bits of p.
[4] Note that this is easily computable, it is in fact enough to check that L and R are empty, and this can be done just by checking the first two bits of the first sequence in the left and in the first sequence on the right.

algorithm from the previous proof we can convert every element in $L_{q+_s q'}$ and in $R_{q+_s q'}$ into a sequence (note that by induction the codes of these cuts are in $\mathrm{dom}(\delta_{\mathrm{Cut}_{\mathbb{Q}_\kappa}})$ so we can use the algorithm). Then by the same method used in the previous lemma, we can compute the code of the sequence representation for $q +_s q'$. Once we have the code of the sequence representation for $q +_s q'$ we can convert it to a code of the cut representation by using the first algorithm from the previous lemma.

Given that $\mathbb{R}_\kappa$ is the Cauchy completion of $\mathbb{Q}_\kappa$, the following is a natural representation of $\mathbb{R}_\kappa$. We let $\delta_{\mathbb{R}_\kappa}(p) = x$ iff $p = (p_\alpha)_{\alpha<\kappa}$, where for each $\alpha < \kappa$ we have $p_\alpha \in \mathrm{dom}(\delta_{\mathbb{Q}_\kappa})$, $\delta_{\mathbb{Q}_\kappa}(p_\alpha) < x +_s \frac{1}{\alpha+1}$, and $x < \delta_{\mathbb{Q}_\kappa}(p_\alpha) +_s \frac{1}{\alpha+1}$. It is routine to check the following.

Theorem 11. *The field operations $+_s$, $-_s$, $\cdot_s$, and $\frac{1}{x}$ are $\delta_{\mathbb{R}_\kappa}$-computable.*

Proof. Let us do the proof for $\cdot_s$, the others being similar. Given codes $p = (p_\alpha)_{\alpha<\kappa}$ and $q = (q_\alpha)_{\alpha<\kappa}$ for $x,y \in \mathbb{R}_\kappa$ respectively, let $x_\alpha = \delta_{\mathbb{Q}_\kappa}(p_\alpha)$ and $y_\alpha = \delta_{\mathbb{Q}_\kappa}(q_\alpha)$. Note that for each α we can compute some α' such that $\frac{1}{\alpha'+1}(x_0 +_s y_0 +_s 3) \leq \frac{1}{\alpha+1}$. We then output $r = (r_\alpha)_{\alpha<\kappa}$, where r_α is a $\delta_{\mathbb{Q}_\kappa}$-name for $x_{\alpha'} y_{\alpha'}$.

We have $xy -_s x_{\alpha'} y_{\alpha'} = x(y -_s y_{\alpha'}) +_s y_{\alpha'}(x -_s x_{\alpha'}) < (x_0 +_s 1)\frac{1}{\alpha'+1} +_s (y_0 +_s 2)\frac{1}{\alpha'+1} \leq \frac{1}{\alpha+1}$, as desired, and likewise we can prove $x_{\alpha'} y_{\alpha'} -_s xy < \frac{1}{\alpha+1}$.

On the other hand, the following is suggested by the definition of $\mathbb{R}_\kappa$ as the collection of Veronese cuts over $\mathbb{Q}_\kappa$. We let $\delta^V_{\mathbb{R}_\kappa}(p) = x$ iff $p = (p_\alpha)_{\alpha<\kappa}$, where for each $\alpha < \kappa$ we have $p_\alpha \in \mathrm{dom}(\delta_{\mathbb{Q}_\kappa})$ and $x = [L \mid R]$, with $L = \{\delta_{\mathbb{Q}_\kappa}(p_\alpha) \; ; \; \alpha < \kappa \text{ is even}\}$; $R = \{\delta_{\mathbb{Q}_\kappa}(p_\alpha) \; ; \; \alpha < \kappa \text{ is odd}\}$; and for each even $\alpha < \kappa$ we have $\delta_{\mathbb{Q}_\kappa}(p_{\alpha+1}) < \delta_{\mathbb{Q}_\kappa}(p_\alpha) +_s \frac{1}{\alpha+1}$.

Theorem 12. $\delta_{\mathbb{R}_\kappa} \equiv \delta^V_{\mathbb{R}_\kappa}$.

Proof. To reduce $\delta^V_{\mathbb{R}_\kappa}$ to $\delta_{\mathbb{R}_\kappa}$, given $p = (p_\alpha)_{\alpha<\kappa}$, we output $q = (q_\alpha)_{\alpha<\kappa}$ by making q_α equal to p_β, where β is the αth even ordinal. It is now easy to see that q is a $\delta_{\mathbb{R}_\kappa}$-name for $\delta^V_{\mathbb{R}_\kappa}(p)$.

For the converse reduction, given $p = (p_\alpha)_{\alpha<\kappa}$, we output $q = (q_\alpha)_{\alpha<\kappa}$ where for each even α we let q_α be a $\delta_{\mathbb{Q}_\kappa}$-name for $\delta_{\mathbb{Q}_\kappa}(p_{2\cdot_s \alpha+2}) -_s \frac{1}{2\cdot_s \alpha+3}$ and $q_{\alpha+1}$ be a $\delta_{\mathbb{Q}_\kappa}$-name for $\delta_{\mathbb{Q}_\kappa}(p_{2\cdot_s \alpha+2}) +_s \frac{1}{2\cdot_s \alpha+3}$. Then letting $L := \{\delta_{\mathbb{Q}_\kappa}(p_\alpha) \; ; \; \alpha < \kappa \text{ is even}\}$ and $R := \{\delta_{\mathbb{Q}_\kappa}(p_\alpha) \; ; \; \alpha < \kappa \text{ is odd}\}$ we have $L < \{x\} < R$ and for each even $\alpha < \kappa$ we have $\delta_{\mathbb{Q}_\kappa}(q_{\alpha+1}) = \delta_{\mathbb{Q}_\kappa}(p_{2\cdot_s \alpha+2}) +_s \frac{1}{2\cdot_s \alpha+3} = \delta_{\mathbb{Q}_\kappa}(q_\alpha) +_s \frac{2}{2\cdot_s \alpha+3} < \delta_{\mathbb{Q}_\kappa}(q_\alpha) +_s \frac{1}{\alpha+1}$, as desired.

7   Generalised Boundedness Principles and the IVT

As shown in, e.g., [1,2], the so-called *boundedness principles* and *choice principles* are important building blocks in characterizing the Weihrauch degrees of interest in computable analysis. In this section we focus on the study of IVT and its relationship with the boundedness principle B_I. In particular we generalise a classical result from Brattka and Gherardi [2], proving that IVT_κ is Weihrauch equivalent to a generalised version of B_I. This strengthens a result from [11], namely that B_I is continuously reducible to IVT_κ.

The theorem IVT_κ as stated in Theorem 5 can be considered as the partial multi-valued function $IVT_\kappa : C_{[0,1]} \rightrightarrows [0,1]$ defined as follows: $IVT_\kappa(f) = \{c \in [0,1] ; f(c) = 0\}$, where $[0,1]$ is represented by $\delta_{\mathbb{R}_\kappa} \restriction [0,1]$ and $C_{[0,1]}$ is endowed with the standard representation of $[[0,1] \to \mathbb{R}_\kappa]$ restricted to $C_{[0,1]}$. By lifting the classical proof to κ it is easy to show that this version of IVT_κ is not continuous, and thus also not computable, relative to these representations.

To introduce the boundedness principle B_I^κ, we will need the following represented spaces. Let $\mathbf{S}_b^\uparrow$ be the space of bounded increasing sequences of κ-rationals, represented by letting p be a name for $\langle x_\alpha \rangle_{\alpha < \kappa}$ iff $p = (p_\alpha)_{\alpha < \kappa}$ where $p_\alpha \in \mathrm{dom}_{\mathbb{Q}_\kappa}$ and $\delta_{\mathbb{Q}_\kappa}(p_\alpha) = x_\alpha$ for each $\alpha < \kappa$. The represented space $\mathbf{S}_b^\downarrow$ is defined analogously, with bounded decreasing sequences of κ-rationals. Note that, unlike the classical case of the real line, not all limits of bounded monotone sequences of length κ exist in $\mathbb{R}_\kappa$. Therefore, although for the real line the spaces $\mathbf{S}_b^\uparrow$ and $\mathbf{S}_b^\downarrow$ naturally correspond to the spaces of *lower reals* $\mathbb{R}_<$ and *upper reals* $\mathbb{R}_>$, respectively, in our generalised setting the correspondence fails. We define B_I^κ as the principle which, given an increasing sequence $\langle q_\alpha \rangle_{\alpha < \kappa}$ and decreasing sequence $\langle q_\alpha' \rangle_{\alpha < \kappa}$ in $\mathbb{Q}_\kappa$ for which there exists $x \in \mathbb{R}_\kappa$ such that $\{q_\alpha ; \alpha < \kappa\} \le \{x\} \le \{q_\alpha' ; \alpha < \kappa\}$, picks one such x. Formally we have the partial multi-valued function $B_I^\kappa : \mathbf{S}_b^\uparrow \times \mathbf{S}_b^\downarrow \rightrightarrows \mathbb{R}_\kappa$ with $x \in B_I^\kappa(s, s')$ iff $\{s(\alpha) ; \alpha < \kappa\} \le \{x\} \le \{s'(\alpha) ; \alpha < \kappa\}$.

Lemma 13. *Let $f : [0,1] \to \mathbb{R}_\kappa$ and $x \in \mathbb{R}_\kappa$. Suppose there exists a sequence $\langle x_\alpha \rangle_{\alpha < \kappa}$ of pairwise distinct elements of $[0,1]$ such that $f(x_\alpha) = x$ if $\alpha < \kappa$ is even and $f(x_\alpha) \ne x$ otherwise, and such that for any odd $\alpha, \beta < \kappa$ there exists an even $\gamma < \kappa$ such that x_γ is between x_α and x_β. Then f is not κ-continuous.*

Proof. If such a sequence exists, then either the preimage of the κ-open set $(x, +\infty)$ or of the κ-open set $(-\infty, x)$ under f must contain x_α for κ-many of the odd $\alpha < \kappa$, and thus cannot be κ-open.

Lemma 14. *Let $f : [0,1] \to \mathbb{R}_\kappa$ be κ-continuous an let $\beta, \beta' < \kappa$, $y \in \mathbb{R}_\kappa$ and let $\langle r_\alpha \rangle_{\alpha < \beta}$ and $\langle r_\alpha' \rangle_{\alpha < \beta'}$ be two sequences in $[0,1]$ such that $\{r_\alpha ; \alpha < \beta\} < \{r_\alpha' ; \alpha < \beta'\}$ and $\{f(r_\alpha) ; \alpha < \beta\} < \{y\} < \{f(r_\alpha') ; \alpha < \beta'\}$. Then there is $x \in [0,1]$ such that $\{r_\alpha ; \alpha < \beta\} < \{x\} < \{r_\alpha' ; \alpha < \beta'\}$ and $f(x) = y$.*

Proof. Assume not. Without loss of generality we can assume that for every x such that $\{r_\alpha ; \alpha < \beta\} < \{x\} < \{r_\alpha' ; \alpha < \beta'\}$ we have $f(x) > y$ (a similar proof works for $f(x) < y$). Note that the set $\{r_\alpha ; \alpha < \beta\}$ has cofinality at most $\beta < \kappa$ and, since $\mathbb{R}_\kappa$ is an η_κ-set, it follows that $R = \{r \in [0,1] ; \forall \alpha < \beta . r_\alpha < r\}$ has coinitiality κ. Therefore R is not κ-open. Now since f is κ-continuous we have that $f^{-1}[(y, +\infty)]$ is κ-open. Therefore $f^{-1}[(y, +\infty)] = \bigcup_{\alpha \in \gamma} (y_\alpha, b_\alpha)$ with $\gamma < \kappa$ and $y_\alpha, b_\alpha \in [0,1]$ for every $\alpha < \gamma$. Now consider the set $I := \{\alpha \in \gamma ; (y_\alpha, b_\alpha) \cap R \ne \emptyset\}$. We have that $R \subset \bigcup_{\alpha \in I} (y_\alpha, b_\alpha)$. Note that since R is not κ-open we have $R \ne \bigcup_{\alpha \in I} (y_\alpha, b_\alpha)$. Now assume $r \in \bigcup_{\alpha \in I} (y_\alpha, b_\alpha) \setminus R$, so that there is $\alpha \in I$ such that $r \in (y_\alpha, b_\alpha)$. Take $r' \in (y_\alpha, b_\alpha) \cap R$. By the fact that $r \notin R$, there is $\alpha' < \beta$ such that $r < r_{\alpha'}$ and by IVT_κ there is a root of f between $r_{\alpha'}$ and r', but this is a contradiction because $(y_\alpha, b_\alpha) \subset f^{-1}[(y, +\infty)]$.

Corollary 15. *Let $f : [0,1] \to \mathbb{R}_\kappa$ be κ-continuous, and let $x \in [0,1]$, $\langle r_\alpha \rangle_{\alpha < \kappa}$ and $\langle r'_\alpha \rangle_{\alpha < \kappa}$ be respectively increasing and decreasing sequences in $[0,1]$ such that for all $\alpha < \kappa$ we have $f(r_\alpha) < x$ and $f(r'_\alpha) > x$. Then there exists $y \in [0,1]$ such that $f(y) = x$ and $\{r_\alpha ; \alpha < \kappa\} < \{y\} < \{r'_\alpha ; \alpha < \kappa\}$.*

Proof. Construct a sequence $\langle x_\alpha \rangle_{\alpha < \gamma}$ for some $\gamma \leq \kappa$ as follows. First let $\delta_0 = 1$. Having constructed $\langle x_\beta \rangle_{\beta < \alpha}$ for some even $\alpha < \kappa$, by Lemma 14 there exists $x_\alpha \in [0,1]$ such that $f(x_\alpha) = x$ and $\{r_\beta ; \beta < \sup_{\nu < \alpha} \delta_\nu\} < \{x_\alpha\} < \{r'_\beta ; \beta < \sup_{\nu < \alpha} \delta_\nu\}$. If $\{r_\beta ; \beta < \kappa\} < \{x_\alpha\} < \{r'_\beta ; \beta < \kappa\}$, then we are done and $\gamma = \alpha$. Otherwise there exists $\beta < \kappa$ such that $r_\beta > x$ or $r'_\beta < x$, so we let $x_{\alpha+1} = r_\beta$ or $x_{\alpha+1} = r'_\beta$ accordingly, and let $\delta_\alpha = \beta + 1$. If the construction goes on for κ steps, then $\langle x_\alpha \rangle_{\alpha < \kappa}$ is as in Lemma 13, a contradiction. Hence the construction ends at some stage $\gamma < \kappa$, and therefore $\{r_\beta ; \beta < \kappa\} < \{x_\gamma\} < \{r'_\beta ; \beta < \kappa\}$.

Theorem 16. *1. If there exists an effective enumeration of a dense subset of $\mathbb{R}_\kappa$, then $\mathrm{IVT}_\kappa \leq_{\mathrm{W}} \mathrm{B}_{\mathrm{I}}^\kappa$.*
2. We have $\mathrm{B}_{\mathrm{I}}^\kappa \leq_{\mathrm{W}} \mathrm{IVT}_\kappa$.
3. We have $\mathrm{IVT}_\kappa \leq_{\mathrm{W}}^{\mathrm{t}} \mathrm{B}_{\mathrm{I}}^\kappa$, and therefore $\mathrm{IVT}_\kappa \equiv_{\mathrm{W}}^{\mathrm{t}} \mathrm{B}_{\mathrm{I}}^\kappa$.

Proof. For item 1, let the κ-continuous function $f : [0,1] \to \mathbb{R}_\kappa$ be given, $\mathbb{D}$ be a dense subset of $\mathbb{R}_\kappa$ and $\langle d_\gamma \rangle_{\gamma < \kappa}$ be an effective enumeration of $[0,1] \cap \mathbb{D}$. Without loss of generality we can assume $f(0) < 0$ and $f(1) > 0$, and start setting $r_0 = 0$ and $r'_0 = 1$. Now assume that for $0 < \alpha < \kappa$ we have already defined an increasing sequence $\langle r_\beta \rangle_{\beta < \alpha}$ and a decreasing sequence $\langle r'_\beta \rangle_{\beta < \alpha}$ of elements of $[0,1] \cap \mathbb{D}$ with $\{r_\beta ; \beta < \alpha\} < \{r'_\beta ; \beta < \alpha\}$ and $\{f(r_\beta) ; \beta < \alpha\} < \{0\} < \{f(r'_\beta) ; \beta < \alpha\}$. By Lemma 14 there is still a root of f between the two sequences. Note that, since $\mathbb{R}_\kappa$ is an η_κ-set and again by applying Lemma 14, there exist $r_L, r_R \in \mathbb{D}$ such that $\{r_\beta ; \beta < \alpha\} < \{r_L\} < \{r_R\} < \{r'_\beta ; \beta < \alpha\}$ and $f(r_L) < 0$, $f(r_R) > 0$. Therefore, by searching in the sequence $\langle d_\gamma \rangle_{\gamma < \kappa}$ and running the corresponding algorithms in parallel, we can find such a pair r_L, r_R in fewer than κ computation steps. Let β, γ, δ be such that $\mathfrak{g}(\beta, \mathfrak{g}(\gamma, \delta)) = \alpha$, where $\mathfrak{g}$ is the Gödel pairing function, which has a computable inverse by Lemma 7. If $r_L < d_\gamma < d_\delta < r_R$, $f(d_\gamma) < 0$, and $f(d_\delta) > 0$, where the last two comparisons are decided in fewer than β steps of computation, then let $r_\alpha = d_\gamma$ and $r'_\alpha = d_\delta$; otherwise let $r_\alpha = r_L$ and $r'_\alpha = r_R$.

By Corollary 15 we have that there exists $x \in [0,1]$ such that $\{r_\alpha ; \alpha < \kappa\} < \{x\} < \{r'_\alpha ; \alpha < \kappa\}$. It remains to be proved that $f(x) = 0$ for any such x. Suppose not, say $f(x) > 0$ for some such x. Then also $f(y) > 0$ for some $y \in \mathbb{D}$ such that $\{r_\alpha ; \alpha < \kappa\} < \{y\} < \{r'_\alpha ; \alpha < \kappa\}$. Now let $\beta, \gamma, \delta < \kappa$ be such that $d_\gamma = y$, $d_\delta = r_\nu$ for some ν such that $\{y -_{\mathrm{s}} r_\nu\} < \{r'_\alpha -_{\mathrm{s}} r_\beta ; \alpha, \beta < \kappa\}$ and $f(y) < 0$, $f(r_\nu) > 0$ are decided in fewer than β computation steps. Then at stage $\alpha = \mathfrak{g}(\beta, \mathfrak{g}(\gamma, \delta))$ of the computation we define a pair r_α, r'_α such that $r'_\alpha -_{\mathrm{s}} r_\alpha \leq y -_{\mathrm{s}} r_\nu$, a contradiction. This ends the proof of 1.

Item 2 can be proved by a straightforward generalisation of the proof of [2, Theorem 6.2], and the proof of item 3 is the same as that of item 1 without the requirement that the enumeration $\langle d_\gamma \rangle_{\gamma < \kappa}$ of the dense subset of $[0, 1] \cap \mathbb{D}$ be effective.

Note that the antecedent of item 1 of Theorem 16 is satisfied, e.g., in the constructible universe $\mathbf{L}$. We leave for future work the task of investigating the set-theoretic properties of that condition more deeply.

Acknowledgments. This research was partially done whilst the authors were visiting fellows at the Isaac Newton Institute for Mathematical Sciences in the programme *Mathematical, Foundational and Computational Aspects of the Higher Infinite*. The research benefited from the Royal Society International Exchange Grant *Infinite games in logic and Weihrauch degrees*. The second author was also supported by the Capes Science Without Borders grant number 9625/13-5. The authors are grateful to Benedikt Löwe and Arno Pauly for the many fruitful discussions and to the Institute for Logic, Language and Computation for the hospitality offered to the first author. Finally, the authors wish to thank the three anonymous referees for the helpful comments which have improved the paper.

References

1. Brattka, V., de Brecht, M., Pauly, A.: Closed choice and a uniform low basis theorem. Ann. Pure Appl. Log. **163**(8), 986–1008 (2012)
2. Brattka, V., Gherardi, G.: Effective choice and boundedness principles in computable analysis. Bull. Symb. Log. **17**(1), 73–117 (2011)
3. Carl, M.: Generalized effective reducibility. In: Beckmann, A., Bienvenu, L., Jonoska, N. (eds.) CiE 2016. LNCS, vol. 9709, pp. 225–233. Springer, Cham (2016). doi:10.1007/978-3-319-40189-8_23
4. Carl, M., Galeotti, L., Löwe, B.: The Bolzano-Weierstraß theorem for the generalised real line. In preparation
5. Cohn, P.M.: Basic Algebra. Groups, Rings and Fields. Springer, Heidelberg (2003)
6. Conway, J.H.: On Numbers and Games. Taylor & Francis, New York (2000)
7. Dawson, B.: Ordinal time Turing computation. Ph.D. thesis, University of Bristol (2009)
8. Dales, H.G., Woodin, W.H.: Super-Real Fields: Totally Ordered Fields with Additional Structure. Clarendon Press, Oxford (1996)
9. Ehrlich, P.: Dedekind cuts of Archimedean complete ordered abelian groups. Algebra Univers. **37**(2), 223–234 (1997)
10. Galeotti, L.: Computable analysis over the generalized Baire space. Master's thesis, Universiteit van Amsterdam (2015)
11. Galeotti, L.: A candidate for the generalised real line. In: Beckmann, A., Bienvenu, L., Jonoska, N. (eds.) CiE 2016. LNCS, vol. 9709, pp. 271–281. Springer, Cham (2016). doi:10.1007/978-3-319-40189-8_28
12. Gonshor, H.: An Introduction to the Theory of Surreal Numbers. Cambridge University Press, Cambridge (1986)
13. Koepke, P.: Turing computations on ordinals. Bull. Symb. Logic **11**(3), 377–397 (2005)

14. Koepke, P., Seyfferth, B.: Ordinal machines and admissible recursion theory. Ann. Pure Appl. Log. **160**(3), 310–318 (2009)

15. Pauly, A.: On the topological aspects of the theory of represented spaces. Preprint, (2015). arXiv:1204.3763v3

16. Rin, B.: The computational strengths of α-tape infinite time Turing machines. Ann. Pure Appl. Log. **165**(9), 1501–1511 (2014)

17. Weihrauch, K.: Computable Analysis: An Introduction. Springer, Heidelberg (2012)

Finite Language Forbidding-Enforcing Systems

Daniela Genova[1]([✉]) and Hendrik Jan Hoogeboom[2]

[1] Department of Mathematics and Statistics, University of North Florida,
Jacksonville, FL 32224, USA
d.genova@unf.edu
[2] LIACS, Leiden University, Leiden, The Netherlands
h.j.hoogeboom@liacs.leidenuniv.nl

Abstract. The forbidding and enforcing paradigm was introduced by
Ehrenfeucht and Rozenberg as a way to define families of languages based
on two sets of boundary conditions. Later, a variant of this paradigm was
considered where an fe-system defines a single language. We investigate
this variant further by studying fe-systems in which both the forbidding
and enforcing sets are finite and show that they define regular languages.
We prove that the class of languages defined by finite fe-systems is strictly
between the strictly locally testable languages and the class of locally
testable languages.

Keywords: Fe-systems · Fe-languages · Regular languages · (Strictly)
locally testable · Natural computing · Biomolecular computing

1 Introduction

Many novel models of computation have been proposed in the field of natural computing, such as, splicing, membrane, and reaction systems [7,14,15,17].
These models were inspired by the information-processing capabilities of molecular interactions, the functioning of the living cell, cell-to-cell communication,
etc. The forbidding-enforcing paradigm, proposed by Ehrenfeucht and Rozenberg [4–6] was motivated by the non-determinism of molecular reactions. There,
forbidding-enforcing systems (fe-systems) were used to define classes of languages, fe-families, based on two sets of constraints. The idea behind using
constraints rather than rules to define classes of languages was that "everything that is not forbidden is allowed", which is orthogonal to the determinism
of grammars and automata where "everything that is not allowed is forbidden".

Variants of fe-systems have been considered in the study of membrane systems [2] and to describe self-assembly of graphs [8]. They, also, have been formulated to define classes of graphs [13] and abstracted to categories in [12].
In [9], a variant of fe-systems was introduced, in which one fe-system defines a
single language (fe-language) and in [11] it was shown how fe-languages characterize DNA codes (conditions on words abstracted from unwanted hybridization
of DNA strands). Normal forms for forbidding sets of language fe-systems were
studied in [10].

© Springer International Publishing AG 2017
J. Kari et al. (Eds.): CiE 2017, LNCS 10307, pp. 258–269, 2017.
DOI: 10.1007/978-3-319-58741-7_25

Forbidding sets of words as subwords has been previously discussed in the literature, although, before forbidding-enforcing systems, the discussions have always been in the sense of "strict" forbidding. For example, local languages are languages, for which subwords from a given finite set of words are strictly forbidden. In [3], the authors define the notion of subword avoidable (resp. subword unavoidable) sets. A set of words is subword avoidable if there is an infinite language which does not have subwords from that set. Otherwise, a set of words is subword unavoidable. The authors use subword unavoidable sets to characterize regular languages. The model from [9] that we consider uses forbidding sets that generalize the notion of sets of strictly forbidden words in the sense that if the forbidders are finitely many and singletons, then the languages coincide with local languages. If the forbidders are singletons but not necessarily finite, they are still forbidding subwords in the strict sense and coincide with avoidable sets. But in general, forbidders may be non-singletons and then they forbid the entire combinations of subwords, while still allowing parts of them as subwords.

The enforcing constraint was originally motivated by the evolution of molecular reactions, where if some molecules are present in a molecular system, then this triggers the presence of other molecules, as well [6]. The enforcing set from the model defined in [9] requires that if certain subwords appear in a word, they are enclosed in larger subwords from pre-specified finite sets. Since DNA molecules can be modeled by strings, this enforcing condition was motivated by imposing restrictions on strands of nucleotides to define, for example, restrictions imposed by DNA codewords [11] and capture the idea of linear splicing, as defined by Head [14] and constants [1].

This paper investigates fe-languages defined by finite fe-systems. After the definitions recalled in Sect. 2, we prove that such languages, as a class, are regular (Sect. 3) and stand strictly between the strictly locally testable and the locally testable languages (Theorem 9). Results showing that (separately and together) finite forbidding sets and finite enforcing sets define locally testable languages are included in Sect. 3 and reinforced in Sect. 4 using the Subword automata. Automata are a natural tool for studying locally testable languages because their finite state memory makes it easy to implement a sliding window that moves over the string to inspects the consecutive substrings of the input string. For the (strictly) locally testable languages membership is determined by the collection of substrings of the string. Finally, Sect. 5 investigates how the classes of strictly locally testable languages and the languages defined by finite fe-systems are related.

2   Forbidding-Enforcing Systems

We use standard formal language concepts and fix our notation here. The length of a word w is denoted by $|w|$. The empty word is denoted by λ, it has length 0. The set of all nonempty words over the alphabet A is denoted by A^+, the set of all words is A^*, and A^m is the set of all words of length m.

The word y is a *subword* of $x \in A^*$, if there exist $s, t \in A^*$, such that $x = syt$. The set of subwords of a word x is denoted by $\text{sub}(x)$ and $\text{sub}(L) = \bigcup_{x \in L} \text{sub}(x)$

is the set of subwords of a language L. Moreover, y is a *prefix* of x if $x = yt$, and a *suffix* of x if $x = sy$. We denote by pref(x) and suff(x) the sets of all prefixes and suffixes of x. These notations are extended to languages.

Forbidding Systems, f-languages. As originally defined [4–6], a forbidding-enforcing system (fe-system) was used to define a *family* of languages, where the conditions imposed by the system were required to hold for every language in the family. Here, we consider the fe-systems model, introduced in [9], in which a fe-system defines a single language. This means that the forbidding mechanism is applied to every single word in that language, rather than to a language in the family.

A *forbidding set* $\mathcal{F}$ is a family of finite nonempty subsets of A^+, each element of which is called a *forbidder*.

A word w is *consistent with a forbidder* F, if $F \nsubseteq$ sub(w). A word w is *consistent with a forbidding set* $\mathcal{F}$ if w is consistent with all $F \in \mathcal{F}$. The set of all words consistent with $\mathcal{F}$ defines the language $L(\mathcal{F})$, which is called a *forbidding language* or an f-*language*.

Note that $L(\mathcal{F}) = A^*$ if and only if $\mathcal{F} = \emptyset$ and that $\lambda \in L(\mathcal{F})$ for every $\mathcal{F}$.

Example 1. Let $\mathcal{F} = \{\{aa, bb\}, \{ab, ba\}\}$. A word w is consistent with the first forbidder $\{aa, bb\}$ if and only if w does not contain both aa and bb as subwords. Thus, words like $ababa, baabaab$, and $abbbb$ are consistent with the first forbidder and words like $aabb$ and $bbaabbaa$ are not consistent with it. Similarly, the second forbidder $\{ab, ba\}$ disallows both ab and ba to be in the subwords of w, but either one of them is allowed by itself. Thus, words like $aaa, aabb$, and $bbbaa$ are allowed, but $aababa$ and $bbab$ are forbidden by the second forbidder. The language defined by the entire forbidding set is the intersection of the sets of words consistent with each forbidder. Thus, $L(\mathcal{F}) = \{a^n, b^n, ab^n, a^n b, ba^n, b^n a \mid n \geq 0\}$.

Example 2. Let $A = \{a, b\}$ and $\mathcal{F} = \{\{ba^k b\} \mid k \text{ even}\}$. Then, $L(\mathcal{F})$ contains words where any two b's are separated by an odd number of a's.

Enforcing Systems, e-languages. An *enforcer* is a pair (x, Y) such that $x \in A^*$ and Y is a finite nonempty subset of A^+, where x is a proper subword of each $y \in Y$. The semantics of enforcers differs depending on whether x is the empty string or not. In the case that $x \in A^+$, a word w *satisfies an enforcer* (x, Y), if every occurrence of x in w is embedded in some word from Y. Formally, if $w = uxv$ for some $u, v \in A^*$ then there exists $y \in Y$ and $u_1, u_2, v_1, v_2 \in A^*$ such that $u = u_1 u_2$, $v = v_2 v_1$, and $y = u_2 x v_2$. If $x \notin$ sub(w) then w satisfies the enforcer (x, Y) trivially. When $x = \lambda$, the enforcer (λ, Y) is called *brute*. In this case, a word w satisfies the enforcer if there exists a word $y \in Y$ such that y is a subword of w.

An *enforcing set* $\mathcal{E}$ is a family of enforcers. A word w *satisfies* $\mathcal{E}$ if and only if w satisfies every enforcer in that set. The language of all words that satisfy $\mathcal{E}$ is denoted by $L(\mathcal{E})$. Such a language is called an *enforcing language* or an e-*language*.

Clearly, if $y \in Y$ then y satisfies (x, Y) and x does not. Also, $L(\mathcal{E}) = A^*$ if and only if $\mathcal{E} = \emptyset$.

In part, this research is motivated by the fact that *infinite* sets of enforcers form a powerful tool. The next example from [9] shows how a context-sensitive languages can be defined. Basically, 'unwanted' strings are forced to be prolonged indefinitely. So, one might say that enforcers are, in a way, used as forbidders here.

Example 3. Let $A = \{a\}$ and $L = \{a^{2^n} \mid n \geq 0\}$. Then, the enforcing set $\mathcal{E} = \{(\lambda, \{a, aa\})\} \cup \{(a^{2^i+1}, \{a^{2^{i+1}}\}) \mid i \geq 1\}$ defines L, i.e., $L = L(\mathcal{E})$.

Forbidding-Enforcing Systems, fe-languages. Finally, the two components are combined into a single system. A *forbidding-enforcing system* is an ordered pair $(\mathcal{F}, \mathcal{E})$, such that $\mathcal{F}$ is a forbidding set and $\mathcal{E}$ is an enforcing set. The language $L(\mathcal{F}, \mathcal{E})$ defined by this system consists of all words that are both consistent with $\mathcal{F}$ and satisfy $\mathcal{E}$, i.e., $L(\mathcal{F}, \mathcal{E}) = L(\mathcal{F}) \cap L(\mathcal{E})$. Such a language L is called a *forbidding-enforcing language* or an *fe-language*. To stress the fact that the fe-system in question is of this kind and defines a single language, rather than a family of languages, we use the term *language forbidding-enforcing system*.

Example 4. Let $\mathcal{F} = \{\{ba\}\}$ and $\mathcal{E} = \{(\lambda, \{a\})\} \cup \{(a^i, \{a^{i+1}, a^i b^i\}) \mid i \geq 1\}$. The forbidding set ensures the language is a subset of $a^* b^*$. The brute enforcer ensures we have at least the symbol a in the strings. Consider the enforcer $(a^i, \{a^{i+1}, a^i b^i\})$ for the string $a^n b^m$. For $i < n$, it is satisfied since a^{i+1} is a subword of $a^n b^m$. If $i > n$ it is satisfied trivially; for $i = n$ it is satisfied if $m \geq n$. Then, $L = L(\mathcal{F}, \mathcal{E}) = \{a^n b^m \mid n \leq m \text{ and } n, m \geq 1\}$.

3 Finite Fe-Systems and Regular Languages

Finite Forbidding Sets. Finite forbidding sets define regular languages. In fact, given a finite forbidding set $\mathcal{F}$, each forbidder $F \in \mathcal{F}$ prohibits the combined presence of all of its elements as subwords, so it defines the regular language $L(\{F\}) = \bigcup_{u \in F} A^* \setminus A^* u A^*$, a union of local languages. Then, by Proposition 1 (5) in [9], $L(\mathcal{F}) = \bigcap_{F \in \mathcal{F}} L(\{F\})$, which is a finite intersection of regular languages and, hence, is regular.

Example 5. Consider $\mathcal{F} = \{\{aa, bb\}, \{ab, ba\}\}$ from Example 1. Then, $L(\mathcal{F}) = (A^* \setminus A^* aa A^* \cup A^* \setminus A^* bb A^*) \cap (A^* \setminus A^* ab A^* \cup A^* \setminus A^* ba A^*) = \{a^n, b^n, ab^n, ba^n, a^n b, b^n a \mid n \geq 0\}$, as previously stated.

We show that finite forbidding sets, actually, define locally testable languages, a subclass of regular languages. We recall the definitions for locally testable and strictly locally testable languages from [18] below. For $k \geq 0$ and $w \in A^*$ such that $|w| \geq k$, i.e. for $w \in A^{\geq k}$, denote by $\text{pref}_k(w)$ and $\text{suff}_k(w)$, respectively, the singleton sets containing the prefix and suffix of w of length k and by $\text{int}_k(w)$ (interior subwords) the set of all subwords of w of length k that occur at other than prefix or suffix positions.

The locally testable languages are those languages whose membership $w \in L$ is determined by the triple $\text{pref}_k(w)$, $\text{suff}_k(w)$, and $\text{int}_k(w)$.

Definition 1. A language $L \subseteq A^*$ is called k-*testable* if and only if for any words $x, y \in A^*$, the conditions $\text{pref}_k(x) = \text{pref}_k(y)$, $\text{suff}_k(x) = \text{suff}_k(y)$, and $\text{int}_k(x) = \text{int}_k(y)$ imply that $x \in L$ if and only if $y \in L$. A language is called *locally testable* if it is k-testable for some integer $k \geq 1$.

A language is strictly locally testable if we can specify sets P, S, I of "allowed" prefixes, suffixes, and internal subwords, respectively, such that a string is in the language if and only if its subwords are within these respective sets.

Definition 2. A language $L \subseteq A^*$ is called *strictly k-testable* if there exist finite sets $P, S, I \subseteq A^*$ such that for all $w \in A^{\geq k}$, $w \in L$ if and only if $\text{pref}_k(w) \subseteq P$, $\text{suff}_k(w) \subseteq S$, and $\text{int}_k(w) \subseteq I$. A language is called *strictly locally testable* if it is strictly k-testable for some integer $k \geq 1$.

Locally testable languages are the boolean closure of strictly locally testable languages. There are languages that are locally testable but not strictly locally testable, such as the language L from the following example.

Example 6. Let $A = \{a, b\}$. The language $L = aaA^*aa \cup bbA^*bb$ is 2-testable and hence, locally testable. A word w belongs to L if and only if $\text{int}_2(w) \neq \emptyset$ and $\text{pref}_2(w) = \text{suff}_2(w) = \{aa\}$ or $\text{pref}_2(w) = \text{suff}_2(w) = \{bb\}$. The language L, however, is not strictly locally testable. Intuitively, the argument is as follows. As both $aauaa$ and $bbvbb$ belong to L for any $u, v \in A^*$, any triple P, S, I defining L must allow both aa and bb as prefix, as well as, suffix. So, $P = S = \{aa, bb\}$. Also, $I = A^2$ since any $u, v \in A^*$ are allowed. Thus, $aazbb$ will be allowed, for any $z \in A^*$, contradicting the definition of L. This shows that L is not strictly 2-testable and similarly, it is not strictly k-testable for all $k \geq 2$. This example will be revisited in the proof of Lemma 1.

We show that finite forbidding sets define locally testable languages.

Theorem 1. *Let $\mathcal{F}$ be a finite forbidding set. Then, $L(\mathcal{F})$ is locally testable.*

Proof. Let $\mathcal{F}$ be a finite forbidding set. If $\mathcal{F} = \emptyset$ then $L(\mathcal{F}) = A^*$. The language A^* is strictly locally testable, and hence, locally testable. Suppose $\mathcal{F} = \{\{u\}\}$, i.e. $\mathcal{F}$ has only one forbidder $F = \{u\}$, which is a singleton. Then, $L(\mathcal{F}) = A^* \setminus A^*uA^*$. Let $|u| = k$ and set $P = S = I = A^k \setminus \{u\}$. Since every word $w \in L(\mathcal{F})$ with $|w| \geq k$ has a k-length prefix and a k-length suffix that are not u and none of its interior subwords are u, it follows that $L(\mathcal{F})$ is strictly locally testable. Consider now a nonempty finite $\mathcal{F}$ with not necessarily singleton forbidders. It follows from the discussion at the beginning of this section that $L(\mathcal{F}) = \bigcap_{F \in \mathcal{F}} (\bigcup_{u \in F} A^* \setminus A^*uA^*)$, i.e., $L(\mathcal{F})$ is a finite intersection of finite unions of strictly locally testable languages and, thus, it is locally testable. □

Remark 1. Note that if $\mathcal{F}$ contains a non-singleton forbidder F, the f-language is not necessarily strictly locally testable, even if $\mathcal{F}$ is a singleton. Consider $\mathcal{F} = \{\{aa, bb\}\}$. Then, the words in $L(\mathcal{F})$ either don't have aa or don't have bb (or don't have both) in their subwords. However, as the language $L(\mathcal{F})$ contains all possible prefixes, suffixes, and subwords, it is impossible to find suitable P, S, I. For example, both aa and bb should be in I, but then words with subwords $aabb$ will be in L, contradicting the definition of $L(\mathcal{F})$.

Remark 2. Observe that this theorem confirms the fact that local languages are a subclass of the locally testable languages. Namely, if a language L is local, then by Proposition 2 in [9] there exists a forbidding set $\mathcal{F}$ such that $L = L(\mathcal{F})$ and $\mathcal{F} = \{\{u_1\}, \{u_2\}, \ldots, \{u_n\}\}$ where $H = \{u_1, u_2, \ldots, u_n\}$ such that $L = A^* \setminus A^* H A^*$. By the proof of the above theorem, $L(\{\{u_i\}\})$ is strictly locally testable for all $i = 1, \ldots, n$ and hence, L, which equals the intersection of these languages is locally testable.

The converse of the above theorem does not hold, since there are languages that are locally testable but cannot be defined by a forbidding set only. The fe-systems defining such languages will need to have enforcers.

Theorem 2. *There exists a locally testable language that is not an f-language.*

Proof. By Theorem 2 in [9], f-languages are factorial (that is the language contains all of its subwords) and vice versa. Consider L from Example 6. It is not factorial, since $ababab \in \mathrm{sub}(L)$ but $ababab \notin L$. Hence, there is no $\mathcal{F}$ (finite or infinite) such that $L = L(\mathcal{F})$. $\quad\square$

Finite Enforcing Sets. It turns out that finite enforcing sets, also, define locally testable languages.

Theorem 3. *Let $\mathcal{E}$ be a finite enforcing set. Then $L(\mathcal{E})$ is locally testable.*

Proof. Assume that $\mathcal{E}$ is finite. If $\mathcal{E} = \emptyset$ then $L(\mathcal{E}) = A^*$, which is strictly locally testable and, hence, locally testable. Otherwise, $\mathcal{E}$ contains at least one enforcer (x, Y). Let k be twice the length of the longest string in any enforcer of $\mathcal{E}$. Thus, $k = 2 \cdot \max\{|y| \mid y \in Y$ for some $(x, Y) \in \mathcal{E}\}$. As $\mathcal{E}$ is finite and nonempty, k is well-defined.

Consider $w \in A^{\geqslant k}$. We show that the sets $\mathrm{pref}_k(w)$, $\mathrm{suff}_k(w)$, and $\mathrm{int}_k(w)$ determine whether $w \in L(\mathcal{E})$, and thus, we show that $L(\mathcal{E})$ is locally testable.

First, we consider enforcing sets without brute enforcers. For every $(x, Y) \in \mathcal{E}$ we have to make sure that every occurrence of x in a word is embedded in some $y \in Y$. We prove that $L(\mathcal{E})$ is strictly locally testable by specifying the sets P, S, and I. The set I consists of all words of length k which either don't have x in the middle or every occurrence of x in the middle is enclosed by some $y \in Y$. The set P consists of all words that either don't contain x in the first half or every occurrence of x in the first half is in some $y \in Y$. Symmetrically, S can be defined.

In the case that $\mathcal{E}$ contains brute enforcers, for every enforcer (λ, Y), we determine membership of w by considering the sets $\mathrm{pref}_k(w)$, $\mathrm{suff}_k(w)$, and $\mathrm{int}_k(w)$ and checking whether there is a word in any of these sets that has some $y \in Y$ as a subword. Such a language is locally testable. $\quad\square$

Example 7. Consider the brute enforcer $E = (\lambda, \{aa, bb\})$. Since a word from $Y = \{aa, bb\}$ has to be in the subwords of any w that satisfies this enforcer, the words that satisfy it will either have aa or bb or both in their subwords. This enforcer specifies the language $L(\{E\}) = A^*\{aa, bb\}A^*$. That is, if the enforcing set $\mathcal{E} = \{E\}$, then $L(\mathcal{E}) = A^*\{aa, bb\}A^*$, which is not strictly locally testable. Indeed, it is not possible to find suitable sets P, S, and I to define $L(\mathcal{E})$, since for any $k \geq 1$, S, P, and I would have to equal A^k. But then, the word $(ab)^k \in L$, for example, which would contradict the definition of $L(\mathcal{E})$.

Lemma 1. *There exists a locally testable language that cannot be defined by a finite enforcing set.*

Proof. Consider the 2-testable language $L = aaA^*aa \cup bbA^*bb$ from Example 6 and assume it can be defined by a finite enforcing set $\mathcal{E}$. Let ℓ be the length of the longest word in Y among all enforcers $(x, Y) \in \mathcal{E}$. Let z be any word that contains all strings of length ℓ at least once. By definition, L contains $aazaa$ and $bbzbb$, so they must satisfy any enforcer $(x, Y) \in \mathcal{E}$. We argue that also aaz^4bb satisfies $\mathcal{E}$. As z contains all words of length ℓ, the string aaz^4bb satisfies any brute enforcer (λ, Y) in $\mathcal{E}$. Otherwise, if $x \neq \lambda$, consider any occurrence of x in aaz^4bb. It is either in the beginning $aazz$, which is a subword of aaz^4aa, or at the end $zzbb \in sub(bbz^4bb)$, or in the middle $zz \in sub(aaz^4aa)$. Since $aaz^4aa, bbz^4bb \in L(\mathcal{E})$, x is embedded in a word in Y. But then, aaz^4bb satisfies (x, Y) and, hence, is in $L(\mathcal{E})$, which is a contradiction.   □

Finite Fe-Systems. Following Theorems 1 and 3 above, we can conclude that finite fe-systems define regular languages and, more specifically, they define locally testable languages, as locally testable languages are closed under intersection.

Theorem 4. *Let $(\mathcal{F}, \mathcal{E})$ be a fe-system where both $\mathcal{F}$ and $\mathcal{E}$ are finite sets. Then, $L(\mathcal{F}, \mathcal{E})$ is a locally testable language.*

Not all locally testable languages can be defined by finite fe-systems.

Theorem 5. *There exists a locally testable language that cannot be defined by a finite fe-system.*

Proof. Let $A = \{a, b\}$. We consider again the 2-testable language $L = aaA^*aa \cup bbA^*bb$ from Example 6. We assume there exists a finite fe-system $(\mathcal{F}, \mathcal{E})$ which defines L and derive a contradiction.

First note that $int(L) = A^*$, so we must have $\mathcal{F} = \emptyset$. Otherwise, if $\mathcal{F}$ contains forbidder $F = \{u_1, \ldots, u_n\}$ then, for example, $u_1u_2 \ldots u_n \notin int(L)$. Then, there should be a finite $\mathcal{E}$ that defines L. However, Lemma 1 shows that such $\mathcal{E}$ does not exist, which is a contradiction.   □

4   Automata Representation of Finite Fe-Systems

In this section we reconsider some of our results using an automata approach to local testability. Locally testable languages can be recognized by a finite state process which slides a window of fixed size over the string. Acceptance is based on the set of substrings that have been seen during the scan. In the case of strictly locally testable languages we verify whether the prefix, interior subwords, and the suffix of a word are within the sets of allowed subwords. For locally testable languages, the exact triple $(\mathrm{pref}_k(w), \mathrm{suff}_k(w), int_k(w))$ of subwords seen determines whether the word w is in the language, based on the set of allowed triples.

Subword automaton. The basis of automata for (strictly) locally testable languages is the *subword automaton*[1], see Fig. 1. For a given k and an alphabet A, the state set of the automaton equals $Q = A^k$, the words of length k over A. The automaton, basically, keeps track of the last k-letters seen, so its transition function equals $\delta(aw, b) = wb$ for $|w| = k - 1$ and $a, b \in A$. Note that this subword automaton is not a proper finite state automaton, as it has no initial and final states. To a given string w, we associate a path, i.e. the automaton starts in $\mathrm{pref}_k(w)$, traces the remaining letters of w by visiting the successive subwords in $int_k(w)$, and ends in $\mathrm{suff}_k(w)$.

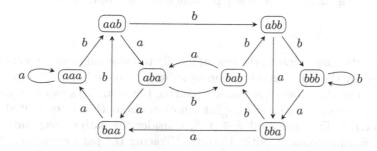

Fig. 1. Subword automaton for $k = 3$ and $A = \{a, b\}$

Strictly locally testable languages. For a specific combination of (P, S, I) in the case of strictly k-testable languages we turn the subword automaton into a finite state automaton by choosing initial and final states, and by deleting transitions. We mark the states of P and S as initial and final, respectively. For states not in $P \cup I$ we delete the outgoing edges, while similarly, for states not in $I \cup S$ we delete the incoming edges, see Example 8 below.

Given a strictly locally testable language L with subsets P, S, I of A^k, the automaton M, such that $L(M) = L$ is specified as $M = (A^k, A, \delta, P, S)$, where $\delta(aw, b) = wb$ for $a, b \in A$ such that $aw \in P \cup I$ and $wb \in I \cup S$.

---

[1] It is called the k-local universal automaton in [16].

This automaton may have several initial states, and also "jumps" to the initial prefix of the input, rather than reading the initial letters one-by-one as we are used to in finite state automata. This can be changed to a more classic representation by adding all prefixes of the strings in P and adding transitions $\delta(w, b) = wb$ for $|w| < k$ and $b \in A$. In this way, we build a tree from the new initial state λ to the words of length k by reading the first k symbols one-by-one.

Example 8. With $P = \{aaa, aab\}$, $I = \{aab, abb, baa, bba, bbb\}$, and $S = \{bab, bbb\}$ we obtain the automaton in Fig. 2. It has *two* initial states, the words in P.

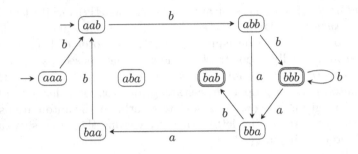

Fig. 2. A strictly locally testable language, see Example 8

Locally testable languages. In the case of locally testable languages, acceptance is determined by the actual combination of subwords (including prefix and suffix) encountered by the subword automaton during the scan. This can be built from the subword automaton by storing information about the states visited in the form of a triple. For each word $w \in A^{\geq k}$, consider the path corresponding to w in the subword automaton with A^k states. Tracing the path associated with w begins with first visiting the state corresponding to $\text{pref}_k(w)$ and then following the transitions $\delta(aw, b) = wb$, with $a, b \in A$, to trace w until $wb = \text{suff}_k(w)$ is reached. This is associated with a sequence of triples, where the first component is always the first state visited (which is $\text{pref}_k(w)$), the second component is the current state visited (which at the end of the path is $\text{suff}_k(w)$), and the third component is the set of visited states (which at the end of the path is $\text{int}_k(w)$). Thus, the state set becomes $A^k \times A^k \times 2^{A^k}$. Hence, for every word w a sequence of triples is recorded along the path tracing w ending in $(\text{pref}_k(w), \text{suff}_k(w), \text{int}_k(w))$. The word w is accepted, if and only if, the last triple is an accepting state, i.e. an allowed triple for the locally testable language.

With this automata terminology it is possible to reconsider the result on finite enforcers, Theorem 3.

Theorem 6. *Finite enforcing sets define locally testable languages. If no enforcer of the form (λ, Y) is present, the language is strictly locally testable.*

Proof. Suppose $\mathcal{E}$ is a singleton, i.e. $\mathcal{E} = \{(x, Y)\}$. We show strict local testability by constructing a finite state automaton that accepts the language based on the subword automaton for certain sets (P, S, I).

We first consider $x \neq \lambda$. Whenever x is present in the string it must be embedded in one of the strings in Y. This can be checked locally, as follows. Choose ℓ to be the length of the longest string in Y. When y in Y contains x, the other letters of y can extend at most $\ell - |x|$ positions in either direction of x. This means that we can check whether (x, Y) is satisfied using a window of length $k = 2\ell - |x|$, but we will use $k = 2\ell$ for simplicity. Thus, I contains all those subwords that, whenever x occurs in the middle of the window, it is covered by a string in Y. For the prefix and suffix P, S, the initial and final states, we similarly consider strings such that all occurrences of x that are within ℓ positions from the start (from the end, respectively) are covered by a string in Y.

When $\mathcal{E}$ consists of finitely many (more than one) non-brute enforcers, then $L(\mathcal{E})$ is the intersection of all $L(\{(x, Y)\})$, that is the P, S, and I will be intersections of all the others, and hence, $L(\mathcal{E})$ will be strictly locally testable.

An enforcing set containing a brute enforcer (λ, Y), due to its special semantics, is not strictly locally testable, as we have seen in Example 7. We must verify that at least one of the strings in Y is present as a substring. This is locally testable: $y \in Y$ occurs as a subword when it occurs as a subword in either $\mathrm{pref}_k(w)$, $\mathrm{suff}_k(w)$, or $\mathrm{int}_k(w)$, for any $k \geq \ell$.   □

5   (Strictly) Locally Testable Languages and Fe-Systems

In Sect. 3 we showed that finite fe-systems define locally testable languages, but the converse does not hold. Here, we investigate the relationship between strictly locally testable languages and languages defined by finite fe-systems.

Theorem 7. *Let L be a strictly locally testable language, then there exists a finite fe-system $(\mathcal{F}, \mathcal{E})$ such that $L = L(\mathcal{F}, \mathcal{E})$.*

Proof. Assume that L is strictly locally testable. Then, there exists $k \geq 1$ and sets $P, S, I \subseteq A^k$ such that $\mathrm{pref}_k(L) \subseteq P$, $\mathrm{suff}_k(L) \subseteq S$, and $\mathrm{int}_k(L) \subseteq I$. We construct a finite fe-system $(\mathcal{F}, \mathcal{E})$ and show that $L = L(\mathcal{F}, \mathcal{E})$. The construction below considers the various intersections and unions among the three sets. If any k-letter word is not in any of the three sets, it must be a forbidden subword $(\mathcal{F}_1)$. The forbidders $\mathcal{F}_2$ ensure that if a k-letter subword is a prefix only and not a suffix or an interior subword then it appears at the beginning of the word only and hence any extension to the left is forbidden. Similarly, $\mathcal{F}_3$ forbids any extension to the right of subwords which are suffixes only. If a k-letter subword is not an interior subword, but it is both a prefix and a suffix, then simultaneous extension on both sides is forbidden by $\mathcal{F}_4$. Construct:

$$\mathcal{F}_1 = \{\{u\} \mid u \in A^k \setminus (P \cup S \cup I)\}$$
$$\mathcal{F}_2 = \{\{au\} \mid u \in P \setminus (I \cup S) \text{ and } a \in A\}$$
$$\mathcal{F}_3 = \{\{ub\} \mid u \in S \setminus (I \cup P) \text{ and } b \in A\}$$

$\mathcal{F}_4 = \{\{aub\} \mid u \in (P \cap S) \setminus I \text{ and } a, b \in A\}$

We construct enforcing sets to ensure that interior subwords which are not prefixes or suffixes are indeed interior and are being extended on both sides ($\mathcal{E}_1$). Similarly, suffixes that are not prefixes must be extended to the left ($\mathcal{E}_2$) and prefixes that are not suffixes must be extended to the right ($\mathcal{E}_3$).

$\mathcal{E}_1 = \{(x, \{axb \mid a, b \in A\}) \mid x \in I \setminus (P \cup S)\}$
$\mathcal{E}_2 = \{(x, \{ax \mid a \in A\}) \mid x \in S \setminus P\}$
$\mathcal{E}_3 = \{(x, \{xb \mid b \in A\}) \mid x \in P \setminus S\}$

Then, let $\mathcal{F} = \bigcup_{i=1}^{4} \mathcal{F}_i$ and let $\mathcal{E} = \bigcup_{i=1}^{3} \mathcal{E}_i$. By the above construction, $L = L(\mathcal{F}, \mathcal{E})$. □

Example 8 was used to illustrate the subword automaton. The same sets are represented here using the construction in the above proof.

Example 9. Let the strictly locally testable language L be determined by $P = \{aaa\}$, $I = \{aab, abb, baa, bba, bbb\}$, and $S = \{bab, bbb\}$. We construct an $(\mathcal{F}, \mathcal{E})$ system which defines L.
Let $\mathcal{F}_1 = \{\{aba\}\}$, $\mathcal{F}_2 = \{\{aaaa\}, \{baaa\}\}$, $\mathcal{F}_3 = \{\{baba\}, \{babb\}\}$, and $\mathcal{F}_4 = \emptyset$.
Let $\mathcal{E}_1 = \{(aab, \{aaaba, aaabb, baaba, baabb\}), (abb, \{aabba, aabbb, babba, babbb\}),$
$(baa, \{abaaa, abaab, bbaaa, bbaab\}), (bba, \{abbaa, abbab, bbbaa, bbbab\})\}$,
$\mathcal{E}_2 = \{(bab, \{abab, bbab\}), (bbb, \{abbb, bbbb\})\}$, and $\mathcal{E}_3 = \{(aaa, \{aaaa, aaab\})\}$.
Then, let $\mathcal{F} = \bigcup_{i=1}^{4} \mathcal{F}_i$ and let $\mathcal{E} = \bigcup_{i=1}^{3} \mathcal{E}_i$.

We have presented two examples, so far, of fe-languages defined by finite systems, which are not strictly locally testable, Remark 1 and Example 7. Thus, we conclude that the converse of the above theorem does not hold.

Theorem 8. *There exists a language defined by a finite fe-system that is not strictly locally testable.*

We conclude with a summary of our results.

Theorem 9. *The class of fe-languages defined by finite fe-systems is strictly between the class of strictly locally testable languages and the class of locally testable languages.*

Proof. By Theorems 7 and 8, the class of strictly locally testable languages is included strictly in the class of fe-languages defined by finite fe-systems. Also, by Theorems 4 and 5, languages defined by finite fe-systems are strictly included in the class of locally testable languages. □

Acknowledgement. This paper was started when the first author was visiting Leiden University, while on sabbatical from UNF. The authors thank both institutions for making this possible.

References

1. Bonizzoni, P., Jonoska, N.: Existence of constants in regular splicing languages. Inf. Comput. **242**, 340–353 (2015)
2. Cavaliere, M., Jonoska, N.: Forbidding and enforcing in membrane computing. Nat. Comput. **2**, 215–228 (2003)
3. Ehrenfeucht, A., Haussler, D., Rozenberg, G.: On regularity of context-free languages. Theoret. Comput. Sci. **27**, 311–332 (1983)
4. Ehrenfeucht, A., Hoogeboom, H.J., Rozenberg, G., van Vugt, N.: Forbidding and enforcing, In: Winfree, E., Gifford, D.K. (eds.) DNA Based Computers V. AMS DIMACS, Providence, RI, vol. 54, pp. 195–206 (2001)
5. Ehrenfeucht, A., Hoogeboom, H.J., Rozenberg, G., van Vugt, N.: Sequences of languages in forbidding-enforcing families. Soft. Comput. **5**(2), 121–125 (2001)
6. Ehrenfeucht, A., Rozenberg, G.: Forbidding-enforcing systems. Theoret. Comput. Sci. **292**, 611–638 (2003)
7. Ehrenfeucht, A., Rozenberg, G.: Reaction systems. Fundamenta Informaticae **76**, 1–18 (2006)
8. Franco, G., Jonoska, N.: Forbidding and enforcing conditions in DNA self-assembly of graphs. In: Chen, J., Jonoska, N., Rozenberg, G. (eds.) Nanotechnology: Science and Computation, Part I. Natural Computing Series, pp. 105–118. Springer, Heidelberg (2006)
9. Genova, D.: Defining languages by forbidding-enforcing systems. In: Löwe, B., Normann, D., Soskov, I., Soskova, A. (eds.) CiE 2011. LNCS, vol. 6735, pp. 92–101. Springer, Heidelberg (2011). doi:10.1007/978-3-642-21875-0_10
10. Genova, D.: Forbidding sets and normal forms for language forbidding-enforcing systems. In: Dediu, A.-H., Martín-Vide, C. (eds.) LATA 2012. LNCS, vol. 7183, pp. 289–300. Springer, Heidelberg (2012). doi:10.1007/978-3-642-28332-1_25
11. Genova, D.: Language forbidding-enforcing systems defining DNA codewords. In: Bonizzoni, P., Brattka, V., Löwe, B. (eds.) CiE 2013. LNCS, vol. 7921, pp. 220–229. Springer, Heidelberg (2013). doi:10.1007/978-3-642-39053-1_25
12. Genova, D., Jonoska, N.: Defining structures through forbidding and enforcing constraints. Phys. B Condens. Matter **394**(2), 306–310 (2007)
13. Genova, D., Jonoska, N.: Forbidding and enforcing on graphs. Theoret. Comput. Sci. **429**, 108–117 (2012)
14. Head, T.: Formal language theory and DNA: an analysis of the generative capacity of specific recombinant behaviors. Bull. Math. Biol. **49**(6), 737–759 (1987)
15. Păun, G.: Membrane Computing. An Introduction. Springer, Berlin (2002)
16. Perrin, D.: Finite automata. In: van Leeuwen, J. (ed.) Handbook of Theoretical Computer Science, vol. B, Formal Models and Semantics, pp. 1–57. Elsevier, Amsterdam (1990)
17. Rozenberg, G., Bäck, T., Kok, J.N. (eds.): Handbook of Natural Computing, 3 vols. Springer, Heidelberg (2012)
18. Yu, S.: Regular languages. In: Rozenberg, G., Saloma, A. (eds.) Handbook of Formal Languages, vol. 1, pp. 41–110. Springer, Heidelberg (1997)

Surjective H-Colouring: New Hardness Results

Petr A. Golovach[1], Matthew Johnson[2], Barnaby Martin[2], Daniël Paulusma[2], and Anthony Stewart[2(✉)]

[1] Department of Informatics, University of Bergen, Bergen, Norway
petr.golovach@ii.uib.no
[2] School of Engineering and Computing Sciences,
Durham University, South Road, Durham DH1 3LE, U.K.
{matthew.johnson2,barnaby.d.martin,daniel.paulusma,a.g.stewart}@dur.ac.uk

Abstract. A homomorphism from a graph G to a graph H is a vertex mapping f from the vertex set of G to the vertex set of H such that there is an edge between vertices $f(u)$ and $f(v)$ of H whenever there is an edge between vertices u and v of G. The H-COLOURING problem is to decide whether or not a graph G allows a homomorphism to a fixed graph H. We continue a study on a variant of this problem, namely the SURJECTIVE H-COLOURING problem, which imposes the homomorphism to be vertex-surjective. We build upon previous results and show that this problem is NP-complete for every connected graph H that has exactly two vertices with a self-loop as long as these two vertices are not adjacent. As a result, we can classify the computational complexity of SURJECTIVE H-COLOURING for every graph H on at most four vertices.

1 Introduction

The well-known COLOURING problem is to decide whether or not the vertices of a given graph can be properly coloured with at most k colours for some given integer k. If we exclude k from the input and assume it is fixed, we obtain the k-COLOURING problem. A *homomorphism* from a graph $G = (V_G, E_G)$ to a graph $H = (V_H, E_H)$ is a vertex mapping $f : V_G \to V_H$, such that there is an edge between $f(u)$ and $f(v)$ in E_H whenever there is an edge between u and v in E_G. We observe that k-COLOURING is equivalent to the problem of asking whether a graph allows a homomorphism to the complete graph K_k on k vertices. Hence, a natural generalization of the k-COLOURING problem is the H-COLOURING problem, which is to decide whether or not a graph allows a homomorphism to an arbitrary fixed graph H. We call this fixed graph H the *target graph*. Throughout the paper we consider undirected graphs with no multiple edges. We assume that an input graph G contains no vertices with self-loops (we call such vertices *reflexive*), whereas a target graph H may contain such vertices. We call H *reflexive* if all its vertices are reflexive, and *irreflexive* if all its vertices are irreflexive.

Supported by the Research Council of Norway via the project "CLASSIS" and the Leverhulme Trust (RPG-2016-258).

J. Kari et al. (Eds.): CiE 2017, LNCS 10307, pp. 270–281, 2017.
DOI: 10.1007/978-3-319-58741-7_26

For a survey on graph homomorphisms we refer the reader to the textbook of Hell and Nešetřil [10]. Here, we will discuss the H-COLOURING problem, a number of its variants and their relations to each other. In particular, we will focus on the *surjective* variant: a homomorphism f from a graph G to a graph H is *(vertex-)surjective* if f is surjective, that is, if for every vertex $x \in V_H$ there exists at least one vertex $u \in V_G$ with $f(u) = x$.

The computational complexity of H-COLOURING has been determined completely. The problem is trivial if H contains a reflexive vertex u (we can map each vertex of the input graph to u). If H has no reflexive vertices, then the Hell-Nešetřil dichotomy theorem [9] tells us that H-Colouring is solvable in polynomial time if H is bipartite and that it is NP-complete otherwise.

The LIST H-COLOURING problem takes as input a graph G and a function L that assigns to each $u \in V_G$ a list $L(u) \subseteq V_H$. The question is whether G allows a homomorphism f to the target H with $f(u) \in L(u)$ for every $u \in V_G$. Feder, Hell and Huang [2] proved that LIST H-COLOURING is polynomial-time solvable if H is a bi-arc graph and NP-complete otherwise (we refer to [2] for the definition of a bi-arc graph). A homomorphism f from G to an induced subgraph H of G is a *retraction* if $f(x) = x$ for every $x \in V_H$, and we say that G *retracts* to H. A retraction from G to H can be viewed as a list-homomorphism: choose $L(u) = \{u\}$ if $u \in V_H$, and $L(u) = V_H$ if $u \in V_G \setminus V_H$. The corresponding decision problem is called H-RETRACTION. The computational complexity of H-RETRACTION has not yet been classified. Feder et al. [3] determined the complexity of the H-RETRACTION problem whenever H is a pseudo-forest (a graph in which every connected component has at most one cycle). They also showed that H-RETRACTION is NP-complete if H contains a connected component in which the reflexive vertices induce a disconnected graph.

We impose a surjective condition on the graph homomorphism. An important distinction is whether the surjectivity is with respect to vertices or edges. Furthermore, the condition can be imposed locally or globally. If we require a graph homomorphism f to be vertex-surjective when restricted to the open neighbourhood of every vertex u of G, we say that f is an H-*role assignment*. The corresponding decision problem is called H-ROLE ASSIGNMENT and its computational complexity has been fully classified [6]. We refer to the survey of Fiala and Kratochvíl [5] for further details on locally constrained homomorphisms and from here on only consider global surjectivity.

It has been shown that deciding whether a given graph G allows a surjective homomorphism to a given graph H is NP-complete even if G and H both belong to one of the following graph classes: disjoint unions of paths; disjoint unions of complete graphs; trees; connected cographs; connected proper interval graphs; and connected split graphs [7]. Hence it is natural, just as before, to fix H which yields the following problem:

SURJECTIVE H-COLOURING
 Instance: a graph G.
 Question: does there exist a surjective homomorphism from G to H?

We emphasize that being vertex-surjective is a different condition than being edge-surjective. A homomorphism from a graph G to a graph H is called *edge-surjective* or a *compaction* if for any edge $xy \in E_H$ with $x \neq y$ there exists an edge $uv \in E_G$ with $f(u) = x$ and $f(v) = y$. Note that the edge-surjectivity condition does not hold for any self-loops $xx \in E_H$. If f is a compaction from G to H, we say that G *compacts* to H. The corresponding decision problem is known as the H-COMPACTION problem. A full classification of this problem is still wide open. However partial results are known, for example when H is a reflexive cycle, an irreflexive cycle, or a graph on at most four vertices [13–15], or when G is restricted to some special graph class [16]. Vikas also showed that whenever H-RETRACTION is polynomial-time solvable, then so is H-COMPACTION [14]. Whether the reverse implication holds is not known. A complete complexity classification of SURJECTIVE H-COLOURING is also still open. Below we survey the known results.

We first consider irreflexive target graphs H. The SURJECTIVE H-COLOURING problem is NP-complete for every such graph H if H is non-bipartite, as observed by Golovach et al. [8]. The straightforward reduction is from the corresponding H-COLOURING problem, which is NP-complete due to the aforementioned Hell-Nešetřil dichotomy theorem. However, the complexity classifications of H-COLOURING and SURJECTIVE H-COLOURING do not coincide: there exist bipartite graphs H for which SURJECTIVE H-COLOURING is NP-complete, for instance when H is the graph obtained from a 6-vertex cycle to each of which vertices we add a path of length 3 [1].

We now consider target graphs with at least one reflexive vertex. Unlike the H-COLOURING problem, the presence of a reflexive vertex does not make the SURJECTIVE H-COLOURING problem trivial to solve. We call a connected graph *loop-connected* if all its reflexive vertices induce a connected subgraph. Golovach, Paulusma and Song [8] showed that if H is a tree (in this context, a connected graph with no cycles of length at least 3) then SURJECTIVE H-COLOURING is polynomial-time solvable if H is loop-connected and NP-complete otherwise. As such the following question is natural:

Is SURJECTIVE H-COLOURING NP-*complete for every connected graph H that is not loop-connected?*

The reverse statement is not true (if P$\neq$ NP): SURJECTIVE H-COLOURING is NP-complete when H is the 4-vertex cycle C_4^* with a self-loop in each of its vertices. This result has been shown by Martin and Paulusma [11] and independently by Vikas, as announced in [16]. Recall also that SURJECTIVE H-COLOURING is NP-complete if H is irreflexive (and thus loop-connected) and non-bipartite.

It is known that SURJECTIVE H-COLOURING is polynomial-time solvable whenever H-COMPACTION is [1]. Recall that H-COMPACTION is polynomial-time solvable whenever H-RETRACTION is [14]. Hence, for instance, the aforementioned result of Feder, Hell and Huang [2] implies that SURJECTIVE H-COLOURING is polynomial-time solvable if H is a bi-arc graph. We also recall that H-RETRACTION is NP-complete whenever H is a connected graph that is not loop-connected [3]. Hence, an affirmative answer to the above question would mean that for these target graphs H the complexities of H-RETRACTION, H-COMPACTION and SURJECTIVE H-COLOURING coincide.

In Fig. 1 we display the relationships between the different problems discussed. In particular, it is a major open problem whether the computational complexities of H-COMPACTION, H-RETRACTION and SURJECTIVE H-COLOURING coincide for each target graph H. Even showing this for specific cases, such as the case $H = C_4^*$, has been proven to be non-trivial. If it is true, it would relate the SURJECTIVE H-COLOURING problem to a well-known conjecture of Feder and Vardi [4], which states that the $\mathcal{H}$-CONSTRAINT SATISFACTION problem has a dichotomy when $\mathcal{H}$ is some fixed finite target structure and which is equivalent to conjecturing that H-RETRACTION has a dichotomy [4]. We refer to the survey of Bodirsky, Kara and Martin [1] for more details on the SURJECTIVE H-COLOURING problem from a constraint satisfaction point of view.

Fig. 1. Relations between SURJECTIVE H-COLOURING and its variants. An arrow from one problem to another indicates that the latter problem is polynomial-time solvable for a target graph H whenever the former is polynomial-time solvable for H. Reverse arrows do not hold for the leftmost and rightmost arrows, as witnessed by the reflexive 4-vertex cycle for the rightmost arrow and by any reflexive tree that is not a reflexive interval graph for the leftmost arrow (Feder, Hell and Huang [2] showed that the only reflexive bi-arc graphs are reflexive interval graphs). It is not known whether the reverse direction holds for the two middle arrows.

1.1   Our Results

We present further progress on the research question of whether SURJECTIVE H-COLOURING is NP-complete for every connected graph H that is not loop-connected. We first consider the case where the target graph H is a connected graph with exactly two reflexive vertices that are non-adjacent. In Sect. 2 we prove that SURJECTIVE H-COLOURING is indeed NP-complete for every such target graph H. In the same section we slightly generalize this result by showing that it holds even if the reflexive vertices of H can be partitioned into two non-adjacent sets of twin vertices. This enables us to classify in Sect. 3 the computational complexity of SURJECTIVE H-COLOURING for every graph H on at most four vertices, just as Vikas [15] did for the H-COMPACTION problem. A classification of SURJECTIVE H-COLOURING for target graphs H on at most four vertices has also been announced by Vikas in [16], and it is interesting to note that NP-hardness proofs for H-COMPACTION of [15] may lift to NP-hardness for SURJECTIVE H-COLOURING. However, this is not true for the reflexive cycle C_4^*, where a totally new proof was required.

1.2 Future Work

To conjecture a dichotomy of SURJECTIVE H-COLOURING between P and NP-complete seems still to be difficult. Our first goal is to prove that SURJECTIVE H-COLOURING is NP-complete for every connected graph H that is not loop-connected. However, doing this via using our current techniques does not seem straightforward and we may need new hardness reductions. Another way forward is to prove polynomial equivalence between the three problems SURJECTIVE H-COLOURING, H-COMPACTION and H-RETRACTION. However, completely achieving this goal also seems far from trivial. Our classification for target graphs H up to four vertices does show such an equivalence for these cases (see Sect. 3).

2 Two Non-adjacent Reflexive Vertices

We say that a graph is 2-*reflexive* if it contains exactly 2 reflexive vertices *that are non-adjacent*. In this section we will prove that SURJECTIVE H-COLOURING is NP-complete whenever H is connected and 2-reflexive. The problem is readily seen to be in NP. Our NP-hardness reduction uses similar ingredients as the reduction of Golovach, Paulusma and Song [8] for proving NP-hardness when H is a tree that is not loop-connected. There are, however, a number of differences. For instance, we will reduce from a factor cut problem instead of the less general matching cut problem used in [8]. We will explain these two problems and prove NP-hardness for the former one in Sect. 2.1. Then in Sect. 2.2 we give our hardness reduction, and in Sect. 2.3 we extend our result to be valid for target graphs H with more than two reflexive vertices as long as these reflexive vertices can be partitioned into two non-adjacent sets of twin vertices.

2.1 Factor Cuts

Let $G = (V_G, E_G)$ be a connected graph. For $v \in V_G$ and $E \subseteq E_G$, let $d_E(v)$ denote the number of edges of E incident with v. For a partition (V_1, V_2) of V_G, let $E_G(V_1, V_2)$ denote the set of edges between V_1 and V_2 in G.

Let i and j be positive integers, $i \leq j$. Let (V_1, V_2) be a partition of V_G and let $M = E_G(V_1, V_2)$. Then (V_1, V_2) is an (i, j)-*factor cut* of G if, for all $v \in V_1$, $d_M(v) \leq i$, and, for all $v \in V_2$, $d_M(v) \leq j$. Observe that if a vertex v exists with degree at most j, then there is a trivial (i, j)-factor cut $(V \setminus \{v\}, \{v\})$. Two distinct vertices s and t in V_G are (i, j)-*factor roots* of G if, for each (i, j)-factor cut (V_1, V_2) of G, s and t belong to different parts of the partition and, if $i < j$, $s \in V_1$ and $t \in V_2$ (of course, if $i = j$, we do not require the latter condition as (V_2, V_1) is also an (i, j)-factor cut). We note that when no (i, j)-factor cut exists, every pair of vertices is a pair of (i, j)-factor roots. We define the following decision problem.

(i, j)-FACTOR CUT WITH ROOTS

  Instance: a connected graph G with (i, j)-factor roots s and t.
  Question: does G have an (i, j)-factor cut?

We emphasize that the (i,j)-factor roots are given as part of the input. That is, the problem asks whether or not an (i,j)-factor cut (V_1, V_2) exists, but we know already that if it does, then s and t belong to different parts of the partition. That is, we actually define (i,j)-FACTOR CUT WITH ROOTS to be a promise problem in which we assume that if an (i,j)-factor cut exists then it has the property that s and t belong to different parts of the partition. The promise class may not itself be polynomially recognizable but one may readily find a subclass of it that is polynomially recognizable and includes all the instances we need for NP-hardness. In fact this will become clear when reading the proof of Theorem 1 but we refer also to [8] where such a subclass is given for the case $(i,j) = (1,1)$. A $(1,1)$-factor cut (V_1, V_2) of G is also known as a *matching cut*, as no two edges in $E_G(V_1, V_2)$ have a common end-vertex, that is, $E_G(V_1, V_2)$ is a *matching*. Similarly $(1,1)$-FACTOR CUT WITH ROOTS is known as MATCHING CUT WITH ROOTS and was proved NP-complete by Golovach, Paulusma and Song [8] (by making an observation about the proof of the result of Patrignani and Pizzonia [12] that deciding whether or not any given graph has a matching cut is NP-complete). The proof of Theorem 1 has been omitted.

Theorem 1. *Let i and j be positive integers, $i \leq j$. Then (i,j)-FACTOR CUT WITH ROOTS is NP-complete.*

2.2   The Hardness Reduction

Let H be a connected 2-reflexive target graph. Let p and q be the two (non-adjacent) reflexive vertices of H. The *length* of a path is its number of edges. The *distance* between two vertices u and v in a graph G is the length of a shortest path between them and is denoted $\mathrm{dist}_G(u,v)$. We define two induced subgraphs H_1 and H_2 of H whose vertex sets partition V_H. First H_1 contains those vertices of H that are closer to p than to q; and H_2 contains those vertices that are at least as close to q as to p (so contains any vertex equidistant to p and q). That is, $V_{H_1} = \{v \in V_H : \mathrm{dist}_H(v,p) < \mathrm{dist}_H(v,q)\}$ and $V_{H_2} = \{v \in V_H : \mathrm{dist}_H(v,q) \leq \mathrm{dist}_H(v,p)\}$. See Fig. 2 for an example. The following lemma follows immediately from our assumption that H is connected.

Lemma 1. *Both H_1 and H_2 are connected. Moreover, $\mathrm{dist}_{H_1}(x,p) = \mathrm{dist}_H(x,p)$ for every $x \in V_{H_1}$ and $\mathrm{dist}_{H_2}(x,q) = \mathrm{dist}_H(x,q)$ for every $x \in V_{H_2}$.*

A *clique* is a subset of vertices of G that are pairwise adjacent to each other. Let ω denote the size of a largest clique in H.

From graphs H_1 and H_2 we construct graphs F_1 and F_2, respectively, in the following way:

1. for each $x \notin \{p, q\}$, create a vertex t_x^1;
2. for p, create ω vertices $t_p^1, \ldots, t_p^\omega$;
3. for q, create ω vertices $t_q^1, \ldots, t_q^\omega$;
4. for $i = 1, 2$, add an edge in F_i between any two vertices t_x^h and t_y^j if and only if xy is an edge of E_{H_i}.

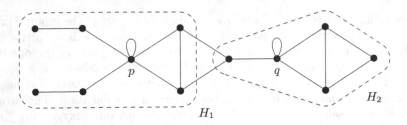

Fig. 2. An example of the construction of graphs H_1 and H_2 from a connected 2-reflexive target graph H with $\omega = 3$.

We observe that since p and q are reflexive, there are edges pp and qq, hence $t_p^1, \ldots, t_p^\omega$ and $t_q^1, \ldots, t_q^\omega$ form cliques of size ω. Note also that F_1 is the graph obtained by taking H_1 and replacing p by a clique of size ω. Similarly, F_2 is the graph obtained by taking H_2 and replacing q by a clique of size ω. We say that $t_p^1, \ldots, t_p^\omega$ are the roots of F_1 and that $t_q^1, \ldots, t_q^\omega$ are the roots of F_2. Figure 3 shows an example of the graphs F_1 and F_2 obtained from the graph H in Fig. 2.

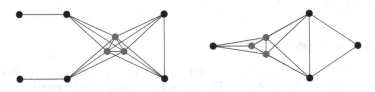

Fig. 3. The graphs F_1 (left) and F_2 (right) resulting from the graph H in Fig. 2.

Let $\ell = \mathrm{dist}_H(p, q) \geq 2$ denote the distance between p and q. Let N_p be the set of neighbours of p that are each on some shortest path (thus of length ℓ) from p to q in H. Let r_p be the size of a largest clique in N_p. We define N_q and r_q similarly. We will reduce from (r_p, r_q)-FACTOR CUT WITH ROOTS, which is NP-complete due to Theorem 1. Hence, consider an instance (G, s, t) of (r_p, r_q)-FACTOR CUT WITH ROOTS, where G is a connected graph and s and t form the (ordered) pair of (r_p, r_q)-factor roots of G. Recall that we assume that G is irreflexive.

We say that we *identify* two vertices u and v of a graph when we remove them from the graph and replace them with a single vertex that we make adjacent to every vertex that was adjacent to u or v. From F_1, F_2, and G we construct a new graph G' as follows:

1. For each edge $e = uv \in E_G$, we do as follows. We create four vertices, $g_{u,e}^r$, $g_{u,e}^b$, $g_{v,e}^r$ and $g_{v,e}^b$. We also create two paths P_e^1 and P_e^2, each of length $\ell - 2$, between $g_{u,e}^r$ and $g_{v,e}^b$, and between $g_{v,e}^r$ and $g_{u,e}^b$, respectively. If $\ell = 2$ we identify $g_{u,e}^r$ and $g_{v,e}^b$ and $g_{v,e}^r$ and $g_{u,e}^b$ to get paths of length 0.

2. For each vertex $u \in V_G$, we do as follows. First we construct a clique C_u on ω vertices. We denote these vertices by $g_u^1, \ldots, g_u^\omega$. We then make every vertex in C_u adjacent to both $g_{u,e}^r$ and $g_{u,e}^b$ for every edge e incident to u; we call $g_{u,e}^r$ and $g_{u,e}^b$ a *red* and *blue* neighbour of C_u, respectively; if $\ell = 2$, then the vertex obtained by identifying two vertices $g_{u,e}^r$ and $g_{v,e}^b$, or $g_{v,e}^r$ and $g_{u,e}^b$ is simultaneously a red neighbour of one clique and a blue neighbour of another one. Finally, for every two edges e and e' incident to u, we make $g_{u,e}^r$ and $g_{u,e'}^r$ adjacent, that is, the set of red neighbours of C_u form a clique, whereas the set of blue neighbours form an independent sets.

3. We add F_1 by identifying t_p^i and g_s^i for $i = 1, \ldots, \omega$, and we add F_2 by identifying t_q^i and g_t^i for $i = 1, \ldots, \omega$. We denote the vertices in F_1 and F_2 in G' by their label t_x^i in F_1 or F_2.

See Fig. 4 for an example of a graph G'.

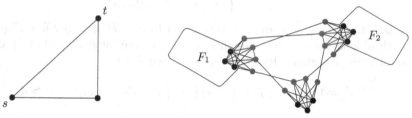

(a) An example of a graph G with a $(1,2)$-factor cut with $(1,2)$-factor roots s and t.

(b) The corresponding graph G' where H is a 2-reflexive target graph with $\ell = 3$ and $\omega = 3$.

Fig. 4. An example of a graph G and the corresponding graph G'. (Color figure online)

The next lemma describes a straightforward property of graph homomorphisms that will prove useful.

Lemma 2. *If there exists a homomorphism $h : G' \to H$ then $\mathrm{dist}_{G'}(u, v) \geq \mathrm{dist}_H(h(u), h(v))$ for every pair of vertices $u, v \in V_{G'}$.*

Here is the key property of our construction (proof omitted).

Lemma 3. *For every homomorphism h from G' to H, there exists at least one clique C_a with $p \in h(C_a)$ and at least one clique C_b with $q \in h(C_b)$.*

Proof Sketch. Since for each $u \in V_G$ and any edge e incident to u, every clique $C_u \cup \{g_{u,e}^r\}$ in G' is of size at least $\omega + 1$, we find that h must map at least two of its vertices to a reflexive vertex, so either to p or q. Hence, for every $u \in V_G$, we find that h maps at least one vertex of C_u to either p or q.

We prove the lemma by contradiction. We will assume that h does not map any vertex of any C_u to q, thus $p \in h(C_u)$ for all $u \in V_G$. We will note later that if instead $q \in h(C_u)$ for all $u \in V_G$ we can obtain a contradiction in the same way.

We consider two vertices $t_p^i \in F_1$ and $t_q^j \in F_2$ such that $h(t_p^i) = h(t_q^j) = p$. Without loss of generality let $i = j = 1$. We shall refer to these vertices as t_p and t_q respectively. We now consider a vertex $v \in V_{F_1} \cup V_{F_2}$. By Lemma 2, $\mathrm{dist}_{G'}(v, t_p) \geq \mathrm{dist}_H(h(v), p)$ and $\mathrm{dist}_{G'}(v, t_q) \geq \mathrm{dist}_H(h(v), p)$. In other words:

$$\min\left(\mathrm{dist}_{G'}(v, t_p), \mathrm{dist}_{G'}(v, t_q)\right) \geq \mathrm{dist}_H(h(v), p).$$

In fact by applying Lemma 2 we can generalize this further to any vertex mapped to p by h:

$$\min_{w \in h^{-1}(p)} \left(\mathrm{dist}_{G'}(v, w)\right) \geq \mathrm{dist}_H(h(v), p). \tag{1}$$

For every $v \in V_{G'}$ we define a value $\mathcal{D}(v)$ as follows:

$$\mathcal{D}(v) = \begin{cases} \mathrm{dist}_{F_1}(v, t_p) & \text{if } v \in F_1 \\ \mathrm{dist}_{F_2}(v, t_q) & \text{if } v \in F_2 \\ \lfloor \ell/2 \rfloor & \text{otherwise} \end{cases}$$

As $h(t_p) = h(t_q) = p$ and any vertex not in $F_1 \cup F_2$ is either in a clique or on a path of length ℓ between two cliques, we can use inequality (1) to prove that the following inequality holds for any distance $d \geq \ell$:

$$\left|\{t_w^1 \in V_{F_1} \cup V_{F_2} : \mathcal{D}(t_w^1) \geq d\}\right| \geq \left|\{w \in V_H : \mathrm{dist}_H(w, p) \geq d\}\right|. \tag{2}$$

In the remainder we only present the intuition behind the final part of the proof. Consider the graphs F_1, F_2 and H in the example shown in Fig. 5. We recall that every vertex v (other than p or q) has a single corresponding vertex t_v in F_1 or F_2. We may naturally want to map the vertices of F_1 onto the vertices of H_1,

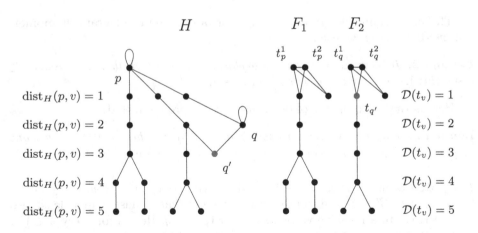

Fig. 5. An example of a graph H with corresponding graphs F_1 and F_2. Vertices in H equidistant from p are plotted at the same vertical position and likewise vertices $t_v \in F_1$ and $t_w \in F_2$ with $\mathcal{D}(t_v) = \mathcal{D}(t_w)$ are plotted at the same vertical position. The vertices $q' \in H$ and corresponding $t_{q'} \in F_2$ are highlighted.

which is possible by definition of F_1. However, when we try to map the vertices of F_2 onto the vertices of H_2, with $h(t_q^i) = p$ (for some i), we will prove that there is at least one vertex q' in H_2 which is further from p in H than it is from q and that cannot be mapped to and thus violates the surjectivity constraint. In Fig. 5 this vertex, which will play a special role in our proof, is shown in red. In the example of this figure, $\ell = 3$ and we observe that there are ten vertices (including q') in H with $\mathrm{dist}_H(p,v) \geq 3$ but only nine vertices (excluding q') in $F_1 \cup F_2$ with $\mathcal{D}(t_v) \geq 3$ which could be mapped to these vertices. This contradicts inequality (2). □

We are now ready to state and prove our main result.

Theorem 2. *For every connected 2-reflexive graph H, the* SURJECTIVE H-COLOURING *problem is* NP-*complete.*

Proof. Let H be a connected 2-reflexive graph with reflexive vertices p and q at distance $\ell \geq 2$ from each other. Let ω be the size of a largest clique in H. We define the graphs H_1, H_2, F_1 and F_2 and values r_p, r_q as above. Recall that the problem is readily seen to be in NP and that we reduce from (r_p, r_q)-FACTOR CUT WITH ROOTS. From F_1, F_2 and an instance (G, s, t) of the latter problem we construct the graph G'. We claim that G has an (r_p, r_q)-factor cut (V_1, V_2) if and only if there exists a surjective homomorphism h from G' to H.

First suppose that G has an (r_p, r_q)-factor cut (V_1, V_2). By definition, $s \in V_1$ and $t \in V_2$. We define a homomorphism h as follows. For every $x \in V_{F_1} \cup V_{F_2}$, we let h map t_x^1 to x. This shows that h is surjective. It remains to define h on the other vertices. For every $u \in V_G$, let h map all of C_u to p if u is in V_1 and let h map all of C_u to q if u is in V_2 (note that this is consistent with how we defined h so far). For each $uv \in E_G$ with $u, v \in V_1$, we map the vertices of the paths P_e^1 and P_e^2 to p. For each $uv \in E_G$ with $u, v \in V_2$, we map the vertices of the paths P_e^1 and P_e^2 to q. We are left to show that the vertices of the remaining paths P_e^1 and P_e^2 can be mapped to appropriate vertices of H.

Note that the red neighbours of each C_u form a clique (whereas all blue vertices of each C_u form an independent set and inner vertices of paths P_e^1 and P_e^2 have degree 2). However, as (V_1, V_2) is an (r_p, r_q)-factor cut of G, all but at most r_p vertices of these red cliques have been mapped to p already if $u \in V_1$ and all but at most r_q vertices have been mapped to q already if $u \in V_2$. By definition of r_p and r_q, this means that we can map the vertices of the paths P_e^1 and P_e^2 with $e = uv$ for $u \in V_1$ and $v \in V_2$ to vertices of appropriate shortest paths between p and q in H, so that h is a homomorphism from G' to H (recall that we already showed surjectivity). In particular, the clique formed by the red neighbours of each C_u is mapped to a clique in $N_p \cup \{p\}$ or $N_q \cup \{q\}$.

Now suppose that there exists a surjective homomorphism h from G' to H. For a clique C_u, we may choose any edge e incident to u, such that $C_u' = C_u \cup \{g_{u,e}^r\}$ is a clique of size $\omega + 1$. Since H contains no cliques larger than ω, we find that h maps each clique C_u' (which has size $\omega + 1$) to a clique in H that contains a reflexive vertex. Note that at least two vertices of C_u' are mapped to a reflexive vertex. Hence we can define the following partition of V_G. We let

$V_1 = \{v \in V_G : p \in h(C_v)\}$ and $V_2 = V_G \setminus V_1 = \{v \in V_G : q \in h(C_v)\}$. Lemma 3 tells us that $V_1 \neq \emptyset$ and $V_2 \neq \emptyset$. We define $M = \{uv \in E_G : u \in V_1, v \in V_2\}$.

Let $e = uv$ be an arbitrary edge in M. By definition, h maps all of C_u to a clique containing p and all of C_v to a clique containing q. Hence, the vertices of the two paths P_e^1 and P_e^2 must be mapped to the vertices of a shortest path between p and q. At most r_p red neighbours of every C_u with $u \in V_1$ can be mapped to a vertex other than p. This is because these red neighbours form a clique. As such they must be mapped onto vertices that form a clique in H. As such vertices lie on a shortest path from p to q, the clique in H has size at most r_p. Similarly, at most r_q red neighbours of every C_u with $u \in V_2$ can be mapped to a vertex other than q. As such, (V_1, V_2) is an (r_p, r_q)-factor cut in G. $\qquad\square$

2.3   A Small Extension

Two vertices u and v in a graph G are *true twins* if they are adjacent to each other and share the same neighbours in $V_G \setminus \{u, v\}$. Let $H^{(i,j)}$ be a graph obtained from a connected 2-reflexive graph H with reflexive vertices p and q after introducing i reflexive true twins of p and j reflexive true twins of q. In the graph G' we increase the cliques C_u to size $\omega + \max(i,j)$. We call the resulting graph G''. Then it is readily seen that there exists a surjective homomorphism from G' to H if and only if there exists a surjective homomorphism from G'' to $H^{(i,j)}$.

Theorem 3. *For every connected 2-reflexive graph H and integers $i, j \geq 0$,* SURJECTIVE $H^{(i,j)}$-COLOURING *is* NP-*complete.*

3   Target Graphs of at Most Four Vertices

For proving Theorem 4 we use Theorem 3 combined with known results [11,15]; we omit the details. In Fig. 6, the three graphs C_4^*, D and paw* are displayed.

Fig. 6. The graphs C_4^*, D and paw*.

Theorem 4. *Let H be a graph with $|V_H| \leq 4$. Then* SURJECTIVE H-COLOURING *is* NP-*complete if some connected component of H is not loop-connected or is an irreflexive complete graph on at least three vertices, or $H \in \{C_4^*, D, \text{paw}^*\}$. Otherwise* SURJECTIVE H-COLOURING *is polynomial-time solvable.*

Theorem 4 corresponds to Vikas' complexity classification of H-COMPACTION for targets graphs H of at most four vertices. Vikas [15] showed that H-COMPACTION and H-RETRACTION are polynomially equivalent for target graphs H of at most four vertices. Thus, we obtain the following corollary.

Corollary 1. *Let H be a graph on at most four vertices. Then the three problems* SURJECTIVE H-COLOURING, H-COMPACTION *and* H-RETRACTION *are polynomially equivalent.*

References

1. Bodirsky, M., Kára, J., Martin, B.: The complexity of surjective homomorphism problems - a survey. Discrete Appl. Math. **160**, 1680–1690 (2012)
2. Feder, T., Hell, P., Huang, J.: Bi-arc graphs and the complexity of list homomorphisms. J. Graph Theory **42**, 61–80 (2003)
3. Feder, T., Hell, P., Jonsson, P., Krokhin, A., Nordh, G.: Retractions to pseudo-forests. SIAM J. Discrete Math. **24**, 101–112 (2010)
4. Feder, T., Vardi, M.Y.: The computational structure of monotone monadic SNP and constraint satisfaction: a study through datalog and group theory. SIAM J. Comput. **28**, 57–104 (1998)
5. Fiala, J., Kratochvíl, J.: Locally constrained graph homomorphisms - structure, complexity, and applications. Comput. Sci. Rev. **2**, 97–111 (2008)
6. Fiala, J., Paulusma, D.: A complete complexity classification of the role assignment problem. Theoret. Comput. Sci. **349**, 67–81 (2005)
7. Golovach, P.A., Lidický, B., Martin, B., Paulusma, D.: Finding vertex-surjective graph homomorphisms. Acta Informatica **49**, 381–394 (2012)
8. Golovach, P.A., Paulusma, D., Song, J.: Computing vertex-surjective homomorphisms to partially reflexive trees. Theoret. Comput. Sci. **457**, 86–100 (2012)
9. Hell, P., Nešetřil, J.: On the complexity of H-colouring. J. Comb. Theory Ser. B **48**, 92–110 (1990)
10. Hell, P., Nešetřil, J.: Graphs and Homomorphisms. Oxford University Press, Oxford (2004)
11. Martin, B., Paulusma, D.: The computational complexity of disconnected cut and $2K_2$-partition. J. Comb. Theory Ser. B **111**, 17–37 (2015)
12. Patrignani, M., Pizzonia, M.: The complexity of the matching-cut problem. In: Brandstädt, A., Le, V.B. (eds.) WG 2001. LNCS, vol. 2204, pp. 284–295. Springer, Heidelberg (2001). doi:10.1007/3-540-45477-2_26
13. Vikas, N.: Computational complexity of compaction to reflexive cycles. SIAM J. Comput. **32**, 253–280 (2002)
14. Vikas, N.: Compaction, retraction, and constraint satisfaction. SIAM J. Comput. **33**, 761–782 (2004)
15. Vikas, N.: A complete and equal computational complexity classification of compaction and retraction to all graphs with at most four vertices and some general results. J. Comput. Syst. Sci. **71**, 406–439 (2005)
16. Vikas, N.: Algorithms for partition of some class of graphs under compaction and vertex-compaction. Algorithmica **67**, 180–206 (2013)

On Higher Effective Descriptive Set Theory

Margarita Korovina[1]([⊠]) and Oleg Kudinov[2]

[1] A.P. Ershov Institute of Informatics Systems, SbRAS, Novosibirsk, Russia
rita.korovina@gmail.com
[2] Sobolev Institute of Mathematics, SbRAS, Novosibirsk, Russia
kud@math.nsc.ru

Abstract. In the framework of computable topology, we propose an approach how to develop higher effective descriptive set theory. We introduce a wide class $\mathbb{K}$ of effective T_0-spaces admitting Borel point recovering. For this class we propose the notion of an (α, m)-retractive morphism that gives a great opportunity to extend classical results from EDST to the class $\mathbb{K}$. We illustrate this by several examples.

Keywords: Higher effective descriptive set theory · Effective topological space · Effective T_0–space admitting Borel point recovering · (α, m)-retractive morphism · Effective Borel and Lusin hierarchies · Suslin-Kleene Theorem · Uniformisation Theorem

1 Introduction

Having a long term history, classical descriptive set theory (DST) has been based on the fundamental idea that all Polish spaces have common properties related to definability of functions and sets as well as to the resulting Borel and Lusin hierarchies [5,13]. The similar idea can be used for the computable Polish spaces under consideration of effective versions of the corresponding hierarchies. One of the approaches to effective descriptive set theory (EDST) has been proposed in [10,12,14] and developed in [3,13,18] among others, where most of results have been obtained for the computable Polish spaces. A comparison of concepts and results from computable analysis and EDST has been presented in [4].

It is worth noting that in the non-effective case a significant progress has been done in [2,17], where a big part of DST has been developed first for ω-continuous domains and then for the wider class of quasi-Polish spaces. Two of the main problems that naturally arise in EDST are the following. The first one is to discover wide classes of effective topological spaces for which the main results of DST hold. Here the class of quasi-Polish spaces looks like a promising candidate. The second problem is to discover wide classes of effective topological spaces admitting results of EDST related to higher levels of the effective Lusin hierarchy.

The research has been partially supported by the DFG grants CAVER BE 1267/14-1 and WERA MU 1801/5-1.

J. Kari et al. (Eds.): CiE 2017, LNCS 10307, pp. 282–291, 2017.
DOI: 10.1007/978-3-319-58741-7_27

The main results of our paper concern this second problem. Informally, in the paper we consider a part of EDST that is nearly Δ_1^1–level or above in the effective Lusin hierarchy and refer to this part as higher effective descriptive set theory (Higher EDST). In order to give a flavor of Higher EDST it is worth noting that an effective version of the Hausdorff theorem does not belong to Higher EDST while the extended Suslin-Kleene theorem does.

The main contributions of the paper are the following. To attack the second problem mentioned above we propose a wide class of effective topological spaces admitting effective Borel point recovering which contains effective quasi-Polish spaces, weakly-effective ω–continuous domains as proper subclasses. For this class we provide a fruitful technique that allows us to effectivise classical results from DST. We illustrate this by the Suslin-Kleene Theorem, Uniformisation Theorem.

2   Notations and Preliminaries

We refer the reader to [14,15] for basic definitions and fundamental concepts of recursion theory, to [3,5,13] for basic definitions and fundamental concepts of DST and EDST. We use bold Greek letters $\alpha, \beta, \gamma, \ldots$ to denote numberings and light Greek letters to denote ordinals. We use only computable ordinals i.e. that are less than ω_1^{CK}. We work with the Baire space $\mathcal{N} = (\omega^\omega, \alpha_\mathcal{N})$, the Cantor space $\mathcal{C} = (2^\omega, \alpha_\mathcal{C})$ with the standard topologies and numberings of the bases, and the space $\mathbb{P} = (\mathcal{P}(\omega), \beta)$, where β is the standard numbering of the topology base generated by all open sets of the type $W_D = \{I \subseteq \omega \mid D \subseteq I\}$, where D is a finite subset of ω.

3   Effective Topological Spaces and Hierarchies

Let (X, τ, α) be a topological space, where X is a non-empty set, $B_\tau \subseteq 2^X$ is a base of the topology τ and $\alpha : \omega \to B_\tau$ is a numbering. In notations we skip τ since it can be recovered by α. Further on we will often abbreviate (X, α) by X if α is clear from a context.

Definition 1. *A topological space (X, α) is effective if the following condition holds.*

- *There exists a computable function $g : \omega \times \omega \times \omega \to \omega$ such that*

$$\alpha(i) \bigcap \alpha(j) = \bigcup_{n \in \omega} \alpha(g(i, j, n)).$$

Now we recall the notion of an effectively enumerable topological space.

Definition 2. [8] *An effective topological space (X, α) is effectively enumerable if the following condition holds.*

- *The set $\{i \mid \alpha(i) \neq \varnothing\}$ is computably enumerable.*

It is worth noting that the similar concept of a weakly computable cb_0–space has been used in [18].

Definition 3. *Let (X, α) be an effective topological space. A set $A \subseteq X$ is effectively open if there exists a computably enumerable set $V \subseteq \omega$ such that*

$$A = \bigcup_{n \in V} \alpha(n).$$

Let $\mathcal{O}_X$ denote the set of all open subsets of X and $\mathcal{O}_X^e$ denote the set of all effectively open subsets of X. The set $\mathcal{O}_X^e$ is closed under intersection and union since the class of effectively enumerable sets is a lattice. The following proposition is a natural corollary of the definition.

Proposition 1. [7] *For every effective topological space X there exists a principal computable numbering α_X^e of $\mathcal{O}_X^e$.*

In this paper we use the notation $f : X \to Y$ for a partial function unless the word *total* is written.

Definition 4. *Let $X = (X, \alpha)$ be an effective topological space and $Y = (Y, \beta)$ be an effective T_0–space. A total function $f : X \to Y$ is called computable if there exists a computable function $H : \omega^2 \to \omega$ such that*

$$f^{-1}(\beta(m)) = \bigcup_{i \in \omega} \alpha(H(m, i)).$$

We adopt the notion of the effective Borel and Lusin hierarchies for computable Polish spaces [13] to effective T_0–spaces. Put for finite ordinals

$\Sigma_1^0 = $ all effecively open sets,

$\Sigma_{n+1}^0 = \exists^\omega$(the set of finite boolean combinations of Σ_n^0),

where $\exists^\omega$ denotes the projection along ω,

$\Pi_n^0 = \neg \Sigma_n^0$,

$\Delta_n^0 = \Sigma_n^0 \bigcap \Pi_n^0.$

Following [13], $\{(\Sigma_n^0, \Pi_n^0)\}_{n<\omega}$ is called the *Kleene hierarchy*. Further on, for $R \subseteq X$ and $R \in \Sigma_n^0$ we write $R \in \Sigma_n^0[X]$.

Remark 1. It is worth noting that the difference between our and Moschovakis's definitions of the Kleene hierarchy [13] is based on the following observation. For the space $\mathbb{P}$, $\Sigma_1^0[\mathbb{P}] \not\subseteq \exists^\omega \Pi_1^0[\mathbb{P}]$. To show that it is sufficient to prove that $A = \{x \subseteq \omega \mid 0 \notin x\} \notin \forall^\omega \Sigma_1^0[\mathbb{P}]$. Assume contrary $A \in \forall^\omega \Sigma_1^0[\mathbb{P}]$. Then $A = \bigcap_{n \in \omega} E_n$, where E_n are effectively open sets. Therefore, for all $n \in \omega$, $E_n \neq \varnothing$ and, by the properties of the topology on $\mathbb{P}$, $\omega \in E_n$. So, $\omega \in A$. This contradicts the definition of A. In fact this definition of the Kleene hierarchy has been proposed in [18].

One way to introduce the *effective Borel hierarchy* $\{(\Sigma^0_\xi, \Pi^0_\xi)\}_{\xi < \omega^{CK}_1}$ considered in [10] is to use effective Borel coding [5,13], i.e., for any B such that $B = B_\alpha$ for some computable $\alpha \in \mathcal{N}$, $\xi(B) = \inf \{\xi(\alpha) \mid \alpha \text{ is computable and } B = B_\alpha\}$, where $\xi(\alpha)$ is introduced in [5]. Another more direct way proposed in [14] for $\mathcal{N}$ and in [3] for computable Polish Spaces can be adopted to effective T_0–spaces [18]. In different situations one of the approaches has its advantages. For inductive considerations the second one is more preferable. In particular, in the framework of the second approach the inclusions $\Sigma^0_\alpha \subseteq \Sigma^0_{\alpha+1}$, $\Pi^0_\alpha \subseteq \Sigma^0_{\alpha+1}$ are just parts of the definition. Therefore, $\Sigma^0_\alpha \cup \Pi^0_\alpha \subseteq \Delta^0_{\alpha+1} = \Sigma^0_{\alpha+1} \cap \Pi^0_{\alpha+1}$.

In [3,18] the effective Lusin hierarchy $\{(\Sigma^1_n[X], \Pi^1_n[X])\}_{n<\omega}$ is defined by induction as follows:

$$\Sigma^1_1[X] = \{\mathrm{pr}_\mathcal{N}(B) \mid B \in \Pi^0_2[\mathcal{N} \times X]\},$$
$$\Sigma^1_{n+1}[X] = \{\mathrm{pr}_\mathcal{N}(B) \mid B \in \Pi^1_n[\mathcal{N} \times X]\},$$
$$\Pi^1_n[X] = \neg\Sigma^1_n[X],$$
$$\Delta^1_n[X] = \Sigma^1_n[X] \cap \Pi^1_n[X].$$

Some properties of the effective Lusin hierarchy on particular spaces (e.g. perfect computable Polish spaces, computable reflective ω-algebraic domains) can be found in [18].

Lemma 1. *For an effective T_0–space X, $\{(x, y) \mid x = y\} \in \Pi^0_2[X \times X]$.*

4   The Class $\mathbb{K}$

Definition 5. *An effective T_0–space (X, α) is said to admit effective Borel point recovering if the following condition holds.*

- *The set $\{A_x \mid x \in X\}$ is a Δ^1_1-subset of $\mathcal{P}(\omega)$, where $A_x = \{n \mid x \in \alpha(n)\}$. Here $\mathcal{P}(\omega)$ is considered as the Cantor space $\mathcal{C}$.*

We denote this class of effective T_0–spaces admitting Borel point recovering by $\mathbb{K}$. Now we show the correspondence between spaces from $\mathbb{K}$ and Δ^1_1-subsets of $\mathbb{P}$.

Proposition 2. $\Delta^1_1[\mathcal{C}] = \Delta^1_1[\mathbb{P}]$.

Proof. The inclusion $\Delta^1_1[\mathcal{C}] \supseteq \Delta^1_1[\mathbb{P}]$ follows from the fact that $\mathcal{C}$ has more Π^0_2-subsets than $\mathbb{P}$. Now us take id : $(\mathbb{P}, \beta) \to (\mathcal{C}, \alpha_\mathcal{C})$. Since $\mathrm{id}^{-1}(\alpha_\mathcal{C}(n)) = \beta(h(n)) \setminus \beta(t(n))$, where h and t are computable functions, $\Pi^0_2[\mathcal{C}] \subseteq \Pi^0_3[\mathbb{P}]$. Therefore, $\Delta^1_1[\mathcal{C}] \subseteq \Delta^1_1[\mathbb{P}]$. $\square$

Definition 6. *We write $(X_0, \beta_0) < (X, \beta)$ if*

1. *X_0 is a subset of X,*
2. *$\beta_0(n) = \beta(n) \cap X$ for any $n \in \omega$.*

Theorem 1. *For any effective (effectively enumerable) T_0–space (X, α) the following assertions are equivalent.*

1. $(X, \alpha) \in \mathbb{K}$.
2. (X, α) *is computably homeomorphic to some* $(X_0, \beta_0) < (\mathbb{P}, \beta)$ *with*

$$X_0 \in \Delta_1^1[\mathbb{P}].$$

Proof. $(2 \to 1)$. Assume $X \in \Delta_1^1[\mathbb{P}]$. Consider (X, α), where α is induced by β. It is easy to see that $\{P_x \mid x \in \mathbb{P}\} \in \Delta_2^0[\mathcal{C}]$, where $P_x = \{n \mid x \in \beta(n)\}$. In particular, $\mathbb{P} \in \mathbb{K}$. Let $A_y = \{n \mid y \in \alpha(n)\}$. Then

$$
\begin{aligned}
&J \in \{A_y \mid y \in X\} \leftrightarrow \\
&J \in \{P_y \mid y \in \mathbb{P}\} \wedge (\exists y \in X) J = P_y \leftrightarrow \\
&J \in \{P_y \mid y \in \mathbb{P}\} \wedge (\exists y \in X)(\forall n \in \omega)(y \in \beta(n) \leftrightarrow n \in J).
\end{aligned}
$$

This is a Σ_1^1-condition on $\mathcal{C}$.

$$
\begin{aligned}
&J \notin \{A_y \mid y \in X\} \leftrightarrow \\
&J \notin \{P_y \mid y \in \mathbb{P}\} \vee (\exists y \in \mathbb{P})(J = P_y \wedge y \notin X).
\end{aligned}
$$

This is also a Σ_1^1-condition on $\mathcal{C}$. Therefore, $\{A_y \mid y \in X\} \in \Delta_1^1[\mathcal{C}]$.

$(1 \to 2)$. Assume now that $\{A_x \mid x \in X\} \in \Delta_1^1[\mathcal{C}]$, where $A_x = \{n \mid x \in \alpha(n)\}$. We take the function $F : X \xrightarrow{1-1} \mathbb{P}$ defined by the rule $F(x) = A_x$. It is clear that F is effectively continuous, i.e. computable. Indeed, $F^{-1}(W_D) = \bigcap_{m \in D} \alpha(m)$. Let $X_0 = F(X)$ and $\beta_0(m) = W_{D_m} \cap X_0$. By definition, (X_0, β_0) is a subspace of $\mathbb{P}$. In order to prove that (X_0, β_0) is computably homeomorphic to (X, α) we show that $F^{-1} : (X_0, \beta_0) \to (X, \alpha)$ is computable. For that we check $F(\alpha(m)) = W_{\{m\}} \cap X_0$. The inclusion $F(\alpha(m)) \subseteq W_{\{m\}} \cap X_0$ is trivial. Let us show the inclusion $F(\alpha(m)) \supseteq W_{\{m\}} \cap X_0$. Let $I \in W_{\{m\}} \cap X_0$, i.e., $m \in I$ and $I \in X_0$. Since $I \in X_0$, by definition, there exists $x \in X$ that $F(x) = I$, so $I = A_x$. Since $m \in A_x$, $x \in \alpha(m)$, so $I \in F(\alpha(m))$.

Now we show that $X_0 \in \Delta_1^1[\mathbb{P}]$. Since (X_0, β_0) is computably homeomorphic to (X, α), $\{P_x \mid x \in X_0\} \in \Delta_1^1[\mathcal{C}] = \Delta_1^1[\mathbb{P}]$. For $x \in \mathbb{P}$, we have

$$
\begin{aligned}
&x \in X_0 \leftrightarrow \\
&(\exists I \in \{P_y \mid y \in X_0\}) I = P_x \leftrightarrow \\
&(\exists I \in \{P_y \mid y \in \mathbb{P}\})(\forall n \in \omega)(x \in \beta(n) \leftrightarrow n \in I)
\end{aligned}
$$

and

$$
\begin{aligned}
&x \notin X_0 \leftrightarrow \\
&(\exists I \in \{P_y \mid y \in \mathbb{P}\})(I = P_x \wedge I \notin \{P_y \mid y \in X_0\}).
\end{aligned}
$$

Therefore $X_0 \in \Delta_1^1[\mathcal{C}] = \Delta_1^1[\mathbb{P}]$. $\qquad\square$

Taking into account results from [2], where the class of quasi-Polish spaces has been introduced, we define in a natural way the notion of a computable quasi-Polish space.

Definition 7. *The space (X, α) is called computable quasi-Polish if it is computably homeomorphic to some $(X_0, \beta_0) < (\mathbb{P}, \beta)$ with*

$$X_0 \in \Pi_2^0[\mathbb{P}].$$

Now we show that there exists $X \in \mathbb{K}$ that is not computable quasi-Polish.

Proposition 3. *Let (X, α) be a computable quasi-Polish space. Then*

$$\{A_x \mid x \in X\} \in \Pi_3^0[\mathcal{C}] \bigcap \Pi_4^0[\mathbb{P}].$$

Proof. It is sufficient to consider $(X_0, \beta_0) < (\mathbb{P}, \beta)$. Let us recall $A_I = \{n \mid D_n \subseteq I\}$, where $I \in X_0$ and D_n is finite. Assume that $X_0 \in \Pi_2^0[\mathbb{P}]$. This means that $X_0 = \bigcap_{n \in \omega} E_n$, where $\{E_n\}_{n \in \omega}$ is an effective sequence of boolean combinations of effectively open sets in $\mathbb{P}$. Let us observe that if we denote $\mathcal{A}_n \rightleftharpoons \{A_x \mid x \in E_n\}$ then $\mathcal{A} \rightleftharpoons \{A_x \mid x \in X\} = \bigcap_{n \in \omega} \mathcal{A}_n$. Therefore, it is sufficient to understand the level of $\{A_x \mid x \in \beta(i)\}$. Since $\mathcal{B}_i \rightleftharpoons \{A_x \mid x \in \beta(i)\} = \{\{n \mid D_n \subseteq x\} \mid x \supseteq D_i\}$, by a routine computation it can be shown that $\{\mathcal{B}_i\}_{i \in \omega}$ is a computable sequence of $\Pi_1^0[\mathcal{C}]$–sets ($\Pi_2^0[\mathbb{P}]$–sets). Finally, $\mathcal{A}_n$ is a boolean combination of $\Sigma_2^0[\mathcal{C}]$–sets ($\Pi_3^0[\mathbb{P}]$–sets). Therefore $\mathcal{A} \in \Pi_3^0[\mathcal{C}]$ ($\mathcal{A} \in \Pi_4^0[\mathbb{P}]$). $\square$

Theorem 2. *1. There exists a space from the class $\mathbb{K}$ which is not computable quasi-Polish.*

2. There exists a perfect space from the class $\mathbb{K}$ which is not computable quasi-Polish.

Proof. We use the notations from the proposition above.

1. Let us consider $(X_0, \beta_0) < (\mathbb{P}, \beta)$ such that $X_0 \in \Delta_1^1[\mathcal{C}] \setminus \Pi_4^0[\mathcal{C}]$. Such space exist since effective hierarchies on $\mathcal{C}$ are strict. We take the function $F : \mathbb{P} \xrightarrow{1\text{-}1} \mathbb{P}$ defined by the rule $F(x) = A_x$. The function F is effectively continuous, i.e. computable. Therefore this function is continuous and computable from $\mathcal{C}$ to $\mathbb{P}$. It is clear that if $I \in \Pi_4^0[\mathbb{P}]$ then $F^{-1}(I) \in \Pi_4^0[\mathcal{C}]$. Now if we assume that X_0 is quasi-Polish then by Proposition 3 $\{A_x \mid x \in X_0\} \in \Pi_4^0[\mathbb{P}]$. So $X_0 = F^{-1}(\{A_x \mid x \in X_0\}) \in \Pi_4^0[\mathcal{C}]$. This leads to a contradiction.

2. In order to construct a perfect space we consider $\tilde{X}_0 = \{2 \cdot I \mid I \in X_0\}$. Take $X_0^* = \{I \oplus J \mid I \in X_0, J \subseteq \omega\}$. It is easy to see that X_0^* is computably homeomorphic to $\tilde{X}_0 \times \mathcal{P}(2 \cdot \omega + 1)$. So $X_0^* \in \mathbb{K}$ is perfect but not computable quasi-Polish. $\square$

5 (α, m)–Retractive Morphisms

In this section we assume that all our spaces belong to the class $\mathbb{K}$. We introduce a useful tool that provides a fruitful technique to get effective versions of classical theorems from DST that hold for all spaces from $\mathbb{K}$.

Definition 8. *Let* $f : X \to Y$ *and* $g : Y \to X$, *where* $(X, \alpha) \in \mathbb{K}$ *and* $(Y, \gamma) \in \mathbb{K}$. *The pair* (f, g) *is called a* (α, m)–*retractive morphism from* X *to* Y *(denoted by* $X \underset{g}{\overset{f}{\rightleftarrows}} Y$ *) if the following conditions hold.*

1. $f \circ g = \mathrm{id}_Y$,
2. $\mathrm{dom}(f) \in \Sigma^0_\alpha[X]$,
3. $f^{-1}(\gamma(m)) = \bigcup_{i \in I_m} \alpha(i) \cap \mathrm{dom}(f)$, *where* I_m *is a computable sequence of c.e. sets,*
4. *For* $\mathcal{O}_n = \alpha^e_X(n)$, $\{g^{-1}(\mathcal{O}_n)\}_{n \in \omega}$ *is a computable sequence of elements of* $\Sigma^0_{m+1}[X]$, *i.e.,* $\bigcup_{n \in \omega} \{n\} \times g^{-1}(\mathcal{O}_n) \in \Sigma^0_{m+1}[\omega \times X]$.

Remark 2. The mappings f and g are effective Borel functions, i.e., their preimages of effectively open sets are effective Borel sets [13,18]. From the first condition it follows that the function f is onto and the function g is a total injection the third condition corresponds to the effective continuity of f.

Remark 3. The notion of a computable sequence of elements of $\Sigma^0_m[X]$ can be defined in terms of good parametrisations for $\Sigma^0_m[X]$ (see e.g. [13]).

Lemma 2. *Let* $X \underset{g_1}{\overset{f_1}{\rightleftarrows}} X_1 \underset{g_2}{\overset{f_2}{\rightleftarrows}} X_2$, *where* (f_1, g_1) *is an* (α, n)–*retractive morphism from* X *to* X_1 *and* (f_2, g_2) *is a* (β, m)–*retractive morphism from* X_1 *to* X_2 *then there exist ordinals* $\gamma < \omega^{\mathrm{CK}_1}$ *and* $k < \omega$ *such that* $(f_2 \circ f_1, g_1 \circ g_2)$ *is a* (γ, k)–*retractive morphism from* X *to* X_2. *The ordinal* k *can be chosen as* $n + m$.

Lemma 3. *Let* $X \underset{g_1}{\overset{f_1}{\rightleftarrows}} Y_0 \underset{\varphi}{\overset{\varphi^{-1}}{\rightleftarrows}} Y$, *where* $Y_0 \subseteq \mathbb{P}$ *is a computable homeomorphic copy of* Y *under* φ, (f, g) *is an* (α, n)–*retractive morphism between* X *and* Y. *Then* $(\varphi^{-1} \circ f, g \circ \varphi)$ *is an* (α, n)–*retractive morphism between* X *and* Y.

Theorem 3. *For any space* $Y \in \mathbb{K}$ *there exists an* $(\alpha, 1)$–*retractive morphism from* $\mathcal{N}$ *to* Y *for some ordinal* $\alpha < \omega^{\mathrm{CK}}_1$.

Proof. By Theorem 1 and Lemma 3, without loss of generality we assume that $Y \in \Delta^1_1[\mathbb{P}]$. A required $(\alpha, 1)$–retractive morphism from $\mathcal{N}$ to Y is the composition of the following $(\alpha, 1)$–retractive morphisms:

$$\mathcal{N} \underset{g_1}{\overset{f_1}{\rightleftarrows}} C \underset{g_2}{\overset{f_2}{\rightleftarrows}} \mathbb{P} \underset{g_3}{\overset{f_3}{\rightleftarrows}} Y,$$

where the functions are defined as follows.

1. The function f_1 and g_1 are standard well known computable mappings: $f(\chi)(n) = \chi(n) \bmod 2$ and $g = \mathrm{id}$.
2. Put $f_2 = \mathrm{id}$ and $g_2 = \mathrm{id}$.
3. Put $f_3 = \begin{cases} x, & x \in Y \\ \uparrow, & x \notin Y \end{cases}$ and $g_3 = \mathrm{id}$.

The functions satisfy the requirements of Definition 8. Let us show non-trivial parts. It is easy to see that

$$g_2^{-1}(\alpha_{\mathcal{C}}(n)) = \{f \mid (\forall j \in D_2)f(j) = 1\} \setminus \{f \mid (\forall i \in D_1)f(i) = 1\},$$

where $\alpha_{\mathcal{C}}(n) = \{f : \omega \to \{0, 1\} \mid (\forall i \in D_1)f(i) = 0 \land (\forall j \in D_2)f(j) = 1\}$ for some finite sets D_1 and D_2 such that $D_1 \cap D_2 = \varnothing$. Therefore, $g_2^{-1}(\alpha_{\mathcal{C}}(n)) \in \Delta_2^0[\mathbb{P}]$ uniformly in n. Since $Y \in \Delta_1^1[\mathbb{P}] = \Delta_1^1[\mathcal{C}]$ and the Suslin-Kleene Theorem holds for all perfect computable Polish spaces (see e.g. [18]), $Y \in \Sigma_\alpha^0[\mathcal{C}]$. So, $Y \in \Sigma_{\alpha'}^0[\mathbb{P}]$ for some $\alpha' \geq \alpha$. $\qquad\square$

Proposition 4. *Let $Y \in \mathbb{K}$ and $\mathcal{N} \xrightarrow[\overleftarrow{g}]{f} Y$ be an $(\alpha, 1)$–retractive morphism from $\mathcal{N}$ to Y. For any $A \in \Pi_1^1[Y]$, $g(A) \in \Pi_1^1[\mathcal{N}]$.*

Proof. By the definition of (f, g), $g(A) = f^{-1}(A) \cap g(Y)$. Let us denote $Y_0 = g(Y)$ and assume $A \in \Pi_1^1[Y]$. Since f is an effective Borel function, $f^{-1}(A) \in \Pi_1^1[\mathcal{N}]$. It is easy to see that since $\{(x, y) \mid x = y\} \in \Pi_2^0[X \times X]$, $Y_0 = \{x \in \mathcal{N} \mid g(f(x)) = x\} \in \Pi_{\alpha'}^0$ for some $\alpha' \geq \alpha$, Therefore, $g(A) \in \Pi_1^1[\mathcal{N}]$. $\qquad\square$

Proposition 5. *Let $X \in \mathbb{K}$ and $B \in \Delta_1^1[X \times X]$. Then $A = \mathrm{pr}_X(B) \in \Sigma_1^1[X]$.*

Proof. Let us fix an $(\alpha, 1)$–retractive morphism $\mathcal{N} \xrightarrow[\overleftarrow{g}]{f} X$. Put $B' = \{(x, y) \in X \times \mathcal{N} \mid B(x, f(y))\}$. It is clear that $B \in \Sigma_\alpha^0[X \times \mathcal{N}]$ for some $\alpha < \omega_1^{\mathrm{CK}}$. Let us show that $x \in A \leftrightarrow (\exists y \in \mathcal{N})B'(x, y)$.
($\to$). Assume $B(x, z)$ for some $z \in X$. Put $y = g(z)$. By definition, $f(y) = z$. We have $B(x, f(y))$, i.e. $B'(x, y)$.
($\leftarrow$). Assume $B'(x, y)$, i.e. $B(x, f(y))$ for some $y \in \mathcal{N}$. Put $z = f(y)$. Then $B(x, z)$, i.e. $x \in A$. So, the projections of Σ_1^1–sets along X are again Σ_1^1–sets. $\qquad\square$

6 Transferring EDST Theorems

In this section we show how (α, m)–retractive morphisms can be used to make an effectivisation of classical results from DST that hold on the spaces from this class $\mathbb{K}$. The following proposition is an extension of the classical Suslin-Kleene Theorem.

Theorem 4. *For any $Y \in \mathbb{K}$,*

$$\Delta_1^1[Y] = \bigcup_{\alpha < \omega_1^{\mathrm{CK}}} \Sigma_\alpha^0[Y].$$

Proof. The inclusion $\Delta_1^1 \supseteq \bigcup_{\alpha < \omega_1^{\mathrm{CK}}} \Sigma_\alpha^0$ follows from the observation that $\Sigma_1^1[Y]$ is closed under effective infinite unions, intersections. In order to show the inclusion $\Delta_1^1 \subseteq \bigcup_{\alpha < \omega_1^{\mathrm{CK}}} \Sigma_\alpha^0$ let us fix an $(\alpha, 1)$–retractive morphism $\mathcal{N} \xrightarrow[\overleftarrow{g}]{f} Y$. Let us denote $Y_0 = g(Y)$ and assume $A \in \Delta_1^1[Y]$. Since f is an effective Borel function, $f^{-1}(A) \in \Delta_1^1[\mathcal{N}]$. By the Suslin-Kleene Theorem for $\mathcal{N}$ there exists a computable ordinal α such that $f^{-1}(A) \in \Delta_\alpha^0[\mathcal{N}]$. Since $f^{-1}(A) \cap g(Y) = g(A)$, $\{(x,y) \mid x = y\} \in \Pi_2^0[X \times X]$, $Y_0 = \{x \in \mathcal{N} \mid g(f(x)) = x\} \in \Pi_{\alpha'}^0[\mathcal{N}]$ for some ordinal $\alpha' \geq \alpha$, Therefore, $f^{-1}(A) \cap g(Y) \in \Sigma_\beta^0[\mathcal{N}]$ for some $\beta \geq \alpha$. Finally, $g^{-1}(f^{-1}(A) \cap g(Y)) = A$. Therefore, $A \in \Sigma_\gamma^0[X]$ for some $\gamma < \omega_1^{\mathrm{CK}}$. $\qquad\square$

The following proposition is an extension of the classical Novikov-Kondo-Addison Uniformisation Theorem.

Theorem 5 (Uniformisation). *Let $X \in \mathbb{K}$. If $Y \in \Pi_1^1[X \times X]$ then there exists a function $F : X \to X$ such that*

1. *The graph Γ_F of the function F is a subset of Y.*
2. $\delta(F) = \delta(Y) = \{x \mid (\exists y \in X)\, (x,y) \in Y\}$,
3. $\Gamma_F \in \Pi_1^1[X \times X]$.

Proof. Let us fix an $(\alpha, 1)$–retractive morphism $\mathcal{N} \underset{g}{\overset{f}{\rightleftarrows}} Y$. Let $Y \in \Pi_1^1[X]$. By Proposition 4, $g(Y) \in \Pi_1^1[\mathcal{N}]$. From the Novikov-Kondo-Addison Theorem for $\mathcal{N}$ it follows that there exists G that is a Π_1^1–uniformisation of $g(Y)$. Put $F = g^{-1}(G)$. Since g is an effective Borel function and a bijection between Y and $g(Y)$, $\Gamma_F \in \Pi_1^1[X]$ and F is a required function. $\qquad\square$

References

1. Becher, V., Heiber, P., Slaman, T.A.: Normal numbers and the Borel hierarchy. Fundamenta Mathematicae **22**, 63–77 (2014)
2. de Brecht, M.: Quasi-Polish spaces. Ann. Pure Appl. Logic **164**, 356–381 (2013)
3. Gao, S.: Invariant Descriptive Set Theory. CRC Press, New York (2009)
4. Gregoriades, V., Kispeter, T., Pauly, A.: A comparison of concepts from computable analysis and effective descriptive set theory. Math. Struct. Comput. Sci. 1–23 (2016). https://doi.org/10.1017/S0960129516000128. (Published online: 23 June 2016)
5. Kechris, A.S.: Classical Descriptive Set Theory. Springer, New York (1995)
6. Korovina, M., Kudinov, O.: Complexity for partial computable functions over computable Polish spaces. Math. Struct. Comput. Sci. (2016). doi:10.1017/S0960129516000438. (Published online: 19 December 2016)
7. Korovina, M., Kudinov, O.: Index sets as a measure of continuous constraint complexity. In: Voronkov, A., Virbitskaite, I. (eds.) PSI 2014. LNCS, vol. 8974, pp. 201–215. Springer, Heidelberg (2015). doi:10.1007/978-3-662-46823-4_17
8. Korovina, M., Kudinov, O.: Towards computability over effectively enumerable topological spaces. Electr. Notes Theor. Comput. Sci. **221**, 115–125 (2008)
9. Korovina, M., Kudinov, O.: Towards computability of higher type continuous data. In: Cooper, S.B., Löwe, B., Torenvliet, L. (eds.) CiE 2005. LNCS, vol. 3526, pp. 235–241. Springer, Heidelberg (2005). doi:10.1007/11494645_30
10. Louveau, A.: Recursivity and compactness. In: Müller, G.H., Scott, D.S. (eds.) Higher Set Theory. LNM, vol. 669, pp. 303–337. Springer, Heidelberg (1978). doi:10.1007/BFb0103106

11. Montalban, A., Nies, A.: Borel structures: a brief survey. Lect. Notes Logic **41**, 124–134 (2013)
12. Moschovakis, Y.N.: Descriptive Set Theory. Studies in Logic Series. North Holland, Amsterdam (1980)
13. Moschovakis, Y.N.: Descriptive Set Theory, 2nd edn. North-Holland, Amsterdam (2009)
14. Rogers, H.: Theory of Recursive Functions and Effective Computability. McGraw-Hill, New York (1967)
15. Soare, R.I.: Recursively Enumerable Sets and Degrees: A Study of Computable Functions and Computably Generated Sets. Springer Science and Business Media, Heidelberg (1987)
16. Spreen, D.: On effective topological spaces. J. Symb. Log. **63**(1), 185–221 (1998)
17. Selivanov, V.L.: Towards a descriptive set theory for domain-like structures. Theor. Comput. Sci. **365**(3), 258–282 (2006)
18. Selivanov, V.: Towards the effective descriptive set theory. In: Beckmann, A., Mitrana, V., Soskova, M. (eds.) CiE 2015. LNCS, vol. 9136, pp. 324–333. Springer, Cham (2015). doi:10.1007/978-3-319-20028-6_33
19. Weihrauch, K.: Computable Analysis. Springer, Heidelberg (2000)

Σ_2^μ is decidable for Π_2^μ

Karoliina Lehtinen[1,3]($\boxtimes$) and Sandra Quickert[2,3]

[1] University of Kiel, Kiel, Germany
kleh@informatik.uni-kiel.de
[2] University of St. Andrews, St. Andrews, UK
sq21@st-andrews.ac.uk
[3] University of Edinburgh, Edinburgh, UK

Abstract. Given a Π_2^μ formula of the modal μ calculus, it is decidable whether it is equivalent to a Σ_2^μ formula.

1 Introduction

The modal μ calculus, L_μ, is a well-established verification logic describing properties of labelled transition systems. It consists of a simple modal logic, augmented with the least fixpoint μ and its dual, the greatest fixpoint ν. Alternations between μ and ν are key for measuring complexity: the fewer alternations, the easier a formula is to model check. We call this the formula's index. For any fixed index, the model-checking problem is in P. However, no fixed index is sufficient to capture all properties expressible in L_μ [1,3,14], and it is notoriously difficult to decide whether a formula can be simplified. So far only properties expressible without fixpoints [21], or with only one type of fixpoint [10] are known to be decidable.

In automata theory, the corresponding index problem is to decide the simplest acceptance condition sufficient to express a property with a specified type of automata. This is often referred to as the Mostowski-Rabin index of a language.

Given a deterministic automaton on labelled binary trees, the minimal index of equivalent deterministic [18], non-deterministic [20,24], and alternating [19] automata are all known to be decidable. In [7] these results were extended to show that the non-deterministic and alternating index problems are also decidable for languages of labelled binary trees recognised by game-automata, a slightly more general model than deterministic automata.

For the case of non-deterministic automata, the index problem reduces to the uniform universality of distance parity automata [5]. In [4] it was shown that given a Büchi definable language $\mathcal{L}$, it is decidable whether it can be described by an alternating co-Büchi automaton. Skrzypczak and Walukiewicz [23] give an alternative proof of the same result and add a topological characterisation of the recognised languages. A Büchi definable language which is co-Büchi is said to be weakly definable: it is definable in weak monadic second order logic [22], and equivalently, by an alternating automaton which is simultaneously both Büchi and co-Büchi. In L_μ terms, this result corresponds to deciding whether a formula in the class Π_2^μ is equivalent on binary trees to a formula in the class Σ_2^μ.

© Springer International Publishing AG 2017
J. Kari et al. (Eds.): CiE 2017, LNCS 10307, pp. 292–303, 2017.
DOI: 10.1007/978-3-319-58741-7_28

This paper provides a novel proof of the same result extended to arbitrary structures: given a Π_2^μ formula, it is decidable whether it is equivalent to a Σ_2^μ formula. The proof defines an n-parametrised game such that the decidability of Σ_2^μ reduces to deciding whether for some n this is the model-checking game for a formula. From this game, we derive a family Ψ^n of Σ_2^μ formulas, such that an input formula Ψ is equivalent to a Σ_2^μ formula if and only if it is equivalent to some formula in this family. To decide the parameter n for Π_2^μ input formulas, we simply argue that the game construction in [23], designed for binary trees, extends to the case of labelled transition systems.

We consider the most interesting contributions of this paper to be the reduction of the decidability of Σ_2^μ to finding the parameter n such that Ψ is equivalent to Ψ^n. With this result, finding a way to generalise the game construction from [23] to arbitrary inputs would suffice to decide Σ_2^μ.

All the proofs omitted here can be found in the technical report [13].

2   Preliminaries

2.1 L_μ

Let us fix, once and for all a finite set of actions $Act = \{a, b, ...\}$, a countably infinite set of propositional variables $Prop = \{P, Q, ...\}$, and fixpoint variables $Var = \{X, Y, ...\}$. A literal is either P or $\neg P$ for $P \in Prop$.

Definition 1 (Labelled transition tree). *A labelled transition tree is a structure $\mathcal{T} = (V, v_r, E, L, P)$ where V is a set of states, v_r is the root, the only node without predecessor, $E \subset V \times V$ is an edge relation, $L : E \to Act$ labels edges with actions and $P : V \to 2^{Prop}$ labels vertices with propositional variables. Furthermore, for each $v \in V$ the set of ancestors $\{w \in V \mid \exists w_1 ... w_k. \, wEw_1E...w_kEv\}$ is finite and well-ordered with respect to the transitive closure of E; the set of successors $\{w \in V \mid vEw\}$ is also finite.*

We can represent repetition in an infinite tree with back edges. Note that we allow more than one successor per label.

Definition 2 (Modal μ). *The syntax of L_μ is given by:*

$$\phi := P \ / \ X \ / \ \neg P \ / \ \phi \wedge \phi \ / \ \phi \vee \phi \ / \ \langle a \rangle \phi \ / \ [a]\phi \ / \ \mu X.\phi \ / \ \nu X.\phi \ / \ \bot \ / \ \top$$

The order of operator precedence is $[a], \langle a \rangle, \wedge, \vee, \mu$ and ν.

The operators $\langle a \rangle$ and $[a]$ are called *modalities*, and formulas $\langle a \rangle \phi$ and $[a]\phi$ are called *modal formulas*. If $\psi = \mu X.\phi$ or $\psi = \nu X.\phi$, we call the formula ϕ the binding formula of X within ψ and denote it by ϕ_X. We say that ϕ' is an immediate subformula of ϕ if either ϕ is built from ϕ' in one step using the syntax rules above, or, in a slight abuse of notation, if $\phi = X$ and ϕ' its binding formula. Hence ϕ is an immediate subformula of the formulas $\phi \vee \psi$, $\langle a \rangle \phi$, $\mu X.\phi$ and also of X in $\nu X.\phi$. A formula is guarded if every fixpoint variable is in the scope of a modality within its binding. Without loss of expressivity [9,15], we restrict ourselves to L_μ in guarded positive form. We will also assume throughout the paper that all fixpoint variables within a formula have distinct names.

The semantics of L_μ are standard, see for example [2]. We now define the *priority assignment* and *index* of a formula, following Niwiński's notion of alternation in [17].

Definition 3 (Priority assignment, index and alternation classes). *A priority assignment Ω is a function assigning an integer value to each fixpoint variable in a formula such that: (a) μ-bound variables receive odd priorities and ν-bound variables receive even priorities, and (b) if X is free in ϕ_Y, the binding formula of Y, then $\Omega(X) \geq \Omega(Y)$. A formula has index $\{q, ..., i\}$ where $i \in \{0,1\}$ if it has a priority assignment with co-domain $\{q, ..., i\}$.*

Formulas without fixpoints form the modal fragment of L_μ. Formulas with one type of fixpoint have index $\{0\}$ or $\{1\}$, corresponding to the alternation classes Π_1^μ and Σ_1^μ, respectively. Then the class Π_i^μ and Σ_i^μ for even i correspond to formulas with indices $\{i, ..., 1\}$ and $\{i - 1, ..., 0\}$, respectively, while for odd i they correspond to formulas with indices $\{i - 1, ..., 0\}$ and $\{i, ..., 1\}$, respectively. A formula has semantic alternation class C if it is equivalent to a formula in C.

Example 1. The formula $\mu X.\nu Y.\Box Y \wedge \mu Z.\Box(X \vee Z)$ accepts the priority assignment $\Omega(X) = 1, \Omega(Y) = 0$ and $\Omega(Z) = 1$, so it has index $\{1,0\}$ and is in the class Σ_2^μ. However, it is equivalent to $\mu X.\Box X$ which holds in structures without infinite paths, and is therefore semantically in Σ_1^μ.

In this paper we present a new proof that Σ_2^μ is decidable for formulas in Π_2^μ: given an arbitrary L_μ formula Ψ with index $\{2,1\}$, it is decidable whether Ψ is equivalent to a formula with index $\{1,0\}$.

2.2   Parity Games

The semantics of L_μ formulas (like that of alternating parity automata) can be described in terms of winning regions of parity games.

Definition 4. *A parity game $G = (V, v_i, E, \Omega)$ consists of a set of vertices V partitioned into those belonging to Even, V_e, and those belonging to Odd, V_o, an initial position $v_i \in V$, and a set of edges $E \in V \times V$. A priority assignment Ω assigns a priority to every vertex.*

At each turn, the player who owns the current position v chooses a successor position from the successors of v via E. A play is a potentially infinite sequence of positions starting at the initial position v_i. A finite play is winning for Even if the final position has even priority, and for Odd otherwise. An infinite play is winning for the player of the parity of the highest priority seen infinitely often.

Parity games are known to be determined and we can restrict ourselves to positional winning strategies [6,16]. It is a standard result that given a structure $\mathcal{M}$ and a formula Ψ, there is a model-checking parity game $\mathcal{M} \times \Psi$ such that Even wins if and only if $\mathcal{M}$ satisfies Ψ [26].

Definition 5 (The model-checking game $\mathcal{M} \times \Psi$). *The parity game $\mathcal{M} \times \Psi$ has for states $s \times \phi$ where s is a state of $\mathcal{M}$ and ϕ is a subformula of Ψ. There is an edge from $s \times \psi$ to $s \times \phi$ if ϕ is an immediate subformula of a non-modal formula ψ; there is an edge from $s \times \langle a \rangle \phi$ and $s \times [a]\phi$ to $s' \times \phi$ for s' an a-successor of s. Positions $s \times \phi$ where ϕ is a disjunction or starting with an existential modality $\langle a \rangle$ belong to Even while those where ϕ is a conjunction or universal modality $[a]$ belong to Odd. Positions with a single successor are given to Even, although the game is deterministic at those. The priority assignment is inherited from the priority assignment Ω_Ψ on Ψ: a fixpoint variable X receives priority $\Omega_\Psi(X)$ while other nodes receive the minimal priority in the co-domain of Ω_Ψ.*

2.3   Disjunctive Form

Disjunctive L_μ is a fragment restricting conjunctions in a way reminiscent of non-deterministic automata [25]. Its use is key to several of our proofs.

Definition 6 (*Disjunctive formulas*). *The set of disjunctive form formulas of L_μ is the smallest set $\mathcal{F}$ satisfying:*

- *$\top, \bot$, fixpoint variables and finite sets (conjunctions) of literals are in $\mathcal{F}$;*
- *If $\psi \in \mathcal{F}$ and $\phi \in \mathcal{F}$, then $\psi \lor \phi \in \mathcal{F}$;*
- *If for each a in Act the set $\mathcal{B}_a \subseteq \mathcal{F}$ is a finite set of formulas, and if A is a finite set of literals, then $A \land \bigwedge_{a \in Act} \overset{a}{\to} \mathcal{B}_a \in \mathcal{F}$ where $\overset{a}{\to} \mathcal{B}_a$ is short for $(\bigwedge_{\psi \in \mathcal{B}_a} \langle a \rangle \psi) \land [a] \bigvee_{\psi \in \mathcal{B}_a} \psi$ – that is to say, every formula in $\mathcal{B}_a$ holds at least one successor and at every successor at least one of the formulas in $\mathcal{B}_a$ holds;*
- *$\mu X.\psi$ and $\nu X.\psi$ are in $\mathcal{F}$ as long as $\psi \in \mathcal{F}$.*

Every formula is known to be equivalent to an effectively computable formula in disjunctive form [25]. The transformation preserves guardedness.

Given an L_μ formula with unrestricted conjunctions, the model-checking parity game requires Even to have a strategy to verify both conjuncts. A strategy for Even will agree with the plays corresponding to each of Odd's choices, leading potentially to several plays on some branches. In contrast, disjunctive form restricts conjunctions, and the only branching in Even's strategies is at a position where the formula is of the form $A \land \bigwedge_{a \in Act} \overset{a}{\to} \mathcal{B}_a \in \mathcal{F}$, called an *Odd-choice formula*.

Disjunctive form guarantees that Even can use strategies which only agree with one play per branch. For further details, see [12].

Lemma 1. *[12] Given a disjunctive formula Ψ, for any structure $\mathcal{M}$ and strategy σ in $\mathcal{M} \times \Psi$, there is a structure $\mathcal{M}'$ bisimilar to $\mathcal{M}$ such that a strategy σ' in $\mathcal{M}' \times \Psi$ induced from σ only agrees with one play per branch. We then say that $\mathcal{M}'$ and σ' are well-behaved.*

Lemma 2. *Given a Π_2^μ formula, the transformation into disjunctive form as presented in [25] yields a disjunctive Π_2^μ formula.*

This can be proved using the concepts of tableau, tableau equivalence, and traces from [25]. The crux of the argument is that the tableau of a disjunctive formula not in Π_2^μ must have an even cycle nested in an odd cycle which in turn implies the existence of a trace on which a μ-fixpoint dominates a ν fixpoint in any equivalent tableau. The proof can be found in the technical report [13].

Note that the dual is not true: a Σ_2^μ formula may yield a formula of arbitrarily large alternation depth when turned into disjunctive form [11]. This is in line with alternating Büchi automata being equivalent to non-deterministic Büchi automata while the same is not necessarily true for co-Büchi automata.

2.4   Automata and L_μ

The relationship between L_μ and automata theory is based on the fact that the automata model that L_μ formulas correspond to is, when restricted to binary trees[1], equivalent to alternating automata with a parity condition [8]. The model-checking problems in these two settings are equivalent: Model-checking a formula ψ on a structure $\mathcal{M}$ reduces to checking an automaton $A(\psi)$ on a binary tree encoding of $\mathcal{M}$. Model checking disjunctive L_μ similarly reduces to model checking non-deterministic automata, albeit one of potentially higher index. For the index problem, the comparison is not as simple and to the best of our knowledge there is no known reduction from the (disjunctive) L_μ index problem to the (non-deterministic) automata index problem. Part of the difficulty is that only considering binary trees affects the semantic complexity of formulas: for example, the formula $\langle a \rangle \psi \wedge \langle a \rangle \bar{\psi}$ (where $\bar{\psi}$ is the negation of ψ) is semantically trivial when interpreted on trees with only one a-successor while in the general case its index depends on ψ. Furthermore, non-deterministic parity automata are weaker than disjunctive L_μ in the sense that some properties of binary trees can be expressed with a lower index using disjunctive form.

3   Deciding Σ_2^μ Reduces to a Bounding Problem

The first part of the proof of our main result defines a parametrised n-challenge game on a parity game arena. For each finite n, the n-challenge game is described by a Σ_2^μ formula Ψ^n which holds in $\mathcal{M}$ if and only if Even wins the n-challenge game on $\mathcal{M} \times \Psi$. We then show that a disjunctive formula Ψ is equivalent to a (not necessarily disjunctive) formula in Σ_2^μ if and only if there is some n such that Ψ is equivalent to Ψ^n. As any formula can be turned into disjunctive form, this reduces the decidability of Σ_2^μ to bounding the parameter n. For the main result of this paper, we will only use this construction for Π_2^μ input formulas to determine equivalence to a Σ_2^μ formula. However, using this more general construction, a generalisation of the second part of our proof beyond Π_2^μ would suffice to decide Σ_2^μ entirely.

---

[1] assuming $|Act| = 2$; otherwise trees with one successor per label.

When restricted to automata on binary trees and two priorities, this construction is equivalent to those found for example in [4, 23].

We fix a disjunctive formula Ψ with index $\{q, ..., 0\}$. Let $I = \{q, ..., 0\}$ if the maximal priority q is even and $\{q + 1, q, ..., 0\}$ otherwise. Write I_e for the even priorities in I. The n-challenge game consists of a normal parity game augmented with a set of challenges, one for each even priority i. A challenge can either be *open* or *met* and has a counter c_i attached to it. Each counter is initialised to n, and decremented when the corresponding challenge is opened. The Odd player can at any point open challenges of which the counter is non-zero, but he must do so in decreasing order: an i-challenge can only be opened if every j-challenge for $j > i$ is opened. When a play encounters a priority greater or equal to j while the j-challenge is open, the challenge is said to be met. All i-challenges for $i < j$ are *reset*. This means that the counters c_i are set back to n.

A play of this game is a play in a parity game, augmented with the challenge and counter configuration at each step. A play with dominant priority d is winning for Even if either d is even or if every opened $d + 1$ challenge is eventually met or reset.

Example 2. The formula $\nu Y.\mu X.(A \wedge \Diamond X) \vee (B \wedge \Diamond Y)$ is true if on some path B always eventually holds. This formula does not hold in this structure:

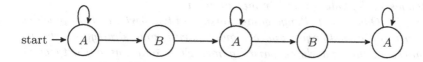

However, Even wins the 1- and 2-challenge games: her strategy is to loop in the current state until Odd opens a 2-challenge, then meet the challenge by moving to the next state, as seeing a B corresponds to seeing 2. Odd will run out of challenges before reaching the last state. Although Odd wins the 3-challenge game in this structure, for any n it is easy to construct a similar structure in which he loses the n-challenge game but wins the parity game. This section argues that this is sufficient to show that $\nu Y.\mu X.(A \wedge \Diamond X) \vee (B \wedge \Diamond Y)$ is not equivalent to any Σ_2^μ formula.

In contrast, in the formula $\nu Y.\mu X.(A \wedge \Box X) \vee (B \wedge \Diamond Y)$, Odd wins the 1-challenge game whenever he wins the parity game: he can open the challenge when his strategy in the parity game reaches the point at which he can avoid B. This formula is therefore equivalent to a Σ_2^μ formula, namely the alternation free formulas $\nu Y.((A \wedge \Box Y) \vee (B \wedge \Diamond Y)) \wedge \mu X.(A \wedge \Box X) \vee B)$.

Definition 7. *A configuration $(v, p, \bar{c}, r)$ of the n-challenge game on a parity game G of index $\{q, ..., 0\}$ where q is even consists of:*

– *a position v in the parity game;*

– an even priority p indicating the least significant priority on which a challenge
is open or $p = q + 2$ if all challenges are currently met;
– $\bar{c} = (c_0, c_2, \ldots, c_q)$ a collection of counter values c_i for each even priority i.
– $r \in \{0, 1\}$ indicating the round of the game: 1 for Odd's turn to open challenges, 0 for a turn in the parity game.

At a configuration $(v, p, \bar{c}, 1)$, corresponding to Odd's turn, he can open challenges up to any $p' \leq p$, as long as $c[i] > 0$ for each i such that $p' \leq i < p$. Then
the configuration becomes $(v, p', \bar{c}', 0)$ where $c'[i] = c[i] - 1$ for all newly opened
challenges i, that is to say i such that $p' \leq i < p$ and $c'[i] = c[i]$ for all other i.

At the configuration $(v, p, \bar{c}, 0)$, the player whose turn it is in the parity
game decides the successor position v' of v and the configuration is updated
to $(v', p', \bar{c}', 1)$ according to the priority i of v' as follows:

– If $i \geq p$ then $p' = i + 2$ if i is even, $p' = i + 1$ otherwise. This indicates which
challenges have been met. Note that if all challenges are met, $p = q + 2$.
– For each $j < i$, the counter value c_j is reset to n.
– If i is even and $c_i = 0$, then the game ends immediately with a win for Even.

A play is a potentially infinite sequence of configurations starting at the ini-
tial configuration $(v_\iota, q + 2, (n, \ldots, n), 1)$, where v_ι is the initial position of the
parity game. An infinite play is winning for Even if the dominant priority on
the sequence of parity game positions is d but the game reaches infinitely many
configurations $(v, p, \bar{c}, 0)$ where $p > d + 1$. This is the case if d is even or if all
$d + 1$ challenges set by Odd are either met or reset.

A strategy for Odd in a challenge game consists of two parts: a strategy which
dictates when to open challenges, and a regular parity-game strategy which dictates his moves in the underlying parity game. Even only has a parity game
strategy. Both players' strategies may of course depend on the challenge configuration as well as the parity game configuration. Given a challenge-game strategy
for even σ, a challenging strategy γ for Odd induces a normal parity game strategy σ_γ for Even which does not depend on the challenge configuration.

We first establish that the winning regions of the n-challenge games for Ψ
can be described by a Σ_2^μ formula Ψ^n.

Lemma 3. *For all Ψ and finite n, there is a formula $\Psi^n \in \Sigma_2^\mu$ which holds in
$\mathcal{M}$ if and only if Even wins the n-challenge game on $\mathcal{M} \times \Psi$.*

This lemma can be proven by describing explicitly an alternating parity
automaton computing winning regions of the above mentioned n-challenge game.
The automaton can be defined in a way that it only uses priorities 0 and 1, and
therefore corresponds to a formula in Σ_2^μ which is the formula Ψ^n as required.
The proof can be found in the technical report [13].

Next we prove our core theorem, reducing the decidability of Σ_2^μ to a boundedness criterion. .

Theorem 1. *If a disjunctive formula Ψ is semantically in Σ_2^μ, then there is a
finite n such that $\Psi \Leftrightarrow \Psi^n$.*

Proof. Assume that Ψ is semantically in Σ_2^μ, *i.e.* equivalent to some Φ of index $\{1,0\}$, and that for all n, $\Psi \not\Leftrightarrow \Psi^n$. Fix n to be larger than $2^{|\Psi|+|\Phi|}$. There is a structure $\mathcal{M}$, such that Odd wins the parity game $\mathcal{M} \times \Psi$ but Even wins the n-challenge game on $\mathcal{M} \times \Psi$. W.l.o.g, take $\mathcal{M}$ to be finitely branching. The overall structure of this proof is to first use a winning strategy τ for Odd in $\mathcal{M} \times \Phi$ to define a challenging strategy γ for him in the n-challenge game on $\mathcal{M} \times \Psi$ (Part I). We then use Even's winning strategy σ to add back edges to $\mathcal{M}$ (Part II), turning it into a new structure $\mathcal{M}'$ which preserves Odd's winning strategy τ in $\mathcal{M}' \times \Phi$ while turning σ_γ into a winning strategy in $\mathcal{M}' \times \Psi$ (Part III). This contradicts the equivalence of Φ and Ψ.

Part I. Let τ be Odd's winning strategy in $\mathcal{M} \times \Phi$. Since $\mathcal{M}$ is finitely branching, for any node v reachable via τ, there is a finite bound i such that any play that agrees with τ sees 1 within i modal steps of any position $v \times \alpha$ that it reaches (König's Lemma). For a branch b of $\mathcal{M}$, on which τ reaches a node v, indicate by $next(b,v)$ the ith node on b from v. This node has the property that any play on the branch b agreeing with τ must see a 1 between v and $next(b,v)$.

If τ does not agree with any plays on the branch b, then let $next(b,v)$ be a node on b which τ does not reach.

Now consider the n-challenge game on $\mathcal{M} \times \Psi$. Let Odd's challenging strategy γ be: to open all challenges at the start of the game, and whenever its counter is reset; if a challenge for a priority i is met at v, and its counter c_i is not at 0, to open the next challenge when the play reaches a node $next(b,v)$ for any branch b, unless the counter is reset before then (*i.e.* a higher priority is seen).

Part II. Even wins the n-challenge game on $\mathcal{M} \times \Psi$, so let σ be her winning strategy. Recall that σ_γ is an Even's strategy for Ψ up to the point where an nth challenge in the original challenge game is met, and undefined thereafter. Since Ψ is disjunctive, we can adjust $\mathcal{M}$ into a bisimilar structure in which the pure parity game strategy σ_γ is well-behaved wherever it is defined – it reaches each position of $\mathcal{M}$ at either one subformula, or none.

The strategy σ_γ is winning in the challenge game against any strategy for Odd which uses the challenging strategy γ. Since Odd always eventually opens the next challenge, the only way for him to lose is that the play either reaches a position that is winning for Even in the parity game, or a position of priority p when $c_p = 0$. Thus, every play is finite.

Since σ_γ is well-behaved, each branch carries at most one play. For every branch b the finite play it may carry must end either in an immediate win for Even, or in a long streak in which the highest priority seen is some even p, and it is seen at least n times, corresponding to every instance of Even meeting a p-challenge. As long as n is sufficiently large, on every such branch there are two nodes v and its descendant w, at which Odd opens challenges on p, which agree on the set of subformulas that σ_γ reaches there in $\mathcal{M} \times \Psi$ and that τ reaches there in $\mathcal{M} \times \Phi$. We now consider the structure $\mathcal{M}'$, which is as $\mathcal{M}$ except that the predecessor of each w-node has an edge to v instead. The strategies σ_γ and τ transfer in the obvious way to $\mathcal{M}'$.

Part III. We now claim that τ is winning in $\mathcal{M}' \times \Phi$ and that σ_γ is winning in $\mathcal{M}' \times \Psi$. Starting with σ_γ, first consider plays that do not go through back edges infinitely often. These are the ones that end in positions immediately winning for Even. Any play in $\mathcal{M} \times \Psi$ that agrees with σ_γ which sees both v and w is dominated by an even priority between v and w. Then, as the w and v agree on which subformula σ_γ reaches them at, an even priority dominates any play that goes through back edges in $\mathcal{M}' \times \Psi$ infinitely many times. The strategy σ_γ is therefore winning in $\mathcal{M}' \times \Psi$.

Now onto τ in $\mathcal{M}' \times \Phi$. If a branch is unchanged by the transformation, then any play on it is still winning for τ, because such a play would be consistent with τ in the original game. If a branch that τ plays on has been changed, then consider in $\mathcal{M}$ the two nodes v and w at which the transformation is done. These both are nodes at which Odd opens challenges according to γ, therefore, from the definition of *next* and γ, the highest priority seen between them by any play agreeing with τ is 1. Since v and w agree on which subformulas τ reaches them at, any play in $\mathcal{M}' \times \Phi$ which goes through a back-edge infinitely often sees 1 infinitely often and is therefore winning for Odd.

This contradicts the equivalence of Ψ and Φ. Therefore, if Ψ is semantically in Σ_2^μ, then for all structures $\mathcal{M}$ the n-challenge game and the parity game on $\mathcal{M} \times \Psi$ have the same winner for $n > 2^{|\Phi|+|\Psi|}$.

Theorem 2. *Let $\Psi \in L_\mu$, and Ψ_d a disjunctive formula equivalent to Ψ. Then Ψ is semantically in Σ_2^μ if and only if there is some finite n such that $\Psi \Leftrightarrow \Psi_d^n$.*

4   Deciding Σ_2^μ for Π_2^μ

To complete the proof of the namesake result, it suffices to show that the parameter n from Theorem 2 can be bounded. If we restrict ourselves to disjunctive $\Psi \in \Pi_2^\mu$, we argue that the tree-building game $\mathcal{F}$ from [23] extends to arbitrary labelled transition systems and delivers such a bound.

Since the $\mathcal{F}$ game is already well-exposed in [23], and the adjustments to cater for disjunctive L_μ and labelled transition systems are relatively straightforward but verbose, we omit the proof of the following theorem. The proof can be found in the technical report [13].

Theorem 3. *Let $\Psi \in \Pi_2^\mu$ be disjunctive. Then there is a constant K_0 computable from Ψ such that the following statements are equivalent:*

(a) There is some n such that $\Psi \Leftrightarrow \Psi^n$.
(b) $\Psi \Leftrightarrow \Psi^{K_0}$

Placing everything together, we obtain our final result.

Theorem 4. *It is effectively decidable whether any given Π_2^μ formula is equivalent to a Σ_2^μ formula. By duality, it is also effectively decidable whether any given Σ_2^μ formula is equivalent to a Π_2^μ formula.*

Proof. Given any Π_2^μ formula Ψ, it can be effectively turned into a disjunctive formula Ψ_d also in Π_2^μ (Lemma 2). Then, Theorem 2 yields that Ψ_d is semantically in Σ_2^μ if and only if it is equivalent to Ψ_d^n for some n. From Theorem 3, $\Psi_d \Leftrightarrow \Psi_d^n$ if and only if $\Psi_d \Leftrightarrow \Psi_d^{K_0}$ where K_0 is computable from Ψ via Ψ_d. Thus, Ψ is semantically in Σ_2^μ if and only if $\Psi \Leftrightarrow \Psi_d^{K_0}$ if and only if $\Psi_d \Leftrightarrow \Psi_d^{K_0}$.

Given any Σ_2^μ formula, it can also be decided whether it is equivalent to a Π_2^μ formula, via checking whether its negation is equivalent to a Σ_2^μ formula.

5 Discussion

We have shown that given any L_μ formula in Π_2^μ, it can be effectively decided whether it is equivalent to a Σ_2^μ formula. This result is the L_μ-theoretic counterpart of the decidability of weak definability of Büchi definable languages [4,23]. The core contribution is the reduction of the decidability of Σ_2^μ for arbitrary L_μ formulas to deciding whether the n-challenge game is equivalent to the model-checking parity game of a formula for any n. We obtain a family of parameterised Σ_2^μ formulas Ψ^n such that Ψ is in Σ_2^μ if an only if Ψ is equivalent to Ψ^n for some n. Unfortunately, the second part of our proof, based on [23], is less general and only admits input formulas in Π_2^μ. If this could also be generalised to arbitrary formulas, this would yield a decidability proof for Σ_2^μ.

The challenge game can be extended to constructions described by more complex L_μ formulas – this may turn out to be the right way to characterize higher alternation classes. However, for Theorem 1, if there are more than two priorities at play, the different plays along one branch become less manageable and it is not clear how they can inform a challenging strategy. Even when restricted to disjunctive formulas, a new technique seems to be required. However, the result of [5] which achieves this for non-deterministic automata on binary trees justifies cautious optimism for the disjunctive case.

Achnowledgements. We thank the anonymous reviewers for their thoughtful comments and corrections. The work presented here has been supported by an EPSRC doctoral studentship at the University of Edinburgh.

References

1. Arnold, A.: The μ-calculus alternation-depth hierarchy is strict on binary trees. RAIRO - Theoretical Informatics and Applications - Informatique Théorique et Applications **33**(4–5), 329–339 (1999)
2. Bradfield, J.C., Stirling, C.: Modal mu-calculi. Handbook of modal logic **3**, 721–756 (2007)
3. Bradfield, J.C.: The modal mu-calculus alternation hierarchy is strict. In: Montanari, U., Sassone, V. (eds.) CONCUR 1996. LNCS, vol. 1119, pp. 233–246. Springer, Heidelberg (1996). doi:10.1007/3-540-61604-7_58
4. Colcombet, T., Kuperberg, D., Löding, C., Vanden Boom, M.: Deciding the weak definability of Büchi definable tree languages. In: LIPIcs-Leibniz International Proceedings in Informatics, vol. 23. Schloss Dagstuhl-Leibniz-Zentrum fuer Informatik (2013)

5. Colcombet, T., Löding, C.: The non-deterministic mostowski hierarchy and distance-parity automata. In: Aceto, L., Damgård, I., Goldberg, L.A., Halldórsson, M.M., Ingólfsdóttir, A., Walukiewicz, I. (eds.) ICALP 2008. LNCS, vol. 5126, pp. 398–409. Springer, Heidelberg (2008). doi:10.1007/978-3-540-70583-3_33

6. Emerson, E.A., Jutla, C.S.: Tree automata, mu-calculus and determinacy. In: Proceedings of the 32nd Annual Symposium on Foundations of Computer Science, FoCS 1991, pp. 368–377. IEEE Computer Society Press (1991)

7. Facchini, A., Murlak, F., Skrzypczak, M.: Rabin-mostowski index problem: a step beyond deterministic automata. In: Proceedings of the 2013 28th Annual ACM/IEEE Symposium on Logic in Computer Science, pp. 499–508. IEEE Computer Society (2013)

8. Janin, D., Walukiewicz, I.: Automata for the modal μ-calculus and related results. In: Wiedermann, J., Hájek, P. (eds.) MFCS 1995. LNCS, vol. 969, pp. 552–562. Springer, Heidelberg (1995). doi:10.1007/3-540-60246-1_160

9. Kupferman, O., Vardi, M.Y., Wolper, P.: An automata-theoretic approach to branching-time model checking. J. ACM (JACM) **47**(2), 312–360 (2000)

10. Küsters, R., Wilke, T.: Deciding the first level of the μ-calculus alternation hierarchy. In: Agrawal, M., Seth, A. (eds.) FSTTCS 2002. LNCS, vol. 2556, pp. 241–252. Springer, Heidelberg (2002). doi:10.1007/3-540-36206-1_22

11. Lehtinen, K.: Disjunctive form and the modal μ alternation hierarchy. In: FICS 2015 The 10th International Workshop on Fixed Points in Computer Science, EPTCS 191, p. 117 (2015)

12. Lehtinen, K., Quickert, S.: Deciding the first levels of the modal mu alternation hierarchy by formula construction. In: LIPIcs-Leibniz International Proceedings in Informatics, vol. 1. Schloss Dagstuhl-Leibniz-Zentrum fuer Informatik (2015)

13. Lehtinen, M.K., Quickert, S.: Σ_2^μ is decidable for Π_2^μ (extended version). Technical report. arXiv:1703.03239

14. Lenzi, G.: A hierarchy theorem for the μ-calculus. In: Meyer, F., Monien, B. (eds.) ICALP 1996. LNCS, vol. 1099, pp. 87–97. Springer, Heidelberg (1996). doi:10.1007/3-540-61440-0_119

15. Mateescu, R.: Local model-checking of modal mu-calculus on acyclic labeled transition systems. In: Katoen, J.-P., Stevens, P. (eds.) TACAS 2002. LNCS, vol. 2280, pp. 281–295. Springer, Heidelberg (2002). doi:10.1007/3-540-46002-0_20

16. Mostowski, A.W.: Games with forbidden sequences and finite machines. Technical report 78, Instytut Matematyki, Uniwersytet Gdański, Poland (1991)

17. Niwiński, D.: On fixed-point clones. In: Kott, L. (ed.) ICALP 1986. LNCS, vol. 226, pp. 464–473. Springer, Heidelberg (1986). doi:10.1007/3-540-16761-7_96

18. Niwiński, D., Walukiewicz, I.: Relating hierarchies of word and tree automata. In: Morvan, M., Meinel, C., Krob, D. (eds.) STACS 1998. LNCS, vol. 1373, pp. 320–331. Springer, Heidelberg (1998). doi:10.1007/BFb0028571

19. Niwiński, D., Walukiewicz, I.: A gap property of deterministic tree languages. Theor. Comput. Sci. **303**(1), 215–231 (2003)

20. Niwiński, D., Walukiewicz, I.: Deciding nondeterministic hierarchy of deterministic tree automata. Electron. Notes Theor. Comput. Sci. **123**, 195–208 (2005)

21. Otto, M.: Eliminating recursion in the μ-calculus. In: Meinel, C., Tison, S. (eds.) STACS 1999. LNCS, vol. 1563, pp. 531–540. Springer, Heidelberg (1999). doi:10.1007/3-540-49116-3_50

22. Rabin, M.O.: Weakly definable relations and special automata. In: Bar-Hillel, Y. (ed.) Mathematical Logic and Foundations of Set Theory, pp. 1–23 (1970)

23. Skrzypczak, M., Walukiewicz, I.: Deciding the topological complexity of büchi languages. In: Rabani, Y., Chatzigiannakis, I., Mitzenmacher, M., Sangiorgi, D. (eds.) 43rd International Colloquium on Automata, Languages, and Programming (ICALP 2016), vol. 55, Leibniz International Proceedings in Informatics (LIPIcs), pp. 99:1–99:13. Dagstuhl, Germany (2016). Schloss Dagstuhl-Leibniz-Zentrum fuer Informatik

24. Urbański, T.F.: On deciding if deterministic rabin language is in büchi class. In: Montanari, U., Rolim, J.D.P., Welzl, E. (eds.) ICALP 2000. LNCS, vol. 1853, pp. 663–674. Springer, Heidelberg (2000). doi:10.1007/3-540-45022-X_56

25. Walukiewicz, I.: Completeness of Kozen's axiomatisation of the propositional μ-calculus. Inf. Comput. **157**(1–2), 142–182 (2000)

26. Wilke, T.: Alternating tree automata, parity games, and modal m-calculus. Bull. Belg. Math. Soc. Simon Stevin **8**(2), 359 (2001)

Dimension Spectra of Lines

Neil Lutz[1] and D.M. Stull[2(✉)]

[1] Department of Computer Science, Rutgers University, Piscataway, NJ 08854, USA
njlutz@rutgers.edu
[2] Department of Computer Science, Iowa State University, Ames, IA 50011, USA
dstull@iastate.edu

Abstract. This paper investigates the algorithmic dimension spectra of lines in the Euclidean plane. Given any line L with slope a and vertical intercept b, the dimension spectrum $\mathrm{sp}(L)$ is the set of all effective Hausdorff dimensions of individual points on L. We draw on Kolmogorov complexity and geometrical arguments to show that if the effective Hausdorff dimension $\dim(a, b)$ is equal to the effective packing dimension $\mathrm{Dim}(a, b)$, then $\mathrm{sp}(L)$ contains a unit interval. We also show that, if the dimension $\dim(a, b)$ is at least one, then $\mathrm{sp}(L)$ is infinite. Together with previous work, this implies that the dimension spectrum of any line is infinite.

1 Introduction

Algorithmic dimensions refine notions of algorithmic randomness to quantify the density of algorithmic information of individual points in continuous spaces. The most well-studied algorithmic dimensions for a point $x \in \mathbb{R}^n$ are the *effective Hausdorff dimension*, $\dim(x)$, and its dual, the *effective packing dimension*, $\mathrm{Dim}(x)$ [1,7]. These dimensions are both algorithmically and geometrically meaningful [3]. In particular, the quantities $\sup_{x \in E} \dim(x)$ and $\sup_{x \in E} \mathrm{Dim}(x)$ are closely related to classical Hausdorff and packing dimensions of a set $E \subseteq \mathbb{R}^n$ [5,8], and this relationship has been used to prove nontrivial results in classical fractal geometry using algorithmic information theory [8,10,12].

Given the pointwise nature of effective Hausdorff dimension, it is natural to investigate not only the supremum $\sup_{x \in E} \dim(x)$ but the entire *(effective Hausdorff) dimension spectrum* of a set $E \subseteq \mathbb{R}^n$, i.e., the set

$$\mathrm{sp}(E) = \{\dim(x) : x \in E\}.$$

The dimension spectra of several classes of sets have been previously investigated. Gu et al. studied the dimension spectra of randomly selected subfractals of self-similar fractals [4]. Dougherty, et al. focused on the dimension spectra of random

N. Lutz—Research supported in part by National Science Foundation Grant 1445755.

D.M. Stull—Research supported in part by National Science Foundation Grants 1247051 and 1545028.

© Springer International Publishing AG 2017
J. Kari et al. (Eds.): CiE 2017, LNCS 10307, pp. 304–314, 2017.
DOI: 10.1007/978-3-319-58741-7_29

translations of Cantor sets [2]. In the context of symbolic dynamics, Westrick has studied the dimension spectra of subshifts [14].

This work concerns the dimension spectra of lines in the Euclidean plane $\mathbb{R}^2$. Given a line $L_{a,b}$ with slope a and vertical intercept b, we ask what $\mathrm{sp}(L_{a,b})$ might be. It was shown by Turetsky that, for every $n \geq 2$, the set of all points in $\mathbb{R}^n$ with effective Hausdorff 1 is connected, guaranteeing that $1 \in \mathrm{sp}(L_{a,b})$. In recent work [10], we showed that the dimension spectrum of a line in $\mathbb{R}^2$ cannot be a singleton. By proving a general lower bound on $\dim(x, ax + b)$, which is presented as Theorem 5 here, we demonstrated that

$$\min\{1, \dim(a, b)\} + 1 \in \mathrm{sp}(L_{a,b}).$$

Together with the fact that $\dim(a, b) = \dim(a, a^2 + b) \in \mathrm{sp}(L_{a,b})$ and Turetsky's result, this implies that the dimension spectrum of $L_{a,b}$ contains both endpoints of the unit interval $[\min\{1, \dim(a, b)\}, \min\{1, \dim(a, b)\} + 1]$.

Here we build on that work with two main theorems on the dimension spectrum of a line. Our first theorem gives conditions under which the entire unit interval must be contained in the spectrum. We refine the techniques of [10] to show in our main theorem (Theorem 8) that, whenever $\dim(a, b) = \mathrm{Dim}(a, b)$, we have

$$[\min\{1, \dim(a, b)\}, \min\{1, \dim(a, b)\} + 1] \subseteq \mathrm{sp}(L_{a,b}).$$

Given any value $s \in [0, 1]$, we construct, by padding a random binary sequence, a value $x \in \mathbb{R}$ such that $\dim(x, ax+b) = s + \min\{\dim(a, b), 1\}$. Our second main theorem shows that the dimension spectrum $\mathrm{sp}(L_{a,b})$ is infinite for every line such that $\dim(a, b)$ is at least one. Together with Theorem 5, this shows that the dimension spectrum of *any* line has infinite cardinality.

We begin by reviewing definitions and properties of algorithmic information in Euclidean spaces in Sect. 2. In Sect. 3, we sketch our technical approach and state our main technical lemmas. In Sect. 4 we prove our first main theorem and state our second main theorem. We conclude in Sect. 5 with a brief discussion of future directions.

2 Preliminaries

2.1 Kolmogorov Complexity in Discrete Domains

The *conditional Kolmogorov complexity* of binary string $\sigma \in \{0,1\}^*$ given a binary string $\tau \in \{0, 1\}^*$ is the length of the shortest program π that will output σ given τ as input. Formally, it is

$$K(\sigma|\tau) = \min_{\pi \in \{0,1\}^*} \{\ell(\pi) : U(\pi, \tau) = \sigma\},$$

where U is a fixed universal prefix-free Turing machine and $\ell(\pi)$ is the length of π. Any π that achieves this minimum is said to *testify* to, or be a *witness* to, the value $K(\sigma|\tau)$. The *Kolmogorov complexity* of a binary string σ is $K(\sigma) = K(\sigma|\lambda)$, where λ is the empty string. These definitions extend naturally to other finite data objects, e.g., vectors in $\mathbb{Q}^n$, via standard binary encodings; see [6] for details.

2.2 Kolmogorov Complexity in Euclidean Spaces

The above definitions can also be extended to Euclidean spaces, as we now describe. The *Kolmogorov complexity* of a point $x \in \mathbb{R}^m$ at *precision* $r \in \mathbb{N}$ is the length of the shortest program π that outputs a *precision-r* rational estimate for x. Formally, it is

$$K_r(x) = \min \{K(p) \, : \, p \in B_{2^{-r}}(x) \cap \mathbb{Q}^m\} \, ,$$

where $B_\varepsilon(x)$ denotes the open ball of radius ε centered on x. The *conditional Kolmogorov complexity* of x at precision r given $y \in \mathbb{R}^n$ at precision $s \in \mathbb{R}^n$ is

$$K_{r,s}(x|y) = \max \big\{ \min\{K_r(p|q) \, : \, p \in B_{2^{-r}}(x) \cap \mathbb{Q}^m\} \, : \, q \in B_{2^{-s}}(y) \cap \mathbb{Q}^n\big\} \, .$$

When the precisions r and s are equal, we abbreviate $K_{r,r}(x|y)$ by $K_r(x|y)$. As the following lemma shows, these quantities obey a chain rule and are only linearly sensitive to their precision parameters.

Lemma 1 (J. Lutz and N. Lutz [8], N. Lutz and Stull [10]). *Let $x \in \mathbb{R}^m$ and $y \in \mathbb{R}^n$. For all $r, s \in \mathbb{N}$ with $r \geq s$,*

1. $K_r(x, y) = K_r(x|y) + K_r(y) + O(\log r)$.
2. $K_r(x) = K_{r,s}(x|x) + K_s(x) + O(\log r)$.

As a matter of notational convenience, if we are given a nonintegral positive real as a precision parameter, we will always round up to the next integer. For example, $K_r(x)$ denotes $K_{\lceil r \rceil}(x)$ whenever $r \in (0, \infty)$.

2.3 Effective Hausdorff and Packing Dimensions

J. Lutz initiated the study of algorithmic dimensions by effectivizing Hausdorff dimension using betting strategies called *gales*, which generalize martingales. Subsequently, Athreya, et al., defined effective packing dimension, also using gales [1]. Mayordomo showed that effective Hausdorff dimension can be characterized using Kolmogorov complexity [11], and Mayordomo and J. Lutz showed that effective packing dimension can also be characterized in this way [9]. In this paper, we use these characterizations as definitions. The *effective Hausdorff dimension* and *effective packing dimension* of a point $x \in \mathbb{R}^n$ are

$$\dim(x) = \liminf_{r \to \infty} \frac{K_r(x)}{r} \quad \text{and} \quad \mathrm{Dim}(x) = \limsup_{r \to \infty} \frac{K_r(x)}{r} \, .$$

Intuitively, these dimensions measure the density of algorithmic information in the point x. Guided by the information-theoretic nature of these characterizations, J. Lutz and N. Lutz [8] defined the *lower* and *upper conditional dimension* of $x \in \mathbb{R}^m$ given $y \in \mathbb{R}^n$ as

$$\dim(x|y) = \liminf_{r \to \infty} \frac{K_r(x|y)}{r} \quad \text{and} \quad \mathrm{Dim}(x|y) = \limsup_{r \to \infty} \frac{K_r(x|y)}{r} \, .$$

2.4   Relative Complexity and Dimensions

By letting the underlying fixed prefix-free Turing machine U be a universal *oracle* machine, we may *relativize* the definition in this section to an arbitrary oracle set $A \subseteq \mathbb{N}$. The definitions of $K^A(\sigma|\tau)$, $K^A(\sigma)$, $K_r^A(x)$, $K_r^A(x|y)$, $\dim^A(x)$, $\mathrm{Dim}^A(x)$ $\dim^A(x|y)$, and $\mathrm{Dim}^A(x|y)$ are then all identical to their unrelativized versions, except that U is given oracle access to A.

We will frequently consider the complexity of a point $x \in \mathbb{R}^n$ *relative to a point* $y \in \mathbb{R}^m$, i.e., relative to a set A_y that encodes the binary expansion of y is a standard way. We then write $K_r^y(x)$ for $K_r^{A_y}(x)$. J. Lutz and N. Lutz showed that $K_r^y(x) \leq K_{r,t}(x|y) + K(t) + O(1)$ [8].

3   Background and Approach

In this section we describe the basic ideas behind our investigation of dimension spectra of lines. We briefly discuss some of our earlier work on this subject, and we present two technical lemmas needed for the proof our main theorems.

The dimension of a point on a line in $\mathbb{R}^2$ has the following trivial bound.

Observation 2. *For all* $a, b, x \in \mathbb{R}$, $\dim(x, ax + b) \leq \dim(x, a, b)$.

In this work, our goal is to find values of x for which the approximate converse

$$\dim(x, ax + b) \geq \dim^{a,b}(x) + \dim(a, b) \tag{1}$$

holds. There exist oracles, at least, relative to which (1) does not always hold. This follows from the point-to-set principle of J. Lutz and N. Lutz [8] and the existence of Furstenberg sets with parameter α and Hausdorff dimension less than $1 + \alpha$ (attributed by Wolff [15] to Furstenberg and Katznelson "in all probability"). The argument is simple and very similar to our proof in [10] of a lower bound on the dimension of generalized Furstenberg sets.

Specifically, for every $s \in [0, 1]$, we want to find an x of effective Hausdorff dimension s such that (1) holds. Note that equality in Observation 2 implies (1).

Observation 3. *Suppose* $ax + b = ux + v$ *and* $u \neq a$. *Then*

$$\dim(u, v) \geq \dim^{a,b}(u, v) \geq \dim^{a,b}\left(\frac{b - v}{u - a}\right) = \dim^{a,b}(x).$$

This observation suggests an approach, whenever $\dim^{a,b}(x) > \dim(a, b)$, for showing that $\dim(x, ax+b) \geq \dim(x, a, b)$. Since (a, b) is, in this case, the unique low-dimensional pair such that $(x, ax + b)$ lies on $L_{a,b}$, one might naïvely hope to use this fact to derive an estimate of (x, a, b) from an estimate of $(x, ax + b)$. Unfortunately, the dimension of a point is not even semicomputable, so algorithmically distinguishing (a, b) requires a more refined statement.

3.1   Previous Work

The following lemma, which is essentially geometrical, is such a statement.

Lemma 4 (N. Lutz and Stull [10]). *Let $a, b, x \in \mathbb{R}$. For all $(u, v) \in \mathbb{R}^2$ such that $ux + v = ax + b$ and $t = -\log \|(a, b) - (u, v)\| \in (0, r]$,*

$$K_r(u, v) \geq K_t(a, b) + K_{r-t}^{a,b}(x) - O(\log r).$$

Roughly, if $\dim(a, b) < \dim^{a,b}(x)$, then Lemma 4 tells us that $K_r(u, v) > K_r(a, b)$ unless (u, v) is very close to (a, b). As $K_r(u, v)$ is upper semicomputable, this is algorithmically useful: We can enumerate all pairs (u, v) whose precision-r complexity falls below a certain threshold. If one of these pairs satisfies, approximately, $ux + v = ax + b$, then we know that (u, v) is close to (a, b). Thus, an estimate for $(x, ax + b)$ algorithmically yields an estimate for (x, a, b).

In our previous work [10], we used an argument of this type to prove a general lower bound on the dimension of points on lines in $\mathbb{R}^2$:

Theorem 5 (N. Lutz and Stull [10]). *For all $a, b, x \in \mathbb{R}$,*

$$\dim(x, ax + b) \geq \dim^{a,b}(x) + \min\{\dim(a, b), \dim^{a,b}(x)\}.$$

The strategy in that work is to use oracles to artificially lower $K_r(a, b)$ when necessary, to essentially force $\dim(a, b) < \dim^{a,b}(x)$. This enables the above argument structure to be used, but lowering the complexity of (a, b) also weakens the conclusion, leading to the minimum in Theorem 5.

3.2   Technical Lemmas

In the present work, we circumvent this limitation and achieve inequality (1) by controlling the choice of x and placing a condition on (a, b). Adapting the above argument to the case where $\dim(a, b) > \dim^{a,b}(x)$ requires refining the techniques of [10]. In particular, we use the following two technical lemmas, which strengthen results from that work. Lemma 6 weakens the conditions needed to compute an estimate of (x, a, b) from an estimate of $(x, ax + b)$.

Lemma 6. *Let $a, b, x \in \mathbb{R}$, $k \in \mathbb{N}$, and $r_0 = 1$. Suppose that $r_1, \ldots, r_k \in \mathbb{N}$, $\delta \in \mathbb{R}_+$, and $\varepsilon, \eta \in \mathbb{Q}_+$ satisfy the following conditions for every $1 \leq i \leq k$.*

1. *$r_i \geq \log(2|a| + |x| + 6) + r_{i-1}$.*
2. *$K_{r_i}(a, b) \leq (\eta + \varepsilon) r_i$.*
3. *For every $(u, v) \in \mathbb{R}^2$ such that $t = -\log \|(a, b) - (u, v)\| \in (r_{i-1}, r_i]$ and $ux + v = ax + b$, $K_{r_i}(u, v) \geq (\eta - \varepsilon) r_i + \delta \cdot (r_i - t)$.*

Then for every oracle set $A \subseteq \mathbb{N}$,

$$K_{r_k}^A(a, b, x \mid x, ax + b) \leq 2^k \left(K(\varepsilon) + K(\eta) + \frac{4\varepsilon}{\delta} r_k + O(\log r_k) \right).$$

Lemma 7 strengthens the oracle construction of [10], allowing us to control complexity at multiple levels of precision.

Lemma 7. *Let $z \in \mathbb{R}^n$, $\eta \in \mathbb{Q} \cap [0, \dim(z)]$, and $k \in \mathbb{N}$. For all $r_1, \ldots, r_k \in \mathbb{N}$, there is an oracle $D = D(r_1, \ldots, r_k, z, \eta)$ such that*

1. *For every $t \leq r_1$, $K_t^D(z) = \min\{\eta r_1, K_t(z)\} + O(\log r_k)$*
2. *For every $1 \leq i \leq k$,*

$$K_{r_i}^D(z) = \eta r_1 + \sum_{j=2}^{i} \min\{\eta(r_j - r_{j-1}), K_{r_j, r_{j-1}}(z \mid z)\} + O(\log r_k).$$

3. *For every $t \in \mathbb{N}$ and $x \in \mathbb{R}$, $K_t^{z,D}(x) = K_t^z(x) + O(\log r_k)$.*

4 Main Theorems

We are now prepared to prove our two main theorems. We first show that, for lines $L_{a,b}$ such that $\dim(a, b) = \mathrm{Dim}(a, b)$, the dimension spectrum $\mathrm{sp}(L_{a,b})$ contains the unit interval.

Theorem 8. *Let $a, b \in \mathbb{R}$ satisfy $\dim(a, b) = \mathrm{Dim}(a, b)$. Then for every $s \in [0, 1]$ there is a point $x \in \mathbb{R}$ such that $\dim(x, ax + b) = s + \min\{\dim(a, b), 1\}$.*

Proof. Every line contains a point of effective Hausdorff dimension 1 [13], and by the preservation of effective dimensions under computable bi-Lipschitz functions, $\dim(a, a^2 + b) = \dim(a, b)$, so the theorem holds for $s = 0$. For $s = 1$, we may choose an $x \in \mathbb{R}$ that is random relative to (a, b). That is, there is some constant $c \in \mathbb{N}$ such that for all $r \in \mathbb{N}$, $K_r^{a,b}(x) \geq r - c$. By Theorem 5,

$$\dim(x, ax + b) \geq \dim^{a,b}(x) + \min\{\dim(a, b), 1\}$$
$$= \min\{\dim(a, b), 1\} + \liminf_{r \to \infty} \frac{K_r(x)}{r}$$
$$= \min\{\dim(a, b), 1\} + 1,$$

and the conclusion holds.

Now let $s \in (0, 1)$ and $d = \dim(a, b) = \mathrm{Dim}(a, b)$. Let $y \in \mathbb{R}$ be random relative to (a, b). Define the sequence of natural numbers $\{h_j\}_{j \in \mathbb{N}}$ inductively as follows. Define $h_0 = 1$. For every $j > 0$, let

$$h_j = \min\left\{ h \geq 2^{h_{j-1}} : K_h(a, b) \leq \left(d + \frac{1}{j}\right) h \right\}.$$

Note that h_j always exists. For every $r \in \mathbb{N}$, let

$$x[r] = \begin{cases} 0 & \text{if } \frac{r}{h_j} \in (s, 1] \text{ for some } j \in \mathbb{N} \\ y[r] & \text{otherwise} \end{cases}$$

where $x[r]$ is the rth bit of x. Define $x \in \mathbb{R}$ to be the real number with this binary expansion. Then $K_{sh_j}(x) = sh_j + O(\log sh_j)$.

We first show that $\dim(x, ax + b) \leq s + \min\{d, 1\}$. For every $j \in \mathbb{N}$,

$$
\begin{aligned}
K_{h_j}(x, ax + b) &= K_{h_j}(x) + K_{h_j}(ax + b \mid x) + O(\log h_j) \\
&= K_{sh_j}(x) + K_{h_j}(ax + b \mid x) + O(\log h_j) \\
&= K_{sh_j}(y) + K_{h_j}(ax + b \mid x) + O(\log h_j) \\
&\leq sh_j + \min\{d, 1\} \cdot h_j + o(h_j).
\end{aligned}
$$

Therefore,

$$
\begin{aligned}
\dim(x, ax + b) &= \liminf_{r \to \infty} \frac{K_r(x, ax + b)}{r} \\
&\leq \liminf_{j \to \infty} \frac{K_{h_j}(x, ax + b)}{h_j} \\
&\leq \liminf_{j \to \infty} \frac{sh_j + \min\{d, 1\}h_j + o(h_j)}{h_j} \\
&= s + \min\{d, 1\}.
\end{aligned}
$$

If $1 > s \geq d$, then by Theorem 5 we also have

$$
\begin{aligned}
\dim(x, ax + b) &\geq \dim^{a,b}(x) + \dim(a, b) \\
&= \dim(x) + d \\
&= \liminf_{r \to \infty} \frac{K_r(x)}{r} + d \\
&= \liminf_{j \to \infty} \frac{K_{h_j}(x)}{h_j} + d \\
&= s + \min\{d, 1\}.
\end{aligned}
$$

Hence, we may assume that $s < d$.

Let $H = \mathbb{Q} \cap (s, \min\{d, 1\})$. Let $\eta \in H$, $\delta = 1 - \eta > 0$, and $\varepsilon \in \mathbb{Q}_+$. We now show that $\dim(x, ax + b) \geq s + \eta - \frac{\alpha \varepsilon}{\delta}$, where α is some constant independent of η and ε. Let $j \in \mathbb{N}$ and $m = \frac{s-1}{\eta-1}$. We first show that

$$
K_r(x, ax + b) \geq K_r(x) + \eta r - c\frac{\varepsilon}{\delta}r - o(r), \tag{2}
$$

for every $r \in (sh_j, mh_j]$. Let $r \in (sh_j, mh_j]$. Set $k = \frac{r}{sh_j}$, and define $r_i = ish_j$ for all $1 \leq i \leq k$. Note that k is bounded by a constant depending only on s and η. Therefore a $o(r_k) = o(r_i)$ for all r_i. Let $D_r = D(r_1, \ldots, r_k, (a, b), \eta)$ be the oracle defined in Lemma 7. We first note that, since $\dim(a, b) = \mathrm{Dim}(a, b)$,

$$
\begin{aligned}
K_{r_i, r_{i-1}}(a, b \mid a, b) &= K_{r_i}(a, b) - K_{r_{i-1}}(a, b) - O(\log r_i) \\
&= \dim(a, b)r_i - o(r_i) - \dim(a, b)r_{i-1} - o(r_{i-1}) - O(\log r_i) \\
&= \dim(a, b)(r_i - r_{i-1}) - o(r_i) \\
&\geq \eta(r_i - r_{i-1}) - o(r_i).
\end{aligned}
$$

Hence, by property 2 of Lemma 7, for every $1 \le i \le k$,

$$|K_{r_i}^{D_r}(a, b) - \eta r_i| \le o(r_k). \tag{3}$$

We now show that the conditions of Lemma 6 are satisfied. By inequality (3), for every $1 \le i \le k$,

$$K_{r_i}^{D_r}(a, b) \le \eta r_i + o(r_k),$$

and so $K_{r_i}^{D_r}(a, b) \le (\eta + \varepsilon)r_i$, for sufficiently large j. Hence, condition 2 of Lemma 6 is satisfied.

To see that condition 3 is satisfied for $i = 1$, let $(u, v) \in B_1(a, b)$ such that $ux + v = ax + b$ and $t = -\log \|(a, b) - (u, v)\| \le r_1$. Then, by Lemmas 4 and 7, and our construction of x,

$$\begin{aligned}
K_{r_1}^{D_r}(u, v) &\ge K_t^{D_r}(a, b) + K_{r_1-t, r_1}^{D_r}(x|a, b) - O(\log r_1) \\
&\ge \min\{\eta r_1, K_t(a, b)\} + K_{r_1-t}(x) - o(r_k) \\
&\ge \min\{\eta r_1, dt - o(t)\} + (\eta + \delta)(r_1 - t) - o(r_k) \\
&\ge \min\{\eta r_1, \eta t - o(t)\} + (\eta + \delta)(r_1 - t) - o(r_k) \\
&\ge \eta t - o(t) + (\eta + \delta)(r_1 - t) - o(r_k).
\end{aligned}$$

We conclude that $K_{r_1}^{D_r}(u, v) \ge (\eta - \varepsilon)r_1 + \delta(r_1 - t)$, for all sufficiently large j.

To see that that condition 3 is satisfied for $1 < i \le k$, let $(u, v) \in B_{2^{-r_{i-1}}}(a, b)$ such that $ux + v = ax + b$ and $t = -\log \|(a, b) - (u, v)\| \le r_i$. Since $(u, v) \in B_{2^{-r_{i-1}}}(a, b)$,

$$r_i - t \le r_i - r_{i-1} = ish_j - (i-1)sh_j \le sh_j + 1 \le r_1 + 1.$$

Therefore, by Lemma 4, inequality (3), and our construction of x,

$$\begin{aligned}
K_{r_i}^{D_r}(u, v) &\ge K_t^{D_r}(a, b) + K_{r_i-t, r_i}^{D_r}(x|a, b) - O(\log r_i) \\
&\ge \min\{\eta r_i, K_t(a, b)\} + K_{r_i-t}(x) - o(r_i) \\
&\ge \min\{\eta r_i, dt - o(t)\} + (\eta + \delta)(r_i - t) - o(r_i) \\
&\ge \min\{\eta r_i, \eta t - o(t)\} + (\eta + \delta)(r_i - t) - o(r_i) \\
&\ge \eta t - o(t) + (\eta + \delta)(r_i - t) - o(r_i).
\end{aligned}$$

We conclude that $K_{r_i}^{D_r}(u, v) \ge (\eta - \varepsilon)r_i + \delta(r_i - t)$, for all sufficiently large j. Hence the conditions of Lemma 6 are satisfied, and we have

$$\begin{aligned}
K_r(x, ax + b) &\ge K_r^{D_r}(x, ax + b) - O(1) \\
&\ge K_r^{D_r}(a, b, x) - 2^k \left(K(\varepsilon) + K(\eta) + \frac{4\varepsilon}{\delta}r + O(\log r) \right) \\
&= K_r^{D_r}(a, b) + K_r^{D_r}(x \mid a, b) \\
&\quad - 2^k \left(K(\varepsilon) + K(\eta) + \frac{4\varepsilon}{\delta}r + O(\log r) \right) \\
&\ge sr + \eta r - 2^k \left(K(\varepsilon) + K(\eta) + \frac{4\varepsilon}{\delta}r + O(\log r) \right).
\end{aligned}$$

Thus, for every $r \in (sh_j, mh_j]$,

$$K_r(x, ax + b) \geq sr + \eta r - \frac{\alpha \varepsilon}{\delta} r - o(r),$$

where α is a fixed constant, not depending on η and ε.

To complete the proof, we show that (2) holds for every $r \in [mh_j, sh_{j+1})$. By Lemma 1 and our construction of x,

$$\begin{aligned} K_r(x) &= K_{r,h_j}(x \mid x) + K_{h_j}(x) + o(r) \\ &= r - h_j + sh_j + o(r) \\ &\geq \eta r + o(r). \end{aligned}$$

The proof of Theorem 5 gives $K_r(x, ax + b) \geq K_r(x) + \dim(x)r - o(r)$, and so $K_r(x, ax + b) \geq r(s + \eta)$.

Therefore, equation (2) holds for every $r \in [sh_j, sh_{j+1})$, for all sufficiently large j. Hence,

$$\begin{aligned} \dim(x, ax + b) &= \liminf_{r \to \infty} \frac{K_r(x, ax + b)}{r} \\ &\geq \liminf_{r \to \infty} \frac{K_r(x) + \eta r - \frac{\alpha \varepsilon}{\delta} r - o(r)}{r} \\ &\geq \liminf_{r \to \infty} \frac{K_r(x)}{r} + \eta - \frac{\alpha \varepsilon}{\delta} \\ &= s + \eta - \frac{\alpha \varepsilon}{\delta}. \end{aligned}$$

Since η and ε were chosen arbitrarily, the conclusion follows.   □

Theorem 9. *Let $a, b \in \mathbb{R}$ such that $\dim(a, b) \geq 1$. Then for every $s \in [\frac{1}{2}, 1]$ there is a point $x \in \mathbb{R}$ such that $\dim(x, ax + b) \in [\frac{3}{2} + s - \frac{1}{2s}, s + 1]$.*

Corollary 10. *Let $L_{a,b}$ be any line in $\mathbb{R}^2$. Then the dimension spectrum $\mathrm{sp}(L_{a,b})$ is infinite.*

Proof. Let $(a, b) \in R^2$. If $\dim(a, b) < 1$, then by Theorem 5 and Observation 2, the spectrum $\mathrm{sp}(L_{a,b})$ contains the interval $[\dim(a, b), 1]$. Assume that $\dim(a, b) \geq 1$. By Theorem 9, for every $s \in [\frac{1}{2}, 1]$, there is a point x such that $\dim(x, ax + b) \in [\frac{3}{2} + s - \frac{1}{2s}, s + 1]$. Since these intervals are disjoint for $s_n = \frac{2n-1}{2n}$, the dimension spectrum $\mathrm{sp}(L_{a,b})$ is infinite.

5   Future Directions

We have made progress in the broader program of describing the dimension spectra of lines in Euclidean spaces. We highlight three specific directions for further progress. First, it is natural to ask whether the condition on (a, b) may be dropped from the statement our main theorem: *Does Theorem 8 hold for arbitrary $a, b \in \mathbb{R}$?*

Second, the dimension spectrum of a line $L_{a,b} \subseteq \mathbb{R}^2$ may *properly* contain the unit interval described in our main theorem, even when $\dim(a,b) = \text{Dim}(a,b)$. If $a \in \mathbb{R}$ is random and $b = 0$, for example, then $\text{sp}(L_{a,b}) = \{0\} \cup [1,2]$. It is less clear whether this set of "exceptional values" in $\text{sp}(L_{a,b})$ might itself contain an interval, or even be infinite. *How large (in the sense of cardinality, dimension, or measure) may* $\text{sp}(L_{a,b}) \cap [0, \min\{1, \dim(a,b)\})$ *be?*

Finally, any non-trivial statement about the dimension spectra of lines in higher-dimensional Euclidean spaces would be very interesting. Indeed, an n-dimensional version of Theorem 5 (i.e., one in which $a, b \in \mathbb{R}^{n-1}$, for all $n \geq 2$) would, via the point-to-set principle for Hausdorff dimension [8], affirm the famous Kakeya conjecture and is therefore likely difficult. The additional hypothesis of Theorem 8 might make it more conducive to such an extension.

References

1. Athreya, K.B., Hitchcock, J.M., Lutz, J.H., Mayordomo, E.: Effective strong dimension in algorithmic information and computational complexity. SIAM J. Comput. **37**(3), 671–705 (2007)
2. Dougherty, R., Lutz, J., Mauldin, R.D., Teutsch, J.: Translating the Cantor set by a random real. Trans. Am. Math. Soc. **366**(6), 3027–3041 (2014)
3. Downey, R., Hirschfeldt, D.: Algorithmic Randomness and Complexity. Springer, New York (2010)
4. Xiaoyang, G., Lutz, J.H., Mayordomo, E., Moser, P.: Dimension spectra of random subfractals of self-similar fractals. Ann. Pure Appl. Logic **165**(11), 1707–1726 (2014)
5. Hitchcock, J.M.: Correspondence principles for effective dimensions. Theory Comput. Syst. **38**(5), 559–571 (2005)
6. Li, M., Vitányi, P.M.B.: An Introduction to Kolmogorov Complexity and Its Applications, 3rd edn. Springer, New York (2008)
7. Lutz, J.H.: The dimensions of individual strings and sequences. Inf. Comput. **187**(1), 49–79 (2003)
8. Lutz, J.H., Lutz, N.: Algorithmic information, plane Kakeya sets, and conditional dimension. In: Vollmer, H., Vallee, B. (eds.) 34th Symposium on Theoretical Aspects of Computer Science (STACS 2017), Leibniz International Proceedings in Informatics (LIPIcs), vol. 66, pp. 53:1–53:13. Schloss Dagstuhl–Leibniz-Zentrum fuer Informatik, Germany (2017)
9. Lutz, J.H., Mayordomo, E.: Dimensions of points in self-similar fractals. SIAM J. Comput. **38**(3), 1080–1112 (2008)
10. Lutz, N., Stull, D.M.: Bounding the dimension of points on a line. In: Gopal, T.V., Jäger, G., Steila, S. (eds.) TAMC 2017. LNCS, vol. 10185, pp. 425–439. Springer, Cham (2017). doi:10.1007/978-3-319-55911-7_31
11. Mayordomo, E.: A Kolmogorov complexity characterization of constructive Hausdorff dimension. Inf. Process. Lett. **84**(1), 1–3 (2002)
12. Reimann, J.: Effectively closed classes of measures and randomness. Ann. Pure Appl. Logic **156**(1), 170–182 (2008)
13. Turetsky, D.: Connectedness properties of dimension level sets. Theor. Comput. Sci. **412**(29), 3598–3603 (2011)

14. Westrick, L.B.: Computability in ordinal ranks and symbolic dynamics. Ph.D. thesis, University of California, Berkeley (2014)
15. Wolff, T.: Recent work connected with the Kakeya problem. In: Prospects in Mathematics, pp. 129–162 (1999)

A Universal Oracle for Signal Machines

Thierry Monteil[✉]

CNRS – LIPN – Université Paris 13, Villetaneuse, France
thierry.monteil@lipn.univ-paris13.fr
http://monteil.perso.math.cnrs.fr

Abstract. We construct two universal oracles for signal machines, one via the binary expansion of irrational numbers, another via their continued fraction expansion, settling a conjecture of Durand-Lose in CiE 2013. This latter is optimal in the number of speeds and irrational parameters involved in the construction (three and one respectively).

Keywords: Signal machine · Geometric computation · Oracle · Linear Blum Shub Smale model · Binary expansion · Continued fractions

1 Introduction: Signal Machines

Signal machines were introduced in [DL03] as a geometric model of computation featuring discrete and deterministic interactions of signals within a continuous space and time. More precisely, a *meta-signal* (c, s) is characterized by its *color* $c \in C$ and *speed* $s \in \mathbb{R}$, where C is a finite set. A *signal* (c, s, p) is a meta-signal together with a *position* $p \in \mathbb{R}$. A *signal machine* is given by:

- a finite set of meta-signals M,
- an *initial configuration*, *i.e.* a finite set of signals $I \subset M \times \mathbb{R}$,
- a set of deterministic *collision rules*, *i.e.* a map $R : 2^M \to 2^M$ whose restriction to sets of meta-signals with the same speed is the identity.

Note that two signals with the same position and speed, but different colors can be superposed. This allows some flexibility in defining signal machines without having to artificially consider products of colors. The condition on meta-signals with the same speed ensures that the non-trivial collisions are discrete (in particular, no collision appears along a single signal). In this paper, by convention, when a collision rule is only partially defined, it is extended by the identity, so that when signals meet in an undefined way, they ignore each other.

The *execution* of a signal machine could be defined from its initial configuration, then considering that signals move along the real line with constant speed, until signals with different speeds meet, in which case the corresponding collision rule is applied, leading to the substitution of collided signals into their images by R. Since a picture is worth a thousand words, let us describe a simple example which will be used in the next section: a signal machine that computes the middle of two positions. The *middle machine* has five meta-signals on three colors:

© Springer International Publishing AG 2017
J. Kari et al. (Eds.): CiE 2017, LNCS 10307, pp. 315–326, 2017.
DOI: 10.1007/978-3-319-58741-7_30

$C = \{\text{black}, \text{red}, \text{green}\}$ $M = \{(\text{black}, 0), (\text{green}, 0), (\text{red}, 1), (\text{red}, 3), (\text{red}, -3)\}$. Initial configuration (with two parameters x_1 and x_2), collision rules and execution are described in Figs. 1, 2, 3, respectively. The representations show the space ($\mathbb{R}$) horizontally, and the time ($\mathbb{R}_+$) vertically, from bottom up.

While the initial condition could be understood as the input of a signal machine and the collision rules could be understood as the transition function, there is purposely no formal definition for the output of a signal machine. Indeed, depending on the situation, some outputs could be the positions of signals of given colors, or distances between such pairs of signals, some Boolean output for decision problems could depend on the apparition of a given color during an execution, or even on the existence of accumulations of collisions in the execution.

$$I = \{(\text{black}, 0, x_1), (\text{black}, 0, x_2), (\text{red}, 1, x_1), (\text{red}, 3, x_1)\}$$

Fig. 1. Middle machine: initial configuration. (Color figure online)

$R(\{(\text{red}, 3), (\text{black}, 0)\}) = \{(\text{red}, -3), (\text{black}, 0)\}$ $R(\{(\text{red}, 1), (\text{red}, -3)\}) = \{(\text{green}, 0)\}$

Fig. 2. Middle machine: collision rules (other collisions lead to no signal). (Color figure online)

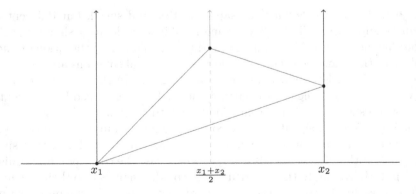

Fig. 3. Middle machine: execution. (Color figure online)

In the present article, a non-trivial oracle never stops, and reacts to queries from a Turing machine, leading to a on demand sequence of outputs.

There are different ways to deal with accumulations of collisions. Allowing accumulations to output some information leads to a stronger model (see [DL09]). In the present article, we consider machines without accumulations, *i.e.* the computation is aborted when an accumulation occurs. It was proved in [DL07] that such an accumulation-free model is equivalent to the linear Blum, Shub and Smale model (denoted Linear-BSS) [BSS89], which can be seen as a Turing machine which can handle real numbers, and where allowed operations on them are addition, comparison, multiplication by constants (in particular it is less powerful than the classical Blum, Shub and Smale model where multiplication between variables is allowed).

2   An Oracle Based on Binary Expansion

A usual way to define the interaction between a Turing machine and an oracle representing a language $L \subseteq \{0, 1\}^*$ is as follows: the Turing machine queries the oracle about a word $w \in \{0, 1\}^*$ and gets an answer 0 or 1 depending on whether the word w belongs to L. Here, our oracle provides an infinite sequence S of bits, whose values can be consulted one after another. Those two versions are equivalent. Indeed, since there exists a computable enumeration $e : \{0, 1\}^* \to \mathbb{N}$, the sequence can S represent the language L if $S \circ e$ is the characteristic function of L. Since a Turing machine has unbounded caching capabilities, it can store all the bits it gets from the oracle. To know if a word w belongs to L, it computes $n = e(w)$, and looks at the nth obtained bit if it was received already, or asks for enough new bits until it gets the nth bit.

In the Turing model of computation, an oracle is a black box whose internals are unspecified. In order to understand the computational power of signal machines, we want to see if *any* oracle can be produced from a concrete signal machine.

The aim of this paper is to construct a *universal oracle* for signal machines, that is, a single explicit signal machine O which depends on an input p, such that every infinite sequence $(x_n) \in \{0, 1\}^{\mathbb{N}}$ can be produced by $O(p)$ for some $p \in \mathbb{R}$.

First, let us describe how the simulation of a Turing machine by a signal machine as described in [DL03] can query an oracle. It should be noticed that the Turing machine simulation uses a bounded space. The oracles we will construct will run in space bounded by 1, and the duration of the computation of one term of the sequence by those oracles will be bounded by 1. In particular, we can concatenate their configuration with the Turing machine on the left and the oracle on the right (see Fig. 4). To avoid possible interference between the Turing machine and the oracle, we add a static signal in-between that blocks any lost signal that is not a query nor an answer (another possibility would be to ensure that the query and answers are the only common colors between the two devices). The oracle computes one term of the sequence and waits for a query with a zero-speed signal whose color (say green for 0 and blue for 1) corresponds to that term. When

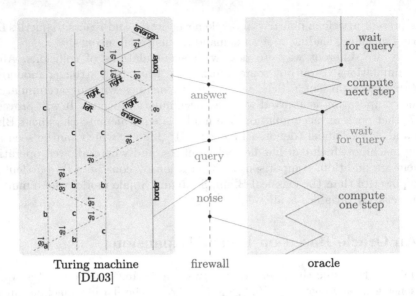

Turing machine
[DL03]   firewall   oracle

Fig. 4. Communication between a Turing machine and an oracle in the signal machine model. (Color figure online)

the Turing machine needs a value from the oracle, it sends an orange signal going to the right (with speed $s > 0$). Once this signal meets the waiting one from the oracle, the answer is returned to the Turing machine as a green or blue signal of speed $-s$. If the execution zones of the Turing machine and the oracle are separated by a distance at least $1/s$, we are sure that the orange query will reach the oracle after this latter finishes the computation of the term.

Since finite and periodic sequences can easily be produced by the (simulated) Turing machine itself, interesting oracles are the ones that produce uncomputable sequences, in particular infinite non-periodic sequences, and we might skip details regarding few finite or periodic sequences if those lead to additional technicalities regarding non-generic collisions.

It is pretty easy to describe such a universal oracle in the Linear-BSS model: given an infinite sequence (x_n) of 0 and 1 which is not ultimately constant, we can construct the real number whose binary representation is $p = 0.x_0x_1x_2x_3\cdots$ and let a Turing machine extract its digits one by one by multiplication by 2 (which corresponds to shifting the sequence) and taking the leading digit. A simple implementation of such a generator could be (using Python syntax as a pseudocode):

```python
def binary_oracle(p):
    while True:
        if p < 1/2:
            yield 0
            p = 2*p
        elif 1/2 < p:
```

```
        yield 1
        p = 2*p-1
  else:
        return
```

Hence, it is possible to simulate such an algorithm with a signal machine, by translating this algorithm into a Linear-BSS machine, and then into a signal machine following the construction of [DL07].

However, such a construction will lose all the geometrical features of the signal-machine model, while we can benefit from it easily. Indeed, constructing the middle of two points is pretty easy in that framework, as we saw in Sect. 1. Let us roughly implement it as a signal machine as follows:

The point $p \in (0, 1)$ is a static (zero speed) signal. We proceed by dichotomy of the unit interval, the last middle is the active signal and knows (from its color) on which side is the point p. When a query from the Turing machine reaches the last middle, the previous result is returned and a new step of the dichotomy is computed for the next query. To this end, two red signals create the next middle, the one which meets the point p gives the right color to the next middle, which knows on which side is p (See Figs. 5, 6, 7).

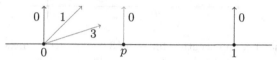

Fig. 5. Binary machine: initial configuration. (Color figure online)

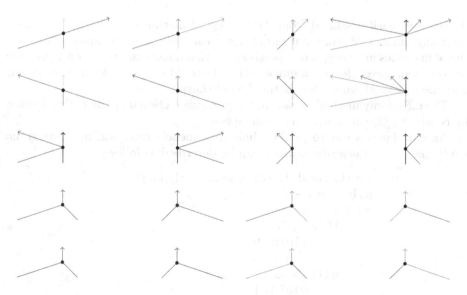

Fig. 6. Binary machine: collision rules. (Color figure online)

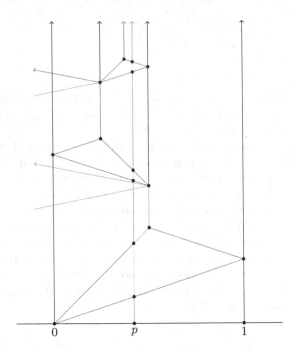

Fig. 7. Binary machine: an execution for some $3/8 < p < 1/2$, answering two queries as $0, 1$ and ready to answer the third as 1. (Color figure online)

3   An Optimal Oracle Based on Continued Fractions

As could be easily deduced from [DL13], signal machines involving two speeds (and any number of irrational parameters) can only produce finite sequences; signal machines involving three speeds and rational parameters can only produce periodic sequences. Hence, an universal oracle would require at least three speeds and one irrational parameter. Is this bound sharp?

The dichotomy method we saw in the previous section uses the multiplication by constant $1/2$, which itself uses four speeds.

Another famous way to produce infinite sequences from real numbers is the *continued fraction algorithm*, which can be described as follows:

```
def continued_fractions_oracle(p):
    a,b = p,1-p
    while True:
        if a < b:
            yield 0
            a,b = b-a,a
        elif b < a:
            yield 1
            a,b = b,a-b
        else:
            return
```

Actually, this algorithm is a slight variant of the classical continued fraction algorithm which iteratively sends (a, b) to $(a, b - a)$ if $a < b$ and (a, b) to (b, a) if $b < a$ and stops when $a = b$ (see, *e.g.* [KE64]). While sharing the features of the classical continued fraction algorithm (see Theorem 1), it is easier to implement geometrically by avoiding swapping values.

One advantage over the dichotomy algorithm is that it performs only subtractions and comparisons. Let us first construct a machine that performs a single step of the algorithm (see Figs. 8, 9, 10).

Let us now turn the subtraction machine into a universal oracle with three speed and a single irrational parameter. We have to log the result of the comparison by specializing the gray color into gray when the comparison is not done yet, green when $a < b$ and blue when $b < a$. When a query arrives from the Turing machine, we have to release this color to the Turing machine, and start a new subtraction for the next round. This leads to the following signal machine $O(p)$ with parameter $p \in [0, 1]$ (see Figs. 11, 12, 13).

Note that the speed s of queries and answers is arbitrary and was set to a high value to avoid signal superposition in the pictures, but it could be set to 1, therefore not artificially adding additional speeds.

The continued fraction machine simulates the `continued_fractions_oracle` algorithm.

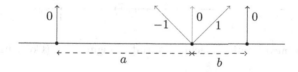

Fig. 8. Subtraction machine: initial configuration with parameters a and b. (Color figure online)

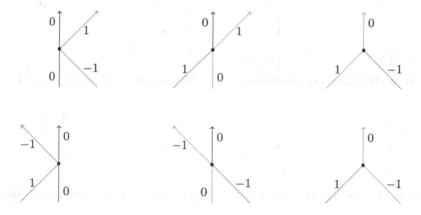

Fig. 9. Subtraction machine: collision rules. (Color figure online)

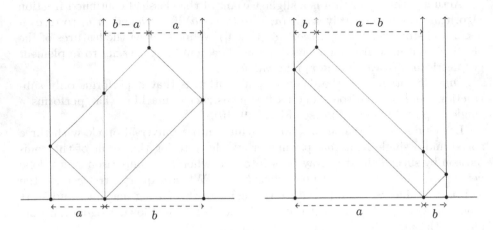

Fig. 10. Subtraction machine: two possible executions depending on whether $a < b$ or $b < a$. (Color figure online)

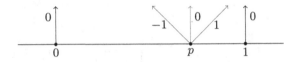

Fig. 11. Continued fraction machine: initial configuration. (Color figure online)

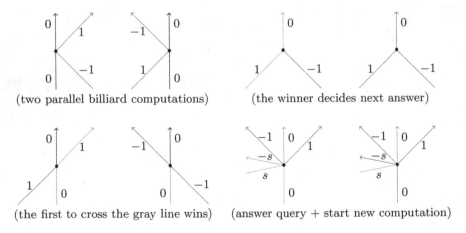

Fig. 12. Continued fraction machine: collision rules. (Color figure online)

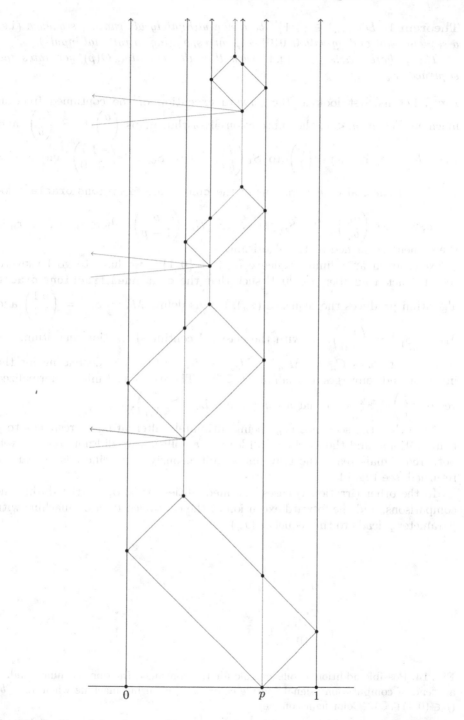

Fig. 13. Continued fraction machine: execution with some parameter $2/3 < p < 5/7$ answering three queries as $1, 0, 0$ and ready to answer the fourth as 1. Note that the speed of queries and answers can be set to ± 1 so that the machine has only 3 speeds. (Color figure online)

Theorem 1. *Let $(x_n) \in \{0,1\}^{\mathbb{N}}$ be a non-ultimately alternating sequence (i.e. a sequence not ending with $010101 \cdots$, corresponding to rational inputs).*

Then, there exists $p \in [0,1]$ such that the machine $O(p)$ generates the sequence (x_n).

Proof. Let us first look at the forward execution of the continued fraction machine. When, $a < b$, the subtraction algorithm maps $\begin{pmatrix} a \\ b \end{pmatrix}$ to $S_0 \begin{pmatrix} a \\ b \end{pmatrix}$, and when $b < a$, it maps $\begin{pmatrix} a \\ b \end{pmatrix}$ to $S_1 \begin{pmatrix} a \\ b \end{pmatrix}$, where $S_0 := \begin{pmatrix} -1 & 1 \\ 1 & 0 \end{pmatrix}$ and $S_1 := \begin{pmatrix} 0 & 1 \\ 1 & -1 \end{pmatrix}$. Hence, after n iterations of the `continued_fractions_oracle` algorithm, we have $\begin{pmatrix} a_n \\ b_n \end{pmatrix} = S_{x_n} S_{x_{n-1}} \cdots S_{x_2} S_{x_1} S_{x_0} \begin{pmatrix} p \\ 1-p \end{pmatrix}$, where $x_0, x_1, \ldots, x_n$ is the sequence returned by the algorithm.

Now, given an infinite sequence $(x_n) \in \{0,1\}^{\mathbb{N}}$, we have to go backward to construct a number $p \in [0,1]$ such that the `continued_fractions_oracle` algorithm produces the sequence (x_n). Let us define $M_0 := S_0^{-1} = \begin{pmatrix} 0 & 1 \\ 1 & 1 \end{pmatrix}$ and $M_1 := S_1^{-1} = \begin{pmatrix} 1 & 1 \\ 1 & 0 \end{pmatrix}$. As with the classical continued fraction algorithm, the sequence of cones $C_n := M_{x_0} M_{x_1} M_{x_2} \cdots M_{x_{n-1}} M_{x_n} \mathbb{R}_+^2$ is decreasing for the inclusion and converges to a half line $L \subseteq \mathbb{R}_+^2$. There exists a (unique) normalized vector $\begin{pmatrix} p \\ q \end{pmatrix} \in \mathbb{R}_+^2$ such that $p + q = 1$ and $L = \mathbb{R}_+ \begin{pmatrix} p \\ q \end{pmatrix}$.

Note that the sequence (x_n) being ultimately alternating corresponds to p being rational and the machine will lead to a collision we did not consider yet: both red signals reach the gray line simultaneously, a specific answer can be returned, see Fig. 14.

In the other (irrational) cases, the inequalities are always strict during the comparisons, and the forward execution of the continued fraction machine with parameter p leads to the sequence (x_n).   □

Fig. 14. Possible additional collision rule for the continued fraction machine: sending an "end of computation" signal (purple color) to the Turing machine when $a_n = b_n$ ($p \in (0,1) \cap \mathbb{Q}$). (Color figure online)

As an illustration of the method, the set of parameters for which (x_n) starts with 1001 is $(2/3, 5/7)$ (see Fig. 13). Indeed, $M_1 M_0 M_0 M_1 \begin{pmatrix} 0 \\ 1 \end{pmatrix} = \begin{pmatrix} 2 \\ 1 \end{pmatrix}$ and $M_1 M_0 M_0 M_1 \begin{pmatrix} 1 \\ 0 \end{pmatrix} = \begin{pmatrix} 5 \\ 2 \end{pmatrix}$, so the normalization leads to the endpoints $2/(2+1) = 2/3$ and $5/(5+2) = 5/7$ respectively.

Three Speeds and One Irrational Parameter for the Whole System

The Turing machine described in [DL13] also works with 3 speeds and a single irrational ratio. The speeds can be chosen identical for the Turing machine and the continued fraction oracle. However, for the Turing machine, the irrational parameter is the golden ratio, which allows to create cells accumulating on a single side. Here, the positions of the vertical strips which are created is not increasing, but oscillating in a way that depends on the continued fraction expansion of the irrational parameter. However, it is possible, still using only 3 speeds, to add some rules to the continued fraction machine that swap a and b before each subtraction step when $a > b$, so that each new vertical strips becomes larger than the previous one. In particular, a single irrational parameter can be first duplicated and used to construct both the oracle (as described in this paper) and the creation of cells for the Turing machine.

4   Conclusion: Further Work

In this paper, we answered a little part of one of the main questions about signal machines: *"what can they compute?"*. We will see in a further paper how signal machines can semi-decide the algebraicity of real numbers, without accumulations, while this is not doable in the Linear-BSS model. We will also propose a new kind of signal whose resulting signal machine model is equivalent to the (classical) BSS model. We will also see how signal machines which are allowed to deal with accumulations can go far beyond BSS: they can compute arbitrary analytic functions, or even *draw* any compact subset of the unit interval.

References

[BSS89] Blum, L., Shub, M., Smale, S.: On a theory of computation and complexity over the real numbers: NP-completeness, recursive functions and universal machines. Bull. Am. Math. Soc. New Ser. **21**(1), 1–46 (1989)

[DL03] Durand-Lose, J.: Calculer géométriquement sur le plan - machines à signaux. Habilitation à Diriger des Recherches, École Doctorale STIC, Université de Nice-Sophia Antipolis (2003) (In French)

[DL07] Durand-Lose, J.: Abstract geometrical computation and the linear Blum, Shub and Smale model. In: Cooper, S.B., Löwe, B., Sorbi, A. (eds.) CiE 2007. LNCS, vol. 4497, pp. 238–247. Springer, Heidelberg (2007). doi:10.1007/978-3-540-73001-9_25

[DL09] Durand-Lose, J.: Abstract geometrical computation 3: black holes for classical and analog computing. Nat. Comput. **8**(3), 455–472 (2009)

[DL13] Durand-Lose, J.: Irrationality is needed to compute with signal machines with only three speeds. In: Bonizzoni, P., Brattka, V., Löwe, B. (eds.) CiE 2013. LNCS, vol. 7921, pp. 108–119. Springer, Heidelberg (2013). doi:10.1007/978-3-642-39053-1_12

[KE64] Khinchin, A.I.A., Eagle, H.: Continued Fractions. Dover Books on Mathematics. Dover Publications, New York (1964)

Game Characterizations and Lower Cones in the Weihrauch Degrees

Hugo Nobrega[1(⊠)] and Arno Pauly[2]

[1] Institute for Logic, Language, and Computation,
University of Amsterdam, Amsterdam, The Netherlands
h.nobrega@uva.nl
[2] Département D'informatique, Université Libre de Bruxelles, Brussels, Belgium
Arno.M.Pauly@gmail.com

Abstract. We introduce generalized Wadge games and show that each lower cone in the Weihrauch degrees is characterized by such a game. These generalized Wadge games subsume (a variant of) the original Wadge game, the eraser and backtrack games as well as Semmes's tree games. In particular, we propose that the lower cones in the Weihrauch degrees are the answer to Andretta's question on which classes of functions admit game characterizations. We then discuss some applications of such generalized Wadge games.

1 Introduction

The use of games in set theory has a well-established tradition, going back to work by Banach, Borel, Zermelo, Kőnig, and others (see [16, Sect. 27] for a thorough historical account of the subject). In particular, games which go on for infinitely many rounds have taken a prominent role in the field especially since the work of Gale and Stewart on the determinacy of certain types of such games [12].

In this paper, we will focus on infinite games which have been used to characterize classes of functions in descriptive set theory. Interest in this particular area began with a re-reading of the seminal work of Wadge [40], who introduced a game in order to analyze a notion of reducibility—*Wadge reducibility*—between subsets of Baire space. In the variant—which by a slight abuse we call the *Wadge game*—two players, **I** and **II**, are given a partial function $f :\subseteq \mathbb{N}^{\mathbb{N}} \to \mathbb{N}^{\mathbb{N}}$ and play with perfect information for ω rounds. In each run of this game, at each round player **I** first picks a natural number and player **II** responds by either

This research was partially done whilst the authors were visiting fellows at the Isaac Newton Institute for Mathematical Sciences in the programme 'Mathematical, Foundational and Computational Aspects of the Higher Infinite'. The research benefited from the Royal Society International Exchange Grant *Infinite games in logic and Weihrauch degrees*. The first author was partially supported by a CAPES Science Without Borders grant (9625/13-5), and the second author was partially supported by the ERC inVEST (279499) project.

J. Kari et al. (Eds.): CiE 2017, LNCS 10307, pp. 327–337, 2017.
DOI: 10.1007/978-3-319-58741-7_31

picking a natural number or passing, although she must pick natural numbers at infinitely many rounds. Thus, in the long run **I** and **II** build elements $x \in \mathbb{N}^{\mathbb{N}}$ and $y \in \mathbb{N}^{\mathbb{N}}$, respectively, and **II** wins the run if and only if $x \notin \mathrm{dom}(f)$ or $f(x) = y$. It can be considered a folklore result, also implicit in Wadge's work, that this game *characterizes* the continuous functions in the following sense.

Theorem 1. *A partial function* $f : \mathbb{N}^{\mathbb{N}} \to \mathbb{N}^{\mathbb{N}}$ *is relatively continuous iff player* **II** *has a winning strategy in the Wadge game for* f.

By adding new possibilities for player **II** at each round, one can obtain games characterizing larger classes of functions. For example, in the *eraser game* (implicit in [11]) characterizing the Baire class 1 functions, player **II** is allowed to erase past moves, the rule being that she is only allowed to erase each position of her output sequence finitely often. In the *backtrack game* (implicit in [44]) characterizing the functions which preserve the class of Σ_2^0 sets under preimages, player **II** is allowed to erase *all* of her past moves at any given round, the rule in this case being that she only do this finitely many times.

In his PhD thesis [39], Semmes introduced the *tree game* characterizing the class of Borel functions in Baire space. Player **I** plays as in the Wadge game and therefore builds some $x \in \mathbb{N}^{\mathbb{N}}$ in the long run, but at each round n player **II** now plays a finite *labeled tree*, i.e., a pair (T_n, ϕ_n) consisting of a finite tree $T_n \subseteq \mathbb{N}^{<\mathbb{N}}$ and a function $\phi_n : T_n \setminus \{\langle\rangle\} \to \mathbb{N}$, where $\langle\rangle$ denotes the empty sequence. The rules are that $T_n \subseteq T_{n+1}$ and $\phi_n \subseteq \phi_{n+1}$ must hold for each n, and that the *final* labeled tree $(T, \phi) = (\bigcup_{n \in \mathbb{N}} T_n, \bigcup_{n \in \mathbb{N}} \phi_n)$ must be an infinite tree with a unique infinite branch. Player **II** then wins if the sequence of labels along this infinite branch is exactly $f(x)$. By providing suitable extra requirements on the structure of the final tree, Semmes was able to obtain a game characterizing the Baire class 2 functions, and although this is not done explicitly in [39], it is not difficult to see that restrictions of the tree game also give his *multitape game* characterizing the classes of functions which preserve Σ_3^0 under preimages and the *multitape eraser game* characterizing the class of functions for which the preimage of any Σ_2^0 set is a Σ_3^0 set.

As examples of applications of these games, Semmes found a new proof of a theorem of Jayne and Rogers characterizing the class of functions which preserve Σ_2^0 under preimages and extended this theorem to the classes characterized by the multitape and multitape eraser games, by performing a detailed analysis of the corresponding game in each case.

Given the success of such game characterizations, in [1] Andretta raised the question of which classes of functions admit a characterization by a suitable game. Significant progress towards an answer was made by Motto Ros in [23]: Starting from a general definition of a reduction game, he shows how to construct new games from existing ones in ways that mirror the typical constructions of classes of functions (e.g., piecewise definitions, composition, pointwise limits). In particular, Motto Ros's results show that all the usual subclasses of the Borel functions studied in descriptive set theory admit game characterizations.

In order to arrive at a full characterization of the classes of functions characterizable by a game, we need to find the appropriate language to formulate such a result. Weihrauch reducibility (in its modern form) was introduced by Gherardi and Marcone [13] and Brattka and Gherardi [2,3] based on earlier work by Weihrauch on a reducibility between sets of functions analogous to Wadge reducibility [41,42].

We will show that game characterizations and Weihrauch degrees correspond closely to each other. We can thus employ the results and techniques developed for Weihrauch reducibility to study function classes in descriptive set theory, and vice versa. In particular, we can use the algebraic structure available for Weihrauch degrees [6,15] to obtain game characterizations for derived classes of functions from game characterizations for the original classes, similar to the constructions found by Motto Ros [23]. As a further feature of our work, we should point out that our results apply to the effective setting firsthand, and are then lifted to the continuous setting via relativization. They thus follow the recipe laid out by Moschovakis in [21].

While the traditional scope of descriptive set theory is restricted to Polish spaces, their subsets, and functions between them, these restrictions are immaterial for the approach presented here. Our results naturally hold for multivalued functions between represented spaces. As such, this work is part of a larger development to extend descriptive set theory to a more general setting, cf., e.g., [7,18,29,33,35].

We shall freely use standard concepts and notation from descriptive set theory, referring to [17] for an introduction.

2   Preliminaries on Represented Spaces and Weihrauch Reducibility

Represented spaces and continuous/computable maps between them form the setting for computable analysis [43]. For a comprehensive modern introduction we refer to [31].

A *represented* space $\mathbf{X} = (X, \delta_{\mathbf{X}})$ is given by a set X and a partial surjection $\delta_{\mathbf{X}} :\subseteq \mathbb{N}^{\mathbb{N}} \to X$. A (multivalued) function between represented spaces is just a (multivalued) function on the underlying sets. We say that a partial function $F :\subseteq \mathbb{N}^{\mathbb{N}} \to \mathbb{N}^{\mathbb{N}}$ is a *realizer* for a multivalued function $f :\subseteq \mathbf{X} \rightrightarrows \mathbf{Y}$ (in symbols: $F \vdash f$) if $\delta_{\mathbf{Y}}(F(p)) \in f(\delta_{\mathbf{X}}(p))$ for all $p \in \mathrm{dom}(f\delta_{\mathbf{X}})$. We call f *computable* (*continuous*), if it admits some computable (continuous) realizer.

Represented spaces and continuous functions (in the sense just defined) do indeed generalize Polish spaces and continuous functions (in the usual sense). Indeed, let (X, τ) be some Polish space, and fix a countable dense sequence $(a_i)_{i\in\mathbb{N}}$ and a compatible metric d. Now define $\delta_{\mathbf{X}}$ by $\delta_{\mathbf{X}}(p) = x$ iff $d(a_{p(i)}, x) < 2^{-i}$ holds for all $i \in \mathbb{N}$. In other words, we represent a point by a sequence of basic points converging to it with a prescribed speed. It is a foundational result in computable analysis that the notion of continuity for the represented space $(X, \delta_{\mathbf{X}})$ coincides with that for the Polish space (X, τ).

Definition 2. *Let f and g be partial multivalued functions between represented spaces. We say that f is* Weihrauch reducible *to g, in symbols $f \leq_{\mathrm{W}} g$, if there* *are computable functions* $K :\subseteq \mathbb{N}^\mathbb{N} \to \mathbb{N}^\mathbb{N}$ *and* $H :\subseteq \mathbb{N}^\mathbb{N} \times \mathbb{N}^\mathbb{N} \to \mathbb{N}^\mathbb{N}$ *such that whenever* $G \vdash g$, *the function* $F := (p \mapsto H(p, G(K(p))))$ *is a realizer for* f.

If there are computable functions $K, H :\subseteq \mathbb{N}^\mathbb{N} \to \mathbb{N}^\mathbb{N}$ such that whenever $G \vdash g$ then $HGK \vdash f$, then we say that f is strongly Weihrauch reducible to g ($f \leq_{sW} f$). We write $f \leq_W^c g$ and $f \leq_{sW}^c g$ for the variations where computable is replaced with continuous.

A multivalued function f *tightens* g, denoted by $f \preceq g$, if $\mathrm{dom}(g) \subseteq \mathrm{dom}(f)$ and $f(x) \subseteq g(x)$ whenever $x \in \mathrm{dom}(g)$, cf. [30,34].

Proposition 3 (cf., e.g., [28, Chapter 4]). *Let* $f :\subseteq \mathbf{A} \rightrightarrows \mathbf{B}$ *and* $g :\subseteq \mathbf{C} \rightrightarrows \mathbf{D}$. *We have*

1. $f \leq_{sW} g$ ($f \leq_{sW}^c g$) *iff there exist computable (continuous)* $k :\subseteq \mathbf{A} \rightrightarrows \mathbf{C}$ *and* $h :\subseteq \mathbf{D} \rightrightarrows \mathbf{B}$ *such that* $hgk \preceq f$; *and*
2. $f \leq_W g$ ($f \leq_W^c g$) *iff there exist computable (continuous)* $k :\subseteq \mathbf{A} \rightrightarrows \mathbb{N}^\mathbb{N} \times \mathbf{C}$ *and* $h :\subseteq \mathbb{N}^\mathbb{N} \times \mathbf{D} \rightrightarrows \mathbf{B}$ *such that* $h(\mathrm{id}_{\mathbb{N}^\mathbb{N}} \times g)k \preceq f$.

Although there are plenty of interesting operations defined on Weihrauch degrees (cf., e.g., the introduction of [4] for a recent overview), here we only require the sequential composition operator $\star$ from [5,6]. Rather than defining it explicitly as in [6], we will make use of the following characterization:

Theorem 4 (Brattka and Pauly [6]). $f \star g \equiv_W \max_{\leq_W}\{f' \circ g' ; f' \leq_W f \wedge g' \leq_W g\}$

3   Transparent Cylinders

We call $f :\subseteq \mathbf{X} \rightrightarrows \mathbf{Y}$ a *cylinder* if $\mathrm{id}_{\mathbb{N}^\mathbb{N}} \times f \leq_{sW} f$. Note that f is a cylinder iff $g \leq_W f$ and $g \leq_{sW} f$ are equivalent for all g. This notion is from [3].

Definition 5. *Call* $T :\subseteq \mathbf{X} \rightrightarrows \mathbf{Y}$ *transparent iff for any computable or continuous* $g :\subseteq \mathbf{Y} \rightrightarrows \mathbf{Y}$ *there is a computable or continuous, respectively,* $f :\subseteq \mathbf{X} \rightrightarrows \mathbf{X}$ *such that* $T \circ f \preceq g \circ T$.

The transparent (singlevalued) functions on Baire space were studied by de Brecht under the name *jump operator* in [8]. One of the reasons for their relevance is that they induce endofunctors on the category of represented spaces, which in turn can characterize function classes in DST [32]. The term *transparent* was coined in [5]. Our extension of the concept to multivalued functions between represented spaces is rather straightforward, but requires the use of the notion of tightening. Note that if $T :\subseteq \mathbf{X} \rightrightarrows \mathbf{Y}$ is transparent, then for every $y \in \mathbf{Y}$ there is some $x \in \mathrm{dom}(T)$ with $T(x) = \{y\}$, i.e., T is *slim* in the terminology of [5, Definition 2.7].

Given $p \in \mathbb{N}^\mathbb{N}$, recall that for each $n \in \mathbb{N}$ we can define $(p)_n \in \mathbb{N}^\mathbb{N}$ by $(p)_n(k) = p(\ulcorner n, k \urcorner)$, where $\ulcorner \cdot, \cdot \urcorner$ is some standard pairing function on natural numbers. Two examples of transparent cylinders which will be relevant in what follows are the functions $\lim$, $\lim_\Delta :\subseteq \mathbb{N}^\mathbb{N} \to \mathbb{N}^\mathbb{N}$ defined by letting $\lim(p) = \lim_{n\in\mathbb{N}}(p)_n$ with $\mathrm{dom}(\lim) = \{p \in \mathbb{N}^\mathbb{N}; \lim_{n\in\mathbb{N}}(p)_n$ exists$\}$ and letting $\lim_\Delta(p)$ be the restriction of $\lim$ to the domain $\{p \in \mathbb{N}^\mathbb{N}; \exists n \in \mathbb{N} \forall k \geq n((p)_k = (p)_n)\}$.

To see an example relating to Semmes's tree game characterizing the Borel functions, one first needs to define the appropriate represented space of labeled trees. For this, it is best to work in a quotient space of labeled trees under the equivalence relation of bisimilarity (see [27] for the details, which we omit). The resulting quotient space can be seen as the space of labeled trees in which the order of the subtrees rooted at the children of a node, and possible repetitions among these subtrees, are abstracted away. Then the function Prune, which removes from (any representative of the equivalence class of) a labeled tree which has one infinite branch all of the nodes which are not part of that infinite branch, is a transparent cylinder.

Theorem 6 (Brattka and Pauly [6]). *For every multivalued function g there is a multivalued function $g^t \equiv_W g$ which is a transparent cylinder.*

Proposition 7. *Let $T :\subseteq \mathbf{X} \rightrightarrows \mathbf{Y}$ and $S :\subseteq \mathbf{Y} \rightrightarrows \mathbf{Z}$ be cylinders. If T is transparent then $S \circ T$ is a cylinder and $S \circ T \equiv_W S \star T$. Furthermore, if S is also transparent then so is $S \circ T$.*

4   Generalized Wadge Games

In order to define our generalization of the Wadge game, first we need the following notion, which is just the dual notion to being an admissible representation as in the approach taken by Schröder in [36].

Definition 8. *A probe for $\mathbf{Y}$ is a computable partial function $\zeta :\subseteq \mathbf{Y} \to \mathbb{N}^\mathbb{N}$ such that for every computable or continuous $f :\subseteq \mathbf{Y} \rightrightarrows \mathbb{N}^\mathbb{N}$ there is a computable or continuous, respectively, $e :\subseteq \mathbf{Y} \rightrightarrows \mathbf{Y}$ such that $\zeta e \preceq f$.*

Note that a probe is always transparent.

The following definition generalizes the definition of a reduction game from [23, Subsect. 3.1], which is recovered as the special case in which all involved spaces are $\mathbb{N}^\mathbb{N}$, the map ζ is the identity on $\mathbb{N}^\mathbb{N}$, and T is a singlevalued function.

Definition 9. *Let $\zeta :\subseteq \mathbf{Y} \to \mathbb{N}^\mathbb{N}$ be a probe, $T :\subseteq \mathbf{X} \rightrightarrows \mathbf{Y}$ and $f :\subseteq \mathbf{A} \rightrightarrows \mathbf{B}$. The (ζ, T)-Wadge game for f is played by two players, $\mathbf{I}$ and $\mathbf{II}$, who take turns in infinitely many rounds. At each round of a run of the game, player $\mathbf{I}$ first plays a natural number and player $\mathbf{II}$ then either plays a natural number or passes, as long as she plays natural numbers infinitely often. Therefore, in the long run player $\mathbf{I}$ builds $x \in \mathbb{N}^\mathbb{N}$ and $\mathbf{II}$ builds $y \in \mathbb{N}^\mathbb{N}$, and player $\mathbf{II}$ wins the run of the game if $x \notin \mathrm{dom}(f\delta_\mathbf{A})$, or $y \in \mathrm{dom}(\delta_\mathbf{B}\zeta T\delta_\mathbf{X})$ and $\delta_\mathbf{B}\zeta T\delta_\mathbf{X}(y) \subseteq f\delta_\mathbf{A}(x)$.*

It is easy to see that the Wadge game is the $(\mathrm{id},\mathrm{id})$-Wadge game, the eraser game is the $(\mathrm{id},\lim)$-Wadge game, and the backtrack game is the $(\mathrm{id},\lim_{\Delta})$-Wadge game. Semmes's tree game for the Borel functions is the $(\mathrm{Label},\mathrm{Prune})$-Wadge game, where Label is the probe extracting the infinite sequence of labels from (any representative of the equivalence class of) a pruned labeled tree consisting of exactly one infinite branch.

Theorem 10. *Let T be a transparent cylinder. Then player **II** has a (computable) winning strategy in the (ζ,T)-Wadge game for f iff $f \leq^c_W T$ ($f \leq_W T$).*

Proof. ($\Rightarrow$) Any (computable) strategy for player **II** gives rise to a continuous (computable) function $k :\subseteq \mathbb{N}^\mathbb{N} \to \mathbb{N}^\mathbb{N}$. If the strategy is winning, then $\delta_{\mathbf{B}}\zeta T\delta_{\mathbf{X}}k \preceq f\delta_{\mathbf{A}}$, which implies $\delta_{\mathbf{B}}\zeta T\delta_{\mathbf{X}}k\delta_{\mathbf{A}}^{-1} \preceq f\delta_{\mathbf{A}}\delta_{\mathbf{A}}^{-1} = f$. Thus the continuous (computable) maps $\delta_{\mathbf{B}} \circ \zeta$ and $\delta_{\mathbf{X}}k\delta_{\mathbf{A}}^{-1}$ witness that $f \leq^c_{sW} T$ ($f \leq_{sW} T$).

($\Leftarrow$) As T is a cylinder, if $f \leq^c_W T$ ($f \leq_W T$), then already $f \leq^c_{sW} T$ ($f \leq_{sW} T$). Thus, there are continuous (computable) h, k with $h \circ T \circ k \preceq f$. As $\delta_{\mathbf{B}} \circ \delta_{\mathbf{B}}^{-1} = \mathrm{id}_{\mathbf{B}}$, we find that $\delta_{\mathbf{B}} \circ \delta_{\mathbf{B}}^{-1} \circ h \circ T \circ k \preceq f$. Now $\delta_{\mathbf{B}}^{-1} \circ h :\subseteq \mathbf{Y} \rightrightarrows \mathbb{N}^\mathbb{N}$ is continuous (computable), so by definition of a probe, there is some continuous (computable) $e :\subseteq \mathbf{Y} \rightrightarrows \mathbf{Y}$ with $\delta_{\mathbf{B}} \circ \zeta \circ e \circ T \circ k \preceq f$. As T is transparent, there is some continuous (computable) g with $e \circ T \succeq T \circ g$, thus $\delta_{\mathbf{B}} \circ \zeta \circ T \circ g \circ k \preceq f$.

As $(g \circ k) :\subseteq \mathbf{A} \rightrightarrows \mathbf{X}$ is continuous (computable), it has some (continuous) computable realizer $K :\subseteq \mathbb{N}^\mathbb{N} \to \mathbb{N}^\mathbb{N}$. By Theorem 1, player **II** has a winning strategy in the Wadge game for K, and it is easy to see that this strategy also wins the (ζ,T)-Wadge game for f for her.

Corollary 11. *Let T and S be transparent cylinders. If the (ζ,T)-Wadge game characterizes the class $\underline{\Gamma}$ and the (ζ',S)-Wadge game characterizes the class $\underline{\Gamma}'$, then the $(\zeta',S \circ T)$-Wadge game characterizes the class $\underline{\Gamma}' \circ \underline{\Gamma} := \{f \circ g; f \in \underline{\Gamma}' \wedge g \in \underline{\Gamma}\}$.*

We thus get game characterizations of many classes of functions, including, e.g., ones not covered by Motto Ros's constructions in [23]. For example, consider the function $\mathrm{Sort} : \{0,1\}^\mathbb{N} \to \{0,1\}^\mathbb{N}$ given by $\mathrm{Sort}(p) = 0^n1^\mathbb{N}$ if p contains exactly n occurrences of 0 and $\mathrm{Sort}(p) = 0^\mathbb{N}$ otherwise. This map was introduced and studied in [26]. From their results it follows that the class $\underline{\Gamma}$ of total functions on $\mathbb{N}^\mathbb{N}$ which are Weihrauch-reducible to Sort is neither the class of pointwise limits of functions in some other class, nor the class of $\mathbf{X}$-measurable functions for any boldface pointclass $\mathbf{X}$ of subsets of $\mathbb{N}^\mathbb{N}$ closed under countable unions and finite intersections. By Theorem 6, Sort is Weihrauch-equivalent to a a transparent cylinder $\mathrm{Sort}^t_{\mathbb{N}^\mathbb{N}}$. Thus, by Theorem 10, $\underline{\Gamma}$ is characterized by the $(\mathrm{id},\mathrm{Sort}^t_{\mathbb{N}^\mathbb{N}})$-Wadge game.

The converse of Theorem 10 is almost true, as well:

Proposition 12. *If the (ζ,T)-Wadge game characterizes a lower cone in the Weihrauch degrees, then it is the lower cone of $\zeta \circ T$, and $\zeta \circ T$ is a transparent cylinder.*

5   Using Game Characterizations

One main advantage of having game characterizations of some properties is realized together with determinacy: either by choosing our set-theoretic axioms accordingly, or by restricting to simple cases and invoking, e.g., Borel determinacy [20], we can conclude that if the property is false, i.e., player **II** has no winning strategy, then player **I** has a winning strategy. Thus, player **I**'s winning strategies serve as explicit witnesses of the failure of a property. Applying this line of reasoning to our generalized Wadge games, we obtain the following corollaries of Theorem 10:

Corollary 13 (ZFC). *Let T be a transparent cylinder and ζ a probe such that $\zeta \circ T$ is single-valued and $\mathrm{dom}(\zeta \circ T)$ is Borel. Then for any $f : \mathbf{A} \rightrightarrows \mathbf{B}$ such that $\mathrm{dom}(\delta_{\mathbf{A}})$ and $f(x)$ are Borel for any $x \in \mathbf{A}$, we find that $f \not\leq^{\mathrm{c}}_{\mathrm{W}} T$ iff player* **I** *has a winning strategy in the (ζ, T)-Wadge game for f.*

Corollary 14 ZF + DC+ AD). *Let T be a transparent cylinder and ζ a probe. Then $f \not\leq^{\mathrm{c}}_{W} T$ iff player* **I** *has a winning strategy in the (ζ, T)-Wadge game for f.*

Unfortunately, as determinacy fails in a computable setting (cf., e.g., [10,19]), we do not retain the computable counterparts. More generally, we are lacking a clear understanding of how these winning strategies of player **I** might look like. As pointed out to the authors by Carroy and Louveau, this holds even for the original Wadge game, i.e., the (id, id)-Wadge game. Here, we already have a notion of explicit witnesses for discontinuity: points of discontinuity. We can thus inquire about their relation:

Question 15. *Let a point of discontinuity of a function $f : \mathbb{N}^{\mathbb{N}} \to \mathbb{N}^{\mathbb{N}}$ be given as a sequence $(a_n)_{n \in \mathbb{N}}$, a point $a \in \mathbb{N}^{\mathbb{N}}$, and a word $w \in \mathbb{N}^{<\mathbb{N}}$ with $w \sqsubseteq f(a)$ such that $\forall n\ d(a_n, a) < 2^{-n} \wedge w \not\sqsubseteq f(a_n)$. Let Point be the multivalued map that takes as input a winning strategy for player* **I** *in the (id, id)-Wadge game for some function $f : \mathbb{N}^{\mathbb{N}} \to \mathbb{N}^{\mathbb{N}}$, and outputs a point of discontinuity for that function. Is Point computable? More generally, what is the Weihrauch degree of Point?*

We can somewhat restrict the range of potential answers for the preceding question:

Theorem 16 ([1]). *Let player* **I** *have a computable winning strategy in the (id, id)-Wadge game for $f : \mathbb{N}^{\mathbb{N}} \to \mathbb{N}^{\mathbb{N}}$. Then f has a computable point of discontinuity.*

A more convenient way of exploiting determinacy of the (ζ, T)-Wadge games could perhaps be achieved if a more symmetric version were found. In this, we could hope for a dual principle S, where for any f either $f \leq^{\mathrm{c}}_{\mathrm{W}} T$ or $S \leq^{\mathrm{c}}_{\mathrm{W}} f$ holds. More generally, we hope that a better understanding of the (ζ, T)-Wadge games would lead to structural results about the Weihrauch lattice, similar to the results obtained by Carroy on the strong Weihrauch reducibility [9].

---

[1] A key lemma for the proof of this theorem goes back to helpful comments by Takayuki Kihara.

6   Generalized Wadge Reductions

As mentioned in the introduction, the Wadge game was introduced not to characterize continuous functions, but in order to reason about a reducibility between sets. Given $A, B \subseteq \mathbb{N}^{\mathbb{N}}$, we say that A is *Wadge-reducible* to B, in symbols $A \leq_w B$, if there exists a continuous $F : \mathbb{N}^{\mathbb{N}} \to \mathbb{N}^{\mathbb{N}}$ such that $F^{-1}[B] = A$. Equivalently, we could consider the multivalued function $\frac{B}{A} : \mathbb{N}^{\mathbb{N}} \rightrightarrows \mathbb{N}^{\mathbb{N}}$ defined by $\frac{B}{A}(x) = B$ if $x \in A$ and $\frac{B}{A}(x) = (\mathbb{N}^{\mathbb{N}} \setminus B)$ if $x \notin A$. Now, $A \leq_w B$ iff $\frac{B}{A}$ is continuous. A famous structural result following from the determinacy of the corresponding Wadge game is that for any Borel $A, B \subseteq \mathbb{N}^{\mathbb{N}}$, either $A \leq_w B$ or $\mathbb{N}^{\mathbb{N}} \setminus B \leq_w A$. In particular, the Wadge hierarchy on the Borel sets is a strict weak order of width 2.

Both definitions immediately generalize to the case where $A \subseteq \mathbf{X}$ and $B \subseteq \mathbf{Y}$ for represented spaces $\mathbf{X}, \mathbf{Y}$. However, they yield different notions, for not every continuous multivalued function has a continuous choice function. As noted, e.g., by Hertling [14], extending the former definition to the reals already introduces infinite antichains in the resulting degree structure. The second generalization was proposed by Pequignot [35] as an alternative.[2]

It is a natural variation to replace *continuous* in the definition of Wadge reducibility by some other class of functions (ideally one closed under composition). Motto Ros has shown that for the typical candidates of more restrictive classes of functions, the resulting degree structures will not share the nice properties of the standard Wadge degrees (they are *bad*) [24]. Larger classes of functions as reduction witnesses have been explored by Motto Ros, Schlicht, and Selivanov [25] in the setting of quasi-Polish spaces—using the generalization of the first definition of the reduction.

Definition 17. *Let T be a Weihrauch degree. We define a relation $\preceq^T$ on subsets of represented spaces as follows: For $A \subseteq \mathbf{X}$, $B \subseteq \mathbf{Y}$ let $A \preceq^T B$ hold iff $\frac{B}{A} \leq_W^c T$.*

Observation 18. *If $T \star T \equiv_W T$, then $\preceq^T$ is a quasiorder.*

The following partially generalizes [23, Theorem 6.10]:

Theorem 19. *Let $A \subseteq \mathbf{X}$ and $B \subseteq \mathbf{Y}$, let $T : \mathbf{U} \rightrightarrows \mathbf{V}$ be a transparent cylinder, and let $\zeta : \subseteq \mathbf{Y} \to \mathbb{N}^{\mathbb{N}}$ be a probe such that the (ζ, T)-Wadge game for $\frac{B}{A}$ is determined. Then either $A \preceq^T B$ or $B \leq_w \mathbb{N}^{\mathbb{N}} \setminus A$.*

Proof. If player **II** has a winning strategy in the (ζ, T)-Wadge game for $\frac{B}{A}$, then by Theorem 10, we find that $\frac{B}{A} \leq_W^c T$, hence $A \preceq^T B$.

Otherwise, player **I** has a winning strategy. This winning strategy induces a continuous function $s : \mathbb{N}^{\mathbb{N}} \to \mathbb{N}^{\mathbb{N}}$, such that if player **II** plays $y \in \mathbb{N}^{\mathbb{N}}$, then

---

[2] While Pequignot only introduces the notion for second countable T_0 spaces, the extension to all represented spaces is immediate. Note that one needs to take into account that for general represented spaces, the Borel sets can show unfamiliar properties, e.g., even singletons can fail to be Borel (cf. also [37,38]).

player **I** plays $s(y) \in \mathbb{N}^{\mathbb{N}}$. As T is a transparent cylinder and ζ a probe, there is a continuous function $t : \mathbb{N}^{\mathbb{N}} \to \mathbb{N}^{\mathbb{N}}$ such that $(\zeta \circ T \circ \delta_{\mathbf{U}} \circ t) = \mathrm{id}_{\mathbb{N}^{\mathbb{N}}}$. Now we consider $s \circ t : \mathbb{N}^{\mathbb{N}} \to \mathbb{N}^{\mathbb{N}}$. If $\delta_{\mathbf{X}}(x) \in A$, then if player **II** plays $t(x)$, player **I** needs to play some $s(t(x))$ such that $\delta_{\mathbf{Y}}(s(t(x))) \notin B$. Likewise, if $\delta_{\mathbf{X}}(x) \notin A$, then for player **I** to win, it needs to be the case that $\delta_{\mathbf{Y}}(s(t(x))) \in B$. Thus, $s \circ t$ is a continuous realizer of $\frac{B}{\mathbb{N}^{\mathbb{N}} \setminus A}$, and $B \leq_w \mathbb{N}^{\mathbb{N}} \setminus A$ follows.

Corollary 20 (ZF + DC + AD). *Suppose* $T \star T \equiv_{\mathrm{W}} T$. *Then* $\prec^T$ *is strict weak order of width at most 2.*

In [22], Motto Ros has identified sufficient conditions on a generalized reduction (although in a different formalism) to ensure that its degree structure is equivalent to the Wadge degrees. We leave for future work the task of determining precisely for which T the degree structure of $\prec^T$ (restricted to subsets of $\mathbb{N}^{\mathbb{N}}$) is equivalent to the Wadge degrees, and which other structure types are realizable.

Acknowledgments. We are grateful to Benedikt Löwe, Luca Motto Ros, Takayuki Kihara and Raphaël Carroy for helpful and inspiring discussions. We would also like to thank the anonymous referees for the many corrections which have significantly improved the paper.

References

1. Andretta, A.: The SLO principle and the Wadge hierarchy. In: Bold, S., Löwe, B., Räsch, T., van Benthem, J. (eds.) Foundations of the Formal Sciences V: Infinite Games, pp. 1–38. College Publications (2007)
2. Brattka, V., Gherardi, G.: Effective choice and boundedness principles in computable analysis. Bull. Symbolic Log. **1**, 73–117 (2011). arXiv:0905.4685
3. Brattka, V., Gherardi, G.: Weihrauch degrees, omniscience principles and weak computability. J. Symbolic Log. **76**, 143–176 (2011). arXiv:0905.4679
4. Brattka, V., Gherardi, G., Hölzl, R.: Probabilistic computability and choice. Inf. Comput. **242**, 249–286 (2015). http://www.sciencedirect.com/science/article/pii/S0890540115000206, arXiv:1312.7305
5. Brattka, V., Gherardi, G., Marcone, A.: The Bolzano-Weierstrass theorem is the jump of weak König's lemma. Ann. Pure Appl. Log. **163**(6), 623–625 (2012)
6. Brattka, V., Pauly, A.: On the algebraic structure of Weihrauch degrees (2016). arXiv 1604.08348, http://arxiv.org/abs/1604.08348
7. de Brecht, M.: Quasi-Polish spaces. Ann. Pure Appl. Log. **164**(3), 354–381 (2013)
8. de Brecht, M.: Levels of discontinuity, limit-computability, and jump operators. In: Brattka, V., Diener, H., Spreen, D. (eds.) Logic, Computation, Hierarchies, pp. 79–108 (2014). de Gruyter, arXiv:1312.0697
9. Carroy, R.: A quasi-order on continuous functions. J. Symbolic Log. **78**(2), 633–648 (2013)
10. Cenzer, D., Remmel, J.: Recursively presented games and strategies. Math. Soc. Sci. **24**(2–3), 117–139 (1992)

11. Duparc, J.: Wadge hierarchy and Veblen hierarchy part I: Borel sets of finite rank. J. Symbolic Log. **66**(1), 56–86 (2001). http://projecteuclid.org/euclid.jsl/1183746360

12. Gale, D., Stewart, F.M.: Infinite games with perfect information. In: Kuhn, H.W., Tucker, A.W. (eds.) Contributions to the Theory of Games, vol. 2, pp. 245–266. Princeton University Press (1953)

13. Gherardi, G., Marcone, A.: How incomputable is the separable Hahn-Banach theorem? Notre Dame J. Formal Log. **50**(4), 393–425 (2009)

14. Hertling, P.: Unstetigkeitsgrade von Funktionen in der effektiven Analysis. Ph.D. thesis, Fernuniversität, Gesamthochschule in Hagen (Oktober 1996)

15. Higuchi, K., Pauly, A.: The degree-structure of Weihrauch-reducibility. Log. Methods Comput. Sci. **9**(2), 1–17 (2013)

16. Kanamori, A.: The Higher Infinite: Large Cardinals in Set Theory from their Beginnings. Springer Monographs in Mathematics, 2nd edn. Springer, Heidelberg (2005)

17. Kechris, A.S.: Classical Descriptive Set Theory. Graduate Texts in Mathematics, vol. 156. Springer, Heidelberg (1995)

18. Kihara, T., Pauly, A.: Point degree spectra of represented spaces (2014). arXiv:1405.6866

19. Le Roux, S., Pauly, A.: Weihrauch degrees of finding equilibria in sequential games. In: Beckmann, A., Mitrana, V., Soskova, M. (eds.) CiE 2015. LNCS, vol. 9136, pp. 246–257. Springer, Cham (2015). doi:10.1007/978-3-319-20028-6_25

20. Martin, D.A.: Borel determinacy. Ann. Math. **102**(2), 363–371 (1975)

21. Moschovakis, Y.N.: Classical descriptive set theory as a refinement of effective descriptive set theory. Ann. Pure Appl. Log. **162**, 243–255 (2010)

22. Ros, L.M.: Borel-amenable reducibilities for sets of reals. J. Symbolic Log. **74**(1), 27–49 (2009). http://dx.doi.org/10.2178/jsl/1231082301

23. Ros, L.M.: Game representations of classes of piecewise definable functions. Math. Log. Q. **57**(1), 95–112 (2011)

24. Ros, L.M.: Bad Wadge-like reducibilities on the baire space. Fundam. Math. **224**(1), 67–95 (2014)

25. Ros, L.M., Schlicht, P., Selivanov, V.: Wadge-like reducibilities on arbitrary quasi-Polish spaces. Mathematical Structures in Computer Science pp. 1–50 (2014). http://journals.cambridge.org/article_S0960129513000339, arXiv:1204.5338

26. Neumann, E., Pauly, A.: A topological view on algebraic computation models (2016). arXiv:1602.08004

27. Nobrega, H., Pauly, A.: Game characterizations and lower cones in the Weihrauch degrees (2015). arXiv:1511.03693

28. Pauly, A.: Computable Metamathematics and its Application to Game Theory. Ph.D. thesis, University of Cambridge (2012)

29. Pauly, A.: The descriptive theory of represented spaces (2014). arXiv:1408.5329

30. Pauly, A.: Many-one reductions and the category of multivalued functions. Mathematical Structures in Computer Science (2015). arXiv:1102.3151

31. Pauly, A.: On the topological aspects of the theory of represented spaces. Computability **5**(2), 159–180 (2016). http://arxiv.org/abs/1204.3763

32. Pauly, A., de Brecht, M.: Towards synthetic descriptive set theory: An instantiation with represented spaces (2013). http://arxiv.org/abs/1307.1850

33. Pauly, A., de Brecht, M.: Descriptive set theory in the category of represented spaces. In: 30th Annual ACM/IEEE Symposium on Logic in Computer Science (LICS), pp. 438–449 (2015)

34. Pauly, A., Ziegler, M.: Relative computability and uniform continuity of relations. J. Log. Anal. **5**, 1–39 (2013)

35. Pequignot, Y.: A Wadge hierarchy for second countable spaces. Arch. Math. Log. **54**(5), 1–25 (2015). http://dx.doi.org/10.1007/s00153-015-0434-y
36. Schröder, M.: Extended admissibility. Theoret. Comput. Sci. **284**(2), 519–538 (2002)
37. Schröder, M., Selivanov, V.: Hyperprojective hierarchy of QCB_0-spaces (2014). arXiv:1404.0297, http://arxiv.org/abs/1404.0297
38. Schröder, M., Selivanov, V.L.: Some hierarchies of QCB_0-spaces. Math. Struct. Comput. Sci. **25**(8), 1–25 (2014). arXiv:1304.1647
39. Semmes, B.: A game for the Borel functions. Ph.D. thesis, University of Amsterdam (2009)
40. Wadge, W.W.: Reducibility and determinateness on the Baire space. Ph.D. thesis, University of California, Berkeley (1983)
41. Weihrauch, K.: The degrees of discontinuity of some translators between representations of the real numbers. Informatik Berichte 129, FernUniversität Hagen, Hagen, July 1992
42. Weihrauch, K.: The TTE-interpretation of three hierarchies of omniscience principles. Informatik Berichte 130, FernUniversität Hagen, Hagen, September 1992
43. Weihrauch, K.: Computable Analysis. Springer, Heidelberg (2000)
44. Wesep, R.: Wadge degrees and descriptive set theory. In: Kechris, A.S., Moschovakis, Y.N. (eds.) Cabal Seminar 76–77. LNM, vol. 689, pp. 151–170. Springer, Heidelberg (1978). doi:10.1007/BFb0069298

Randomness Deficiencies

Gleb Novikov[✉]

Lomonosov Moscow State University, Moscow, Russia
novikov.g.e@gmail.com

Abstract. The notion of random sequence was introduced by Martin-Löf in [3]. In the same article he defined the so-called randomness deficiency function that shows how close are random sequences to non-random (in some natural sense). Other deficiency functions can be obtained from the Levin-Schnorr theorem, that describes randomness in terms of Kolmogorov complexity. The difference between all of these deficiencies is bounded by a logarithmic term (Proposition 1). In this paper we show (Theorems 1 and 2) that the difference between some deficiencies can be as large as possible.

1 Introduction

Classical probability theory cannot deal with individual random objects, such as binary sequences or points on the real line: each sequence or point has measure zero (with respect to the uniform measure). However, our intuition says that the sequence of zeros (and any other computable sequence) is not random, while the result of tossing a coin is. Martin-Löf in [3] tried to formalize this statement. He used an algorithmic approach to define random binary sequences.

Martin-Löf random sequences have many nice properties: adding, deleting or changing finitely many bits doesn't change randomness; random sequences satisfy the law of large numbers; computable permutations preserve randomness. So if the sequence ω is random, the sequence $\omega' = 0^{1000000000}\omega$ (billion of zeros concatenated with ω) is also random. But intuitively ω' is "less random". We can make this argument formal using a randomness deficiency function d: this function is finite on random sequences and infinite on non-random sequences. If $d(\omega') \geq d(\omega)$ we say that ω' is less random than ω. It turns out that there are some natural types of deficiency functions that have similar properties to the so-called finite deficiency (the difference between the length of the string and its Kolmogorov complexity). For example, adding n zeros to the sequence increases randomness deficiency by $n + O(\log n)$. Using this fact one can reformulate statements about random sequences in terms of the deficiency functions to look for the connections between algorithmic randomness and Kolmogorov complexity theories.

In this paper we consider several deficiency functions: the first was introduced by Martin-Löf (Definition 3), the others appear from the Levin-Schnorr's criterion of randomness in terms of different types of Kolmogorov complexity: the prefix-free complexity and the a priori complexity. The difference between all

© Springer International Publishing AG 2017
J. Kari et al. (Eds.): CiE 2017, LNCS 10307, pp. 338–350, 2017.
DOI: 10.1007/978-3-319-58741-7_32

of the deficiencies is not greater than $(1+\varepsilon)\log d$ (up to a constant, for all $\varepsilon > 0$) (Proposition 1), where d is one of the deficiency functions. We show that the difference between some of the deficiencies can be greater than $\log d$. For example, some of the deficiency functions (given in exponential scale) are integrable, while the others are not and that is the reason of the difference (Theorem 1). To differ the non-integrable deficiencies we construct a special rarefied set of intervals in the Cantor space (Theorem 2).

1.1 Notation

The set of all infinite binary sequences is called the Cantor space and is denoted by Ω. An interval in the Cantor space is a set of extensions of some string x, it is denoted by $[x]$.

The set of all binary strings is denoted by $\mathbb{B}^*$. The length of the string x is denoted by $|x|$. We write $y \prec x$ if y is a prefix of x.

$\mathbb{I}_S$ is the indicator function of the set S.

In this paper, log means binary logarithm.

Notation $f <^+ g$ ($f <^* g$) means that there exists a constant c such that for all x, $f(x) < c + g(x)$ ($f(x) < cg(x)$).

2 Preliminaries

One can find all of the notions and statements of this section in [1,2].

Definition 1. *A measure μ over Ω is called computable, if there exists a Turing machine that from each string x and rational $\varepsilon > 0$ returns an ε-approximation of the value $\mu([x])$.*

The collection of intervals in the Cantor space forms a base for its standard topology. We will talk about closed and open sets relative to this topology.

Definition 2. *Let μ be a computable measure. A nested sequence of open sets $\{V_n\}$ is called a Martin-Löf test with respect to μ if:*

(1) $\{V_n\}$ is uniformly effectively open, that is there exists a Turing machine that for each input k enumerates the set V_k.

(2) $\mu(V_n) \leq 2^{-n}$ for each n.

Definition 3. *Let $\{V_n\}$ be a Martin-Löf test with respect to a computable measure μ. Function $d_{\mu;\{V_n\}}(\omega) = \max\{k : \omega \in V_k\}$ is called a randomness deficiency of ω with respect to the test $\{V_n\}$.*

Lemma 1. *For every computable measure μ there exists a Martin-Löf test $\{U_n\}$ with respect to μ such that for any Martin-Löf test $\{V_n\}$ with respect to μ there exist a constant c such that for all sequences ω*

$$d_{\mu;\{U_n\}}(\omega) \geq d_{\mu;\{V_n\}}(\omega) - c \tag{1}$$

The deficiency function $\mathbf{d}_\mu$ was defined by Martin-Löf in [3]. In the same article he introduced the following notion of randomness:

Definition 4. *Let μ be a computable measure. A sequence $\omega \in \Omega$ is called Martin-Löf random with respect to μ if $\mathbf{d}_\mu(\omega) < \infty$.*

There are some other types of deficiency functions. To show the relations between them, we need to reformulate the definition of $\mathbf{d}_\mu$. First we define the so-called lower semicomputable functions.

Definition 5. *A function $t : \Omega \to \mathbb{R}$ is called lower semicomputable if there exists a machine that given a rational r enumerates the set of intervals $\{\omega : t(\omega) > r\}$ (so this set should be open).*

Let's note the following property of $\mathbf{d}_\mu$: the function $\mathbf{t}_\mu = 2^{\mathbf{d}_\mu}$ is probability bounded, that is

$$\mu\{\mathbf{t}_\mu(\omega) > c\} \le \frac{1}{c} \tag{2}$$

for rational numbers c. Moreover, $\mathbf{t}_\mu$ is the largest (up to a multiplicative constant) among all lower semicomputable probability bounded functions (the sets $V_n = \{t(\omega) > 2^n\}$ form a Martin-Löf test). Therefore we can define the function $\mathbf{d}_\mu$ as logarithm of the largest lower semicomputable probability bounded function and from now we denote this function as $\mathbf{d}_\mu^P$ (and $\mathbf{t}_\mu$ as $\mathbf{t}_\mu^P$).

To define other deficiency functions we need the following notion:

Definition 6. *Function $f : \Omega \to \mathbb{Q}$ is called basic if its value on every sequence ω is determined by some finite prefix of ω.*

By compactness of Ω there exist finitely many intervals where basic function is constant, and the union of these intervals is Ω. Therefore basic functions are constructive objects and we can consider computable sequences of basic functions.

The following lemma gives the equivalent definition of lower semicomputable functions.

Lemma 2. *Function $t : \Omega \to \mathbb{R}$ is lower semicomputable iff it is a limit of increasing computable sequence of basic functions.*

If the function is integrable and its integral is less than 1 it is probability bounded (by Markov's inequality). We call these functions expectation bounded. There exists maximal (up to a multiplicative constant) lower semicomputable expectation bounded function $\mathbf{t}_\mu^E$: we can enumerate all probability bounded functions (with respect to μ); the integral of such function is a limit of integrals of basic functions, so if it is greater than 1 we always know it after finitely many steps of computation. If the integral is greater than 1, we decrease the values of basic functions to make it less than 1. The sum of these new functions with weights 2^{-n} is the maximal lower semicomputable expectation bounded function.

Definition 7. *Let μ be a computable measure. The expectation bounded deficiency is the function*

$$d_\mu^E(\omega) = \log t_\mu^E(\omega) \tag{3}$$

The following proposition shows that the difference between d_μ^P and d_μ^E is not large.

Proposition 1. *Let μ be a computable measure and $\varepsilon > 0$. Then*

$$d_\mu^E \leq^+ d_\mu^P \leq^+ d_\mu^E + (1+\varepsilon) \log d_\mu^E \tag{4}$$

Proof. The first part follows from Markov's inequality. To prove the second part, let's consider a function $t_\mu^P \log^{-1-\varepsilon} t_\mu^P$. Its integral does not exceed

$$\sum_n \int_{A_n} t_\mu^P(\omega) \log^{-1-\varepsilon} t_\mu^P(\omega) d\mu(\omega) \leq \sum_n 2n^{-1-\varepsilon} \tag{5}$$

where $A_n = \{2^n \leq t_\mu^P < 2^{n+1}\}$, so this integral is finite. Therefore

$$d_\mu^P \leq^+ d_\mu^E + (1+\varepsilon) \log d_\mu^P \leq^+ d_\mu^E + (1+\varepsilon) \log d_\mu^E \tag{6}$$

$\square$

The deficiency function d_μ^E can be described in terms of prefix-free Kolmogorov complexity (see, for example, [2]). We will briefly describe this construction. At first we define the discrete analogues of basic and lower semicomputable functions.

Definition 8. *Function $f : \mathbb{B}^* \to \mathbb{Q}$ is called basic if its support is finite.*

Definition 9. *Function $f : \mathbb{B}^* \to \mathbb{R}$ is called lower semicomputable if it is a limit of a computable sequence of increasing basic functions.*

Definition 10. *A lower semicomputable function $m : \mathbb{B}^* \to [0, \infty)$ such that $\sum_x m(x) \leq 1$ is called a discrete lower semicomputable semimeasure.*

Let's denote the prefix-free Kolmogorov complexity of a string x as $K(x)$. The function $m(x) = 2^{-K(x)}$ is called the discrete a priori probability. The famous coding theorem (see, for example, [2]) states that this function is the largest (up to a multiplicative constant) among all discrete lower semicomputable semimeasures.

It can be shown (see, for example, [1]) that

$$t_\mu^E(\omega) =^* \sum_n \frac{m(\omega_{1...n})}{\mu([\omega_{1...n}])} =^* \sup_n \frac{m(\omega_{1...n})}{\mu([\omega_{1...n}])} \tag{7}$$

In logarithmic scale:

$$d_\mu^E(\omega) =^+ \sup_n \{-\log \mu([\omega_{1...n}]) - K(\omega_{1...n})\} \tag{8}$$

This result is due to Gacs (see [4]). The value in the right part of 8 is finite iff the sequence is random. It was first shown by Schnorr and Levin independently in [5,6]. Informally, the sequence is random iff its initial segments are incompressible. The Eq. 8 also shows that if one adds n zeros to the sequence then the randomness deficiency (probability or expectation bounded) increases by at most $n + O(\log n)$.

The Schnorr-Levin theorem can be formulated in terms of the so-called a priori complexity. To define it we need the notion of continuous a priori probability.

Definition 11. *A lower semicomputable function* $a : \mathbb{B}^* \to [0, \infty)$ *such that* $\sum_{x \in S} a(x) \le 1$ *for every prefix-free set* S *is called a continuous lower semicomputable semimeasure.*

We can enumerate all continuous lower semicomputable semimeasures and consider a semimeasure $\mathbf{a}(x) = \sum_j a_j(x) 2^{-j}$. This semimeasure is also continuous and lower semicomputable, and it is the largest (up to a multiplicative constant) in this class of semimeasures. We will call $\mathbf{a}(x)$ the continuous a priori probability.

Definition 12. *The value* $KM(x) = - \log \mathbf{a}(x)$ *is called the a priori complexity of* x.

The Schnorr-Levin theorem for the a priori complexity states that the sequence ω is random iff $\sup_n \{ - \log \mu([\omega_{1...n}]) - KM(\omega_{1...n}) \}$ is finite. Moreover, supremum can be replaced by lim sup or lim inf. Using this theorem we can define other types of deficiency functions.

Definition 13. *Let* μ *be a computable measure. We will consider functions*

$$d_\mu^M (\omega) = \sup_n \{ - \log \mu([\omega_{1...n}]) - KM(\omega_{1...n}) \} \tag{9}$$

$$d_\mu^{\lim \sup M} (\omega) = \lim_n \sup \{ - \log \mu([\omega_{1...n}]) - KM(\omega_{1...n}) \} \tag{10}$$

$$d_\mu^{\lim \inf M} (\omega) = \lim_n \inf \{ - \log \mu([\omega_{1...n}]) - KM(\omega_{1...n}) \} \tag{11}$$

and call them a priori randomness deficiencies.

Each continuous lower semicomputable semimeasure can be represented as a probability distribution on the initial segments of outputs of some probabilistic machine that prints bits one after another and does not have to stop (see, for example, [2]). That is for each $a(x)$ there exists a machine A such that

$$a(x) = \mathbb{P}\{\text{the output of } A \text{ begins on the string } x\} \tag{12}$$

Informally, the Schnorr–Levin theorem states that the sequence ω is random iff the probability of getting the initial segments $\omega_{1...n}$ using a probabilistic machine cannot be much greater than getting it from a random generator (with the distribution μ). The deficiency functions from the Definition 13 show the difference between logarithms of these probabilities.

One can use supermartingales to define the deficiencies d_μ^M, $d_\mu^{\lim \sup M} (\omega)$, $d_\mu^{\lim \inf M} (\omega)$.

Definition 14. *Let μ be a measure on Ω and let M be a function of binary strings.*

If $\mu([x])M(x) = \mu([x0])M(x0) + \mu([x1])M(x1)$ the function M is called a martingale.

If $\mu([x])M(x) \geq \mu([x0])M(x0) + \mu([x1])M(x1)$ the function M is called a supermartingale.

If $\mu([x])M(x) \leq \mu([x0])M(x0) + \mu([x1])M(x1)$ the function M is called a submartingale.

If martingale (or sub/supermartingale) is not bounded on the initial segments of the sequence ω we say that it wins on ω.

If μ is computable, the supermartingale $\mathbf{M}(x) = \frac{\mathbf{a}(x)}{\mu([x])}$ is the largest (up to a multiplicative constant) among all lower semicomputable supermartingales. Supermartingale $\mathbf{M}(x)$ wins on all non-random sequences and does not win on random sequences.

The deficiency $\mathbf{d}_{\mu}^{M}(\omega)$ is a supremum of $\mathbf{M}(\omega_{1...n})$, the deficiencies $\mathbf{d}_{\mu}^{\lim \sup M}(\omega)$ and $\mathbf{d}_{\mu}^{\lim \inf M}(\omega)$ are respectively limsup and liminf of $\mathbf{M}(\omega_{1...n})$.

Now we are going to show the relations between the deficiencies.

Proposition 2.

$$\mathbf{d}_{\mu}^{E} \leq^{+} \mathbf{d}_{\mu}^{\lim \inf M} \tag{13}$$

Proof. We need to construct some continuous lower semicomputable semimeasure a. Once the approximation to $\mathbf{m}(x)$ increases by ε we do the following:

(1) Increase the value of a by ε on prefixes of x
(2) Increase the value of a by $\varepsilon \mu([y])/\mu([x])$ on the extensions y of x.

If $\mathbf{d}_{\mu}^{E} = R$ there exists a string x such that

$$-\log \mu([x]) - K(x) =^{+} R \tag{14}$$

and ω is the extension of x. If $n > |x|$, the logarithm of a is:

$$\log a(\omega_{1...n}) \geq -K(x) + \log \mu([\omega_{1...n}]) - \log \mu([x]) \tag{15}$$

Therefore

$$\mathbf{d}_{\mu}^{\lim \inf M}(\omega) \geq^{+} \liminf_{n}\{-\log \mu([\omega_{1...n}]) + \log a(\omega_{1...n})\} \geq$$

$$\geq \liminf_{n}\{-\log \mu([x]) - K(x)\} = -\log \mu([x]) - K(x) =^{+} \mathbf{d}_{\mu}^{E} \tag{16}$$

The case $\mathbf{d}_{\mu}^{E} = \infty$ can be considered in the same way. $\square$

Proposition 3.

$$\mathbf{d}_{\mu}^{M} \leq^{+} \mathbf{d}_{\mu}^{P} \tag{17}$$

Proof. It is sufficient to show that $\mu\{2^{\mathbf{d}_\mu^M}(\omega) > 2^c\} \le 2^{-c}$ for all rational c. Let's fix c and consider a set of strings

$$S = \{x : \frac{a(x)}{\mu([x])} > 2^c, \forall y \prec x \ \frac{a(y)}{\mu([y])} \le 2^c\} \tag{18}$$

It is evident that $\omega \in \cup_{x \in S}[x]$ iff $\mathbf{d}_\mu^M(\omega) > c$. The set S is prefix-free, so

$$\mu\{2^{\mathbf{d}_\mu^M}(\omega) > 2^c\} = \sum_{x \in S} \mu([x]) < \sum_{x \in S} \frac{a(x)}{2^c} \le 2^{-c} \tag{19}$$

$\square$

Combining the results of Propositions 1, 2 and 3 we can write down the following chain of inequalities:

$$\mathbf{d}_\mu^E \le^+ \mathbf{d}_\mu^{\liminf M} \le^+ \mathbf{d}_\mu^{\limsup M} \le^+ \mathbf{d}_\mu^M \le^+ \mathbf{d}_\mu^P \le^+ \mathbf{d}_\mu^E + (1+\varepsilon)\log \mathbf{d}_\mu^E \tag{20}$$

The natural question is about the difference between these deficiencies. To show the difference between some of them we will need the following lemma from calculus:

Lemma 3. *If $c_k \ge 0$ and $\sum_{k=1}^\infty c_k < \infty$ and $R_k := \sum_{n=k+1}^\infty c_n > 0$, then*

$$\sum_{k=1}^\infty \frac{c_k}{R_k \log \frac{1}{R_K}} = \infty \tag{21}$$

Proof. At first we will prove that the series $\sum_{k=1}^\infty \frac{c_k}{R_k}$ does not converge. Denote $z_k = \frac{c_k}{R_k}$. It is evident that

$$z_k = \frac{R_{k-1} - R_k}{R_k} = \frac{R_{k-1}}{R_k} - 1 \implies \frac{1}{R_k} = \frac{1}{R_0} \prod_{n=1}^k (1 + z_n) \tag{22}$$

If we take the logarithm from both parts, we get

$$\log \frac{1}{R_k} = \log \frac{1}{R_0} + \sum_{n=1}^k \log(1 + z_n) \le^* \sum_{n=1}^k z_n \tag{23}$$

The left part tends to infinity, so the sum $\sum_{n=1}^\infty z_n$ is infinite. To prove the lemma we need to show that $\sum_{k=1}^\infty \frac{z_k}{\log \frac{1}{R_k}} = \infty$. Using 23 we get:

$$\sum_{k=1}^\infty \frac{z_k}{\log \frac{1}{R_k}} \ge^* \sum_{k=1}^\infty \frac{z_k}{\sum_{n=1}^k z_n} \tag{24}$$

Denote $S_k = \sum_{n=1}^k z_n$ and $b_k = \frac{z_k}{S_k}$. It is sufficient to show that if the series $\sum_{n=1}^\infty z_n$ does not converge then the series $\sum_{n=1}^\infty b_n$ also does not converge. We will do it in the same way as the first part of the proof of the lemma:

$$b_k = \frac{S_{k+1} - S_k}{S_k} = \frac{S_{k+1}}{S_k} - 1 \implies S_{k+1} = S_1 \prod_{n=1}^k (1 + b_n) \tag{25}$$

If we take the logarithm from both parts we get

$$\log S_k = \log S_1 + \sum_{n=1}^{k} \log(1 + b_n) \leq^* \sum_{n=1}^{k} b_n \tag{26}$$

The left part tends to infinity, so the sum $\sum_{n=1}^{\infty} b_n$ is infinite. □

3   New Results

Now we are going to show the relations between deficiency functions. Proposition 4 is an effective version of Doob's martingale convergence theorem (see, for example, [7]) and can be easily obtained from it. Theorems 1 and 2 require Lemma 3.

Definition 15. *If the sequence ω is random relative to the oracle $0'$ it is called 2-random.*

Proposition 4. *Let μ be a computable measure. If ω is 2-random (with respect to μ), then $d_{\mu}^{\lim \sup M}(\omega) = d_{\mu}^{\lim \inf M}(\omega)$.*

Proof. Given rational numbers $\beta > \alpha > 0$ we can construct a $0'$-computable supermartingale M_{α}^{β} that wins on sequences ω such that the supermartingale $\mathbf{M}$ infinitely many times becomes smaller than α and greater than β on the initial segments of ω. Using the oracle we compute the values of $\mathbf{M}$ and if $\mathbf{M}(x) < \alpha$ the values $M_{\alpha}^{\beta}(z)$ are equal to $\mathbf{M}(z)$ on extensions z of x such that $\mathbf{M}(z) \leq \beta$. When we find extension y such that $\mathbf{M}(y) > \beta$ we just save the capital $(M_{\alpha}^{\beta}(yw) = M_{\alpha}^{\beta}(y))$ until we find some new string x with small $\mathbf{M}(x)$. On the segments from x to y the value of M_{α}^{β} increases by $\frac{\beta}{\alpha}$ times. The sum of all M_{α}^{β} with weights $\mathbf{m}(\alpha, \beta)$ is a $0'$-lower semicomputable supermartingale, so it is finite on 2-random sequences. □

Corollary 1. *Let μ be a computable measure. Then $2^{d_{\mu}^{\lim \sup M}}$ is the integrable function with respect to μ.*

Proof. By Fatou's lemma:

$$\int_{\Omega} \liminf_{n} \mathbf{M}(\omega_{1...n}) d\mu(\omega) \leq \liminf_{n} \int_{\Omega} \mathbf{M}(\omega_{1...n}) d\mu(\omega) = \liminf_{n} \sum_{|x|=n} \mathbf{a}(x) \leq 1 \tag{27}$$

$d_{\mu}^{\lim \sup M} = d_{\mu}^{\lim \inf M}$ almost everywhere, therefore $2^{d_{\mu}^{\lim \sup M}}$ is integrable. □

The greater deficiencies are not integrable (in exponential scale). To show that $2^{d_{\mu}^{M}}$ is not integrable we need Lemma 3.

Recall the definition of atomic measures.

Definition 16. *If the measure μ on Ω is positive on some sequence, we will say that μ is an atomic measure.*

The following theorems show that the difference between $\mathbf{d}_\mu^M$ and other deficiencies may be greater then $\log \mathbf{d}_\mu$ (here we write $\mathbf{d}_\mu$ without index because difference between logarithms of all of the deficiencies is bounded by some constant).

Theorem 1. *Let μ be a computable non-atomic measure. For all c there exists ω such that*

$$\mathbf{d}_\mu^{\lim \sup M}(\omega) < \mathbf{d}_\mu^M(\omega) - \log \mathbf{d}_\mu^M(\omega) - c \tag{28}$$

Proof. It is sufficient to prove that the function $q = 2^{\mathbf{d}_\mu^M - \log \mathbf{d}_\mu^M}$ is not integrable with respect to μ. We will construct some deterministic (but formally probabilistic) machine f. At each step, after f has printed the string of bits x of length k, f computes measures of $[x0]$ and $[x1]$, and then prints a bit b if $\mu[xb] > \frac{1}{3}\mu[x]$ (if the both bits are suitable, let f print 0). Denote the interval $[xb] = B_k$ if at the k-th step f prints a bit b, and $C_k = B_{k-1} \setminus B_k$. The measure μ is non-atomic, hence

$$\mu B_k = \sum_{n=k+1}^{\infty} \mu C_n \tag{29}$$

The intervals C_k are disjoint, so $\sum_k C_k \leq 1$. By Lemma 3

$$\sum_{k=1}^{\infty} \frac{\mu C_k}{\mu B_k \log \frac{1}{\mu B_k}} = \infty \tag{30}$$

Let's denote

$$t_f(\omega) = \sup_n \frac{\mathbb{P}\{\text{the output of } f \text{ begins on the string } \omega_{1\ldots n}\}}{\mu([\omega_{1\ldots n}])} \tag{31}$$

The function $\frac{x}{\log x}$ is monotone for large enough x, therefore by the universality

$$q \geq^* \frac{t_f}{\log t_f} \tag{32}$$

It is easy to see that

$$\frac{t_f}{\log t_f}(\omega) = \sum_{k=1}^{\infty} \frac{\mathbb{I}_{C_{k+1}}(\omega)}{\mu B_k \log \frac{1}{\mu B_k}} \tag{33}$$

Recall that $\mu B_k \geq \mu B_{k+1} > \frac{1}{3}\mu B_k$

$$\int_\Omega q(\omega)d\omega \geq^* \int_\Omega \frac{t_f}{\log t_f}(\omega)d\omega \geq \sum_{k=1}^{\infty} \frac{\mu C_{k+1}}{\mu B_k \log \frac{1}{\mu B_k}} > \frac{1}{3}\sum_{k=1}^{\infty} \frac{\mu C_{k+1}}{\mu B_{k+1} \log \frac{1}{\mu B_{k+1}}} = \infty \tag{34}$$

$\square$

The following theorem requires some technical constructions in general case, so at first we will prove it in the case of the uniform measure to show the idea.

Theorem 2. *Let μ be a computable non-atomic measure. For all c there exists ω such that*

$$d_\mu^M(\omega) < d_\mu^P(\omega) - \log d_\mu^P(\omega) - c \qquad (35)$$

Proof (Uniform case). The main idea is that one cannot win 50$ after 5 tosses of a coin if he starts with 1$. Let's consider a function $g = \sum_k 2^{2k-1} \mathbb{I}_{[0^k 1^k]}(\omega)$. It is a lower semicomputable probability bounded function. Let's prove the theorem by contradiction. Assume that there exists a constant c such that for all ω

$$t_\mu^M(\omega) \geq 2^{-c} \frac{g}{\log g}(\omega) \qquad (36)$$

That means that there exists a prefix-free set of binary strings w_l^k such that $\cup_l [w_l^k] \supset 0^k 1^k$ and

$$\mathbf{a}(w_l^k) 2^{|w_l^k|} \geq 2^{-c} \frac{2^{2k-1}}{2k-1} \qquad (37)$$

For k large enough

$$|w_l^k| \geq -c - \log(2k-1) + 2k - 1 + KM(w_l^k) > k + 1 \qquad (38)$$

So $[w_l^k] \subset [0^k 1]$. Hence the set $\{w_l^k\}_{k,l}$ is prefix-free. Consider the following chain of inequalities:

$$1 \geq \sum_k \sum_l \mathbf{a}(w_l^k) \geq \sum_k \sum_l 2^{-c-|w_l^k|} \frac{2^{2k-1}}{2k-1} \geq^*$$
$$\geq^* \sum_k 2^{-|0^k 1^k|} \frac{2^{2k-1}}{2k-1} = \sum_k \frac{1}{2(2k-1)} = \infty \qquad (39)$$

This contradiction proves the theorem.   □

Proof (General case). Now we replace the intervals $[0^k 1]$ and $[0^k 1^k]$ by C_k and D_k (see below) respectively. We cannot make the measures of D_k very small, because it decreases g, but they also cannot be large, because g should be probability bounded. We will find suitable sets $\{C_k\}$ and $\{D_k\}$ that satisfy all of the conditions.

Let's consider the intervals B_k and C_k from Theorem 1. The series $\sum \mu(C_k)$ computably converges, hence for every computable sequence of positive rational numbers ε_k that tends to 0, the one-to-one function $\tau : \mathbb{N} \to \mathbb{N}$ such that $\mu(C_{\tau(k)}) \geq (1 - \varepsilon_k)\mu(C_m)$ for all $m \notin \{\tau(1), \ldots, \tau(k-1)\}$ is also computable. We will choose suitable sequence ε_k later. Denote $C_{\tau(k)} = \mathbf{C}_k$ and consider $z_k = -\frac{3}{\log \mu \mathbf{C}_k}$. The sequence $S_k = 1 + \sum_{j \leq k} z_j$ is computable. Let's show that $S_k \to \infty$:

Recall that

$$\sum_k \frac{\mu C_{k+1}}{\mu B_k \log \frac{1}{\mu B_k}} = \infty \qquad (40)$$

The function $\frac{x}{\log x}$ is monotone for large enough x, therefore

$$\sum_k z_k = \sum_k \frac{3}{\log \frac{1}{\mu C_k}} = 3\sum_k \frac{\mu C_k}{\mu C_k \log \frac{1}{\mu C_k}} \geq 3\sum_k \frac{\mu C_{k+1}}{\mu B_k \log \frac{1}{\mu B_k}} = \infty \qquad (41)$$

If $\mu(C_k) \neq 0$, then $k = \tau(j)$ for some j, so $\sum_k z_k = \sum_k \frac{3}{\log \frac{1}{\mu C_k}}$.

For all $\mathbf{C}_k$, we can construct an interval $D_k \subset \mathbf{C}_k$ with such property:

$$\frac{1}{3}(\mu\mathbf{C}_k)^{S_k} < \mu D_k < (\mu\mathbf{C}_k)^{S_k} \qquad (42)$$

Let x_k be a string such that $[x_k] = \mathbf{C}_k$. We compute $\mu([x_k0])$ and $\mu([x_k1])$ and choose the next bit b if $\mu[x_kb] > \frac{1}{3}\mu[x_k]$ (if the both bits are suitable, let's choose 0). After that we repeat this procedure with a string x_kb and so on. We stop when the condition 42 holds for the interval D_k (the set of the extensions of the latest string). It always happens, because the measure is non-atomic (so $\mu[x_kb_1 \ldots b_m]$ tends to 0), and $\mu[x_kb_1 \ldots b_{m-1}] < 3\mu[x_kb_1 \ldots b_m]$.

Consider a function

$$g(\omega) = \sum_k \frac{\mathbb{I}_{D_k}(\omega)}{2\mu D_k} \qquad (43)$$

It is lower semicomputable. To prove that it is probability bounded it is sufficient to show that

$$\mu D_j \geq \sum_{i:\mu D_i < \mu D_j} \mu D_i \qquad (44)$$

Indeed, consider the set $\{g(\omega) > C\}$:

$$\mu\{g(\omega) > C\} = \sum_{i:\mu D_i < \frac{1}{2C}} \mu D_i \leq 2\max\{\mu D_i : \mu D_i < \frac{1}{2C}\} < \frac{1}{C} \qquad (45)$$

The sequence $\mu\mathbf{C}_j^{S_j}$ is exponentially decreasing:

$$\frac{\mu\mathbf{C}_j^{S_j}}{\mu\mathbf{C}_{j+1}^{S_{j+1}}} \geq (1-\varepsilon_j)^{S_j}\mu\mathbf{C}_{j+1}^{S_j-S_{j+1}} = (1-\varepsilon_j)^{1+\sum z_j}\mu\mathbf{C}_{j+1}^{-z_{j+1}} \geq$$
$$\geq (1-\varepsilon_j)^{Mj}2^{-z_{j+1}\log\mathbf{C}_{j+1}} \geq \frac{3}{4}8 = 6 \qquad (46)$$

Where $M = -\frac{3}{\log \frac{2}{3}} \geq \max z_k$. Here one can see how to choose the sequence ε_k: $(1-\varepsilon_j)^{Mj}$ should be not less than $\frac{3}{4}$. This inequality shows that

$$\sum_{i:\mu D_i < \mu D_j} \mu D_i = \sum_{i>j} \mu D_i \leq \sum_{k\geq 1}(\frac{1}{2})^{-k}\mu D_j \leq \mu D_j \qquad (47)$$

Therefore the function g is probability bounded.

Assume that there exists a constant c such that for all ω

$$\mathbf{t}_\mu^M(\omega) \geq 2^{-c}\frac{g}{\log g}(\omega) \tag{48}$$

Where $\mathbf{t}_\mu^M = 2^{\mathbf{d}_\mu^M(\omega)}$. If $\omega \in D_k$, then for this k there exists a prefix-free set of strings w_l^k such that $\cup_l[w_l^k] \supset D_k$ and

$$\frac{\mathbf{a}(w_l^k)}{\mu([w_l^k])} \geq 2^{-c}\frac{1}{2\mu D_k \log\frac{1}{\mu D_k}} \tag{49}$$

Using the Property 42 for large enough k we get:

$$\mu([w_l^k]) \leq 2^{c+1}\mathbf{a}(w_l^k)\mu D_k \log\frac{1}{\mu D_k} < \sqrt{\mu D_k} < \mu C_k \tag{50}$$

Therefore $w_l^k \subset C_k$ and the set $\{w_l^k\}_{k,l}$ is prefix-free.
Consider the following chain of inequalities:

$$1 \geq \sum_{k,l}\mathbf{a}(w_l^k) \geq \sum_{k,l}\mu([w_l^k])2^{-c-1}\frac{1}{\mu D_k \log\frac{1}{\mu D_k}} \geq^*$$
$$\geq^* \sum_k \mu D_k\frac{1}{\mu D_k \log\frac{1}{\mu D_k}} = \sum_k \frac{1}{\log\frac{1}{\mu D_k}} =^* \sum_k \frac{1}{S_k \log\frac{1}{\mu C_k}} \tag{51}$$

In the proof of Lemma 3 we showed that if the series $\sum_n z_n$ does not converge, then the series $\frac{z_n}{S_n}$ where $S_n = \sum_{k \leq n} z_k$ does not converge either, so the right part of the chain of inequalities is ∞. □

Now we can rewrite the chain of inequalities 20 as follows:

$$\mathbf{d}_\mu^E \leq^+ \mathbf{d}_\mu^{\lim\inf M} \overset{\mathrm{a.e.}}{=\!=\!=} \mathbf{d}_\mu^{\lim\sup M} \ll \mathbf{d}_\mu^M \ll \mathbf{d}_\mu^P \leq^+ \mathbf{d}_\mu^E + (1+\varepsilon)\log\mathbf{d}_\mu^E \tag{52}$$

where the symbol $\ll$ means that the difference may be greater than $\log\mathbf{d}_\mu$.

Acknowledgements. This research was supported in part by RaCAF ANR-15-CE40-0016-01. The author thanks Alexander Shen and Mikhail Andreev for their help.

References

1. Bienvenu, L., Gacs, P., Hoyrup, M., Rojas, C., Shen, A.: Algorithmic tests and randomness with respect to a class of measures. Proc. Steklov Inst. Math. **274**, 41–102 (2011)
2. Uspensky, V.A., Vereshchagin, N.K., Shen, A.: Kolmogorov complexity and algorithmic randomness, MCCME (2013) (in Russian)
3. Martin-Lof, P.: The definition of random sequences. Inf. Control **9**, 602–619 (1966)
4. Gacs, P.: Exact expressions for some randomness tests. Z. Math. Log. Grdl. M. **26**, 385–394 (1980). Short version: Springer. Lecture Notes in Computer Science, vo. 67, pp. 124–131 (1979)

5. Schnorr, C.P.: Process complexity and effective random tests. J. Comput. Syst. Sci. **7**(4), 376–388 (1973). Conference version: STOC 1972, pp. 168–176
6. Levin, L.A.: On the notion of a random sequence. Sov. Math. Dokl. **14**(5), 1413–1416 (1973)
7. Williams, D.: Probability with Martingales. Cambridge University Press, Cambridge (1991)

McShane-Whitney Pairs

Iosif Petrakis[✉]

University of Munich, Munich, Germany
petrakis@math.lmu.de

Abstract. We present a constructive version of the classical McShane-Whitney theorem on the extendability of real-valued Lipschitz functions defined on a subset of a metric space. Through the introduced notion of a McShane-Whitney pair we study some abstract properties of this extension theorem showing how the behavior of a Lipschitz function defined on the subspace of the pair affect its McShane-Whitney extensions on the space of the pair. As a consequence, a Lipschitz version of the theory around the Hahn-Banach theorem is formed. We work within Bishop's informal system of constructive mathematics BISH.

1 Introduction

According to the classical extension theorem of McShane and Whitney that first appeared in [12,19], a real-valued Lipschitz function defined on *any* subset A of a metric space X is extended to a Lipschitz function defined on X. To determine metric spaces X and Y such that a similar extension theorem for Y-valued Lipschitz functions defined on a subset A of X holds is a non-trivial problem under active current study (see [1,4,14]). Although the McShane-Whitney theorem has a highly ineffective proof similar to the proof of the Hahn-Banach theorem (see [17], pp.16–17), it also admits a proof based on an explicit definition of two such extension functions. This definition, which involves the notions of infimum and supremum of a bounded subset of reals, can be carried out constructively only if we restrict to certain subsets A of a metric space X.

We define a McShane-Whitney subset A of a metric space X in order to constructively realize the McShane-Whitney explicit definition of the extension functions. A pair (X, A), where X is a metric space and A is a subset of X on which the McShane-Whitney explicit definition is carried out constructively is called here a McShane-Whitney pair. The importance of the McShane-Whitney extension lies in the possibility to relate properties of a given Lipschitz function on A to properties of its extension functions on X in such a way that a Lipschitz-version of the theory around the Hahn-Banach theorem is formed. We present here the first basic results in this direction. We work within Bishop's informal system of constructive mathematics BISH (see [2,3,6]). The constructive reconstruction of the general theory of Lipschitz functions is quite underdeveloped. Some first results on constructive Lipschitz analysis are found in [8,10,13]. All proofs that are not included here due to space restrictions are left to the reader.

© Springer International Publishing AG 2017
J. Kari et al. (Eds.): CiE 2017, LNCS 10307, pp. 351–363, 2017.
DOI: 10.1007/978-3-319-58741-7_33

2   Basic Notions and Facts

Definition 1. *Let $A \subseteq \mathbb{R}$ and $b, l, \lambda, m, \mu \in \mathbb{R}$. If A is bounded above, we define $b \geq A :\leftrightarrow \forall_{a \in A}(b \geq a)$, $[A] := \{b \in \mathbb{R} \mid b \geq A\}$, $l = \sup A :\leftrightarrow l \geq A \wedge \forall_{\epsilon > 0} \exists_{a \in A}(a > l - \epsilon)$, and $\lambda = \text{lub}A :\leftrightarrow \lambda \geq A \wedge \forall_{b \in [A]}(b \geq \lambda)$. If A is bounded below, $b \leq A$, $(A]$, $m = \inf A$, and $\mu = \text{glb}A$ are defined in a dual way.*

In [11], pp. 24–25, Mandelkern gave a necessary and sufficient condition for the existence of lub$A$ and glb$A$ and proved the following remark: If $A \subseteq \mathbb{R}$ bounded and glb$A$ exists, then $\sup(A]$ exists and $\sup(A] = \text{glb}A$, while if lub$A$ exists, then $\inf[A)$ exists and $\inf[A) = \text{lub}A$.

Definition 2. *We denote by $\mathbb{F}(X, Y)$ the set of functions of type $X \to Y$ and by $\mathbb{F}(X)$ the set of functions of type $X \to \mathbb{R}$. If $a \in \mathbb{R}$, then $\overline{a}_X$ denotes the constant map in $\mathbb{F}(X)$ with value a, and $\text{Const}(X)$ is the set of constant maps. If $(X, d), (Y, \rho)$ are metric spaces, then $C_u(X, Y)$ denotes the uniformly continuous functions from X to Y, and $C_u(X)$ denotes the uniformly continuous functions from X to $\mathbb{R}$, where $\mathbb{R}$ is equipped with its standard metric. The metric d_{x_0} at the point $x_0 \in X$ is defined by $d_{x_0}(x) := d(x_0, x)$, for every $x \in X$, and $U_0(X) := \{d_{x_0} \mid x_0 \in X\}$. The set X_0 of the d-distinct pairs of X is defined by $X_0 := \{(x, y) \in X \times X \mid d(x, y) > 0\}$.*

Definition 3. *If A is a subset of a metric space X, $x \in X$, and $\Delta(x, A) := \{d(x, a) \mid a \in A\}$, A is located if $d(x, A) := \inf \Delta(x, A)$ exists, for every $x \in X$, and A is colocated, if $\delta(x, A) := \sup \Delta(x, A)$ exists, for every $x \in X$.*

If A is inhabited and colocated, then A is bounded; if a_0 inhabits A, then $d(a, b) \leq d(a, a_0) + d(a_0, b) \leq 2\delta(a_0, A)$, for every $a, b \in A$. Unless otherwise stated, for the rest X and Y are equipped with metrics d and ρ, respectively.

Definition 4. *The set of Lipschitz functions $\text{Lip}(X, Y)$ from X to Y is*

$$\text{Lip}(X, Y) := \bigcup_{\sigma \geq 0} \text{Lip}(X, Y, \sigma),$$

$$\text{Lip}(X, Y, \sigma) := \{f \in \mathbb{F}(X, Y) \mid \forall_{x, y \in X}(\rho(f(x), f(y)) \leq \sigma d(x, y))\}.$$

If $Y = \mathbb{R}$, we use the notations $\text{Lip}(X)$ and $\text{Lip}(X, \sigma)$, respectively.

Clearly, $\text{Lip}(X, Y) \subseteq C_u(X, Y)$. If $A \subseteq X$ and $f \in \text{Lip}(X, \sigma)$, for some $\sigma \geq 0$, then $f_{|A} \in \text{Lip}(A, \sigma)$. An element of $\text{Lip}(X, Y)$ sends a bounded subset of X to a bounded subset of Y, which is not generally the case for elements of $C_u(X, Y)$; the identity function id : $\mathbb{N} \to \mathbb{R}$, where $\mathbb{N}$ is equipped with the discrete metric, is in $C_u(\mathbb{N}) \setminus \text{Lip}(\mathbb{N})$ and id$(\mathbb{N}) = \mathbb{N}$ is unbounded in $\mathbb{R}$. In [13], p. 370, it is shown constructively that if X is totally bounded, then $\text{Lip}(X)$ is uniformly dense in $C_u(X)$.

Proposition 1. *The set $\text{Lip}(X)$ includes the sets $U_0(X)$, $\text{Const}(X)$, and it is closed under addition and multiplication by reals. If every element of $C_u(X)$ is a bounded function, then $\text{Lip}(X)$ is closed under multiplication.*

Definition 5. *If $f \in \mathbb{F}(X,Y)$, we define the following sets:*

$$\Lambda(f) := \{\sigma \geq 0 \mid \forall_{x,y \in X}(\rho(f(x), f(y)) \leq \sigma d(x,y))\},$$

$$\Xi(f) := \{\sigma \geq 0 \mid \forall_{x,y \in X}(\rho(f(x), f(y)) \geq \sigma d(x,y))\},$$

$$M_0(f) := \{\sigma_{x,y}(f) \mid (x,y) \in X_0\},$$

$$\sigma_{x,y}(f) := \frac{\rho(f(x), f(y))}{d(x,y)}.$$

Proposition 2. *If $f \in \mathbb{F}(X,Y)$, then $\Lambda(f) = [M_0(f))$ and $\Xi(f) = (M_0(f)]$.*

Classically one can prove that if $f \in \mathrm{Lip}(X,Y)$ such that $\inf \Lambda(f)$ exists, then $\sup M_0(f)$ exists and $\sup M_0(f) = \inf \Lambda(f)$. The classical argument in that proof is avoided, if $\sup M_0(f)$ exists.

Proposition 3. *Let $f \in \mathrm{Lip}(X,Y)$.*

(i) *If $\sup M_0(f)$ exists, then $\inf \Lambda(f)$ exists and $\inf \Lambda(f) = \min \Lambda(f) = \sup M_0(f)$.*
(ii) *If $\inf \Lambda(f)$ exists, then $\mathrm{lub}M_0(f)$ exists and $\mathrm{lub}M_0(f) = \inf \Lambda(f)$.*
(iii) *If $\mathrm{lub}M_0(f)$ exists, then $\inf \Lambda(f)$ exists and $\inf \Lambda(f) = \mathrm{lub}M_0(f)$.*

In constructive analysis one usually works with the stronger notions of infima or suprema of sets and not with greatest lower bounds or least upper bounds of sets. An important exception is found in the work of Mandelkern (see his comment in [11], p. 24). Here we also find useful to keep both notions at work.

Definition 6. *Let $f \in \mathrm{Lip}(X,Y)$. We call f L-pseudo-normable, if $\sup M_0(f)$ exists, and we write $L(f) := \sup M_0(f)$. We call f weakly L-pseudo-normable, or L^*-pseudo-normable, if $\mathrm{lub}M_0(f) = \inf \Lambda(f)$ exists, and $L^*(f) := \mathrm{lub}M_0(f)$.*

In general a Lipschitz function need not be L-pseudo-normable. Note that in the case of a linear function f between normed spaces X and Y the boundedness condition is equivalent to the Lipschitz condition and the existence of its norm $\|f\|$ is equivalent to the existence of $L(f)$. If f is L-pseudo-normable, and since $L(f) \geq M_0(f)$, by Proposition 2 we get $\forall_{x,y \in X}(\rho(f(x), f(y)) \leq L(f)d(x,y))$, or $f \in \mathrm{Lip}(X, Y, L(f))$. If f is L^*-pseudo-normable, we work similarly.

Proposition 4. *Let $A \subseteq X$ and $f \in \mathrm{Lip}(A,Y)$ such that*

$$\exists_{g \in \mathbb{F}(X,Y)}(g_{|A} = f \ \wedge \ \forall_{\sigma \geq 0}(f \in \mathrm{Lip}(A,Y,\sigma) \rightarrow g \in \mathrm{Lip}(X,Y,\sigma))).$$

(i) *If f is L-pseudo-normable, g is L-pseudo-normable and $L(g) = L(f)$.*
(ii) *If f is L^*-pseudo-normable, g is L^*-pseudo-normable and $L^*(g) = L^*(f)$.*

Note that if $f \in \mathrm{Lip}(A,Y), g \in \mathrm{Lip}(X,Y)$ such that $L(f), L(g)$ exist and $L(f) = L(g)$, then it is immediate that $\forall_{\sigma \geq 0}(f \in \mathrm{Lip}(A,Y,\sigma) \rightarrow g \in \mathrm{Lip}(X,Y,\sigma))$. Next follows the Lipschitz-version of the extendability of a uniformly continuous function defined on a dense subset of a metric space with values in a complete metric space.

Proposition 5. *Let $D \subseteq X$ be dense in X, Y complete, and $f \in \mathrm{Lip}(D, Y)$.*

(i) $\exists_{!g \in \mathbb{F}(X,Y)}(g_{|A} = f \; \wedge \; \forall_{\sigma \geq 0}(f \in \mathrm{Lip}(D, Y, \sigma) \to g \in \mathrm{Lip}(X, Y, \sigma)))$.
(ii) If f is L-pseudo-normable, then g is L-pseudo-normable and $L(g) = L(f)$.
(iii) If f is L^*-pseudo-normable, g is L^*-pseudo-normable and $L^*(g) = L^*(f)$.

Corollary 1. *Let $D \subseteq X$ be dense in X, let Y be complete, and $g \in \mathrm{Lip}(X, Y)$. If g is L^*-pseudo-normable, $f = g_{|D}$ is L^*-pseudo-normable and $L^*(f) = L^*(g)$.*

3   McShane-Whitney Subsets and Pairs

We formulate a property on the subsets of a metric space so that the McShane-Whitney extension can be carried out on them constructively.

Definition 7. *If $A \subseteq X$ is inhabited, $x \in X$, $\lambda \in \mathbb{R}$, and $g \in \mathrm{Lip}(A)$, the set $\mathrm{MW}_g(A, \lambda, x)$ is defined by*

$$\mathrm{MW}_g(A, \lambda, x) := \{g(a) + \lambda d(x, a) \mid a \in A\}.$$

The set A is called a McShane-Whitney subset of X, if for every $\sigma > 0$, $g \in \mathrm{Lip}(A)$ and $x \in X$ the $\inf \mathrm{MW}_g(A, \sigma, x)$ exists.

A McShane-Whitney subset A of X is located, colocated and bounded. Since $\{d(x, a) \mid a \in A\} = \mathrm{MW}_{\overline{0}_X}(A, 1, x)$, A is located. Since $\mathrm{MW}_{-2d_x}(A, 1, x) = \{-2d(x, a) + d(x, a) \mid a \in A\} = \{-d(x, a) \mid a \in A\} = -\Delta(x, A)$, we get[1] $\delta(x, A) = \sup[-(-\Delta(x, A))] = -\inf(-\Delta(x, A)) = -\inf \mathrm{MW}_{-2d_x}(A, 1, x)$ i.e., A is colocated, and since A is inhabited, A is also bounded.

Proposition 6. *A is a McShane-Whitney subset of X if and only if for every $\sigma > 0$, $g \in \mathrm{Lip}(A)$ and $x \in X$ the $\sup \mathrm{MW}_g(A, -\sigma, x)$ exists.*

Proof. If $\sigma > 0$, $g \in \mathrm{Lip}(A)$ and $x \in X$, then $\mathrm{MW}_g(A, -\sigma, x) = \{g(a) - \sigma d(x, a) \mid a \in A\} = \{-(-g(a) + \sigma d(x, a)) \mid a \in A\} = -\{(-g)(a) + \sigma d(x, a) \mid a \in A\} = -\mathrm{MW}_{-g}(A, \sigma, x)$. Since $-g \in \mathrm{Lip}(A)$, we get $\sup \mathrm{MW}_g(A, -\sigma, x) = \sup(-\mathrm{MW}_{-g}(A, \sigma, x)) = -\inf \mathrm{MW}_{-g}(A, \sigma, x)$. For the converse implication we use the equality $\inf(-B) = -\sup B$, where $B \subseteq \mathbb{R}$ such that $\sup B$ exists, and the similarly shown equality $\mathrm{MW}_g(A, \sigma, x) = -\mathrm{MW}_{-g}(A, -\sigma, x)$. Hence $\inf \mathrm{MW}_g(A, \sigma, x) = \inf(-\mathrm{MW}_{-g}(A, -\sigma, x)) = -\sup \mathrm{MW}_{-g}(A, -\sigma, x)$.

---

[1] If $B \subseteq \mathbb{R}$ is bounded and $\inf B$ exists, then $\sup(-B)$ exists and $\sup(-B) = -\inf B$; if $m = \inf B$, then by definition m is a lower bound of B and $\forall_{\epsilon > 0}\exists_{b \in B}(b < m + \epsilon)$, therefore $-m$ is an upper bound of $-B$ and $\forall_{\epsilon > 0}\exists_{-b \in -B}(-b > -m - \epsilon)$. The following constructively provable properties are used in this paper: if $A, B \subseteq \mathbb{R}$ are inhabited and bounded such that $\sup A, \inf A, \sup B, \inf B$ exist, then $\sup(A + B)$ exists and $\sup(A + B) = \sup A + \sup B$, $\inf(A + B)$ exists and $\inf(A + B) = \inf A + \inf B$, if $\lambda > 0$, then $\sup(\lambda A), \inf(\lambda A)$ exist and $\sup(\lambda A) = \lambda \sup A$, $\inf(\lambda A) = \lambda \inf A$, if $\lambda < 0$, then $\sup(\lambda A), \inf(\lambda A)$ exist and $\sup(\lambda A) = \lambda \inf A$, and $\inf(\lambda A) = \lambda \sup A$.

The next proposition provides examples of McShane-Whitney subsets. A locally compact (totally bounded) metric space X is one every bounded subset of which is included in a compact (totally bounded) subset of X (see [5], p. 46).

Proposition 7. *Let $A \subseteq X$ be inhabited.*

(i) *If A is totally bounded, then A is a McShane-Whitney subset of X.*

(ii) *If X is totally bounded and A is located, then A is a McShane-Whitney subset of X.*

(iii) *If X is locally compact (totally bounded), then A is a McShane-Whitney subset of X if and only if A is bounded and located.*

(iv) *If $X = \mathbb{R}^n$, then A is a McShane-Whitney subset of $\mathbb{R}^n$ if and only if A is totally bounded.*

Proof

(i) If $\sigma > 0$, $g \in \mathrm{Lip}(A)$ and $x \in X$, then $g + \sigma d_x \in \mathrm{Lip}(A) \subseteq \mathrm{C_u}(A)$, and $\inf \mathrm{MW}_g(\sigma, x)$ exists, since A is totally bounded (see [3], Corollary 4.3, p. 94).

(ii) A located subset of X is also totally bounded (see [3], p. 95), and we use (i).

(iii) If A is bounded and located, there is compact (totally bounded) $K \subseteq X$ such that $A \subseteq K$. Since A is located in X, it is a located in K, hence A is totally bounded, and we use (i). For the converse see our remark after Definition 7.

(iv) A is totally bounded if and only if it is located and bounded (see [3], p. 95), and $\mathbb{R}^n$ is locally compact as a finite product of $\mathbb{R}$ (see [3], p. 111). The required equivalence follows from (iii).

Definition 8. *Let $A \subseteq X$. We call (X, A) a McShane-Whitney pair, if for all $\sigma > 0$ and $g \in \mathrm{Lip}(A, \sigma)$ the functions $g^*, {}^*g : X \to \mathbb{R}$, the smallest and largest McShane-Whitney extension of g, defined by*

$$g^*(x) = \inf M_g(A, \sigma, x), \quad {}^*g(x) = \sup M_g(A, -\sigma, x),$$

for every $x \in X$, are well-defined and satisfy the following properties:

(i) $g^*, {}^*g \in \mathrm{Lip}(X, \sigma)$.

(ii) $g^*_{|A} = ({}^*g)_{|A} = g$.

(iii) $\forall_{f \in \mathrm{Lip}(A, \sigma)}(f_{|A} = g \to g^* \le f \le {}^*g)$.

The extensions $g^*, {}^*g$ of g are unique. Let $h^*, {}^*h$ satisfy conditions (i)-(iii) of Definition 8. Since $h^*_{|A} = ({}^*h)_{|A} = g$, $g^* \le h^* \le {}^*g$ and $g^* \le {}^*h \le {}^*g$. Since $g^*_{|A} = ({}^*g)_{|A} = g$, $h^* \le g^* \le {}^*h$ and $h^* \le {}^*g \le {}^*h$, hence $h^* = g^*$ and ${}^*h = {}^*g$.

Theorem 1 (McShane-Whitney). *If A is a McShane-Whitney subset of X, then (X, A) is a McShane-Whitney pair.*

Proof. By Proposition 6, the functions $^*g, g^*$ are well-defined. First we show that $^*g$ extends g. If $a_0 \in A$, then $^*g(a_0) = \inf\{g(a) + \sigma d(a_0, a) \mid a \in A\} \leq g(a_0) + \sigma d(a_0, a_0) = g(a_0)$. If $a \in A$, then $g(a_0) - g(a) \leq |g(a_0) - g(a)| \leq \sigma d(a_0, a)$, hence $g(a) + \sigma d(a_0, a) \geq g(a_0)$. Since a is arbitrary, $^*g(a_0) \geq g(a_0)$. To show $^*g \in \mathrm{Lip}(X, \sigma)$ let $x_1, x_2 \in X$ and $a \in A$. Then $d(x_1, a) \leq d(x_2, a) + d(x_2, x_1)$ and $\sigma d(x_1, a) \leq \sigma d(x_2, a) + \sigma d(x_1, x_2)$, therefore

$$g(a) + \sigma d(x_1, a) \leq (g(a) + \sigma d(x_2, a)) + \sigma d(x_1, x_2) \rightarrow$$
$$^*g(x_1) \leq (g(a) + \sigma d(x_2, a)) + \sigma d(x_1, x_2) \rightarrow$$
$$^*g(x_1) - \sigma d(x_1, x_2) \leq g(a) + \sigma d(x_2, a) \rightarrow$$
$$^*g(x_1) - \sigma d(x_1, x_2) \leq {}^*g(x_2) \rightarrow$$
$$^*g(x_1) - {}^*g(x_2) \leq \sigma d(x_1, x_2).$$

If we start from the inequality $d(x_2, a) \leq d(x_1, a) + d(x_2, x_1)$ and work as above, we get $^*g(x_2) - {}^*g(x_1) \leq \sigma d(x_1, x_2)$, therefore $|{}^*g(x_1) - {}^*g(x_2)| \leq \sigma d(x_1, x_2)$. Working similarly we get that g^* is an extension of g which is in $\mathrm{Lip}(X, \sigma)$. If $f \in \mathrm{Lip}(X, \sigma)$ such that $f_{|A} = g$, $x \in X$ and $a \in A$ we have that

$$f(x) - g(a) = f(x) - f(a) \leq |f(x) - f(a)| \leq \sigma d(x, a) \rightarrow$$
$$f(x) \leq g(a) + \sigma d(a, x) \rightarrow$$
$$f(x) \leq {}^*g(x),$$
$$g(a) - f(x) = f(a) - f(x) \leq |f(a) - f(x)| \leq \sigma d(x, a) \rightarrow$$
$$g(a) - \sigma d(a, x) \leq f(x) \rightarrow$$
$$g^*(x) \leq f(x).$$

Proposition 8. *Let* (X, A) *be a McShane-Whitney pair and* $g \in \mathrm{Lip}(A, \sigma)$.

(i) The set A is located.

(ii) If $\inf g$ and $\sup g$ exist, then $\inf {}^*g$, $\sup g^*$ exist and

$$\inf_{x \in X} {}^*g = \inf_{a \in A} g, \qquad \sup_{x \in X} g^* = \sup_{a \in A} g.$$

Proof

(i) Let $r \in \mathbb{R}$ and $\sigma > 0$. Since $\bar{r}_A \in \mathrm{Lip}(A, \sigma)$, by hypothesis $^*\bar{r}_A$ is well-defined, where $^*\bar{r}_A(x) = \inf\{r + \sigma d(x, a) \mid a \in A\}$, for every $x \in X$. If $x \in X$ and $a \in A$, then $d(x, a) = \frac{1}{\sigma}(r + \sigma d(x, a) - r)$, and $\Delta(x, A) = \{\frac{1}{\sigma}(r + \sigma d(x, a) - r) \mid a \in A\}$. Hence $d(x, A) = \inf\{\frac{1}{\sigma}(r + \sigma d(x, a) - r) \mid a \in A\} = \frac{1}{\sigma}(\inf\{r + \sigma d(x, a) \mid a \in A\} - r) = \frac{1}{\sigma}(^*\bar{r}_A(x) - r)$.

(ii) We show that $m := \inf\{g(a) \mid a \in A\}$ satisfies the properties of $\inf\{^*g(x) \mid x \in X\}$. It suffices to show that $m \leq {}^*g(X)$, since the other definitional condition of inf follows immediately; if $\epsilon > 0$, then there exists $a \in A \subseteq X$ such that $g(a) = {}^*g(a) < m + \epsilon$. If $x \in A$, then $m \leq g(x) = {}^*g(x)$, since $m = \inf g$. Since A is located, the set $-A := \{x \in X \mid d(x, A) > 0\}$ is well-defined. If $x \in -A$, then $d(x, a) \geq d(x, A) > 0$, for every $a \in A$. Hence

$$g(a) + \sigma d(x, a) > g(a) \geq \inf_{a \in A} g \rightarrow \inf_{a \in A}(g(a) + \sigma d(x, a)) \geq \inf_{a \in A} g \leftrightarrow {}^*g(x) \geq m.$$

Since A is located, the set $A \cup (-A)$ is dense in X (see [3], p.88). If $x \in X$, there is some sequence $(d_n)_{n \in \mathbb{N}} \subseteq A \cup (-A)$ such that $d_n \xrightarrow{n} x$. By the continuity of $^*g$ we have that $^*g(d_n) \xrightarrow{n} {}^*g(x)$. Suppose that $^*g(x) < m$. Since $^*g(d_n) \geq m$, for every $n \in \mathbb{N}$, if $\epsilon := (m - {}^*g(x)) > 0$, there is some n_0 such that for every $n \geq n_0$ we have that $|{}^*g(d_n) - {}^*g(x)| = {}^*g(d_n) - {}^*g(x) < m - {}^*g(x) \leftrightarrow {}^*g(d_n) < m$, which is a contradiction. Hence $^*g(x) \geq m$. For the existence of $\sup g^*$ and the equality $\sup_{x \in X} g^* = \sup_{a \in A} g$ we work similarly.

If $g = \overline{r}_A$ and $\sigma > 0$, then $\overline{r}_A^* = \overline{r}_A - \sigma d_A$ and $^*\overline{r}_A = \overline{r}_A + \sigma d_A$. If $g \in \mathrm{Lip}(A, 0)$, it is immediate that $g = \overline{r}_A$, for some $r \in \mathbb{R}$, and $^*g = g^* = \overline{r}_X$. If D is dense in X and (X, D) is a McShane-Whitney pair, then by Proposition 5 there is a unique σ-Lipschitz extension on X of some $g \in \mathrm{Lip}(D)$, hence $^*g = g^*$, a fact which is also shown by the definition of $^*g$ and g^*. A weaker property on A that suffices for the McShane-Whitney extension is that for every $\sigma > 0, g \in \mathrm{Lip}(A, \sigma)$ and $x \in X$ the $\inf \mathrm{MW}_g(A, \sigma, x)$ exists, but since all our examples of McShane-Whitney subsets satisfy the stronger property of Definition 7, we avoid it. The next proposition expresses a "step-invariance" of the McShane-Whitney extension. If $A \subseteq B \subseteq X$ such that $(X, A), (X, B), (B, A)$ are McShane-Whitney pairs and $g \in \mathrm{Lip}(A)$, then g^{*X} is the $(A-X)$ extension of g, g^{*B*X} is the $(B-X)$ extension of the $(A - B)$ extension g^{*B} of g, and similarly for $^{*X}g$ and $^{*X*B}g$.

Proposition 9. *If $A \subseteq B \subseteq X$ such that $(X, A), (X, B), (B, A)$ are McShane-Whitney pairs and $g \in \mathrm{Lip}(A, \sigma)$, for some $\sigma > 0$, then*

$$g^{*X} = g^{*B*X}, \quad {}^{*X}g = {}^{*X*B}g.$$

Proof. We show only the first equality and for the second we work similarly. By definition $g^{*B} : B \to \mathbb{R} \in \mathrm{Lip}(B, \sigma)$ and $g^{*B}(b) = \sup\{g(a) - \sigma d(b, a) \mid a \in A\}$, for every $b \in B$. Moreover, $g^{*B*X} : X \to \mathbb{R} \in \mathrm{Lip}(X, \sigma)$ and $g^{*B*X}(x) = \sup\{g^{*B}(b) - \sigma d(x, b) \mid b \in B\}$, for every $x \in X$. For the $(A-X)$ extension of g we have that $g^{*X} : X \to \mathbb{R} \in \mathrm{Lip}(X, \sigma)$ and $g^{*X}(x) = \sup\{g(a) - \sigma d(x, a) \mid a \in A\}$, for every $x \in X$. Since $(g^{*B*X})_{|B} = g^{*B}$, we have that $(g^{*B*X})_{|A} = (g^{*B})_{|A} = g$. Therefore $g^{*X} \leq g^{*B*X} \leq {}^{*X}g$, and $(g^{*X})_{|B} \leq (g^{*B*X})_{|B} = g^{*B} \leq (^{*X}g)_{|B}$. Since $(g^{*X})_{|A} = g$, we get that $((g^{*X})_{|B})_{|A} = g$, therefore $g^{*B} \leq ((g^{*X})_{|B}) \leq {}^{*B}g$. Since $(g^{*X})_{|B} \leq g^{*B}$ and $g^{*B} \leq (g^{*X})_{|B}$, we get $(g^{*X})_{|B} = g^{*B}$. Hence $g^{*B*X} \leq g^{*X} \leq {}^{*X*B}g$ i.e., we have shown both $g^{*X} \leq g^{*B*X}$ and $g^{*B*X} \leq g^{*X}$.

Proposition 10. *Let (X, A) be a McShane-Whitney pair and $g \in \mathrm{Lip}(A)$ such that $L(g)$ exists.*

(i) $g \in \mathrm{Lip}(A, L(g))$.
(ii) If f is an $L(g)$-Lipschitz extension of g, then $L(f)$ exists and $L(f) = L(g)$.
(iii) $L(^*g), L(g^*)$ exist and $L(^*g) = L(g) = L(g^*)$.

Proof

(i) Since $L(g) = \sup M_0(g)$, we have that $L(g) \geq M_0(g)$ and by Proposition 2 we get $L(g) \in \Lambda(g)$, therefore $g \in \mathrm{Lip}(A, L(g))$.

(ii) Since $f \in \mathrm{Lip}(X, L(g))$, we get $L(g) \in \Lambda(f)$ and $L(g) \geq M_0(f)$. Let $\epsilon > 0$. Since $L(g) = \sup M_0(g)$, there exists $(a, b) \in A_0 \subseteq X_0$ such that $\sigma_{a,b}(g) > L(g) - \epsilon$. Since f extends g, $\sigma_{a,b}(g) = \sigma_{a,b}(f)$.

(iii) By definition ${}^{*}g, g^{*} \in \mathrm{Lip}(X, L(g))$ and they extend g. Hence we use (ii).

Definition 9. *Let $(X, ||.||)$ be a normed space. A subset C of X is called convex, if $\forall_{x,y \in C} \forall_{\lambda \in (0,1)}(\lambda x + (1 - \lambda)y \in C)$. If $C \subseteq X$ is convex, a function $g : C \to \mathbb{R}$ is called convex, if $\forall_{x,y \in C} \forall_{\lambda \in (0,1)}(g(\lambda x + (1 - \lambda)y) \leq \lambda g(x) + (1 - \lambda)g(y))$, and g is called concave, if $\forall_{x,y \in C} \forall_{\lambda \in (0,1)}(g(\lambda x + (1 - \lambda)y) \geq \lambda g(x) + (1 - \lambda)g(y))$. A function $f : X \to \mathbb{R}$ is called sublinear if it is subadditive and positive homogeneous i.e., if $f(x + y) \leq f(x) + f(y)$, and $f(\lambda x) = \lambda f(x)$, for every $x, y \in X$ and $\lambda > 0$, respectively. Similarly, f is called superlinear, if it is superadditive i.e., if $f(x + y) \geq f(x) + f(y)$, for every $x, y \in X$, and positive homogeneous.*

Proposition 11. *Let $(X, ||.||)$ be a normed space, $C \subseteq X$ convex and inhabited, (X, C) a McShane-Whitney pair, and $g \in \mathrm{Lip}(C, \sigma)$, for some $\sigma > 0$.*

(i) If g is convex, then ${}^{*}g$ is convex.
(ii) If g is concave, then g^{*} is concave.

Proof. We show only (i), and for (ii) we work similarly. Let $x, y \in X$, and $\lambda \in (0, 1)$. If we consider the sets $C_x = \{\lambda(g(c) + \sigma||x - c||) \mid c \in C\}$ and $C_y = \{(1 - \lambda)(g(c) + \sigma||y - c||) \mid c \in C\}$, an element of $C_x + C_y$ has the form $\lambda(g(c) + \sigma||x - c||) + (1 - \lambda)(g(d) + \sigma||y - d||)$, for some $c, d \in C$. We show that ${}^{*}g(\lambda x + (1-\lambda)y) \leq \lambda(g(c) + \sigma||x - c||) + (1 - \lambda)(g(d) + \sigma||y - d||)$, where $c, d \in C$. Since C is convex, $c' := \lambda c + (1 - \lambda)d \in C$, and by the convexity of g we get

$$
\begin{aligned}
{}^{*}g(\lambda x + (1 - \lambda)y) &\leq g(c') + \sigma||\lambda x + (1 - \lambda)y - c'|| \\
&\leq \lambda g(c) + (1 - \lambda)g(d) + \lambda \sigma||x - c|| + (1 - \lambda)\sigma||y - d|| \\
&= \lambda(g(c) + \sigma||x - c||) + (1 - \lambda)(g(d) + \sigma||y - d||).
\end{aligned}
$$

Since the element of $C_x + C_y$ considered is arbitrary, we get that ${}^{*}g(\lambda x + (1 - \lambda)y) \leq \inf(C_x + C_y) = \inf C_x + \inf C_y = \lambda {}^{*}g(x) + (1 - \lambda){}^{*}g(y)$.

Proposition 12. *Let (X, A) be McShane-Whitney pair, $g_1 \in \mathrm{Lip}(A, \sigma_1), g_2 \in \mathrm{Lip}(A, \sigma_2)$ and $g \in \mathrm{Lip}(A, \sigma)$, for some $\sigma_1, \sigma_2, \sigma > 0$.*

(i) $(g_1 + g_2)^{*} \leq g_1^{*} + g_2^{*}$ and ${}^{*}(g_1 + g_2) \geq {}^{*}g_1 + {}^{*}g_2$.
(ii) If $\lambda > 0$, then $(\lambda g)^{*} = \lambda g^{*}$ and ${}^{*}(\lambda g) = \lambda {}^{*}g$.
(iii) If $\lambda < 0$, then $(\lambda g)^{*} = \lambda {}^{*}g$ and ${}^{*}(\lambda g) = \lambda g^{*}$.

Proof. In each case we show only one of the two facts.

(i) We have that $g_1 + g_2 \in \mathrm{Lip}(A, \sigma_1 + \sigma_2)$ and

$$
\begin{aligned}
(g_1 + g_2)^{*}(x) &= \sup\{g_1(a) + g_2(a) - (\sigma_1 + \sigma_2)d(x, a) \mid a \in A\} \\
&= \sup\{(g_1(a) - \sigma_1 d(x, a)) + (g_2(a) - \sigma_2 d(x, a)) \mid a \in A\} \\
&\leq \sup\{g_1(a) - \sigma_1 d(x, a) \mid a \in A\} + \sup\{g_2(a) - \sigma_2 d(x, a) \mid a \in A\} \\
&= g_1^{*}(x) + g_2^{*}(x).
\end{aligned}
$$

(ii) If $\lambda \in \mathbb{R}$, then $\lambda g \in \mathrm{Lip}(A, |\lambda|\sigma)$ and if $\lambda > 0$, then

$$\begin{aligned}(\lambda g)^*(x) &= \sup\{\lambda g(a) - |\lambda|\sigma d(x,a) \mid a \in A\}\\ &= \lambda \sup\{g(a) - \sigma d(x,a) \mid a \in A\}\\ &= \lambda g^*(x).\end{aligned}$$

(iii) If $\lambda < 0$, then

$$\begin{aligned}(\lambda g)^*(x) &= \sup\{\lambda g(a) - |\lambda|\sigma d(x,a) \mid a \in A\}\\ &= \sup\{\lambda g(a) - (-\lambda)\sigma d(x,a) \mid a \in A\}\\ &= \sup\{\lambda(g(a) + \sigma d(x,a)) \mid a \in A\}\\ &= \lambda \inf\{g(a) + \sigma d(x,a) \mid a \in A\}\\ &= \lambda {}^*g(x).\end{aligned}$$

Proposition 13. *Let $(X, \|.\|)$ be a normed space, A a non-trivial subspace of X such that (X, A) is a McShane-Whitney pair, and let $g \in \mathrm{Lip}(A, \sigma)$, for some $\sigma > 0$, be linear. Moreover, let $x_1, x_2, x \in X$ and $\lambda \in \mathbb{R}$.*

(i) $g^*(x_1 + x_2) \geq g^*(x_1) + g^*(x_2)$ and ${}^*g(x_1 + x_2) \leq {}^*g(x_1) + {}^*g(x_2)$.
(ii) If $\lambda > 0$, then $g^*(\lambda x) = \lambda g^*(x)$ and ${}^*g(\lambda x) = \lambda {}^*g(x)$.
(iii) If $\lambda < 0$, then $g^*(\lambda x) = \lambda {}^*g(x)$ and ${}^*g(\lambda x) = \lambda g^*(x)$.

Proof. In each case we show only one of the two facts.

(i) If $a_1, a_2 \in A$, then

$$\begin{aligned}{}^*g(x_1 + x_2) &= \inf\{g(a) + \sigma\|(x_1 + x_2) - a\| \mid a \in A\}\\ &\leq g(a_1 + a_2) + \sigma\|x_1 + x_2 - (a_1 + a_2)\|\\ &\leq g(a_1) + \sigma\|x_1 - a_1\| + g(a_2) + \sigma\|x_2 - a_2\|,\end{aligned}$$

therefore

$$\begin{aligned}{}^*g(x_1 + x_2) &\leq \inf\{g(a_1) + \sigma\|x_1 - a_1\| + g(a_2) + \sigma\|x_2 - a_2\| \mid a_1, a_2 \in A\}\\ &= \inf\{g(a_1) + \sigma\|x_1 - a_1\| \mid a_1 \in A\} + \inf\{g(a_2) + \sigma\|x_2 - a_2\| \mid a_2 \in A\}\\ &= {}^*g(x_1) + {}^*g(x_2).\end{aligned}$$

(ii) First we show that ${}^*g(\lambda x) \leq \lambda {}^*g(x)$. If $a \in A$, then

$$\begin{aligned}{}^*g(\lambda x) &= \inf\{g(a) + \sigma\|\lambda x - a\| \mid a \in A\}\\ &\leq g(\lambda a) + \sigma\|\lambda x - \lambda a\|\\ &= \lambda g(a) + |\lambda|\sigma\|x - a\|\\ &= \lambda(g(a) + \sigma\|x - a\|),\end{aligned}$$

therefore

$$\begin{aligned}{}^*g(\lambda x) &\leq \inf\{\lambda(g(a) + \sigma\|x - a\|) \mid a \in A\}\\ &= \lambda \inf\{g(a) + \sigma\|x - a\| \mid a \in A\}\\ &= \lambda {}^*g(x).\end{aligned}$$

For the inclusion $^*g(\lambda x) \geq \lambda^* g(x)$ we work as follows.

$$\lambda^* g(x) = \lambda \inf\{g(a) + \sigma ||x - a|| \mid a \in A\}$$
$$\leq \lambda(g(\frac{1}{\lambda}a) + \sigma ||x - \frac{1}{\lambda}a||)$$
$$= g(a) + \sigma |\lambda| ||x - \frac{1}{\lambda}a||$$
$$= g(a) + \sigma ||\lambda(x - \frac{1}{\lambda}a)||$$
$$= g(a) + \sigma ||\lambda x - a||,$$

therefore

$$\lambda^* g(x) \leq \inf\{g(a) + \sigma ||\lambda x - a|| \mid a \in A\} = {}^*g(\lambda x).$$

(iii) First we show that $^*g(\lambda x) \leq \lambda g^*(x)$. If $a \in A$, then

$$^*g(\lambda x) = \inf\{g(a) + \sigma ||\lambda x - a|| \mid a \in A\}$$
$$\leq g(\lambda a) + \sigma ||\lambda x - \lambda a||$$
$$= \lambda g(a) + |\lambda|\sigma ||x - a||$$
$$= \lambda(g(a) - \sigma ||x - a||),$$

therefore

$$^*g(\lambda x) \leq \inf\{\lambda(g(a) - \sigma ||x - a||) \mid a \in A\}$$
$$= \lambda \sup\{g(a) - \sigma ||x - a|| \mid a \in A\}$$
$$= \lambda g^*(x).$$

For the inclusion $^*g(\lambda x) \geq \lambda g^*(x)$ we work as follows. Since

$$g^*(x) = \sup\{g(a) - \sigma ||x - a|| \mid a \in A\}$$
$$\geq g(\frac{1}{\lambda}a) - \sigma ||x - \frac{1}{\lambda}a||$$

and $\lambda < 0$, we get

$$\lambda g^*(x) \leq \lambda(g(\frac{1}{\lambda}a) - \sigma ||x - \frac{1}{\lambda}a||)$$
$$= g(a) - \lambda \sigma ||x - \frac{1}{\lambda}a||$$
$$= g(a) + \sigma |\lambda| ||x - \frac{1}{\lambda}a||$$
$$= g(a) + \sigma ||\lambda(x - \frac{1}{\lambda}a)||$$
$$= g(a) + \sigma ||\lambda x - a||,$$

therefore

$$\lambda g^*(x) \leq \inf\{g(a) + \sigma ||\lambda x - a|| \mid a \in A\} = {}^*g(\lambda x).$$

Proposition 13 says that $^*g$ is sublinear and g^* is superlinear. If X is a normed space and $x_0 \in X$, then it is not generally the case that $\mathbb{R}x_0 := \{\lambda x_0 \mid \lambda \in \mathbb{R}\}$ is a located subset of X. If $X = \mathbb{R}$, this is equivalent to LPO, the limited principle of omniscience[2] (see [3], p. 122). Things change, if $||x_0|| > 0$. In this case $\mathbb{R}x_0$ is a 1-dimensional subspace of X i.e., a closed and located linear subset of X of dimension one (see [3], p. 307). Of course, $\mathbb{R}x_0$ is a convex subset of X. A standard corollary of the classical Hahn-Banach theorem is that if $x_0 \neq 0$, there is a bounded linear functional u on X such that $||u|| = 1$ and $u(x_0) = ||x_0||$. Its proof is based on the extension of the obvious linear map on $\mathbb{R}x_0$ to X through the Hahn-Banach theorem. Next follows a first approach to the translation of this corollary in Lipschitz analysis. First we need a simple lemma.

Lemma 1. If $(X, ||.||)$ is a normed space and $x_0 \in X$ such that $||x_0|| > 0$, then $\mathbb{I}x_0 := \{\lambda x_0 \mid \lambda \in [-1, 1]\}$ is a compact subset of X.

Proposition 14. If $(X, ||.||)$ is a normed space and $x_0 \in X$ such that $||x_0|| > 0$, there exists $f \in \mathrm{Lip}(X)$ such that $f(x_0) = ||x_0||$ and $L(f) = 1$.

Proof. The function $g : \mathbb{I}x_0 \to \mathbb{R}$, defined by $g(\lambda x_0) = \lambda ||x_0||$, for every $\lambda \in [-1, 1]$, is in $\mathrm{Lip}(\mathbb{I}x_0)$ and $L(g) = 1$; if $\lambda, \mu \in [-1, 1]$, then $|g(\lambda x_0) - g(\mu x_0)| = |\lambda||x_0|| - \mu||x_0||| = |\lambda - \mu|||x_0|| = ||(\lambda - \mu)x_0|| = ||\lambda x_0 - \mu x_0||$, and since

$$M_0(g) = \{\sigma_{\lambda x_0, \mu x_0}(g) = \frac{|g(\lambda x_0) - g(\mu x_0)|}{||\lambda x_0 - \mu x_0||} = 1 \mid (\lambda, \mu) \in [-1, 1]_0\},$$

we get that $L(g) = \sup M_0(g) = 1$. Since $\mathbb{I}x_0$ is inhabited and totally bounded, since by Lemma 1 it is compact, by Proposition 7(i) and Theorem 1 the extension $^*g$ of g is in $\mathrm{Lip}(X)$, while by Proposition 10 we have that $L(^*g) = L(g) = 1$.

Theorem 2. Let $(X, ||.||)$ be a normed space and $x_0 \in X$ such that $||x_0|| > 0$. If $(X, \mathbb{R}x_0)$ is a McShane-Whitney pair, there exist a sublinear Lipschitz function f on X such that $f(x_0) = ||x_0||$ and $L(f) = 1$, and a superlinear Lipschitz function h on X such that $h(x_0) = ||x_0||$ and $L(h) = 1$.

Proof. As in the proof of Proposition 14 the function $g : \mathbb{R}x_0 \to \mathbb{R}$, defined by $g(\lambda x_0) = \lambda ||x_0||$, for every $\lambda \in \mathbb{R}$, is in $\mathrm{Lip}(\mathbb{R}x_0)$ and $L(g) = 1$. Since $(X, \mathbb{R}x_0)$ is a McShane-Whitney pair, the extension $^*g$ of g is a Lipschitz function, and by Proposition 10 $L(^*g) = L(g) = 1$. Since g is linear, by Proposition 13 we get that $^*g$ is sublinear. Similarly, the extension g^* of g is a Lipschitz function, and by Proposition 10 $L(g^*) = L(g) = 1$. Since g is linear, by Proposition 13 we get that g^* is superlinear.

---

[2] From this we can explain why it is not constructively acceptable that any pair (X, A) is McShane-Whitney. If $x_0 \in \mathbb{R}$ and $(\mathbb{R}, \mathbb{R}x_0)$ is a McShane-Whitney pair, then by Proposition 8(i) we have that $\mathbb{R}x_0$ is located, which implies LPO.

4   Concluding Remarks

Similarly to Theorem 1, one can prove an extension theorem for Hölder continuous functions, or for functions which are continuous with respect to a given
modulus of continuity λ i.e., a function of type $[0, +\infty) \rightarrow [0, +\infty)$, which is
subadditive, strictly increasing, uniformly continuous on every bounded sub-
set of $[0, +\infty)$, and $\lambda(0) = 0$ (see also [3], p. 102). Note that one could have
defined a McShane-Whitney pair such that the functions g^* and $^*g$ are given
by $g^*(x) = \mathrm{glb}M_g(A, \sigma, x)$ and $^*g(x) = \mathrm{lub}M_g(A, -\sigma, x)$, for every $x \in X$,
respectively, since only the properties of glb and lub are used in the proof of
Theorem 1.

Some open problems related to the material presented here are the following:

a. To find necessary and sufficient conditions on X, Y and $f \in \mathrm{Lip}(X, Y)$ for
the L-pseudo-normability of f.
b. To find conditions on $(X, ||.||)$ under which one can show constructively that
$(X, \mathbb{R}x_0)$ is a McShane-Whitney pair, if $||x_0|| > 0$. A similar attitude is taken
by Ishihara in his constructive proof of the Hahn-Banach theorem, where the
property of Gâteaux differentiability of the norm is added (see [5,7], p.126).
c. To elaborate the Lipschitz version of the theory of the Hahn-Banach theorem.
d. If $(\mathbb{R}^n, A)$ is a McShane-Whitney pair and $g \in \mathrm{Lip}(A, \mathbb{R}^m, \sigma)$, then by
Theorem 1 there are extensions g^* and $^*g$ of g in $\mathrm{Lip}(\mathbb{R}^n, \mathbb{R}^m, \sqrt{m}\sigma)$.
According to the classical Kirszbraun theorem there is an extension of g in
$\mathrm{Lip}(\mathbb{R}^n, \mathbb{R}^m, \sigma)$ (see [9,15,16,18]). The constructive study of the Kirszbraun
theorem is a non-trivial enterprise.

References

1. Benyamini, Y., Lindenstrauss, J.: Geometric nonlinear functional analysis, vol. 1,
   American Mathematical Society Colloquium Publications, 48. American Mathematical Society, Providence, RI (2000)
2. Bishop, E.: Foundations of Constructive Analysis. McGraw-Hill, New York (1967)
3. Bishop, E., Bridges, D.: Constructive Analysis, Grundlehren der mathematischen
   Wissenschaften. 279. Springer, Heidelberg (1985)
4. Brudnyi, A., Brudnyi, Y.: Methods of geometric analysis in extension and trace
   problems, Volume 1. Monographs in Mathematics, vol. 102. Birkhäuser/Springer,
   Basel (2012)
5. Bridges, D.S., Vîţă, L.S.: Techniques of Constructive Analysis. Universitext.
   Springer, New York (2006)
6. Bridges, D.S., Richman, F.: Varieties of Constructive Mathematics. Cambridge
   University Press, Cambridge (1987)
7. Ishihara, H.: On the constructive Hahn-Banach theorem. Bull. London. Math. Soc.
   21, 79–81 (1989)
8. Julian, W., Philips, K.: Constructive bounded sequences and lipschitz functions.
   J. London Math. Soc. **s2–31**(3), 385–392 (1985)
9. Kirszbraun, M.D.: Über die zusammenziehende und Lipschitzsche Transformatio-
   nen. Fundam. Math. **22**, 77–108 (1934)

10. Loeb, I.: Lipschitz functions in constructive reverse mathematics. Logic J. IGPL **21**(1), 28–43 (2013). (special issue on Non-Classical Mathematics)
11. Mandelkern, M.: Constructive continuity, Mem. Amer. Math. Soc. 277 (1983)
12. McShane, E.J.: Extension of range of functions. Bull. Amer. Math. Soc. **40**(12), 837–842 (1934)
13. Petrakis, Iosif: A direct constructive proof of a stone-weierstrass theorem for metric spaces. In: Beckmann, Arnold, Bienvenu, Laurent, Jonoska, Nataša (eds.) CiE 2016. LNCS, vol. 9709, pp. 364–374. Springer, Cham (2016). doi:10.1007/978-3-319-40189-8_37
14. Tuominen, H.: Analysis in Metric Spaces, Lecture notes (2014)
15. Valentine, F.A.: On the extension of a vector function so as to preserve a Lipschitz condition. Bull. Amer. Math. Soc. **49**(2), 100–108 (1943)
16. Valentine, F.A.: A Lipschitz condition preserving extension for a vector function. Amer. J. Math. **67**, 83–93 (1945)
17. Weaver, N.: Lipschitz Algebras. World Scientific, Singapore (1999)
18. Wells, J.H., Williams, L.R.: Embeddings and Extensions in Analysis. Springer, Heidelberg (1975)
19. Whitney, H.: Analytic extensions of differentiable functions defined in closed sets. Trans. Amer. Math. Soc. **36**(1), 63–89 (1934)

Total Nondeterministic Turing Machines and a p-optimal Proof System for SAT

Zenon Sadowski[(✉)]

Institute of Informatics, University of Białystok,
ul. Ciołkowskiego 1 M, 15-245 Białystok, Poland
sadowski@math.uwb.edu.pl

Abstract. We show that the open problem of the existence of a p-optimal proof system for SAT can be characterized in terms of total nondeterministic Turing machines. We prove that there exists a p-optimal proof system for SAT if and only if there exists a proof system h for SAT such that for any total nondeterministic Turing machine working in polynomial time its totality is provable with short proofs in h and these proofs can be efficiently constructed. We prove that the standard proof system for SAT (a satisfying truth assignment is a proof of a satisfiable propositional formula) is p-optimal if and only if for any total nondeterministic Turing machine working in polynomial time its totality is provable with short proofs in the standard proof system for SAT and these proofs can be efficiently constructed.

Additionally we show that the problem of the existence of an optimal proof system for $TAUT$ can be characterized in terms of pairs of nondeterministic Turing machines which are disjoint (do not accept the same strings). We prove that there exists an optimal proof system for $TAUT$ if and only if there exists a proof system f for $TAUT$ such that for any pair of disjoint nondeterministic Turing machines working in polynomial time their disjointness is provable in f with short proofs.

Keywords: p-optimal proof system for SAT · An optimal propositional proof system · Total Turing machines · Disjoint **NP**-pairs

1 Introduction

Stephen Cook and Robert Reckhow were the first to precisely answer the question of what exactly is a propositional proof system. They introduced the notion of an abstract propositional proof system (see [7]) as a function f mapping in polynomial time the set of all strings over a certain fixed finite alphabet ("proofs") onto $TAUT$ (the set of all propositional tautologies). All propositional proof systems from logic textbooks fall under the concept of an abstract propositional proof system (a proof system for $TAUT$)

There are two tools for comparing the efficiency of proof systems for $TAUT$: the notion of p-simulation and its nondeterministic counterpart the notion of

© Springer International Publishing AG 2017
J. Kari et al. (Eds.): CiE 2017, LNCS 10307, pp. 364–374, 2017.
DOI: 10.1007/978-3-319-58741-7_34

simulation. A proof system for $TAUT$ is optimal (p-optimal) if and only if it simulates (p-simulates) any other proof system for $TAUT$.

The notion of an abstract proof system can be considered not only for $TAUT$, but also for any language L, in particular for SAT (the set of all satisfiable boolean formulas). To compare different proof systems for SAT we can use the notion of p-simulation introduced by Stephen Cook and Robert Reckhow (see [7]). Intuitively a proof system h for SAT p-simulates a second proof system g for SAT if there is a polynomial-time computable function t translating proofs in g into proofs in h. A proof system for SAT is p-optimal if and only if it p-simulates any other proof system for SAT. There are two main open problems concerning proof systems for SAT: the problem of the existence of a p-optimal proof system for SAT posed by Johannes Köbler and Jochen Messner (see [4,9, 10]) and the problem of whether the standard proof system for SAT (a satisfying truth assignment is a proof of a formula α) is p-optimal, posed by Pavel Pudlák (see [4,13]).

Both these problems were intensively studied and it seems impossible that we can prove or disprove them with the currently available means. In such a situation new characterizations of these problems can make some progress in understanding them.

A disjoint NP-pair is a pair of nonempty disjoint sets both belonging to the complexity class **NP** (see [14]). Olaf Beyersdorff proposed to express the disjointness of any disjoint NP-pair on a propositional level, as a sequence of propositional tautologies (see [2]). Any such a sequence he called a propositional representation of this disjoint NP-pair. Any disjoint NP-pair has many propositional representations. In this setting Olaf Beyersdorff formulated an unexpected characterization of the problem of the existence of an optimal proof system for $TAUT$. He proved that the existence of an optimal proof system for $TAUT$ is equivalent to the existence of a proof system for $TAUT$ which proves with short proofs the disjointness of any disjoint NP-pair with respect to all representations (see [2,3]). This result was a starting point to our research.

In this paper we characterize the problem of the existence of a p-optimal proof system for SAT, the problem of whether the standard proof system for SAT (a satisfying truth assignment is a proof of a satisfiable boolean formula) is p-optimal, and the problem of the existence of an optimal proof system for $TAUT$ in a similar, but slightly modified, manner. We provide the new characterization of the problem of the existence of a p-optimal proof system for SAT and the new characterization of the problem of whether the standard proof system for SAT is p-optimal. We also reformulate Olaf Beyersdorff's characterization of the problem of the existence of an optimal proof system for $TAUT$. The last section of the paper is devoted to this issue.

Our characterizations of these problems are in terms of nondeterministic polynomial-time Turing machines which obey special conditions, so called promises (see [5]). In case of a p-optimal proof system for SAT this promise is the totality of a nondeterministic polynomial-time Turing machine. This promise can be expressed on a propositional level in a nonuniform manner as a sequence of satisfiable boolean formulas. A nondeterministic polynomial-time

Turing machine is total if and only if it accepts the set of all strings over a certain fixed finite alphabet. The crucial observation is that deterministic polynomial-time transducers computing proof systems for SAT correspond to certain total nondeterministic polynomial-time Turing machines.

For a given nondeterministic polynomial-time Turing machine we construct, using the methods from the proof of Cook's Theorem (see [6]), the family of propositional formulas which in sum express the totality of this machine. Namely, all formulas from this family are satisfiable if and only if the machine is total.

Having such a family of propositional formulas we use proof systems for SAT to verify that a given nondeterministic polynomial-time Turing machine is total. For a given proof system h for SAT and for a total nondeterministic polynomial-time Turing machine N we say that the totality of N is strongly representable in h if and only if formulas expressing the totality of N are provable with short proofs in h and these proofs can be efficiently constructed. We prove that there exists a p-optimal proof system for SAT if and only if there exists a proof system for SAT such that the totality of any total nondeterministic polynomial-time Turing machine is strongly representable in it.

The hypothesis that in polynomial time one can find accepting computations of total nondeterministic polynomial-time Turing machines is on the list of propositions proved to be equivalent by Steven Fenner, Lance Fortnow, Ashish Naik and John Rogers (see [8]). They used the notation Q to represent the property that any or all of the hypotheses from this list are true. Olaf Beyersdorff, Johannes Köbler and Jochen Messner [4] proved that Q is equivalent to the p-optimality of the standard proof system for SAT.

In this paper we prove that the standard proof system for SAT is p-optimal if and only if the totality of any total nondeterministic polynomial-time Turing machine is strongly representable in it, so we add one more proposition to the list of Q-hypotheses.

2   Preliminaries

We assume some familiarity with basic complexity theory and refer the reader to [1] for standard notions and for definitions of complexity classes appearing in the paper.

The symbol Σ denotes a certain fixed finite alphabet throughout the paper. The set of all strings over Σ is denoted by $\Sigma^\star$. For a string x, $|x|$ denotes the length of x.

We use Turing machines (acceptors and transducers) as our basic computational model. We will not distinguish between a machine and its code. For a Turing machine M the symbol $L(M)$ denotes the language accepted by M.

We consider deterministic polynomial-time clocked Turing transducers with uniformly attached standard clocks that stop their computations in polynomial time (see [1]). We impose some restrictions on our encoding of these machines. From the code of any polynomial-time clocked Turing machine N we can easily detect (in polynomial time) the polynomial p_N which is its polynomial-time bound.

We consider only languages over the alphabet Σ (this means that, for example, boolean formulas have to be suitably encoded). The symbol $TAUT$ denotes the set (of encodings) of all propositional tautologies over a fixed adequate set of connectives, SAT denotes the set of all satisfiable boolean (propositional) formulas. Finally, $\langle ., \ldots, . \rangle$ denotes some standard polynomial-time computable tupling function.

3 Propositional Proof Systems

The concept of an abstract propositional proof system, subsuming all propositional proof systems used in practice, was introduced by S. Cook and R. Reckhow [7] in the following way:

Definition 1. *A proof system for $TAUT$ is a polynomial-time computable function $f : \Sigma^\star \xrightarrow{onto} TAUT$.*

A string w such that $f(w) = \alpha$ we call an f-proof of a formula α. We write $f \vdash^* \alpha_n$ if and only if $\{\alpha_n : n \geq 1\}$ is a sequence of tautologies with polynomial-size f-proofs. A polynomially bounded proof system for $TAUT$ (which allows short proofs to all tautologies) exists if and only if $\mathbf{NP} = \mathbf{co\text{-}NP}$ (see [7]).

Proof systems are compared according to their strength using the notion of simulation and the presumably stronger notion of p-simulation.

Definition 2 *(Krajíček, Pudlák). Let f, f' be two proof systems for $TAUT$. We say that f simulates f' if there exists a polynomial p such that for any $\alpha \in TAUT$, if α has a proof of length n in f', then α has a proof of length $\leq p(n)$ in f.*

The notions of an optimal proof system for $TAUT$ and a p-optimal proof system for $TAUT$ were introduced by J. Krajíček and P. Pudlák [11].

Definition 3. *A proof system for $TAUT$ is optimal (p-optimal) if and only if it simulates (p-simulates) any proof system for $TAUT$.*

The notion of a proof system can be considered not only for $TAUT$, but also for any language L, in particular for SAT.

Definition 4. *A proof system for SAT is a polynomial-time computable function $h : \Sigma^\star \xrightarrow{onto} SAT$.*

It follows from the above definition that for any proof system h for SAT there exists a polynomial-time clocked transducer M_h which computes h. To classify proof system for SAT we can use the notion of p-simulation introduced by Stephen Cook and Robert Reckhow (see [7]).

Definition 5 *(Cook, Reckhow). Let h, h' be two proof systems for SAT. We say that h p-simulates h' if there exists a polynomial-time computable function $\gamma : \Sigma^\star \longrightarrow \Sigma^\star$ such that for every $\alpha \in SAT$ and every $w \in \Sigma^\star$, if w is a proof of α in h', then $\gamma(w)$ is a proof of α in h.*

In other words, γ translates h'-proofs into h-proofs of the same formula.

Definition 6. *A proof system for SAT is p-optimal if and only if it p-simulates any proof system for SAT.*

We will study the problem of the existence of an p-optimal proof system for SAT and the problem of the existence of an optimal proof system for $TAUT$ from a computational-complexity perspective.

Proving our characterization of the problem of the existence of a p-optimal proof system for SAT we will use the following theorem proved by Jochen Messner (see [12], Proposition 16). It says that a p-optimal proof system for SAT is polynomial-time invertible on any easy (polynomial-time recognizable) subset of SAT.

Theorem 1 *(Jochen Messner). Let $S \subset SAT$ and $S \in \mathbf{P}$. If f is a p-optimal proof system for SAT then there exists a polynomial-time algorithm A_S such that for any $\alpha \in S$ this algorithm produces an $f - proof$ of α.*

4 Formulas Expressing the Totality of Nondeterministic Turing Machines

Let N be a nondeterministic Turing machine.

Definition 7. *We say that N is total if and only if N accepts Σ^*.*

To any nondeterministic polynomial-time Turing machine N and any $x \in \Sigma^*$ we shall assign the boolean formula β_x^N which will be used for verifying that N accepts x. The formula β_x^N will satisfy the following condition: β_x^N is satisfiable if and only if there exists an accepting computation N on input x. In sum, $\{\beta_x^N : x \in \Sigma^*\}$ is a set of satisfiable boolean formulas if and only if N is total.

Our construction of the formula β_x^N comes from Cook's proof that SAT is **NP**-complete (see [6]). The formula β_x^N is just Cook's formula for the pair $\langle N, x \rangle$. It follows from this construction that our formulas possess the following property.

Local uniformity property.

Let M be a fixed nondeterministic polynomial-time Turing machine. There exists a polynomial-time computable function f_M such that for any $x \in \Sigma^*$
$$f_M(x) = \beta_x^M.$$

5 Properties of Nondeterministic Turing Machines Representable in Proof Systems

Considering the problem of the existence of a p-optimal proof system for SAT we are concerned with the property of nondeterministic Turing machines called the totality. Using the expressibility of the property of being total as a sequence of satisfiable boolean formulas we can verify that a given nondeterministic Turing machine is total with the help of short proofs in some proof system for SAT. This leads to the following concept:

Let h be a proof system for SAT and let N be a polynomial-time nondeterministic Turing machine which is total.

Definition 8. *We say that the totality of N is strongly representable in h if and only if there exists a polynomial-time algorithm that on input x produces an h-proof of β_x^N for any string x.*

The key thing in this definition is the possibility to construct in polynomial time the proofs of the formulas stating that definite input strings are acceptable.

6   A New Characterization of the Problem of the Existence of a p-optimal Proof System for SAT

Polynomial-time clocked transducers producing all satisfiable boolean formulas (completeness) and only satisfiable boolean formulas (correctness) are machines computing proof systems for SAT. The polynomial-time clocked transducers producing only satisfiable boolean formulas correspond in a natural way to total nondeterministic Turing machines working in polynomial time. It can be explained in the following way:

Definition 9. *We say that a polynomial-time clocked deterministic transducer M behaves well on input x if and only if M on x outputs a satisfiable boolean formula.*

Let S be a fixed nondeterministic Turing machine working in polynomial time and accepting SAT. Let M be a polynomial-time clocked transducer. By $M \circ S$ we denote the nondeterministic Turing machine which on any input x runs M and then runs S on the output produced by M. We can see that $M \circ S$ accepts input string x if and only if M on input x outputs a satisfiable boolean formula, so M behaves well on any input if and only if $M \circ S$ is total.

Now we can formulate our characterization of the problem of the existence of a p-optimal proof system for SAT.

Theorem 2. *Statements (i) – (ii) are equivalent:*

(i) *There exists a p-optimal proof system for SAT.*
(ii) *There exists a proof system for SAT such that the totality of any total nondeterministic Turing machine working in polynomial time is strongly representable in it.*

Proof. $(i) \rightarrow (ii)$

Let Opt be a p-optimal proof system for SAT and let M be a total nondeterministic Turing machine working in polynomial-time. Let us define $A_M = \{\beta_x^M : x \in \Sigma^*\}$. We have $A_M \subset SAT$ because M is total. From the structure of Cook's reduction (as β_x^M clearly displays M and x) it follows that $A_M \in \mathbf{P}$.

According to Theorem 1, there exists a polynomial-time algorithm that on input β_x^M produces an Opt-proof of β_x^M for any $\beta_x^M \in A_M$. From this and from the local uniformity property of β_x^M formulas we conclude that there exists a polynomial-time algorithm that on input x produces an Opt-proof of β_x^M for any $x \in \Sigma^\star$, so the totality of the machine M is strongly representable in Opt.

$(ii) \rightarrow (i)$

Let h be a proof system for SAT such that the totality of any total nondeterministic Turing machine working in polynomial time is strongly representable in it. We say that a string $v \in \Sigma^\star$ is in *good form* if and only if

$$v = \langle M, x, Proof - \beta_x^{M \circ S}, 0^{p_M(|x|)} \rangle$$

where M is a polynomial-time clocked transducer with p_M time bound, $x \in \Sigma^\star$, $Proof - \beta_x^{M \circ S}$ is an h-proof of the formula $\beta_x^{M \circ S}$ expressing the fact that the machine $M \circ S$ accepts x, $0^{p_M(|x|)}$ is the sequence of zeros (padding).

Let us notice that if v is in good form, then M on input x produces a satisfiable boolean formula. We can prove it as follows. The formula $\beta_x^{M \circ S}$ is satisfiable as a formula possessing an h-proof. From this it follows that there is an accepting computation of $M \circ S$ on input x. From the definition of the machine $M \circ S$ we conclude that M on input x produces a satisfiable boolean formula.

Let α_0 be a certain satisfiable boolean formula. We define $Opt : \Sigma^\star \xrightarrow{onto} SAT$ in the following way: $Opt(v) = \alpha$ if v is in good form and α is a satisfiable boolean formula produced by M on input x, otherwise $Opt(v) = \alpha_0$. Clearly $Opt : \Sigma^\star \xrightarrow{onto} SAT$. Padding appearing in the definition of a string in a good form assures that Opt can be computed in polynomial time.

To prove the p-optimality of Opt, let g be a proof system for SAT computed by a polynomial-time clocked transducer K with time bound p_k. Since $K \circ S$ is a total nondeterministic polynomial-time Turing machine it follows that the totality of $K \circ S$ is strongly representable in h and in consequence there exists a polynomial-time algorithm that on input x produces an h-proof of $\beta_x^{K \circ S}$ for any x. The following polynomial-time computable function $t : \Sigma^\star \longrightarrow \Sigma^\star$ defined by

$$t(x) = \langle K, x, Proof - \beta_x^{K \circ S}, 0^{p_K(|x|)} \rangle$$

translates g-proofs into Opt-proofs. The symbol $Proof - \beta_x^{K \circ S}$ denotes an h-proof of the formula $\beta_x^{K \circ S}$. This completes the proof that Opt is a p-optimal proof system for SAT.

7   A New Characterization of the Problem of Whether the Standard Proof System for SAT Is p-optimal

By the standard proof system for SAT we mean the proof system for SAT in which the satisfying assignment is a proof of any satisfiable boolean formula α (see [4]).

The following hypothesis is on the list of equivalent complexity-theoretic conjectures presented by Stephen Fenner, Lance Fortnow, Ashish Naik and John Rogers in [8]:

For any total nondeterministic polynomial-time Turing machine N there exists a polynomial-time computable function g_N such that for all x, g_N outputs an accepting computation of N on x. We name this hypothesis as the main hypothesis on the list of Q-hypotheses.

Olaf Beyersdorff, Johannes Köbler and Jochen Messner proved that the p-optimality of the standard proof system for SAT is equivalent to the assertion Q, so in other words, the p-optimality of the standard proof system for SAT is equivalent to any of the Q-hypotheses from the list in [8]. Specifically, it concerns the main hypothesis on the list of Q-hypotheses.

Theorem 3 *(Olaf Beyersdorff, Johannes Köbler, Jochen Messner). Statements (i) – (ii) are equivalent:*

(i) The standard proof system for SAT is p-optimal.
(ii) For any total nondeterministic polynomial-time Turing machine N there exists a polynomial-time computable function g_N such that for all x, $g_N(x)$ outputs an accepting computation of N on x.

The following theorem presents our characterization of the problem of whether the standard proof system for SAT is p-optimal.

Theorem 4. *Statements (i) – (ii) are equivalent:*

(i) The standard proof system for SAT is p-optimal.
(ii) The totality of any total nondeterministic polynomial-time Turing machine is strongly representable in the standard proof system for SAT.

Proof. From Theorem 3 it follows that it is sufficient to show that (ii) is equivalent to the main hypothesis on the list of Q-hypotheses.

For the proof of the forward implication ((ii) → Q) observe, recalling the proof of Cook's theorem, that β_x^N is a propositional formula that encodes a computation of N on x, so given a satisfying assignment to β_x^N, some accepting computation of N on x can be reconstructed from it in polynomial time.

For the converse implication, observe, again recalling the proof of Cook's theorem that if there exists a polynomial-time computable function g_N such that for all x, $g_N(x)$ outputs an accepting computation of N on x, the satisfying assignment of β_x^N can be easily produced from this accepting computation.

Note that we have actually proved that the statement (ii) from the last theorem can be added to the list of Q-hypotheses.

8   A Characterization of the Problem of the Existence of an Optimal Proof System for TAUT

O. Beyersdorff addressed the problem of the existence of an optimal proof system for $TAUT$ from a completely logic oriented perspective. Defining propositional representations of **NP**-sets he has not referred to nondeterministic Turing machines accepting this sets. We reformulate his characterization of the problem of the existence of an optimal proof system for $TAUT$. Our characterization of this problem is in terms of pairs of disjoint nondeterministic Turing machines working in polynomial time, so the disjointness is the promise that, in our characterization, pairs of nondeterministic Turing machines should obey. This promise can be expressed on a propositional level as a sequence of propositional tautologies. Using nondeterministic Turing machines we replaced a logic-oriented perspective connected with Olaf Beyersdorff's characterization with a complexity theoretic-perspective.

Let (M, N) be a pair of nondeterministic polynomial-time Turing machines.

Definition 10. *We say that the machines M and N are disjoint if and only if there does not exist any input that is accepted by both machines M and N.*

Any pair of disjoint nondeterministic polynomial-time Turing machines accepts a disjoint **NP**-pair. So we can say that pairs of disjoint nondeterministic polynomial-time Turing machines is the machine model of the promise complexity class Disjoint **NP**-Pairs.

Definition 11. *Let (M, N) be a pair of nondeterministic polynomial-time Turing machines and let n be a natural number. We say that the machines M and N are n-disjoint if and only if for any input x of length n there does not exist simultaneously a computation of M accepting x and a computation of N accepting x.*

To any pair (M, N) of nondeterministic polynomial-time Turing machines and any n natural we shall assign the boolean formula $Disj_n^{M,N}$ satisfying the following condition: $Disj_n^{M,N}$ is a propositional tautology if and only if the pair (M, N) is n-disjoint.

Adapting the methods from the proof of Cook's Theorem we can construct the propositional formula α_n^M which satisfies the condition: α_n^M is satisfiable if and only if there exists a string w of length n such that M accepts w. Similarly, we can construct the propositional formula α_n^N which is satisfiable if and only if there exists a string w of length n such that N accepts w. Let G_0^n be the propositional formula satisfying the condition: G_0^n is satisfiable if and only if at the first step of the computation of M, the first n cells of the tape of this machine contain the same symbols as the respective cells of the machine N at the first step of its computation. We define $Disj_n^{M,N} \iff \neg(G_0^n \wedge \alpha_n^M \wedge \alpha_n^N)$. For details of the construction of these formulas see [15]. In sum $\{Disj_n^{M,N} : n \in \mathbf{N}\}$ is a set of propositional tautologies if and only if the machines M and N are disjoint.

Using the expressibility of the property of being disjoint as a sequence of propositional tautologies we can verify that a pair of nondeterministic machines is disjoint with the help of short proofs in some proof system for $TAUT$. This leads to the following definition:

Definition 12. *Let f be a proof system for $TAUT$ and let (M, N) be a pair of disjoint nondeterministic polynomial-time Turing machines. We say that the disjointness of M and N is representable in f if and only if $f \vdash^* Disj_n^{M,N}$.*

Our reformulation of Olaf Beyersdorff's characterization of the problem of the existence of an optimal proof system for $TAUT$ is as follows:

Theorem 5. *Statements (i) – (ii) are equivalent:*

(i) There exists an optimal proof system for $TAUT$.

(ii) There exists a proof system for $TAUT$ such that the disjointness of any pair of disjoint nondeterministic polynomial-time Turing machines is representable in it.

It seems that this characterization corresponds to Krajíček - Pudlák's characterization of the problem of the existence of an optimal proof system for $TAUT$ in terms of finitistic consistency statements (see [11]).

Acknowledgements. I would like to thank the anonymous referees for all their helpful comments on how to improve the paper.

References

1. Balcazar, J., Díaz, J., Gabarró, J.: Structural Complexity I. Springer, Heidelberg (1995)
2. Beyersdorff, O.: Classes of representable disjoint NP-pairs. Theoret. Comput. Sci. **377**, 93–109 (2007)
3. Beyersdorff, O.: Tuples of disjoint NP-sets. Theory Comput. Syst. **43**(2), 118–135 (2008)
4. Beyersdorff, O., Köbler, J., Messner, J.: Nondeterministic functions and the existence of optimal proof systems. Theoret. Comput. Sci. **410**, 3839–3855 (2009)
5. Beyersdorff, O., Sadowski, Z.: Do there exist complete sets for promise classes? Math. Logic Q. **57**(6), 535–550 (2011)
6. Cook, S.: The complexity of theorem proving procedures. In: Proceedings of the 3rd ACM Symposium on Theory of Computing, pp. 151–158 (1971)
7. Cook, S., Reckhow, R.: The relative efficiency of propositional proof systems. J. Symbolic Logic **44**, 36–50 (1979)
8. Fenner, S., Fortnow, L., Naik, A., Rogers, J.: Inverting onto functions. Inf. Comput. **186**, 90–103 (2003)
9. Köbler, J., Messner, J.: Complete problems for promise classes by optimal proof systems for test sets. In: Proceedings of the 13th IEEE Conference on Computational Complexity CCC 1998, pp. 132–140 (1998)
10. Köbler, J., Messner, J., Torán, J.: Optimal proof systems imply complete sets for promise classes. Inf. Comput. **184**(1), 71–92 (2003)
11. Krajíček, J., Pudlák, P.: Propositional proof systems, the consistency of first order theories and the complexity of computations. J. Symbolic Logic **54**, 1063–1079 (1989)
12. Messner, J.: On optimal algorithms and optimal proof systems. In: Meinel, C., Tison, S. (eds.) STACS 1999. LNCS, vol. 1563, pp. 541–550. Springer, Heidelberg (1999). doi:10.1007/3-540-49116-3_51
13. Pudlák, P.: Incompleteness in the finite domain. The Czech Academy of Science, Institute of Mathematics, Preprint No. 5–2016, Praha (2016)

14. Razborov, A.: On provably disjoint NP-pairs. Technical report 94–006, ECCC (1994)
15. Sadowski, Z.: On an optimal quantified propositional proof system and a complete language for NP ∩ co-NP. In: Chlebus, B.S., Czaja, L. (eds.) FCT 1997. LNCS, vol. 1279, pp. 423–428. Springer, Heidelberg (1997). doi:10.1007/BFb0036203

A One-Dimensional Physically Universal Cellular Automaton

Ville Salo and Ilkka Törmä[✉]

Department of Matematics and Statistics,
University of Turku, Turku, Finland
{vosalo,iatorm}@utu.fi

Abstract. Physical universality of a cellular automaton was defined by Janzing in 2010 as the ability to implement an arbitrary transformation of spatial patterns. In 2014, Schaeffer gave a construction of a two-dimensional physically universal cellular automaton. We construct a one-dimensional version of the automaton and a reversibly universal automaton.

Keywords: Cellular automaton · Physical universality · Reversibility

1 Introduction

A cellular automaton (CA) is a finite or infinite lattice of deterministic finite state machines with identical interaction rules, which, at discrete time steps, update their states simultaneously based on those of their neighbors. They are an idealized model of massively parallel computation. From another point of view, the local updates can be seen as particle interactions, and the CA is then a kind of physical law, or dynamics, governing the universe of all state configurations.

We study the notion of *physical universality* of cellular automata, introduced by Janzing in [4], which combines the two viewpoints in a novel and interesting way. Intuitively, a cellular automaton is physically universal if, given a finite subset D of the lattice and a function h on the shape-D patterns over the states of the CA, one can build a 'machine' in the universe of the CA that, under the dynamics of the CA, decomposes any given pattern P and replaces it by $h(P)$.

A crucial point in this definition is that we need to perform arbitrary computation on all patterns, not only carefully constructed ones. This has quite serious implications: The machine M that takes apart an arbitrary pattern of shape D and replaces it by its image under an arbitrary function h is not in any way special, and in particular we are not allowed to have separate 'machine states' and 'data states' with the former operating on the latter. Instead, we can also think of M as a pattern, and must be able to construct a larger machine M' that takes M apart and reassembles it in an arbitrary way.

Research supported by the Academy of Finland Grant 131558.

V. Salo was partially supported by CONICYT Proyecto Anillo ACT 1103.

© Springer International Publishing AG 2017
J. Kari et al. (Eds.): CiE 2017, LNCS 10307, pp. 375–386, 2017.
DOI: 10.1007/978-3-319-58741-7_35

This notion differs essentially from most existing notions of universality for CA such as *intrinsic universality* [7], universality in terms of traces as discussed by Delvenne et al. in [3], and the several different versions of computational universality [2,6]. In these notions, one can usually implement the computations and simulations in a well-behaved subset of configurations. Physical universality bears more resemblance to the *universal constructor machines* of Von Neumann [12], which construct copies of themselves under the dynamics of a particular cellular automaton, and were the initial motivation for the definition of CA. Another property of CA with a similar flavor is *universal pattern generation* as discussed in [5], meaning the property of generating all finite patterns from a given simple initial configuration.

In Janzing's work [4] some results were already proved about physically universal CA, but it was left open whether such an object actually exists. A two-dimensional physically universal CA was constructed by Schaeffer in [8] (see also [9,10]), but it was left open whether this CA can be made one-dimensional. We construct such a CA, solving the question in the positive. We also outline the construction of a *reversibly physically universal* CA, solving another open problem of [8].

2   Definitions

We now define the terms and notation used in this article. Let A be a finite set, called the *state set*, and $\mathbb{Z}^d$ a d-dimensional grid; we will mostly restrict our attention to the case $d = 1$. We call $A^{\mathbb{Z}^d}$ the *d-dimensional full shift over A*, whose elements are *configurations*.

A *cellular automaton* (CA for short) is a map $f : A^{\mathbb{Z}^d} \to A^{\mathbb{Z}^d}$ defined by a finite *neighborhood* $\{n_1, \ldots, n_k\} \subset \mathbb{Z}^d$ and a *local function* $F : A^k \to A$, so that $f(x)_v = F(x_{v+n_1}, \ldots, x_{v+n_k})$ holds for all $x \in A^{\mathbb{Z}^d}$ and $v \in \mathbb{Z}^d$. It is *reversible* if there is another CA $g : A^{\mathbb{Z}^d} \to A^{\mathbb{Z}^d}$ such that $f \circ g = g \circ f = \mathrm{id}$. An example of a reversible CA is the *shift by $n \in \mathbb{Z}^d$*, defined by $\sigma^n(x)_v = x_{v+n}$.

Other examples of reversible CA can be constructed as follows. Let the state set A be a Cartesian product $\prod_{i=1}^k A_i$ with projection maps $\pi_i : A \to A_i$, let $n_1, \ldots, n_k \in \mathbb{Z}^d$ be arbitrary vectors, and let $\gamma : A \to A$ be a bijection. Then the CA f defined by $f(x)_v = \gamma(\pi_1(x_{v+n_1}), \ldots, \pi_k(x_{v+n_k}))$ is reversible. We call f a *partitioned CA*. The components A_i are called the *tracks* of f, and the numbers n_i are *shifts* associated to the tracks. In the CA, the tracks are first shifted individually by the vectors n_i, and then the bijection γ is applied to every cell.

A CA f is *physically universal* if the following condition holds. For all finite domains $D, E \subset \mathbb{Z}^d$, and all functions $h : A^D \to A^E$, there exists a partial configuration $x \in A^{\mathbb{Z}^d \setminus D}$ and a time step $t \in \mathbb{N}$ such that for all $P \in A^D$, we have $f^t(x \cup P)_E = h(P)$. We think of the partial configuration x as a 'gadget' that implements the function h: if any pattern $P \in A^D$ is placed in the unspecified part of x and the CA f is applied exactly t times, the image $h(P)$ appears on the domain E. We say f is *efficiently physically universal* if t is polynomial in the diameter of $D \cup E$ and the computational complexity of h according to some 'reasonable' complexity measure. In this article, we use circuit complexity, or more precisely, the number of binary NAND gates needed to implement h. One could reasonably require also that the configuration x is computed efficiently from the circuit presentation of the function h, which can be seen as a uniformity condition in the sense of circuit complexity. Our proof gives a polynomial time algorithm for this. See Sect. 9 for a discussion.

3   The Cellular Automaton

Our physically universal automaton is a partitioned CA f defined as follows.

- The state set is $A = \{0,1\}^4$.
- The shifts are 2, 1, −1 and −2, and we denote $S = \{2,1,-1,-2\}$.
- For each $a, b \in \{0,1\}$ bijection γ maps the state $(1, a, b, 1)$ to $(1, b, a, 1)$, and $(a, 1, 1, b)$ to $(b, 1, 1, a)$. Everything else is mapped to itself.

Alternatively, f is the CA with neighborhood $\{-2, -1, 1, 2\}$ and local rule

$$(a, b, c, d) \mapsto \begin{cases} (\pi_4(d), 1, 1, \pi_1(a)), & \text{if } \pi_2(b) = \pi_3(c) = 1, \\ (1, \pi_3(c), \pi_2(b), 1), & \text{if } \pi_1(a) = \pi_4(d) = 1, \\ (\pi_1(a), \pi_2(b), \pi_3(c), \pi_4(d)), & \text{otherwise} \end{cases}$$

where $a, b, c, d \in A$ are length-4 Boolean vectors.

Intuitively, in the CA f there are four kinds of particles, which are represented by 1-symbols on the four tracks: fast right (track 1, speed 2), slow right (track 2, speed 1), slow left (track 3, speed −1) and fast left (track 4, speed −2). At most one particle of each kind can be present in a cell. On each step, every particle jumps from its position $n \in \mathbb{Z}$ to $n + s$, where $s \in S$ is its speed. After that, if two fast or two slow particles are present in the same cell, then the direction of every particle of the other speed is reversed. Such interactions are depicted in Fig. 1. This resembles the two-dimensional CA of Schaeffer, where particles move in four directions (NE, NW, SE, SW) with speed one, and the head-on collision of two particles causes other particles in the same cell to make a u-turn.

4   The Logical Cellular Automaton

For the proof of physical universality, we define another CA on an infinite state set. Denote the *ternary conditional operator* by $p(a, b, c) = (a \wedge b) \vee (\neg a \wedge c)$ for $a, b, c \in \{0,1\}$. That is, $p(a, b, c)$ is equal to b if $a = 1$, and to c otherwise. In many programming languages, $p(a, b, c)$ is denoted by $a \,?\, b : c$.

Definition 1. *Let $\mathcal{V} = \{\alpha_1, \alpha_2, \ldots\}$ be an infinite set of variables, and denote by $\mathcal{F}$ the set of Boolean functions over finitely many variables of $\mathcal{V}$. The* logical extension *of f is the CA-like function $\hat{f} : \hat{A}^{\mathbb{Z}} \to \hat{A}^{\mathbb{Z}}$ on the infinite state set $\hat{A} = \mathcal{F}^4$, where the four tracks are first shifted as in f, and then the function*

$$(a, b, c, d) \mapsto (p(b \wedge c, d, a), p(a \wedge d, c, b), p(a \wedge d, b, c), p(b \wedge c, a, d))$$

is applied to each coordinate. A valuation *is a function $v : \mathcal{V} \to \{0,1\}$. It extends to $\mathcal{F}$ and then into a function $v : \hat{A}^{\mathbb{Z}} \to A^{\mathbb{Z}}$ in the natural way.*

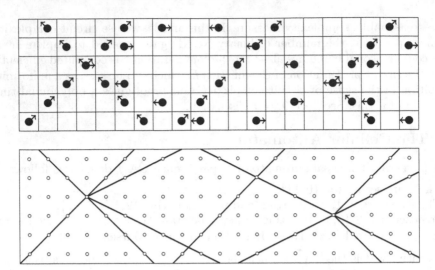

Fig. 1. A sample spacetime diagram of f showing different particle interactions, and a schematic version below it. Time increases upward. Particles are represented by arrowed bullets. For example, a bullet with arrows to the east, northeast and northwest represents three particles moving at speeds 2, 1 and -1, respectively.

The logical extension simulates multiple spacetime diagrams of f: one can see that the definition of $\hat{f}$ is equal to that of f, except that each particle is replaced by a Boolean formula that corresponds to the conditional presence or absence of a particle. We think of A as a subset of $\hat{A}$ containing the constant 0 or constant 1 function in each track. Note that $\hat{f}$ is also reversible, and we denote by $\hat{f}^{-1}$ its inverse function. See Fig. 2 for a spacetime diagram of $\hat{f}$.

The following result holds basically by construction.

Lemma 1. *Let $x \in \hat{A}^{\mathbb{Z}}$ be a configuration, and let $v : \mathcal{V} \to \{0,1\}$ be a valuation, so that $v(x) \in A^{\mathbb{Z}}$. Then for all $t \in \mathbb{Z}$ we have $f^t(v(x)) = v(\hat{f}^t(x))$.*

The idea of the proof of physical universality of f using this new CA is the following. We may assume that $D = E = [0, n-1]$ in the definition of physical universality, for some $n \in \mathbb{N}$. Then, we construct a spacetime diagram of $\hat{f}$ with the following properties. First, in the initial configuration $x \in \hat{A}^{\mathbb{Z}}$, the cells of the interval $[0, n-1]$ contain $4n$ distinct variables from $\mathcal{V}$. All other cells of x contain either 0 or 1. There also exists $t > 0$ such that $\hat{f}^t(x)_{[0,n-1]}$ contains the Boolean functions computing the function h in the definition of physical universality. In the course of the construction, we define which cells of x contain a 1.

The construction proceeds in five stages, which are depicted in Fig. 3:

– *Diffusion Stage*, where the Boolean particles in the input area D disperse into the environment. This stage does not require any auxiliary particles, as Lemma 2 in Sect. 5 will show.

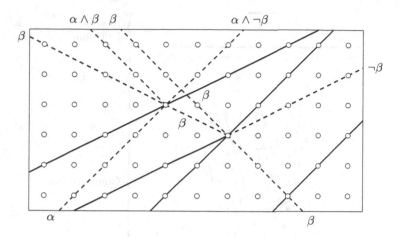

Fig. 2. A schematic spacetime diagram of $\hat{f}$, where $\alpha, \beta \in \mathcal{F}$. Particles are represented by solid lines, and Boolean particles by dotted lines. Note that Boolean particles can be created, even though f itself conserves the number of particles.

- *Collection Stage*, where the dispersed particles are rerouted to travel in the same direction at the same speed. This stage is implemented in Sect. 6.2.
- *Computation Stage*, where the NAND-gates of the circuit representing h are applied one by one to the travelling Boolean particles. This stage is implemented in Sect. 6.3.
- *Assembly Stage*, where the Boolean particles resulting from the computation are rerouted again toward the output area E.
- *Reverse Diffusion Stage*, where the particles converge on the output area and produce the desired pattern encoding h exactly at time t.

Definition 2. *We introduce the following terminology for the construction.*

- *The* configuration of interest, *denoted by* $x \in \hat{A}^{\mathbb{Z}}$, *initially contains the 'fully general' state* $(\alpha_{4i}, \alpha_{4i+1}, \alpha_{4i+2}, \alpha_{4i+3})$ *in every cell* $i \in [0, n-1]$, *and 0 everywhere else. During the construction, we change the cells of* x *from 0 to 1, but keep referring to it as* x, *so some of the definitions below depend on the stage of the proof. The* spacetime diagram of interest *is defined similarly.*
- *A* (spacetime) position *is an element of* $\mathbb{Z} \times S$ *($\mathbb{Z} \times \mathbb{Z} \times S$, respectively), representing a (spacetime) point that may contain a particle of certain speed. Note that time is bi-infinite, since our cellular automata are reversible.*
- *There is a* Boolean particle *at spacetime position* (i, t, s) *if* $\pi_s(\hat{f}^t(x)_i)$ *is not the constant 0 function, and a* particle *if it is the constant 1 function.*
- *There is a* collision *at coordinate* $(i, t) \in \mathbb{Z} \times \mathbb{Z}$ *if* $\hat{f}^t(x)_i$ *contains at least three Boolean particles, and a* crossing *if there are at least two Boolean particles.*
- *The* input *is the pattern* $x_{[0,n-1]} \in \hat{A}^n$.
- *The* gadget *is the contents of* x *outside* $[0, n-1]$, *an element of* $A^{\mathbb{Z} \setminus [0,n-1]}$.

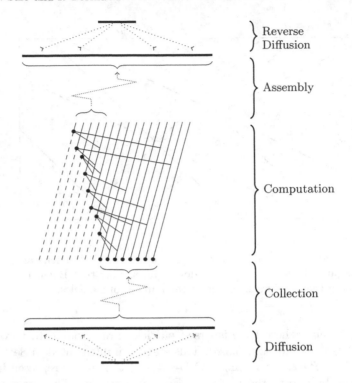

Fig. 3. A schematic diagram of the construction, not drawn to scale.

- A line *is a subset of* $\mathbb{Z} \times \mathbb{N}$ *of the form* $L = L(i,t,s) = \{(i+st,t) \mid t \in \mathbb{N}\}$ *for some speed* $s \in S$ *and* $i \in \mathbb{Z}$. *It is* occupied *(in a time interval* $I \subset \mathbb{Z}$*) if one of its coordinates (in the region* $\mathbb{Z} \times I$*) contains a crossing or a Boolean particle of speed* s. *We denote* $L^{(t)} = i + st$ *and* $L^t = (L^{(t)}, t, s)$. *The set of occupied lines in the spacetime diagram of interest is denoted* $\mathcal{L}_{\mathrm{occ}}$.

For example, there are three crossings in Fig. 2, two of which are collisions. The highest intersection of two dashed lines is not a crossing, as it does not take place at an actual coordinate (one of the white circles). Every line segment in the figure defines an infinite occupied line.

5   The Diffusion Lemma

As stated above, we initialize the gadget to the all-0 partial configuration, in which situation we have the following lemma. It states that any finite set of particles in the CA f eventually stop interacting and scatter to infinity. The corresponding result is proved for the physically universal CA of [8] by considering an *abstract CA* over the state set $\{0, \frac{1}{2}, 1\}$, with the interpretation that $\frac{1}{2}$ can be either 0 or 1, and this lack of information is suitably propagated in collisions. In our version, the role the new state $\frac{1}{2}$ is played by Boolean particles.

Lemma 2 (Diffusion Lemma). *Let $x \in \hat{A}^{\mathbb{Z}}$ be such that $x_i = 0$ for all $i \notin [0, n-1]$. Then there are $O(n^2)$ crossings in the two-directional spacetime diagram of x under $\hat{f}$ that all happen in a time window of length $O(n)$. For all other times $t \in \mathbb{Z}$, there are $O(n)$ Boolean particles in $\hat{f}^t(x)$, and they are contained in the interval $[-2|t|, n + 2|t|]$.*

Proof. We prove the claim in the positive direction of time. By induction, one sees that after any $t \geq 0$ steps, there can be no right-going Boolean particles in the cells $(-\infty, t-1]$, and no left-going particles in the cells $[n-t, \infty)$. After these sets intersect at time $\lceil n/2 \rceil$, there are no collisions, so no new Boolean particles are created. Thus the number of Boolean particles is at most $6n$, that is, twice the length of the segment of $\hat{f}^{\lceil n/2 \rceil}(x)$ that may contain Boolean particles. We may also have $O(n^2)$ crossings between Boolean particles going in the same direction with different speeds. Thus there are $O(n^2)$ crossings in total (Fig. 4).   □

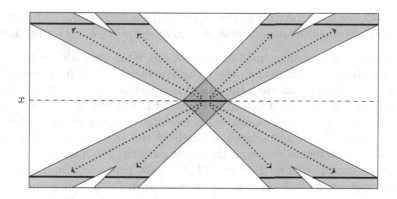

Fig. 4. An illustration of Lemma 2. The dashed line is the configuration x, and the thick segment is the interval $[0, n-1]$. All collisions take place in the dark gray area, and Boolean particles may occur in the light gray area. The eight horizontal line segments are the scattering Boolean particles, grouped by speed.

At this point, in the configuration of interest, we have an empty gadget, and the spacetime diagram contains $O(n)$ Boolean particles at any given time. Since the CA $\hat{f}$ is reversible, the values of the corresponding Boolean expressions determine the values of the original variables.

6   Manipulating the Spacetime Diagram

6.1   Controlled Modifications

In this section, we introduce new particles in the gadget that will collide with the existing Boolean particles and create new ones. This is called *scheduling collisions.* We never schedule a collision on an occupied line, and never add a Boolean particle on an existing crossing. This is formalized in the following definition.

Definition 3. *We say a modification of the gadget is* $(m, n, \mathcal{L}, \mathcal{T}, t, I)$-controlled *for numbers* $m, n \in \mathbb{N}$*, sets of lines* $\mathcal{L}$ *and* $\mathcal{T}$*, time* $t \in \mathbb{Z}$*, and interval* $I \subset \mathbb{Z}$*, if the following conditions hold:*

1. *the modification consists of adding at most* m *particles to the gadget,*
2. *at most* m *new occupied lines and* n *new crossings are introduced,*
3. *no existing crossings become collisions,*
4. *no line in* $\mathcal{L} \cup \mathcal{L}_{\mathrm{occ}}$ *is occupied by a new Boolean particle or a new crossing,*
5. *no spacetime position in the set*

$$F(\mathcal{T}, t, I) = \{(i, t, s') \mid i \in I, s' \in D\} \setminus \mathcal{T}^t \tag{1}$$

gets a new Boolean particle, and
6. *no line in* $\mathcal{T}$ *gets a collision after time* t*.*

If the conditions hold in the time interval $(-\infty, t]$ *(in particular, condition 6 need not hold at all), then the modification is* weakly $(m, n, \mathcal{L}, \mathcal{T}, t, I)$-controlled*.*

In practice, a controlled modification is one where we add to the gadget a finite number of particles that affect the spacetime diagram of interest only where we want it to be affected: the spacetime positions in $\mathcal{T}^t$. The numbers m and n control the amount of new objects added to the diagram (recall that our goal is the keep its size polynomial). An existing crossings should not become a collision, since that could affect the labels of the crossing Boolean particles, which may store information about the input pattern. The lines in $\mathcal{L} \cup \mathcal{L}_{\mathrm{occ}}$ and the positions near $\mathcal{T}$, that is, those in $F(\mathcal{T}, t, I)$, are 'protected' from accidentally obtaining any auxiliary Boolean particles created in the modification. The role of $\mathcal{L}$ and I is to guarantee an empty area in the spacetime diagram where we may reroute more Boolean particles at a later stage of the construction.

The notion of controlled modifications allows us to prove simple lemmas that affect one Boolean particle at a time, and use them repeatedly as 'black boxes' to perform more complex modifications without explicitly specifying the path of every particle in the spacetime diagram. This makes our proof more flexible, but also somewhat technical.

Definition 4. *Let* $j, t \in \mathbb{Z}$*. The* positive cone *rooted at* (j, t) *is the set of spacetime coordinates* $\mathcal{C}(j, t) = \{(j + j', t + t') \mid t' \in \mathbb{N}, j' \in [-t', t']\}$*.*

The following lemmas are parametrized by the numbers m_i for $i = 1, 2, 3, 4$. They denote, intuitively, the number of existing crossings, the number of occupied lines, the size of the additional 'forbidden area,' and the number of particles involved, respectively. Also, the expression 'if $t' = \Omega(N(m_1, m_2, m_3, m_4))$ is large enough, then $P(t')$ holds' for $t' \in \mathbb{N}$, a function $N : \mathbb{N}^4 \to \mathbb{N}$ and a property P means that there exists a number $T \leq K \cdot N(m_1, m_2, m_3, m_4) + K'$, where $K, K' > 0$ are constants independent of the m_i, such that $P(t')$ holds for all $t' \geq T$.

6.2   Moving the Boolean Particles

We now prove that we can add a finite number of particles to the gadget, so that after some number of steps, a collection of Boolean particles is 'moved' onto any desired lines, at the same time. We do this one particle at a time, and without interfering with the trajectories of any other existing particles. The statements of the following lemmas refer to labels of spacetime positions instead of Boolean particles, since we want to also handle 'Boolean particles' whose label happens to be 0. However, in the intuitive explanations we still refer to Boolean particles.

Lemma 3. *Suppose we have a spacetime diagram of interest with m_1 crossings and m_2 occupied lines, let L_p be a line that contains no collisions after time t, and let $\beta \in \mathcal{F}$ be the label of the spacetime position L_p^t. Let $\mathcal{L}$ be a collection of m_3 lines not containing L_p. Let $L \notin \mathcal{L}$ be an unoccupied line that passes through some spacetime coordinate $(j', t') \in \mathcal{C}(j, t)$ with $t' > t$. Let $\mathcal{T}$ be a set of $O(m_3)$ lines containing L, and let $I \subset \mathbb{Z}$ be an interval of length $O(m_3)$. If $t' = t + \Omega(m_1 + m_2 + m_3)$ is large enough, then there is an $(O(1), O(m_2), \mathcal{L}, \mathcal{T}, t', I)$-controlled modification after which the spacetime position $L^{t'}$ has label β. The same holds if the line L is unoccupied only in the time interval $(-\infty, t']$, but the modification is weakly controlled.*

The lemma intuitively states that we can almost completely freely change the position and direction of a Boolean particle by introducing a constant number of auxiliary particles, while preserving the already constructed gadget and introducing relatively few new crossings. The point of the set of forbidden lines $\mathcal{L}$ and positions I is that we can in fact move an arbitrary number of Boolean particles at the same time. The idea is that we move them by applying Lemma 3 repeatedly to one Boolean particle at a time, always adding the target lines of the remaining ones into the protected set $\mathcal{L}$. This guarantees that the lines are not accidentally occupied too early.

Corollary 1. *Suppose we have a spacetime diagram of interest with m_1 crossings and m_2 occupied lines, and for each $k = 1, \ldots, m_4$, a line L_{p_k} that contains no collisions after time t such that the spacetime position $L_{p_k}^t$ has label $\beta_k \in \mathcal{F}$. Let $\mathcal{L}$ be a collection of m_3 lines not containing any of the L_{p_k}. Let $L_k \notin \mathcal{L}$ be unoccupied and mutually disjoint lines that pass through some spacetime coordinates $(j_k', t') \in \mathcal{C}(L_{p_k}^{(t)}, t)$ with $t' > t$. Denote $\mathcal{T} = \{L_1, \ldots, L_{m_4}\}$, and let $I \subset \mathbb{Z}$ be an interval of length $O(m_3)$. If $t' = t + \Omega(m_1 + m_3 + m_4(m_2 + m_4))$ is large enough, then there is an $(O(m_4), O(m_4(m_2 + m_4)), \mathcal{L}, \mathcal{T}, t', I)$-controlled modification after which each spacetime position $L_k^{t'}$ has label β_k. The same holds if the lines L_k are unoccupied only in the time interval $(-\infty, t']$, but the modification is weakly controlled.*

6.3   Computing with the Boolean Particles

Next, we will do some computation with the Boolean particles. Namely, we show that the NAND of two Boolean particles can be computed nondestructively in any spacetime position, as long as we have enough time, and the target line is in the intersection of the cones rooted at the input particles.

Lemma 4. *Suppose we have a spacetime diagram of interest with m_1 crossings and m_2 occupied lines, and two distinct lines L_1 and L_2 of direction 1 containing no collisions after time t such that the spacetime coordinates L_1^t, L_2^t have labels $\beta_1, \beta_2 \in \mathcal{F}$. Let $\mathcal{L}$ be a set of m_3 lines not containing L_1 and L_2, and let $L \notin \mathcal{L}$ be an unoccupied line of slope 1 to the left of L_1 and L_2 that passes through some spacetime coordinate $(j', t') \in \mathcal{C}(L_1^{(t)}, t) \cap \mathcal{C}(L_2^{(t)}, t)$ with $t' > t$. Let $\mathcal{T} = \{L_1, L_2, L\}$. If $t' = t + \Omega(m_1 + m_2 + m_3)$ is large enough, then there is an $(O(1), O(m_2), \mathcal{L}, \mathcal{T}, t', \emptyset)$-controlled modification after which the spacetime positions $L_1^{t'}, L_2^{t'}$ and $L^{t'}$ have labels β_1, β_2 and $\neg(\beta_1 \wedge \beta_2)$, respectively.*

Similarly to the fact that any number of particles can be moved, we can now compute an arbitrary Boolean function, given enough time and space.

Corollary 2. *Suppose we have a spacetime diagram of interest with m_1 collisions and m_2 occupied lines. Let $j \in \mathbb{Z}$, and for every $k = 1, \ldots, m_4$, let $\beta_k \in \mathcal{F}$ be the label of the spacetime position $p_k = (j + k - 1, t, 1)$. Suppose further that the lines L_{p_k} passing through the p_k do not contain collisions after time t. Let $H : \{0, 1\}^{m_4} \to \{0, 1\}^m$ with $m = O(m_4)$ be a Boolean function realizable with C NAND-gates. Let $\mathcal{L}$ be a set of m_3 lines not containing the L_{p_k}, let $L_1, \ldots, L_{C+m} \notin \mathcal{L}$ be unoccupied lines of slope 1 that pass through the spacetime coordinates $(j - 1, t), \ldots, (j - C - m, t)$, and let $t' > t$. Let also $\mathcal{T} = \{L_C, \ldots, L_{C+m}\}$. If*

$$t' = t + \Omega((C + m_4)(C^2 + m_1 + m_3 + (C + m_4)(m_2 + m_4)))$$

is large enough, there is an $(O(C + m_4), O((C + m_2 + m_4)(C + m_4)), \mathcal{L}, \mathcal{T}, t', \emptyset)$-controlled modification after which the spacetime positions $L_{C+\ell}^{t'}$ have labels $H(\beta_1, \ldots, \beta_m)_\ell$.

7   Physical Universality

Combining the lemmas of the previous section, we obtain our main result, the efficient physical universality of f.

Theorem 1. *The cellular automaton f is efficiently physically universal.*

8   A Reversibly Physically Universal CA

In [8], other open questions were also posed, and here we answer one of them.

Call a CA $f : A^{\mathbb{Z}^d} \to A^{\mathbb{Z}^d}$ *reversibly physically universal*, if for any finite domain $D \subset \mathbb{Z}^d$ and any bijection $\theta : A^D \to A^D$, there exist partial configurations $x, y \in A^{\mathbb{Z}^d \setminus D}$ and a time t such that for all patterns $P \in A^D$ we have

$f^t(x \cup P) = y \cup \theta(P)$. In other words, the input cannot permanently affect the gadget during the computation.

The intuition behind this alternative notion is that after the CA has produced the pattern $\theta(P)$, the state of the 'computational machinery' should not depend on P: the entropy present in P does not permanently leak out of the domain D. It was conjectured by Schaeffer in [8] that a two-dimensional reversibly universal CA exists; we construct a one-dimensional one.

Our reversibly physically universal CA is based on the same idea as the physically universal one, but computation must be performed with reversible gates [11]. This, together with additional collision rules for modifying the speeds of particles, allows us to prevent information from spreading beyond the computation zone.

Theorem 2. *There exists a one-dimensional reversibly physically universal CA.*

9   Final Remarks

Our proof of the physical universality of f can be turned into a polynomial time algorithm that, given a circuit computing the function $h : A^D \to A^E$ in the definition of physical universality, computes the corresponding gadget and the polynomially bounded number t. The main issue with this is that occupied lines and crossings cannot be enumerated in polynomial time unless $P = NP$, since it requires checking whether an arbitrary NAND-expression is satisfiable. This issue can be avoided by keeping track of a set of lines $\mathcal{L}_{occ+}$ that contains at least all occupied lines, as well as a set of coordinates $X \subset \mathbb{Z}^2$ that contains at least all crossings. When invoking Lemmas 3 and 4 to modify the configuration of interest, we add to the protected set $\mathcal{L}$ all lines in $\mathcal{L}_{occ+}$ and all lines that pass through a coordinate in X. The set $\mathcal{L}_{occ+}$ is then amended with those lines that 'locally' appear to be occupied based on the added particles, and similarly for X. It can be verified that this approach does not change the asymptotic bounds in the construction, so the resulting configuration is polynomial in size.

The above algorithm will need polynomial space, as it compares the new positions of auxiliary particles to all existing ones. To construct the gadget in logarithmic space, it might be necessary to fix particular choices of where the auxiliary particles are put. We have chosen the more abstract route in the hope that our methods generalize more directly to a larger class of CA.

The existence of a physically universal CA was asked in [4] without fixing the number of states. Our CA has 16 states and radius 2. It would be interesting to find the minimal number of states and the minimal radii for physically universal CA. Of course, one can make any physically universal CA have radius 1 by passing to a blocking presentation, but this increases the number of states. From our CA, one obtains a physically universal radius-1 CA with 256 states.

Question 1. Are there physically universal CA on two states? Which combinations of state set and radius allow physical universality?

A long list of open questions about physical universality is also given in [8]. Finally, it would be interesting to explore the connections between physical universality and the other types of universalities mentioned in Sect. 1. For example, a reversible cellular automaton is not intrinsically universal, since it cannot simulate a non-reversible automaton, but could a physically universal CA be able to simulate all reversible automata?

Acknowledgments. We are thankful to Charalampos Zinoviadis for introducing this problem to us, and for many fruitful discussions on the proof, and Luke Schaeffer for his Golly implementation of our physically universal CA. We would also like to thank Scott Aaronson for popularizing the concept in his blog [1].

References

1. Aaronson, S.: Shtetl-Optimized - the blog of Scott Aaronson. http://www.scottaaronson.com/blog/?p=1896. Accessed 17 Sept 2014
2. Cook, M.: Universality in elementary cellular automata. Complex Syst. **15**(1), 1–40 (2004)
3. Delvenne, J.-C., Kurka, P., Blondel, V.D.: Decidability and universality in symbolic dynamical systems. Fundam. Inform. **74**(4), 463–490 (2006)
4. Janzing, D.: Is there a physically universal cellular automaton or Hamiltonian? ArXiv e-prints, September 2010
5. Kari, J.: Universal pattern generation by cellular automata. Theor. Comput. Sci. **429**, 180–184 (2012). Magic in Science
6. Neary, T., Woods, D.: P-completeness of cellular automaton Rule 110. In: Bugliesi, M., Preneel, B., Sassone, V., Wegener, I. (eds.) ICALP 2006. LNCS, vol. 4051, pp. 132–143. Springer, Heidelberg (2006). doi:10.1007/11786986_13
7. Ollinger, N.: The intrinsic universality problem of one-dimensional cellular automata. In: Alt, H., Habib, M. (eds.) STACS 2003. LNCS, vol. 2607, pp. 632–641. Springer, Heidelberg (2003). doi:10.1007/3-540-36494-3_55
8. Schaeffer, L.: A physically universal cellular automaton. In: Electronic Colloquium on Computational Complexity (ECCC), vol. 21, p. 84 (2014)
9. Schaeffer, L.: A physically universal cellular automaton. In: ITCS 2015– Proceedings of the 6th Innovations in Theoretical Computer Science, pp. 237–246 (2015)
10. Schaeffer, L.: A physically universal quantum cellular automaton. In: Kari, J. (ed.) AUTOMATA 2015. LNCS, vol. 9099, pp. 46–58. Springer, Heidelberg (2015). doi:10.1007/978-3-662-47221-7_4
11. Toffoli, T.: Reversible computing. In: de Bakker, J., van Leeuwen, J. (eds.) ICALP 1980. LNCS, vol. 85, pp. 632–644. Springer, Heidelberg (1980). doi:10.1007/3-540-10003-2_104
12. von Neumann, J.: Theory of Self-Reproducing Automata. University of Illinois Press, Champaign (1966)

Extending Wadge Theory to k-Partitions

Victor L. Selivanov[1,2(✉)]

[1] A.P. Ershov Institute of Informatics Systems SB RAS, Novosibirsk, Russia
vseliv@iis.nsk.su
[2] Kazan (Volga Region) Federal University, Kazan, Russia

Abstract. We extend some results about Wadge degrees of Borel subsets of Baire space to finite partitions of Baire space. A typical new result is the characterization up to isomorphism of the Wadge degrees of k-partitions with $\mathbf{\Delta}_3^0$-components.

Keywords: Baire space · Wadge reducibility · Lipschitz reducibility · Backtrack reducibility · k-partition · h-preorder · Well preorder · Infinite game

1 Introduction

For subsets A, B of the Baire space $\mathcal{N} = \omega^\omega$, A is *Wadge reducible* to B ($A \leq_W B$), if $A = f^{-1}(B)$ for some continuous function f on $\mathcal{N}$. The quotient-poset of the preorder $(P(\mathcal{N}); \leq_W)$ under the induced equivalence relation $\equiv_W$ on the power-set of $\mathcal{N}$ is called *the structure of Wadge degrees* in $\mathcal{N}$. W. Wadge [15,16] characterized the Wadge degrees of Borel sets up to isomorphism, in particular this poset is well-founded and has no 3 pairwise incomparable elements.

Let $2 \leq k < \omega$. By a *k-partition of $\mathcal{N}$* we mean a function $A : \mathcal{N} \to k = \{0, \ldots, k-1\}$ often identified with the sequence $(A_0, \ldots, A_{k-1})$ where $A_i = A^{-1}(i)$ are the components of A. Obviously, 2-partitions of $\mathcal{N}$ can be identified with the subsets of $\mathcal{N}$ using the characteristic functions. The set of all k-partitions of $\mathcal{N}$ is denoted $k^{\mathcal{N}}$, thus $2^{\mathcal{N}} = P(\mathcal{N})$. The Wadge reducibility on subsets of $\mathcal{N}$ is naturally extended to k-partitions: for $A, B \in k^{\mathcal{N}}$, $A \leq_W B$ means that $A = B \circ f$ for some continuous function f on $\mathcal{N}$. In this way, we obtain the preorder $(k^{\mathcal{N}}; \leq_W)$. For any pointclass $\mathbf{\Gamma} \subseteq P(\mathcal{N})$, let $\mathbf{\Gamma}(k^{\mathcal{N}})$ be the set of k-partitions of $\mathcal{N}$ with components in $\mathbf{\Gamma}$.

In contrast with the Wadge degrees of sets, the structure $(\mathbf{\Delta}_1^1(k^{\mathcal{N}}); \leq_W)$ for $k > 2$ has antichains of any finite size. Nevertheless, a basic property of the Wadge degrees of sets may be lifted to k-partitions, as the following very particular case of Theorem 3.2 in [4] shows:

Proposition 1. *For any $2 \leq k < \omega$, the structure $(\mathbf{\Delta}_1^1(k^{\mathcal{N}}); \leq_W)$ is a well preorder, i.e. it has neither infinite descending chains nor infinite antichains.*

Although this result gives an important information about the Wadge degrees of Borel k-partitions, it is far from a characterization. Our aim is to obtain

© Springer International Publishing AG 2017
J. Kari et al. (Eds.): CiE 2017, LNCS 10307, pp. 387–399, 2017.
DOI: 10.1007/978-3-319-58741-7_36

such a characterization, continuing a series of earlier partial results (see e.g. [5,11,12,14]). Our approach is to characterize the initial segments $(\mathbf{\Delta}^0_\alpha(k^\mathcal{N});$ $\leq_W)$ for bigger and bigger ordinals $2 \leq \alpha < \omega_1$. In [11] we have done this for $\alpha = 2$, here we treat the case $\alpha = 3$ (a finitary version of this case was considered in [12]), a general case was outlined in Sect. 5 of [14]. Our original contribution is the discovery of useful properties of natural operations on the k-partitions and of the structures of labeled forests.

Let $\bigoplus_i A_i$ be the disjoint union of a sequence of elements $A_0, A_1, \ldots$ of $k^\mathcal{N}$. Let $\mathcal{N}^+ := \{1, 2, \ldots\}^\omega$ and for $x \in \mathcal{N}^+$ let $x^- := \lambda i.x(i) - 1$, so $x^- \in \mathcal{N}$. Define the binary operation $+$ on $k^\mathcal{N}$ as follows: $(A + B)(x) := A(x^-)$ if $x \in \mathcal{N}^+$, otherwise $(A + B)(x) := B(y)$ where y is the unique element of $\mathcal{N}$ such that $x = \sigma 0 y$ for a unique finite sequence σ of positive integers. For any $i < k$, define a unary operation p_i on $k^\mathcal{N}$ by $p_i(A) := \mathbf{i} + A$ where $\mathbf{i} := \lambda x.i$ are the constant k-partitions (which are precisely the distinct minimal elements of $(k^\mathcal{N}; \leq_W)$). For any $i < k$, define a unary operation q_i on $k^\mathcal{N}$ (for $k = 2$, q_0 and q_1 coincide with the Wadge's operations $\sharp$ and $\flat$ from Sect. III.E of [16]) as follows: $q_i(A)(x) := i$ if x has infinitely many zeroes, $q_i(A)(x) := A(x^-)$ if x has no zeroes, and $q_i(A)(x) := A(y^-)$ otherwise where y is the unique element of $\mathcal{N}^+$ such that $x = \sigma 0 y$ for a string σ of non-negative integers. The introduced operations are correctly defined on Wadge degrees.

Our first result, which is proved with a heavy use of Proposition 1, characterizes some subalgebras of the Wadge degrees generated from the minimal degrees $\{0\}, \ldots, \{k - 1\}$. The item (1) below follows from results in [11] but the proof here is slightly different and easier to generalize.

Theorem 1

(1) *The quotient-poset of* $(\mathbf{\Delta}^0_2(k^\mathcal{N}); \leq_W)$ *is generated from the degrees* $\{0\}, \ldots, \{k - 1\}$ *by the operations* $\bigoplus, p_0, \ldots, p_{k-1}$.

(2) *The quotient-poset of* $(\mathbf{\Delta}^0_3(k^\mathcal{N}); \leq_W)$ *is generated from* $\{0\}, \ldots, \{k - 1\}$ *by the operations* $\bigoplus, +, q_0, \ldots, q_{k-1}$.

Our second result characterizes the structures above in terms of the homomorphism preorder on labeled forests [11,12,14]. Let $(Q; \leq)$ be a preorder. A Q-*poset* is a triple $(P, \leq, c)$ consisting of a countable nonempty poset $(P; \leq)$ without infinite chains, and a labeling $c : P \to Q$. A *morphism* $f : (P, \leq, c) \to (P', \leq', c')$ of Q-posets is a monotone function $f : (P; \leq) \to (P'; \leq')$ satisfying $\forall x \in P(c(x) \leq c'(f(x)))$. Let $\widetilde{\mathcal{F}}_Q$ and $\widetilde{\mathcal{T}}_Q$ denote the sets of all countable Q-forests and Q-trees without infinite chains, respectively. The h-*preorder* $\leq_h$ on $\widetilde{\mathcal{F}}_Q$ is defined as follows: $P \leq_h P'$, if there is a morphism from P to P'. If $Q = \bar{k}$ of the antichain with k elements $0, \ldots, k - 1$, we obtain the Q-preorders denoted by $\widetilde{\mathcal{F}}_k$ and $\widetilde{\mathcal{T}}_k$, respectively. We also need the preorder $\widetilde{\mathcal{F}}_{\widetilde{\mathcal{T}}_k}$.

Theorem 2

(1) *The quotient-posets of* $(\mathbf{\Delta}^0_2(k^\mathcal{N}); \leq_W)$ *and of* $(\widetilde{\mathcal{F}}_k; \leq_h)$ *are isomorphic.*

(2) *The quotient-posets of* $(\mathbf{\Delta}^0_3(k^\mathcal{N}); \leq_W)$ *and of* $(\widetilde{\mathcal{F}}_{\widetilde{\mathcal{T}}_k}; \leq_h)$ *are isomorphic.*

Again, item (1) above was established in [11] and is now extended to (2) (a "finitary" version of (2) is described in [12]). The proof of Theorem 2 generalizes those in [11,12]. It takes the h-quasiorders as natural naming systems for the subalgebras in Theorem 1, providing natural homomorphisms from the forest structures onto the corresponding degree structures. From the properties of the operations in Theorem 1 it follows by induction on the rank of the forests that these homomorphisms are in fact isomorphisms. Because of space bounds, we omit some proofs having published versions for the case of sets.

We do believe that the results above maybe extended to larger segments $(\Delta_\alpha^0(k^{\mathcal{N}}); \leq_W)$, $4 \leq \alpha < \omega_1$. Using the Kuratowski relativization technique [2,8,16], we can define for any $1 \leq \beta < \omega_1$ the binary operation $+_\beta$ on $k^{\mathcal{N}}$ such that $+_1$ coincides with $+$ and, for any $2 \leq \alpha < \omega_1$, the quotient-poset of $(\Delta_\alpha^0(k^{\mathcal{N}}); \leq_W)$ is generated from $\{\mathbf{0}\}, \ldots, \{\mathbf{k-1}\}$ by the operations $\bigoplus$ and $+_\beta$ for all $1 \leq \beta < \alpha$. The extension of Theorem 2 is obtained by defining suitable iterated versions of the h-quasiorder in the spirit of [14]. Since $\Delta_1^1(k^{\mathcal{N}}) = \bigcup_{\alpha<\omega_1} \Delta_\alpha^0(k^{\mathcal{N}})$, we obtain the characterization of Wadge degrees of Borel k-partitions. Note that in [9] we considered a classification of hyperarithmetical k-partitions of ω modulo m-reducibility which is in a precise sense the effective version of the Wadge degrees of k-partitions. Our "algebraic" approach to Wadge theory was motivated by the similar approach in [9].

2 Preliminaries

We use the standard set-theoretic notation. We identify the set of natural numbers with the first infinite ordinal ω. The first uncountable ordinal is denoted by ω_1. Let $\mathcal{N} = \omega^\omega$ be the set of all infinite sequences of natural numbers (i.e., of functions $x: \omega \to \omega$). Let ω^* be the set of finite sequences of elements of ω, including the empty sequence ε. For $\sigma, \tau \in \omega^*$ and $x \in \mathcal{N}$, we write $\sigma \sqsubseteq \tau$ (resp. $\sigma \sqsubseteq x$) to denote that σ is an initial segment of τ (resp. of x). By $\sigma x = \sigma \cdot x$ we denote the concatenation of σ and x, and by $\sigma \cdot \mathcal{N}$ the set of all extensions of σ in $\mathcal{N}$. For $x \in \mathcal{N}$, we write $x = x(0)x(1)\cdots$ where $x(i) \in \omega$ for each $i < \omega$. For $x \in \mathcal{N}$ and $n < \omega$, $x[n] = x(0) \ldots x(n-1)$ is the initial segment of x of length n.

By endowing $\mathcal{N}$ with the product of the discrete topologies on ω, we obtain the so-called *Baire space*. The product topology coincides with the topology generated by the collection of sets of the form $\sigma \cdot \mathcal{N}$ for $\sigma \in \omega^*$. We recall the well-known (see e.g. [6]) relation of closed subsets of $\mathcal{N}$ to trees. A *tree* is a non-empty set $T \subseteq \omega^*$ which is closed downwards under $\sqsubseteq$. A *leaf* of T is a maximal element of $(T; \sqsubseteq)$. By $\partial(T)$ we denote the set of minimal elements in $(\omega^* \setminus T; \sqsubseteq)$. A *pruned tree* is a tree without leafs. A *path through* a tree T is an element $x \in \mathcal{N}$ such that $x[n] \in T$ for each $n \in \omega$. For any tree T, the set $[T]$ of paths through T is closed in $\mathcal{N}$. For any non-empty closed set $A \subseteq \mathcal{N}$ there is a unique pruned tree T with $A = [T]$ and, moreover, there is a Lipschitz surjection $t : \mathcal{N} \to A$ which is constant on A (such a surjection is called a *retraction* onto A). Therefore, there is a bijection between the pruned trees and the non-empty closed sets. Note that the well founded trees T (i.e., trees with $[T] = \emptyset$) and non-empty well founded forests of the form $F := T \setminus \{\varepsilon\}$ are sufficient for defining the h-preorders in the Introduction.

We mention some reducibilities on the k-partitions of Baire space. Since these are many-one reducibilities, they are closely related to classes of functions $\mathcal{F}$ on Baire space that are closed under composition and contain the identity function. Any such a class $\mathcal{F}$ induces a "reducibility" (i.e., a preorder) $\leq_{\mathcal{F}}$ on $k^{\mathcal{N}}$: $A \leq_{\mathcal{F}} B$, if $A = B \circ f$ for some $f \in \mathcal{F}$. If $\mathcal{F}$ is the class of continuous functions we obtain Wadge reducibility. If $\mathcal{F}$ is the class of Lipschitz functions we obtain Lipschitz reducibility which is denoted $\leq_L$ and plays an important role in the Wadge theory. Recall that a *Lipschitz function* (resp. a *strong Lipschitz function*) may be defined as a function f on $\mathcal{N}$ satisfying $f(x)(n) = \phi(x[n+1])$ (resp. $f(x)(n) = \phi(x[n])$), for some $\phi : \omega^* \to \omega$. Every strong Lipschitz function is Lipschitz and every Lipschitz function is continuous, so $\leq_L$ is contained in $\leq_W$ (but not vice versa). For a Lipschitz function f and a string σ, $f(\sigma)$ denotes the obvious initial segment of $f(x)$ for each $x \sqsupseteq \sigma$.

For any pointclass $\Gamma \subseteq P(\mathcal{N})$, by Γ-*function* we mean a function f on $\mathcal{N}$ such that $f^{-1}(A) \in \Gamma$ for each $A \in \Gamma$. Since the Γ-functions are closed under composition and contain the identity function, we obtain the corresponding Γ-reducibility $\leq_{\Gamma}$. Among such reducibilities are Δ_{α}^{0}-reducibilities, for each non-zero countable ordinal α. Note that Δ_{α}^{0}-reducibility coincides with Σ_{α}^{0}-reducibility and Δ_{1}^{0}-reducibility coincides with Wadge reducibility. We usually shorten the notation $\leq_{\Delta_{\alpha}^{0}}$ to $\leq_{\alpha}$, so $\leq_1$ coincides with $\leq_W$. Note that the relation $\leq_{\alpha}$ is contained in $\leq_{\beta}$ for all $1 \leq \alpha < \beta < \omega_1$. The Δ_{α}^{0}-functions and Δ_{α}^{0}-reducibilities were investigated in [1,2,8]. By Jayne-Rogers theorem, the Δ_{2}^{0}-functions coincide with the functions f on $\mathcal{N}$ for which there is a partition $\{A_n\}$ of $\mathcal{N}$ to closed sets A_n such that $f|_{A_n}$ is continuous for each $n < \omega$.

Many reducibilities are closely related to infinite games. Relate to any $A, B \in k^{\mathcal{N}}$ the *Lipschitz game* $G_L(A, B)$ for players I and II as follows. Player I chooses a natural number $x(0)$, Player II responses with his number $y(0)$, I responses with $x(1)$, and so on; every player knows all moves of the opponent. After ω steps, I has produced some $x \in \mathcal{N}$ while II has produced some $y \in \mathcal{N}$; we say that II won this round if $A(x) = B(y)$, otherwise I won the round. A *winning strategy* for II (resp. for I) in the game $G_L(A, B)$ is identified with a Lipschitz function f (resp. a strong Lipschitz function g) such that $A(x) = B(f(x))$ for each $x \in \mathcal{N}$ (resp. $A(g(y)) \neq B(y)$ for each $y \in \mathcal{N}$). It is easy to see that II has a winning strategy in $G_L(A, B)$ iff $A \leq_L B$. As it follows from Martin determinacy theorem, for any Borel A, B the game $G_L(A, B)$ is determined, i.e. one of the players has a winning strategy. Versions of this fact are crucial for applications of infinite games to Wadge theory. In particular, define the *Wadge game* $G_W(A, B)$ as the modification of $G_L(A, B)$ in which Player II is allowed to pass (i.e., not choose a number) at any step. Player II wins a given round if he responses with infinitely many numbers during this round (thus producing some $y \in \mathcal{N}$) and $A(x) = B(y)$. Then II has a winning strategy in $G_W(A, B)$ iff $A \leq_W B$.

There are many other ingenious modifications of $G_L(A, B)$ of which we mention one introduced by R. Van Wesep. The *backtrack game* $G_{bt}(A, B)$ is the modification of $G_W(A, B)$ by giving to Player II the additional ability to backtrack (i.e., delete all his previous moves and start the construction of y from scratch) at any step. Player II wins a given round if he makes only finitely many backtracks during this round (thus producing again some $y \in \mathcal{N}$) and $A(x) = B(y)$. The winning strategies for II in the backtrack games are known as the *backtrack functions*. As shown in Theorem 21 of [1], the backtrack functions coincide with the $\mathbf{\Delta}_2^0$-functions, so II has a winning strategy in $G_{bt}(A, B)$ iff $A \leq_2 B$.

3   Operations on k-partitions

We use some obvious properties of the ω-ary operation $\bigoplus$ on $S^{\mathcal{N}}$ (S is a nonempty set) defined by $\bigoplus_n A_n(i \cdot x) := A_i(x)$. For any $2 \leq m < \omega$, define the m-ary operation on $S^{\mathcal{N}}$ by $B_0 \oplus \cdots \oplus B_{m-1} := \bigoplus_n A_n$ where $A_{mq+r} := B_r$ for all $q \geq 0, 0 \leq r < m$. For any $\sigma \in \omega^*, A \in S^{\mathcal{N}}$, define $A^\sigma \in S^{\mathcal{N}}$ by $A^\sigma(x) := A(\sigma x)$.

The next result is a straightforward extension to k-partitions of some properties of $+$ established in Sect. III.C of [16] for sets.

Proposition 2

(1) If $A \leq_1 A'$ and $B \leq_1 B'$ then $A + B \leq_1 A' + B'$.
(2) $(A + B) + C \equiv_1 A + (B + C)$.
(3) $(\bigoplus_n A_n) + B \equiv_1 \bigoplus_n (A_n + B)$.
(4) $A + B \equiv_2 A \oplus B$.
(5) For any $2 \leq \alpha < \omega_1$, the set $\mathbf{\Delta}_\alpha^0(k^{\mathcal{N}})$ is closed under $+$.

The next result is known (see Theorem 7.6 from [11] and references therein).

Proposition 3. *The structure* $(k^{\mathcal{N}}; \bigoplus, \leq_1, p_0, \ldots, p_{k-1})$ *is a σ-semilattice with discrete closures which by definition means that any p_i is a closure operation (i.e., $A \leq_1 p_i(A)$, $A \leq_1 B$ implies $p_i(A) \leq_1 p_i(B)$, and $p_i(p_i(A)) \leq_1 p_i(A)$), $p_i(A) \leq_1 \bigoplus_n B_n$ implies that $p_i(A) \leq_1 B_n$ for some $n < \omega$, and $p_i(A) \leq_1 p_j(B)$, $i \neq j$ imply that $p_i(A) \leq_1 B$.*

Relate to any k-partition A the tree $T(A) := \{\sigma \in \omega^* \mid A \leq_1 A^\sigma\}$ which is known to be closely related to the Wadge theory, in particular the following relation to the operations $p_0, \ldots, p_{k-1}$ is straightforward (for related facts on sets see e.g. Sect. 2 of [1]):

Proposition 4

(1) If $x \in [T(A)]$ then $p_{A(x)}(A) \leq_1 A$.
(2) For any $i < k$, $p_i(A) \leq_1 A$ iff $i = A(x)$ for some $x \in [T(A)]$.

A k-partition A is α-*irreducible* (a more precise but more complicated term would be σ-join-irreducible w.r.t. $\leq_\alpha$, cf. [11]) if there do not exist k-partitions $A_n <_\alpha A$, $n < \omega$, with $A \equiv_\alpha \bigoplus_n A_n$. If A is not α-irreducible then we call it α-*reducible*. We collect some characterizations of 1-irreducible k-partitions. For the last one see e.g. [3,7] (where the k-partitions A with the property from item (8) are called non-self-dual), the others are rather straightforward.

Proposition 5. *For any $A \in \mathbf{\Delta}_1^1(k^{\mathcal{N}})$, the following are equivalent:*

(1) *The k-partition A is 1-irreducible.*
(2) *There is no $\mathbf{\Delta}_1^0$-partitions $\{D_n\}$ of $\mathcal{N}$ with $A \circ d_n <_1 A$ where d_n is a continuous retraction from $\mathcal{N}$ onto D_n.*
(3) *There are no 1-irreducible k-partitions $A_n <_1 A$ with $A \equiv_1 \bigoplus_n A_n$.*
(4) *If $A \leq_1 \bigoplus_n B_n$ then $A \leq_1 B_n$ for some $n < \omega$.*
(5) *The tree $T(A)$ is not well founded.*
(6) *The tree $T(A)$ is pruned.*
(7) *There is $i < k$ with $p_i(A) \leq_1 A$.*
(8) *Any continuous function f on $\mathcal{N}$ has an A-fixed point (i.e., $A(x) = A(f(x))$ for some $x \in \mathcal{N}$).*

We state a corollary of (8) which extends to k-partitions a well known property of Wadge degrees (in the proof we use the idea of Lemma 29 in [7]).

Proposition 6. *Let $A, B \in \mathbf{\Delta}_1^1(k^{\mathcal{N}})$ and A be 1-irreducible. Then $B \leq_W A$ implies $B \leq_L A$, $A \leq_W B$ implies $A \leq_L B$, and $B \equiv_W A$ implies $B \equiv_L A$.*

Proof. The third assertion follows from the first two. Let $B \leq_W A$ via f. Consider the game $G_{diag}(A, B)$ where players I and II construct x, y as in the Lipschitz game and II wins iff $A(x) \neq B(y)$. Player II does not have a winning strategy in this game since if s were such a strategy then $A(x) \neq B(s(x)) = Af(s(x))$ for all $x \in \mathcal{N}$, contradicting Proposition 5(8). Thus, player I has a winning strategy t satisfying $\forall y A(t(y)) = B(y)$. Since t is Lipschitz, $B \leq_L A$.

Let now $A \leq_W B$ via f. Then player II does not have a winning strategy in $G_{diag}(B, A)$ since if s were such a strategy then $A(x) = B(f(x)) \neq As(f(x))$ for all $x \in \mathcal{N}$, contradicting Proposition 5(8). Thus, player I has a winning strategy t satisfying $B(t(y)) = A(y)$ for all $y \in \mathcal{N}$. Since t is Lipschitz, $A \leq_L B$.   □

Relate to any 1-irreducible $A \in k^{\mathcal{N}}$ the k-partitions $A' \in k^{\mathcal{N}}$ and $\tilde{A} \in k^{\mathcal{N}} \cup \{\bot\}$ (here $\bot$ is a new element strictly $\leq_1$-below any k-partition) as follows: $A' = A \circ t$ where $t : \mathcal{N} \to [T(A)]$ is a Lipschitz retraction onto $[T(A)]$ (modulo $\equiv_1$, A' does not depend on the choice of t), and $\tilde{A} = \bigoplus \{A^\sigma \mid \sigma \in \partial(T(A))\}$ (modulo $\equiv_1$, $\tilde{A}$ does not depend on the choice of a numbering of $\partial(T(A))$ if the last set is non-empty, and if it is empty we set $\tilde{A} = \bot$).

Proposition 7. *For all 1-irreducible $A, B \in k^{\mathcal{N}}$ we have:*

(1) $\tilde{A} = \bot$ *iff* $T(A) = \omega^*$;
(2) $A' \leq_1 A$, $\tilde{A} <_1 A$ *and* $A \equiv_1 A' + \tilde{A}$;

(3) *If $A \equiv_1 B$ then $A' \equiv_1 B'$ (but, in general, $\tilde{A} \not\equiv_1 \tilde{B}$);*
(4) $T(A') = \omega^*$.

Proof. Item (1) obvious. The first assertion in (2) is obvious, the second follows from 1-irreducibility of A, the third is straightforward.

(3) Let $A \equiv_1 B$. By symmetry, it suffices to show that $A \circ t \leq_1 B \circ t_B$ where t_B is a continuous retraction onto $[T(B)]$. Let f witness $A \leq_1 B$, then $f(x) \in [T(B)]$ for each $x \in [T(A)]$. (Suppose not, then $\tau \sqsubseteq f(x)$ for a unique $\tau \in \partial(T(B))$. By continuity of f, $f(\sigma \cdot \mathcal{N}) \subseteq \tau \cdot \mathcal{N}$ for some $\sigma \sqsubseteq x$, hence $A^\sigma \leq_1 B^\tau$. Since $\sigma \in T(A)$, we have $B \leq_1 A \leq_1 A^\sigma \leq_1 B^\tau$, so $\tau \in T(B)$. A contradiction.) Since t_B is a retraction, $f \circ t = t_B \circ f \circ t$, so $A \circ t = B \circ f \circ t = (B \circ t_B) \circ (f \circ t)$, hence $A \circ t \leq_1 B \circ t_B$ via $f \circ t$.

(4) Let $\sigma \in \omega^*$ and $B := A^{t(\sigma)}$. Since $t(\sigma) \in T(A)$, $A \equiv_1 B$. Note that $T(B) = \{\tau \in \omega^* \mid t(\sigma) \cdot \tau \in T(A)\}$. Since t is a Lipschitz retraction onto $T(A)$, $B' \equiv_1 A'^\sigma$. Since $A \equiv_1 B$, by (3) we get $A' \equiv_1 B'$. Therefore, $A' \equiv_1 A'^\sigma$, $\sigma \in T(A')$ and $T(A') = \omega^*$.   □

The next properties of the operations $+, q_0, \ldots, q_{k-1}$ are straightforward:

Proposition 8

(1) *For $3 \leq \alpha < \omega_1$, $\mathbf{\Delta}_\alpha^0(k^{\mathcal{N}})$ is closed under $q_0, \ldots, q_{k-1}$.*
(2) *Structure $(k^{\mathcal{N}}; \bigoplus, \leq_2, q_0, \ldots, q_{k-1})$ is a σ-semilattice with discrete closures.*
(3) *For all $A \in k^{\mathcal{N}}$ and $i < k$, $B \mapsto q_i(A) + B$ is a closure operator on $(k^{\mathcal{N}}; \leq_1)$.*
(4) *For all $A, A_1, B, B_1 \in k^{\mathcal{N}}$ and $i, j < k$, if $q_i(A) + B \leq_1 q_j(A_1) + B_1$ and $q_i(A) \not\leq_1 q_j(A_1)$ then $q_i(A) + B \leq_1 B_1$.*

We proceed with some characterizations of 2-irreducible k-partitions.

Proposition 9. *For any $A \in \mathbf{\Delta}_1^1(k^{\mathcal{N}})$, the following are equivalent:*

(1) *The k-partition A is 2-irreducible.*
(2) *There is no $\mathbf{\Pi}_1^0$-partition $\{D_n\}$ of $\mathcal{N}$ with $A \circ d_n <_2 A$ where d_n is a continuous retraction from $\mathcal{N}$ onto D_n.*
(3) *There are no 2-irreducible k-partitions $A_n <_2 A$ with $A \equiv_2 \bigoplus_n A_n$.*
(4) *If $A \leq_2 \bigoplus_n B_n$ then $A \leq_2 B_n$ for some $n < \omega$.*
(5) *Any $\mathbf{\Delta}_2^0$-function f on $\mathcal{N}$ has an A-fixed point.*
(6) *There is $B \equiv_2 A$ with $T(B) = \omega^*$.*
(7) *There is $i < k$ with $q_i(A) \leq_1 A$.*

Proof. The equivalence of (1)–(4) is rather straightforward.

(5)→(1) Let A be 2 reducible, so $A \equiv_2 \bigoplus_n A_n$ for some $A_n <_2 A$. We have to find a $\mathbf{\Delta}_2^0$-function without A-fixed points. Clearly, it suffices to find such a function without $\bigoplus_n A_n$-fixed points. For any $n < \omega$, choose $n' < \omega$ with $A_{n'} \not\leq_2 A_n$, then Player II does not win $G_{bt}(A_{n'}, A_n)$, hence Player I wins via a strategy s_n, so $A_{n'}(s_n(y)) \neq A_n(y)$ for each $y \in \mathcal{N}$. Define a Lipschitz function f by $f(n \cdot y) := n' \cdot s_n(y)$. Then f has no $\bigoplus_n A_n$-fixed points.

(2)→(5) Suppose for a contradiction that A satisfies (2) but some $\mathbf{\Delta}_2^0$-function has no A-fixed points. Then the following version of Claim 5.3.1 in [8] holds: for any closed set D and any continuous retraction d onto D, if $A \equiv_2 A \circ d$ then there is a $\mathbf{\Delta}_2^0$-function $g : \mathcal{N} \to D$ without $(A \circ d)$-fixed points. Indeed, since $A \equiv_2 A \circ d$, there is a $\mathbf{\Delta}_2^0$-function $f : \mathcal{N} \to \mathcal{N}$ without $(A \circ d)$-fixed points. Since d is a retraction, we can take $g = d \circ f$. With this version at hand, we can repeat (with $A \cap D$ replaced by $A \circ d$) the proof of Theorem 5.3 in [8] (which is based on the proof of Theorem 16 in [2]) and obtain a contradiction.

(5)→(6) Consider the representation $A \equiv_1 A' + \tilde{A}$ from Proposition 7. By this proposition, $A \equiv_2 A' \oplus \tilde{A}$. Repeating the proof of Proposition 6 (with $\leq_W$ replaced by $\leq_2$) we obtain that if any $\mathbf{\Delta}_2^0$-function has an A-fixed point then the $\mathbf{\Delta}_2^0$-degree of A coincides with the Lipshitz degree of A. Therefore, $A \equiv_1 A' \oplus \tilde{A}$. By (5), 2-irreducibility implies 1-irreducibility. Thus, A is 1-irreducible, hence $A \leq_1 A'$ or $A \leq_1 \tilde{A}$. But $A \not\leq_1 \tilde{A}$, hence $A \equiv_1 A'$ and we can take $B = A'$.

(6)→(7). It suffices to show that if $T(A) = \omega^*$ then $q_i(A) \leq_1 A$ for some $i \in rng(A)$. By Proposition 6, for any $\sigma \in \omega^*$ there is a strong Lipschitz function f_σ with $A = A^\sigma \circ f_\sigma$. To simplify notation a bit, we consider the particular case $rng(A) = \{0, 1, 2\}$, it will be clear that the proof works for the general case. Towards a contradiction, suppose that $q_i(A) \not\leq_1 A$ for each $i < 3$, then there are strong Lipschitz functions s_i such that $q_i(A)(s_i(z)) \neq A(z)$ for all $i < 3, z \in \mathcal{N}$. We construct $y \in \mathcal{N}$ as follows.

Let x_0 be the first number in the sequence $s_0(y)$ (it does not depend on y because s_0 is strong Lipschitz). If $x_0 > 0$, we set $y(0) = x_0 - 1$, find the second number x_1 in the sequence $s_0(y)$ and set $y(1) = x_1 - 1$ if $x_1 > 0$, and continue this process until we find the first zero in $s_0(y)$. Such a zero exists because otherwise we would get $y = s_0(y)^-$, hence $A(y) = A(s_0(y)^-) = q_0(A)(s_0(y))$, contradicting the property of s_0. Thus, we have $\tau_0 0 \sqsubseteq s_0(y)$ for a unique string τ_0 without zeroes. We proceed to construct y by concatenating the consecutive numbers of the sequence $f_{\sigma_0}((x_0 - 1)(x_1 - 1) \cdots)$ to $\sigma_0 := \tau_0^-$ where this time $x_0, x_1, \ldots$ are the consecutive non-zero numbers in the sequence $s_1(y)$; we continue until the first 0 in the last sequence is discovered. This 0 exists because otherwise we get $y = \sigma_0 f_{\sigma_0}(s_1(y)^-)$, hence $A(y) = A^{\sigma_0} f_{\sigma_0}(s_1(y)^-) = A(s_1(y)^-) = q_1(A)(s_1(y))$, contradicting the property of s_1. Thus, we have $\tau_1 0 \sqsubseteq s_1(y)$ for a unique string τ_1 without zeroes. We proceed to construct y by concatenating the consecutive numbers of the sequence $f_{\sigma_1}((x_0 - 1)(x_1 - 1) \cdots)$ to $\sigma_1 := \sigma_0 f_{\sigma_0}(\tau_1^-)$ where this time $x_0, x_1, \ldots$ are the consecutive non-zero numbers in the sequence $s_2(y)$. Again we will find the first zero in $s_2(y)$ and the corresponding τ_2 and σ_2. Note that $\sigma_0 \sqsubseteq \sigma_1 \sqsubseteq \sigma_2$ because we work with the strong Lipschitz functions.

At this point, we proceed to construct y by concatenating the consecutive numbers of the sequence $f_{\sigma_2}((x_0 - 1)(x_1 - 1) \cdots)$ to σ_2 where this time $x_0, x_1, \ldots$ are the consecutive non-zero numbers in the unique sequence $z_0 \in \mathcal{N}$ satisfying $s_0(y) = \tau_0 0 z_0$; we continue until the first 0 in the sequence z_0 is discovered. This 0 exists because otherwise we get $y = \sigma_2 f_{\sigma_2}(z_0^-)$, hence $A(y) = A^{\sigma_2} f_{\sigma_2}(z_0^-) = A(z_0^-) = q_i(A)(s_0(y))$, contradicting the property of s_0. Next we work in the same way with $s_1(y) = \tau_1 0 z_1$, $s_2(y) = \tau_2 0 z_2$, and so on.

By the construction, we obtain an infinite sequence $\sigma_0 \sqsubset \sigma_1 \sqsubset \sigma_2 \sqsubset \cdots$ with $y = \bigcup_n \sigma_n$, and any of $s_0(y), s_1(y), s_2(y)$ has infinitely many zeroes. Let $i := A(y), i < 3$. By the definition of $q_i(A)$ we have $q_i(A)(s_i(y)) = i = A(y)$, contradicting the property of s_i.

(7)→(6) Take $B = q_i(A)$. By the definition of $q_i(A)$, $T(B) = \omega^*$.

(6)→(4) Choose $B \equiv_2 A$ with $T(B) = \omega^*$. First we show that $B \leq_2 C$ implies $B \leq_1 C$. Let $B \leq_2 C$ via f such that for some $\mathbf{\Pi}_1^0$-partition $\{D_n\}$ of $\mathcal{N}$ the function f is continuous on D_n for each $n < \omega$. By Baire category theorem, $\sigma \cdot \mathcal{N} \subseteq D_n$ for some n. Thus, $B^\sigma \leq_1 C$ via $\lambda x.f(\sigma x)$. Since $\sigma \in T(B)$, $B \leq_1 B^\sigma$ and therefore $B \leq_1 C$.

Let now $A \leq_2 \bigoplus_n B_n$, hence $B \leq_2 \bigoplus_n B_n$ and therefore $B \leq_1 \bigoplus_n B_n$. Since $T(B) = \omega^*$, B is 1-irreducible by Proposition 5(6), hence $B \leq_1 B_n$ for some $n < \omega$. Therefore, $A \leq_2 B_n$ and A is 2-irreducible. $\qquad\square$

4   Generating Degree Structures

The next proposition coincides with item (1) of Theorem 1.

Proposition 10. *The quotient-poset of* $(\mathbf{\Delta}_2^0(k^\mathcal{N}); \leq_W)$ *is generated from the minimal degrees* $\{\mathbf{0}\}, \ldots, \{\mathbf{k-1}\}$ *by the operations* $\bigoplus, p_0, \ldots, p_{k-1}$ *(and also by the operations* $\bigoplus, +$ *).*

Proof. Let $\mathcal{S}$ (resp. $\mathcal{S}_1$) be the subalgebra of $(k^\mathcal{N}; \bigoplus, p_0, \ldots, p_{k-1})$ (resp. of $(k^\mathcal{N}; \bigoplus, +)$) generated by the set $\{\mathbf{0}, \ldots, \mathbf{k-1}\}$, then $\mathcal{S} \subseteq \mathcal{S}_1 \subseteq \mathbf{\Delta}_2^0(k^\mathcal{N})$ by Proposition 2(5). It remains to show that any $A \in \mathbf{\Delta}_2^0(k^\mathcal{N})$ is Wadge equivalent to some $B \in \mathcal{S}$. This is checked by the rank $rk(A)$ of A in the well founded preorder $(\mathbf{\Delta}_2^0(k^\mathcal{N}); \leq_W)$. If $rk(A) = 0$ then $A \in \{\mathbf{0}, \ldots, \mathbf{k-1}\}$ and there is nothing to prove, so let A be non-constant. If A is 1-reducible then $A \equiv_1 \bigoplus_n A_n$ for some $A_n <_1 A$. By induction, $A_n \equiv_1 B_n$ for some $B_n \in \mathcal{S}$, $n \in \omega$. Then $A \equiv_1 \bigoplus_n B_n \in \mathcal{S}$, as desired.

Finally, let A be 1-irreducible. Consider the representation $A \equiv_1 A' + \tilde{A}$ from Proposition 7. Then A' is constant. (Otherwise, $i, j \in rng(A')$ for some distinct i, j, hence $p_i(A) \equiv_1 p_j(A) \equiv_1 A$ by Proposition 7(4). Since the Wadge degrees of $\mathbf{\Delta}_2^0$-sets are generated by $\bigoplus, p_0, p_1$ from $\{\emptyset\}, \{\mathcal{N}\}$ (see Sect. III.C of [16]), any such degree is Wadge reducible to A_i, hence $A \notin \mathbf{\Delta}_2^0(k^\mathcal{N})$. A contradiction.) If $\tilde{A} = \perp$ then $A \equiv_1 A'$ is constant and we are done. Finally, let $\tilde{A} \neq \perp$. Since $\tilde{A} <_1 A$, by induction $\tilde{A} \equiv_1 B$ for some $B \in \mathcal{S}$. Then $A \equiv_1 p_i(B) \in \mathcal{S}$ for some $i < k$, as desired. $\qquad\square$

The next proposition is interesting in its own right.

Proposition 11. *The quotient-poset of* $(\mathbf{\Delta}_3^0(k^\mathcal{N}); \leq_2)$ *is generated from the minimal degrees* $\{\mathbf{0}\}, \ldots, \{\mathbf{k-1}\}$ *by the operations* $\bigoplus, q_0, \ldots, q_{k-1}$.

Proof. Let $\mathcal{S}$ be the subalgebra of $(k^\mathcal{N}; \bigoplus, q_0, \ldots, q_{k-1})$ generated by the set $\{\mathbf{0}, \ldots, \mathbf{k-1}\}$, then $\mathcal{S} \subseteq \mathbf{\Delta}_3^0(k^\mathcal{N})$ by Proposition 8(1). It remains to show that any $A \in \mathbf{\Delta}_3^0(k^{\mathcal{N}})$ is 2-equivalent to some $B \in \mathcal{S}$. This is checked by the rank $rk(A)$ of A in the well founded preorder $(\mathbf{\Delta}_3^0(k^{\mathcal{N}}); \leq_2)$. If $rk(A) = 0$ then $A \in \{0, \ldots, \mathbf{k} - 1\}$ and there is nothing to prove, so let A be non-constant. If A is 2-reducible then $A \equiv_2 \bigoplus_n A_n$ for some $A_n <_2 A$. By induction, $A_n \equiv_2 B_n$ for some $B_n \in \mathcal{S}$, $n \in \omega$. Then $A \equiv_2 \bigoplus_n B_n \in \mathcal{S}$, as desired.

Finally, let A be 2-irreducible. By Proposition 9(7), $q_i(A) \equiv_2 A$ for some $i < k$. Such i is in fact unique. (Otherwise, $q_i(A) \equiv_1 q_j(A) \equiv_1 A$ for some distinct i, j. Since any $\mathbf{\Delta}_3^0$-set is Wadge reducible to a set generated by $\bigoplus, q_0 = \sharp, q_1 = \flat$ from $\{\emptyset\}, \{\mathcal{N}\}$ by Sect. III.E of [16], any $\mathbf{\Delta}_3^0$-set is Wadge reducible to A_i, hence $A \notin \mathbf{\Delta}_3^0(k^{\mathcal{N}})$. A contradiction.) Moreover, $A \equiv_2 q_i(\tilde{A})$ for some $\tilde{A} <_2 A$ as it follows from the structure of the difference hierarchy of k-partitions over $\mathbf{\Sigma}_2^0$ [14] (A will be Wadge complete in a non-self-dual level of this hierarchy similarly to Sect. 7 of [11], with p_i replaced by q_i; in fact, the set of $\mathbf{\Delta}_2^0$-degrees strictly below A is at most countable, hence we can take $\tilde{A}$ as the disjoint union of all such degrees, then the $\mathbf{\Delta}_2^0$-degree of $\tilde{A}$ is the largest $\mathbf{\Delta}_2^0$-degree strictly below A). Since $\tilde{A} <_2 A$, by induction $\tilde{A} \equiv_2 B$ for some $B \in \mathcal{S}$. Then $A \equiv_2 q_i(B) \in \mathcal{S}$, as desired.   □

The next proposition coincides with item (2) of Theorem 1.

Proposition 12. *The quotient-poset of* $(\mathbf{\Delta}_3^0(k^{\mathcal{N}}); \leq_W)$ *is generated from the minimal degrees* $\{\mathbf{0}\}, \ldots, \{\mathbf{k} - 1\}$ *by the operations* $\bigoplus, +, q_0, \ldots, q_{k-1}$.

Proof. Let $\mathcal{S}$ be the subalgebra of $(k^{\mathcal{N}}; \bigoplus, +, q_0, \ldots, q_{k-1})$ generated by the set $\{\mathbf{0}, \ldots, \mathbf{k} - 1\}$, then $\mathcal{S} \subseteq \mathbf{\Delta}_3^0(k^{\mathcal{N}})$ by Propositions 2(5) and 8. It remains to show that any $A \in \mathbf{\Delta}_3^0(k^{\mathcal{N}})$ is Wadge equivalent to some $B \in \mathcal{S}$. This is checked by the rank $rk(A)$ of A in the well founded preorder $(\mathbf{\Delta}_3^0(k^{\mathcal{N}}); \leq_W)$. If $rk(A) = 0$ then $A \in \{\mathbf{0}, \ldots, \mathbf{k} - 1\}$ and there is nothing to prove, so let A be non-constant. If A is 1-reducible then $A \equiv_1 \bigoplus_n A_n$ for some $A_n <_1 A$. By induction, $A_n \equiv_1 B_n$ for some $B_n \in \mathcal{S}$, $n \in \omega$. Then $A \equiv_1 \bigoplus_n B_n \in \mathcal{S}$, as desired.

Finally, let A be 1-irreducible. Consider the representation $A \equiv_1 A' + \tilde{A}$ from Proposition 7. Since $T(A') = \omega^*$, A' is 2-irreducible by Proposition 9(6). By Proposition 11, $A' \equiv_2 C$ for some $C \in \mathcal{S}$, hence also $A' \equiv_1 C$ by the proof of $(6) \rightarrow (4)$ in Proposition 9. If $\tilde{A} = \bot$ then $A \equiv_1 C \in \mathcal{S}$ and we are done. Finally, let $\tilde{A} \neq \bot$. Since $\tilde{A} <_1 A$, by induction $\tilde{A} \equiv_1 B$ for some $B \in \mathcal{S}$. Then $A \equiv_1 C + B \in \mathcal{S}$, as desired.   □

5   Operations on Labeled Forests

Let us briefly recall from [12–14] some operations of labeled forests and collect their properties used in the sequel (all these operations respect the h-equivalence). Recall that our labeled trees $(T; c)$ in $\widetilde{\mathcal{T}_k}$ (resp. in $\widetilde{\mathcal{T}_{\widetilde{\mathcal{T}_k}}}$) consist of a well founded tree T and a labeling $c : T \to k$ (resp. $c : T \to \widetilde{\mathcal{T}_k}$). The non-empty labeled forests in $\widetilde{\mathcal{F}_{\widetilde{\mathcal{T}_k}}}$ are obtained from such trees by removing the root ε. The ω-ary operation $\bigoplus$ of disjoint union on $\widetilde{\mathcal{F}_{\widetilde{\mathcal{T}_k}}}$ is defined in the obvious way.

For any $i < k$ and $F \in \tilde{\mathcal{F}}_{\tilde{\mathcal{T}}_k}$, let $p_i(F)$ be the tree in $\tilde{\mathcal{T}}_{\tilde{\mathcal{T}}_k}$ obtained from F by adjoining the empty string labeled by i. Note that the set $\tilde{\mathcal{F}}_k$ is closed under the operation p_i (we may think that $\tilde{\mathcal{F}}_k \subseteq \tilde{\mathcal{F}}_{\tilde{\mathcal{T}}_k}$ by identifying labels $i < k$ with the singleton tree $\mathbf{i}$ carrying the label i). Define the binary operation $+$ on $\tilde{\mathcal{F}}_{\tilde{\mathcal{T}}_k}$ as follows: $F+G$ is obtained by adjoining a copy of G below any leaf of F. One easily checks that $\mathbf{i} + F \equiv_h p_i(F)$, $F \leq_h F + G$, $G \leq_h F + G$, $F \leq_h F_1 \to F + G \leq_h F_1 + G$, $G \leq_h G_1 \to F + G \leq_h F + G_1$, $(F + G) + H \equiv_h F + (G + H)$. Note that the set $\tilde{\mathcal{F}}_k$ is closed under the operation $+$. Define the function $s : \tilde{\mathcal{T}}_k \to \tilde{\mathcal{T}}_{\tilde{\mathcal{T}}_k}$ as follows: $s(F)$ is the singleton tree carrying the label F. Note that $s(\mathbf{i}) = \mathbf{i}$ for each $i < k$, and $T \leq_h S$ iff $s(T) \leq_h s(S)$, for all $S, T \in \tilde{\mathcal{T}}_k$. One easily checks the following properties:

Proposition 13

(1) $(\tilde{\mathcal{F}}_k; \bigoplus, \leq_h, p_0, \dots, p_{k-1})$ and $(\tilde{\mathcal{F}}_{\tilde{\mathcal{T}}_k}; \bigoplus, \leq_h, p_0, \dots, p_{k-1})$ are σ-semilattices with discrete closures.
(2) For any $T \in \tilde{\mathcal{T}}_k$, $F \mapsto s(T) + F$ is a closure operator on $\tilde{\mathcal{F}}_{\tilde{\mathcal{T}}_k}$.
(3) For all $T, T_1 \in \tilde{\mathcal{T}}_k$ and $F, F_1 \in \tilde{\mathcal{F}}_{\tilde{\mathcal{T}}_k}$, if $s(T) + F \leq_h s(T_1) + F_1$ and $T \not\leq_h T_1$ then $s(T) + F \leq_h F_1$.

6 Characterizing Degree Structures

The next proposition coincides with item (1) of Theorem 2.

Proposition 14. *The quotient-posets of* $(\Delta_2^0(k^{\mathcal{N}}); \leq_W)$ *and of* $(\tilde{\mathcal{F}}_k; \leq_h)$ *are isomorphic.*

Proof. Let $(T; c) \in \tilde{\mathcal{T}}_k$. Relate to any node $\sigma \in T$ the k-partition $\mu_T(\sigma)$ by induction on the rank $rk(\sigma)$ of σ in $(T; \sqsubseteq)$ as follows: if $rk(\sigma) = 0$, i.e. σ is a leaf of T then $\mu_T(\sigma) := \mathbf{i}$ where $i = c(\sigma)$; otherwise, $\mu_T(\sigma) := p_i(\bigoplus\{\mu_T(\sigma n) \mid n < \omega, \sigma n \in T\})$.

Now we define a function $\mu : \tilde{\mathcal{T}}_k \to k^{\mathcal{N}}$ by $\mu(T) := \mu_T(\varepsilon)$. Then $T \leq_h S$ iff $\mu(T) \leq_W \mu(S)$, for all $T, S \in \tilde{\mathcal{T}}_k$. This is checked using Propositions 13(1) and 3 by induction on $(rk_T(\varepsilon), rk_S(\varepsilon))$ in the lexicographic order of pairs of countable ordinals (for details see the proof of Theorem 5.1 in [10], although the induction there was not on the tree ranks of T, S but rather on their ranks in the well poset $(\tilde{\mathcal{T}}_k; \leq_h)$).

Next we extend μ to $\tilde{\mathcal{F}}_k$ by $\mu(F) := \bigoplus\{\mu_T(n) \mid n < \omega, (n) \in T\}$ where $T := \{\varepsilon\} \cup F$. Again, it is easy to see that $F \leq_h G$ iff $\mu(F) \leq_W \mu(G)$, for all $F, G \in \tilde{\mathcal{F}}_k$.

Repeating the proof of Proposition 10, we check by induction on $rk(A)$ that for each $A \in \Delta_2^0(k^{\mathcal{N}})$ there is $F \in \tilde{\mathcal{F}}_k$ with $\mu(F) \equiv_W A$. Thus, the function μ induces the desired isomorphism of quotient-posets. $\qquad\qquad\square$

The next proposition is interesting in its own right.

Proposition 15. *The quotient-posets of* $(\boldsymbol{\Delta}_3^0(k^{\mathcal{N}}); \leq_2)$ *and of* $(\widetilde{\mathcal{F}}_k; \leq_h)$ *are isomorphic.*

Proof. The proof is a minor modification of the previous one. For any $(T; c) \in \widetilde{\mathcal{T}}_k$, define $\nu_T : T \to k^{\mathcal{N}}$ as follows: if $rk(\sigma) = 0$ then $\nu_T(\sigma) := \mathbf{i}$ where $i = c(\sigma)$, otherwise $\nu_T(\sigma) := q_i(\bigoplus\{\nu_T(\sigma n) \mid n < \omega, \sigma n \in T\})$. Define $\nu : \widetilde{\mathcal{T}}_k \to k^{\mathcal{N}}$ by $\nu(T) := \nu_T(\varepsilon)$. Then $T \leq_h S$ iff $\nu(T) \leq_2 \nu(S)$, for all $T, S \in \widetilde{\mathcal{T}}_k$, because the operations q_i have the same algebraic properties as p_i. Finally, extend ν to $\widetilde{\mathcal{F}}_k$ by $\nu(F) := \bigoplus\{\nu_T(n) \mid n < \omega, (n) \in T\}$ where $T := \{\varepsilon\} \cup F$. Again, it is easy to see that $F \leq_h G$ iff $\nu(F) \leq_2 \nu(G)$, for all $F, G \in \widetilde{\mathcal{F}}_k$.

Repeating the proof of Proposition 11, we check by induction on $rk(A)$ that for each $A \in \boldsymbol{\Delta}_3^0(k^{\mathcal{N}})$ there is $F \in \widetilde{\mathcal{F}}_k$ with $\nu(F) \equiv_2 A$. Thus, the function ν induces the desired isomorphism of quotient-posets. □

The next proposition coincides with item (2) of Theorem 2.

Proposition 16. *The quotient-posets of* $(\boldsymbol{\Delta}_3^0(k^{\mathcal{N}}); \leq_W)$ *and of* $(\widetilde{\mathcal{F}}_{\widetilde{\mathcal{T}}_k}; \leq_h)$ *are isomorphic.*

Proof. Let $(T; c) \in \widetilde{\mathcal{T}}_{\widetilde{\mathcal{T}}_k}$. Relate to any node $\sigma \in T$ the k-partition $\rho_T(\sigma)$ by induction on the rank $rk(\sigma)$ of σ in $(T; \sqsupseteq)$ as follows: if $rk(\sigma) = 0$ then $\rho_T(\sigma) := \nu(Q)$ where $Q = c(\sigma) \in \widetilde{\mathcal{T}}_k$; otherwise, $\rho_T(\sigma) := \nu(Q) + (\bigoplus\{\rho_T(\sigma n) \mid n < \omega, \sigma n \in T\})$. Now define a function $\rho : \widetilde{\mathcal{T}}_{\widetilde{\mathcal{T}}_k} \to k^{\mathcal{N}}$ by $\rho(T) := \rho_T(\varepsilon)$. Then $T \leq_h S$ iff $\rho(T) \leq_W \rho(S)$, for all $T, S \in \widetilde{\mathcal{T}}_{\widetilde{\mathcal{T}}_k}$. This is checked using Propositions 13(2,3) and 2 by induction on $(rk_T(\varepsilon), rk_S(\varepsilon))$. Next we extend ρ to $\widetilde{\mathcal{F}}_{\widetilde{\mathcal{T}}_k}$ by $\rho(F) := \bigoplus\{\rho_T(n) \mid n < \omega, (n) \in T\}$ where $T := \{\varepsilon\} \cup F$. Again, it is easy to see that $F \leq_h G$ iff $\rho(F) \leq_W \rho(G)$, for all $F, G \in \widetilde{\mathcal{F}}_{\widetilde{\mathcal{T}}_k}$.

Repeating the proof of Proposition 12, we check by induction on $rk(A)$ that for each $A \in \boldsymbol{\Delta}_3^0(k^{\mathcal{N}})$ there is $F \in \widetilde{\mathcal{F}}_{\widetilde{\mathcal{T}}_k}$ with $\rho(F) \equiv_W A$. Thus, the function ρ induces the desired isomorphism of quotient-posets. □

References

1. Andretta, A.: More on Wadge determinacy. Ann. Pure Appl. Logic **144**(1–3), 2–32 (2006)
2. Andretta, A., Martin, D.A.: Borel-Wadge degrees. Fund. Math. **177**(2), 175–192 (2003)
3. Block, A.C.: Operations on a Wadge-type hierarchy of ordinal-valued functions. Masters thesis, Universiteit van Amsterdam (2014)
4. van Engelen, F., Miller, A., Steel, J.: Rigid Borel sets and better quasiorder theory. Contemp. Math. **65**, 199–222 (1987)
5. Hertling, P.: Topologische Komplexitätsgrade von Funktionen mit endlichem Bild. Informatik-Berichte 152, Fernuniversität Hagen (1993)
6. Kechris, A.S.: Classical Descriptive Set Theory. Springer, New York (1995)
7. Kihara, T., Montalban, A.: The uniform Martin's conjecture for many-one degrees. arXiv:1608.05065v1 [Math.LO], 17 August 2016

8. Motto Ros, L.: Borel-amenable reducibilities for sets of reals. J. Symbolic Logic **74**(1), 27–49 (2009)
9. Selivanov, V.L.: Hierarchies of hyperarithmetical sets and functions. Algebra Logic **22**, 473–491 (1983)
10. Selivanov, V.L.: The quotient algebra of labeled forests modulo h-equivalence. Algebra Logic **46**, 120–133 (2007)
11. Selivanov, V.L.: Hierarchies of $\mathbf{\Delta}^0_2$-measurable k-partitions. Math. Logic Q. **53**, 446–461 (2007)
12. Selivanov, V.: A fine hierarchy of ω-regular k-partitions. In: Löwe, B., Normann, D., Soskov, I., Soskova, A. (eds.) CiE 2011. LNCS, vol. 6735, pp. 260–269. Springer, Heidelberg (2011). doi:10.1007/978-3-642-21875-0_28
13. Selivanov, V.L.: Fine hierarchies via Priestley duality. Ann. Pure Appl. Logic **163**, 1075–1107 (2012)
14. Selivanov, V.L.: Towards a descriptive theory of cb0-spaces. Mathematical Structures in Computer Science, September 2014. doi:10.1017/S0960129516000177. Earlier version in: arXiv:1406.3942v1 [Math.GN], 16 June 2016
15. Van Wesep, R.: Wadge degrees and descriptive set theory. In: Kechris, A.S., Moschovakis, Y.N. (eds.) Cabal Seminar 76–77. LNM, vol. 689, pp. 151–170. Springer, Heidelberg (1978). doi:10.1007/BFb0069298
16. Wadge, W.: Reducibility and determinateness in the Baire space. Ph.D. thesis, University of California, Berkely (1984)

Erratum to: Counting Substrate Cycles in Topologically Restricted Metabolic Networks

Robert D. Barish[✉] and Akira Suyama

Graduate School of Arts and Sciences, University of Tokyo,
Meguro-ku Komaba 3-8-1, Tokyo 153-8902, Japan
rbarish@genta.c.u-tokyo.ac.jp

Erratum to:
Chapter "Counting Substrate Cycles in Topologically Restricted Metabolic Networks" in: J. Kari et al. (Eds.), Unveiling Dynamics and Complexity, LNCS 10307, DOI: 10.1007/978-3-319-58741-7_14

The original version of the book was inadvertently published with the terms "undirected" and "graph" in full capitalization in Chapter 14. The erratum chapter and the book has been updated with the change.

The updated online version of this chapter can be found at
http://dx.doi.org/10.1007/978-3-319-58741-7_14

© Springer International Publishing AG 2017
J. Kari et al. (Eds.): CiE 2017, LNCS 10307, p. E1, 2017.
DOI: 10.1007/978-3-319-58741-7_37

Author Index

Printed in the United States
By Bookmasters